T0257898

IET POWER AND ENERGY SERIES 95

Communication, Control and Security Challenges for the Smart Grid

Other volumes in this series:

Communication, Control and Security Challenges for the Smart Grid

Edited by
S. M. Muyeen and Saifur Rahman

The Institution of Engineering and Technology

Published by The Institution of Engineering and Technology, London, United Kingdom

The Institution of Engineering and Technology is registered as a Charity in England & Wales (no. 211014) and Scotland (no. SC038698).

© The Institution of Engineering and Technology 2017

First published 2017

The Institution of Engineering and Technology
Michael Faraday House
Six Hills Way, Stevenage
Herts, SG1 2AY, United Kingdom

www.theiet.org

British Library Cataloguing in Publication Data
A catalogue record for this product is available from the British Library

ISBN 978-1-78561-142-1 (hardback)
ISBN 978-1-78561-143-8 (PDF)

Typeset in India by MPS Limited
Printed in the UK by CPI Group (UK) Ltd, Croydon

This book is dedicated to
S. M. Wazed Ali and Nazma Murshida
Serajur Rahman and Sahara Rahman

Contents

9 Smart consumer system 219

Abdul R. Beig

12   Multi-agent based control of smart grid   **321**
K.T.M.U. Hemapala, S.L. Jayasinghe and A.L. Kulasekera

13   Compressive sensing for smart-grid security and reliability   **347**
Mohammad Babakmehr, Marcelo Godoy Simões and Ahmed Al-Durra

Preface

The concept and formation of power system has changed drastically in recent years which is newly introduced as Smart Grid. The changes appear in generation, transmission, and distribution levels and contributions are coming from multi-disciplinary fields such as electrical, electronic, computer, communication, mechanical, aerospace engineering, and many other science disciplines. Each discipline has its own beauty and the combined efforts from scientists from different disciplines are the secret of the success of smart-grid industry.

In this book, the present and future development status and trends of smart grid including generation, transmission, and distribution system are depicted based on the contribution from many renowned scientists and engineers from different disciplines. The prime focus is given in the communication, control, and security aspects of Smart Grid. A wide verity of research results are merged together to make this book useful for students and researchers.

There are 16 chapters in this book. In the first few chapters, the smart-grid architecture, communication features and standards, smart sensing, and measurements are discussed in detail. In the next phase, the smart transmission, distribution, energy management, and control aspects of smart grid are presented. In the last few chapters, the stability, security, and policy and regulation issues are discussed elaborately. A general overview and essence of the chapters can be obtained from Chapter 1 of the book.

Editors

S.M. Muyeen
Department of Electrical and Computer Engineering
Faculty of Science and Engineering, Curtin University
Perth, Australia

and

Saifur Rahman
Electrical and Computer Engineering Department
Virginia Tech, USA

July 31, 2016

Notes on contributors

Dr. S. M. Muyeen received his B.Sc. Eng. Degree from Rajshahi University of Engineering and Technology (RUET), Bangladesh formerly known as Rajshahi Institute of Technology, in 2000 and M. Eng. and Ph.D. Degrees from Kitami Institute of Technology, Japan, in 2005 and 2008, respectively, all in Electrical and Electronic Engineering. At the present, he is working as an Associate Professor in the Electrical and Computer Engineering Department at Curtin University, Perth, Australia. His research interests are power system stability and control, electrical machine, FACTS, energy storage system (ESS), Renewable Energy, HVDC system, and Smart Grid. Dr. Muyeen has developed two research laboratories at the Petroleum Institute, Abu Dhabi, UAE. One of those is the first operational lab among 32 research laboratories under the Petroleum Institute Research Center (PIRC). He has been a Keynote Speaker and an Invited Speaker at many international conferences, workshops, and universities. He has published over 150 articles in different journals and international conferences. He has also published six books as an author or editor. Dr. Muyeen is the Senior Member of IEEE and Fellow of Engineers Australia.

Saifur Rahman is the founding director of the Advanced Research Institute (www.ari.vt.edu) at Virginia Tech where he is the Joseph R. Loring professor of electrical and computer engineering. He also directs the Center for Energy and the Global Environment (www.ceage.vt.edu). He is a Fellow of the IEEE and an IEEE Millennium Medal winner. He is the president-elect of the IEEE Power and Energy Society (PES) for 2016 and 2017 and the President for 2018 and 2019. He was the founding editor-in-chief of the *IEEE Electrification Magazine* and *IEEE Transactions on Sustainable Energy*. He has served on the editorial board of the *Proceedings of the IEEE*, and is now a member of the editorial board of the *IEEE Transactions on Sustainable Energy*. He is a member of the IEEE-USA Energy Policy Committee. In 2006, he served on the IEEE Board of Directors as the vice president for publications. He served as the chair of the US National Science Foundation Advisory Committee for International Science and Engineering from 2010 to 2013. He is a Distinguished Lecturer for the IEEE PES, and has lectured on smart grid, energy efficiency, renewable energy, demand response, distributed generation, and critical infrastructure protection topics in over 30 countries on all six continents. He has published over 500 technical papers in different Journals and International conferences.

Sebastian Lehnhoff is a Full Professor for Energy Informatics at the University of Oldenburg. He received his doctorate at the TU Dortmund University in 2009. He is a member of the executive board of the OFFIS – Institute for Information Technology and speaker of its Energy R&D division. He is speaker of the section 'Energy Informatics' within the German Informatics Society (GI), assoc. editor of the IEEE Computer Society's Computing and Smart Grid Special Technical Community as well as an active member of numerous committees and working groups focusing on ICT in future Smart Grids. He is an honorary professor of the School of Information Technology and Electrical Engineering at the University of Queensland. He is elected chair of the IEEE CA4EPI Working Group P2030.4, and chairman of the architecture and quality committee of the openKONSEQUENZ industry consortium for the development of open source software in power system operation. He is author of over 100 refereed and peer-reviewed scientific publications. His research interests focus on the integration of a large number of decentralized, renewable energy sources into the electricity supply system in combination with the politically motivated reorganization of corporate structures and business processes, which represents a huge challenge for stakeholders in energy.

Mathias Uslar has studied computer science with a minor in legal informatics at the University of Oldenburg, Germany from 1999 until 2004. In October 2004, he started working as a scientific assistant at OFFIS – Institute for Information Systems in Oldenburg, later on working there as project leader and now as group manager in the Energy division of the institute. Since 2008, he is head of the CISE, the Centre for IT Standards in the Energy Sector. In October 2009, he successfully defended his PhD thesis on the integration of heterogeneous standards in the electric utility domain and smart grids. He is leading OFFIS' national and international work packages with the scope of standardisation and interoperability. He is member of the German GI, IEEE, ACM, and IEC German mirror committee member DKE K 952, 952.0.10, 952.0.17 and international member of IE TC 57 WG 14 and 16. His research interests are with semantic modelling and technical interoperability in smart grid architectures. Currently, he is working on modelling DER like CHP or PHEV using the CIM (IEC 61968) and creating control structures for virtual power plants.

Noman Bashir received the B.S. degree in electrical engineering from the University of Engineering and Technology Lahore, in 2013, and the M.S. degree in Energy Systems Engineering from National University of Science and Technology, Islamabad in 2016. He worked as a Research Engineer at Systems and Networks (SysNet) lab on a Microsoft Research Redmond funded project in the domain of smart grid and smart homes from 2013 to 2015. He is currently working as a Research Assistant at Advanced Communication Laboratory at Lahore University of Management Sciences Lahore (LUMS) since August 2015. His research interests include smart homes, smart grid, demand-side management, and solar PV systems.

Naveed Ul Hassan received the B.E. degree in avionics engineering from the College of Aeronautical Engineering, Risalpur, Pakistan, in 2002, and the M.S. and Ph.D. degrees in electrical engineering, with a specialization in digital and wireless communications, from the Ecole Superieure d'Electricite, Gif-sur-Yvette, France, in 2006 and 2010, respectively. In 2011, he joined as an Assistant Professor with the Department of Electrical Engineering, Lahore University of Management Sciences, Lahore, Pakistan. Since 2012, he has been a Visiting Assistant Professor with the Singapore University of Technology and Design, Singapore. He has several years of research experience and has authored/co-authored numerous research papers in refereed international journals and conference proceedings. His major research interests include cross-layer design and resource optimization in wireless networks, demand response management, and integration of renewable energy sources in smart grids, indoor localization, and heterogeneous networks.

Chau Yuen received the B.Eng. and Ph.D. degrees from Nanyang Technological University, Singapore, in 2000 and 2004, respectively. In 2005, he was a Post-Doctoral Fellow with Lucent Technologies Bell Labs, Murray Hill, NJ, USA. In 2008, he was a Visiting Assistant Professor with Hong Kong Polytechnic University, Hong Kong. From 2006 to 2010, he was a Senior Research Engineer with the Institute for Infocomm Research, Singapore, where he was involved in an industrial project developing an 802.11n wireless local area network system and actively participated in the third-generation Partnership Project Long-Term Evolution (LTE) and LTE-A standardization. In 2010, he joined the Singapore University of Technology and Design, Singapore, as an Assistant Professor. He has authored over 200 research papers in international journals or conferences. He holds two U.S. patents. He received the IEEE Asia-Pacific Outstanding Young Researcher Award in 2012. He serves as an Associate Editor of the IEEE Transactions on Vehicular Technology, and was awarded as Top Associate Editor from 2009 to 2014.

Wayes Tushar received the B.Sc. degree in electrical and electronic engineering from the Bangladesh University of Engineering and Technology, Bangladesh, in 2007, and the Ph.D. degree in engineering from Australian National University, Australia, in 2013. He was a Visiting Researcher with National ICT Australia, ACT, Australia. He was also a Visiting Student Research Collaborator with the School of Engineering and Applied Science, Princeton University, NJ, USA, during summer 2011. He is currently a Research Scientist with the SUTD-MIT International Design Center, Singapore University of Technology and Design, Singapore. His research interests include signal processing for distributed networks, game theory, and energy management for smart grids. He was a recipient of two best paper awards, both as the author, in the Australian Communications Theory Workshop in 2012 and the IEEE International Conference on Communications in 2013.

Ahmed S. Musleh (S'11) earned his B.Sc. (summa cum laude) in Electrical Engineering from Abu Dhabi University, Abu Dhabi, UAE, in 2014. Currently, he is a graduate research and teaching assistant at The Petroleum Institute, Abu Dhabi, UAE, where he also pursues his M.Sc. degree in Electrical Engineering. His area of interests include smart-grid technologies, wide-area monitoring and control, FACTS devices, and power quality issues.

Ahmed Al-Durra (S'07-M'10-SM'14) received the B.S., M.S., and PhD in Electrical and Computer Engineering from the Ohio State University in 2005, 2007, and 2010, respectively. For his M. Sc. degree, he investigated the application of several nonlinear control techniques on automotive traction PEM fuel cell systems. He conducted his Ph.D. research at the Center for Automotive Research in the Ohio State University on the applications of modern estimation and control theories to automotive propulsion systems. At the present, he is working as Associate Professor in Electrical Engineering Department and the Downstream Research Coordinator of the PI Research Center at the Petroleum Institute, Abu Dhabi. His research interests are application of estimation and control theory in power system stability, Micro and Smart Grids, renewable energy, and process control. He has published over 60 scientific articles in Journals and International Conferences. He has successfully accomplished several research projects at international and national levels. He is the co-founder of Renewable Energy Laboratory at the Petroleum Institute, Senior Member in IEEE, and a Member in ASME.

Mohammed A. Abou-Khousa (S'02-M'09-SM'10) received the B.S.E.E. (magna cum laude) degree from the American University of Sharjah, Sharjah, UAE, in 2003, the M.S.E.E. degree from Concordia University, Montréal, QC, Canada, in 2004, and the Ph.D. degree in electrical engineering from the Missouri University of Science and Technology, MO, USA, in 2009. He was a Research Scientist with the Robarts Research Institute, London, ON, Canada. He is currently with the Petroleum Institute, Abu Dhabi, UAE. His research and development efforts are focused on RF, microwave, and millimeter wave instrumentation, sensors, and imaging.

Yong Liu received his Ph.D. degree in electrical engineering (power system direction) from the University of Tennessee, Knoxville, in 2013. He received his M.S. and B.S. degrees in electrical engineering from Shandong University, China, in 2007 and 2010, respectively. He is currently a research Assistant Professor in the DOE/NSF-cofunded engineering research center CURENT and Department of Electrical Engineering and Computer Science at the University of Tennessee, Knoxville. His research interests are wide-area power system measurement, power system dynamic analysis, and renewable energy integration.

Shutang You received the B.S. and M.S. degrees from Xian Jiaotong University in 2011 and 2014. He is currently working towards his Ph.D. degree in electrical engineering at the University of Tennessee, Knoxville. His major research interests include power system monitoring and system dynamics.

Yilu Liu received her M.S. and Ph.D. degrees from the Ohio State University, Columbus, in 1986 and 1989. She received the B.S. degree from Xian Jiaotong University, China. She is currently the Governor's Chair at the University of Tennessee, Knoxville and Oak Ridge National Laboratory (ORNL). She is also the Deputy Director of the DOE/NSF-cofunded engineering research center CURENT and a member of the U.S. National Academy of Engineering (NAE). Prior to joining UTK/ORNL, she was a Professor at Virginia Tech. She led the effort to create the North American power grid Frequency Monitoring Network (FNET) at

Virginia Tech, which is now operated at UTK and ORNL as GridEye. Her current research interests include power system wide-area monitoring and control, large interconnection-level dynamic simulations, electromagnetic transient analysis, and power transformer modelling and diagnosis.

Haris M. Khalid (M'13) received his B.S. (Hons.) degree in Mechatronics and Control Systems Engineering from University of Engineering and Technology (UET), Lahore, Pakistan, in 2007, and the M.S. and Ph.D. degrees in Control Systems Engineering from King Fahd University of Petroleum and Minerals (KFUPM), Dhahran, Kingdom of Saudi Arabia, in 2009 and 2012, respectively. He then joined Distributed Control Research Group (DCRG) at KFUPM, as a Research Fellow. In 2013, he joined as a Research Fellow with the Power Systems Research Laboratory (PSRL) at Department of Electrical Engineering and Computer Science, Institute Center for Energy, Masdar Institute of Science and Technology (MI), Masdar City, UAE, which is an MI-MIT Cooperative Program with Massachusetts Institute of Technology (MIT), Cambridge, MA, USA. Currently, he is a Visiting Scholar with Petroleum Institute, Abu Dhabi, UAE.

His research interests are power systems, cyber-physical systems, electric vehicles, signal processing, applied mathematics, fault diagnostics, filtering, estimation, performance monitoring, and battery management systems. He has authored 40+ peer-reviewed publications, which includes 1 Monograph, 6 IEEE Transactions, 3 IET Journals, 2 Elsevier Journals, 3 Springer Journals, and 15+ peer-reviewed International Conferences.

Eleonora Riva Sanseverino received the Doctor and Ph.D. degrees in electrical engineering from the University of Palermo, Palermo, Italy, in 1995 and 2000, respectively. From 2000 to 2001, she worked as a Limited Time Researcher with the Research Group of Electrical Power Systems, University of Palermo. From December 2001 to October 2002, she worked as a Researcher with the National Council of Research, Palermo, in the field of computer systems and computer networks. Since 2002, she has been an Assistant Professor in Power Systems with the University of Palermo. Her research interests include optimization methods for electrical distribution systems design, operation, and planning.

Maria Luisa Di Silvestre received the Masters and Ph.D. degrees in electrical engineering (electric power systems) from the University of Palermo, Palermo, Italy, in 1993 and 1998, respectively. Since 1999, she has been with the Department of Electrical Engineering, University of Palermo, as a Grant Holder from 1999 to 2000, and as a University Researcher teaching electrotechnics since 2000. Her research interests include analysis and modelling of complex grounding systems, power systems design, operation and planning.

Gaetano Zizzo received the Master's and Doctorate degrees in electrical engineering from the University of Palermo, Palermo, Italy, in 2002 and 2006, respectively. He is a Researcher with the Department of Energy, Information Engineering and Mathematical Models, University of Palermo, in the Power

System Research Group. His main research interests include electrical safety, risk analysis, earthing systems, power systems design, and storage.

Ninh Nguyen Quang received his Engineer degree in electrical engineering from Hanoi University of Science and Technology (HUST), Vietnam, in July 2006. Then, he joined the Institute of Energy Science (IES) – Vietnam Academy of Science and Technology (VAST) and focused his research on power flow and optimal power flow in power system since then. In late 2007, he went back to HUST for a Master course and received the Master Degree in electrical engineering in December 2009. Nguyen started his career at Institute of Energy Science (IES) in 2010 and specialized in the areas of Transmission and Distribution (T&D) Engineering, System Operations, Transmission Planning, Smart Grid Planning, and System Integration. He earned his Ph.D. in Electrical Engineering at Dipartimento di Energia, ingegneria dell'Informazione modelli Matematici (DEIM) in the University of Palermo, Italia, in March 2016. His Ph.D. thesis focused on optimal power flow in islanded microgrid. Since returning to IES early this year, he has covered researching on solar photovoltaics renewable energy integration assessment, optimal power flow in three phases microgrids and implementing renewable energy-related projects.

Adriana Carolina Luna Hernandez received the B.S. degree in electronic engineering in 2006 and M.S. degree in Industrial Automation in 2011, both from Universidad Nacional de Colombia. She is currently working toward the Ph.D. degree in the Department of Energy Technology at Aalborg University. Her research work focuses on energy management systems of microgrids, and specifically on architectures and algorithms for scheduling and optimization for operation level in microgrids.

Josep M. Guerrero (S'01-M'04-SM'08-F'15) received the B.S. degree in telecommunications engineering, the M.S. degree in electronics engineering, and the Ph.D. degree in power electronics from the Technical University of Catalonia, Barcelona, Spain, in 1997, 2000, and 2003, respectively. Since 2011, he has been a Full Professor with the Department of Energy Technology, Aalborg University, Aalborg, Denmark, where he is responsible for the Microgrid Research Program. His research interests include different microgrid aspects, power electronics, distributed energy-storage systems, hierarchical and cooperative control, energy management systems, and optimization of microgrids and islanded minigrids.

Jasrul Jamani Bin Jamian obtained his B.Eng. (Hons) in Electrical Engineering (2008), M.Eng. (2010), and Ph.D. (2013) in Electrical (Power) from Universiti Teknologi Malaysia (UTM), Johor Bahru, Malaysia. He is currently a Senior Lecturer in Electrical Power Engineering department, Faculty of Electrical Engineering, Universiti Teknologi Malaysia. He is the author and co-author of more than 50 publications in the area of Power Systems and Energy since 2013. He also involves actively as reviewer for major international journal of IET, Elsevier, Wiley, and several international conferences. His current research interests include power system optimization, power system stability, renewable energy application, and their control method.

Muhammad Ariff Baharudin completed his Master of Engineering (Electrical-Communination) at Universiti Teknologi Malaysia and Shibaura Institute of Technology (SIT), Tokyo Japan in 2011 and Ph.D. (2015) in communication network from Shibaura Institute of Technology (SIT), Tokyo Japan. He is currently a Senior Lecturer in Department of Communication Engineering, Faculty of Electrical Engineering, Universiti Teknologi Malaysia. He is the author and co-author of more than 30 publications in the area of Communication and Power Systems and Energy since 2013. His research interests are in communication, renewable energy systems, and power systems engineering.

Mohd Wazir Mustafa received his B.Eng. degree (1988), M. Sc. (1993), and PhD (1997) from the University of Strathclyde, UK. He is currently a Professor and Deputy Dean (Research and Graduate Studies) at Faculty of Electrical Engineering, Universiti Teknologi Malaysia. He is the author and co-author of more than 200 publications in the area of Power Systems and Energy. He is actively involved in research as a principle investigator and member with a total amount of research grant worth more than 1.5 million (Ringgit Malaysia). He also involves actively as reviewer for major international journal of IET, IEEE, Elsevier, and several international conferences. His research interest includes power system stability, FACTS, and power system distribution automation.

Hazlie Mokhlis received B.Eng. degree in electrical engineering and M.Eng.Sc. degree from the University of Malaya, Malaysia, in 1999 and 2002, respectively. He obtained his Ph.D. degree from the University of Manchester, UK, in 2009. He is currently Associate Professor and Head of Department for Electrical Engineering department, University of Malaya. He was a Deputy Dean for Research and Higher Degree at the Faculty of Engineering. He is actively involved in research as a principle investigator with a total amount of research grant worth more than 2 million (Ringgit Malaysia). He is the author and co-author of more than 200 publications in the area of Power Systems and Energy. He also involves actively as reviewer for major international journal of IET, IEEE, Elsevier, and several international conferences. Besides involve with research, he is also active in the development of Malaysian Standard for Malaysian's Power System Analysis and Studies. His research interest includes fault location, network reconfiguration, islanding operation, islanding detection, and renewable energy.

Abdul R. Beig received the Bachelor's degree in Electrical Engineering from the National Institute of Technology Karnataka, Suratkal, India, in 1989, and the Masters and Ph.D. degrees from the Indian Institute of Science, Bangalore, India, in 1998 and 2004, respectively. He is currently an Associate Professor with the Department of Electrical Engineering, Petroleum Institute, Abu Dhabi, UAE. His research interests include ac drives, multilevel inverters, power quality, power electronics applications to power systems, and smart-grid technology. He has excellent industrial experience especially in the design of inverters; DSP/FPGA based embedded controllers and development of control algorithms for electric drives. From 1989 to 1992, he was with M/S Kirloskar Electric Company, Ltd, Mysore, India, as a R&D Engineer with the design team of BLDC drive. He later joined the Department of Electrical Engineering, National Institute of Technology Karnataka Suratkal. He is the recipient of Research award and Young Teacher Award, The Petroleum Institute, Abu Dhabi, UAE, National award for the Ph.D. thesis by Indian National academy of Engineers (INAE), India, and L&T-ISTE National award: Best M.E. Thesis in Electrical/Electronic Engineering.

Faisal Mumtaz received the B.Sc. degree in Electrical Power Engineering from the COMSATS Institute of Information Technology, Pakistan, in 2012 and the M.Sc. degree in Electrical Power Engineering from Masdar Institute of Science and Technology (MIST), Abu Dhabi, UAE, in 2015. He also worked as a research assistant during his M.Sc. at MIST. He is currently pursuing his Ph.D. degree in Sustainable Energy from Hamad Bin Khalifa University, Doha, Qatar. His research interests include distributed generation, renewable energy generation, demand side management, intelligent load shedding, smart girds, microgrids, stability and protection of microgrids.

Islam Safak Bayram received the B.S. degree in electrical and electronics engineering from Dokuz Eylul University, Izmir, Turkey, in June 2007, the M.S. degree in telecommunications from the University of Pittsburgh in August 2010, and the Ph.D. degree in computer engineering from North Carolina State University, in December 2013. He received the Best Paper Award at the Third IEEE International Conference on Smart Grid Communications. Between January and December 2014, he worked as a postdoctoral research scientist at Texas A&M University at Qatar. Currently, he is working as an assistant professor at College of Science and Engineering at Hamad Bin Khalifa University and a scientist at Qatar Environment and Energy Research Institute and his research interests include stochastic modelling and control of communications and power networks.

Ali Y. Elrayyah, received his B.Sc. in Electrical Engineering from University of Khartoum, Sudan, 2003, M.Sc. in Systems Engineering from King Fahd University of Petroleum and Minerals, Saudi Arabia, 2009 and Ph.D. in Electrical Engineering from The University of Akron, OH, USA, 2013. He has conducted a number of research work related to microgrids control, solar photo-voltaic systems integration with the grid, and electric motors drives. He also has conducted research work related to modelling and control of solar thermal energy systems. His research interests are renewable energy sources, energy management systems and demand side response.

Zhile Yang is pursuing his Ph.D. degree at the School of Electrical, Electronics and Computer Science, Queen's University Belfast (QUB). He received the B.Eng. in Electrical Engineering and the M.Sc. degree in Control Theory and Control Engineering from Shanghai University (SHU) in 2010 and 2013, respectively. His research interests focus on computational intelligence especially evolutionary computation methods and their applications on smart-grid integration with renewable energy and electric vehicles. He is founding chair of IEEE QUB student branch and an active student member of IEEE PES, CIS, and SMC societies. He is the author or co-author of more than 20 articles in peer-reviewed international journals and conferences.

Kang Li received the B.Sc. degree in industrial automation from Xiangtan University, Hunan, China, in 1989; the M.Sc. degree in control theory and applications from Harbin Institute of Technology, Harbin, China, in 1992; and the Ph.D. degree in control theory and applications from Shanghai Jiaotong University, Shanghai, China, in 1995. Between 1995 and 2002, he was with Shanghai Jiaotong University, Shanghai, China; Delft University of Technology, Delft, The Netherlands; and Queen's University Belfast, Belfast, UK, as a Research Fellow. Since 2002, he was a Lecturer, a Senior Lecturer (2007), and a Reader (2009) with the School of Electronics, Electrical Engineering, and Computer Science, Queen's University Belfast, and he became a Professor in 2011. He is a Visiting Professor with Harbin Institute of Technology and Ningbo Institute of Technology, Zhejiang University, Hangzhou, China. His research interests include nonlinear system modelling, identification, and control, and bio-inspired computational intelligence, with applications to energy and power systems, smart grid, electric vehicles, and polymer processing. He is the author or co-author of more than 300 articles, and has edited or coedited over 10 conference proceedings.

K.T.M. Udayanga Hemapala received the B.Sc. (Eng.) degree from University of Moratuwa, Sri Lanka, in 2004 and the PhD degree from University of Genova, Italy. He is a Senior Lecturer in the Department of Electrical Engineering, University of Moratuwa, Sri Lanka. His research interests are in industrial robotics, distributed generation, power system control and smart grid.

Sarinda Lahiru Jayasinghe received the B.Sc. degree in electrical engineering from the University of Moratuwa, Sri Lanka, in 2014, and currently he is an M.Eng. candidate in electrical engineering of Memorial University of Newfoundland, Canada. His research interests include Internet of Things (IoT), SCADA systems, Multi-Agent System (MAS)-based Smart Grids, Digital Displacement, Fuzzy Logic and Neural Control and Electric Vehicles.

Asitha L. Kulasekera is a lecturer in the department of Mechanical Engineering, University of Moratuwa, Sri Lanka specializing in Mechatronics Systems. He completed his bachelor's degree in Engineering (2010) and M.Sc. (2012) from the University of Moratuwa. His main research areas include Intelligent Systems, Multi-Agent Systems, Energy Conservation, Renewable Energy Systems, MEMS-based sensor design, and cognition.

Mohammad Babakmehr (S'14) received the B.S. degree in electrical engineering in 2008 from Central Tehran University and the M.Sc. degree in Biomedical-Bioelectric engineering in 2011 from the Amirkabir University of Technology, Tehran, Iran. He is currently a Ph.D. degree candidate in the Department of Electrical Engineering and Computer Science, Colorado School of Mines (CSM), Golden. He has been with the Center for the Advanced Control of Energy and Power Systems (CSM), since 2013. His research interests include smart-grid technologies, compressive sensing, advance signal processing, and control theory.

Marcelo G. Simões (S'89-M'95-SM'98-F'15) Marcelo Godoy Simões received a B.Sc. degree from the University of São Paulo, Brazil, an M.Sc. degree from the University of São Paulo, Brazil, and a Ph.D. degree from The University of Tennessee, USA, in 1985, 1990, and 1995, respectively. He received his D.Sc. degree (Livre-Docência) from the University of São Paulo in 1998. He was a US Fulbright Fellow for AY 2014–2015, working for Aalborg University, Institute of Energy Technology (Denmark). He is currently with Colorado School of Mines. He has been elevated to the grade of IEEE Fellow, Class of 2016, with the citation: 'for applications of artificial intelligence in control of power electronics systems'.

Srinivasa Rao Kamala received his Bachelor of Technology (B.Tech.) in Electrical and Electronics Engineering from Acharya Nagarjuna University, India. in the year 2007. He completed his M.S./Research in 2011 from Indian Institute of Technology Madras, India. He worked as Associate Programmer at Cognizant Technology Solutions, India, during 2011–2014. He also worked as Research Associate at National University of Singapore during 2014–2016. He is currently a Ph.D. Research Scholar at National University of Singapore, supervised by Prof. Sanjib Kumar Panda. His research interests include FACTS controllers, power system analysis and evolutionary computation techniques, and power quality. He is currently working on electrical network modelling, simulations, and harmonic mitigation techniques

Charalambos Konstantinou received the five-year diploma degree in electrical and computer engineering from the National Technical University of Athens (NTUA), Athens, Greece. He is currently pursuing the Ph.D. degree in electrical engineering at the Tandon School of Engineering, New York University (NYU), Brooklyn, NY, USA. His interests include hardware security with particular focus on embedded systems and smart-grid technologies.

Michail Maniatakos received the B.Sc. degree in computer science and the M.Sc. degree in embedded systems from the University of Piraeus, Piraeus, Greece, in 2006 and 2007, and the M.Sc. and M.Phil. degrees and the Ph.D. degree in electrical engineering from Yale University, New Haven, CT, USA, in 2009, 2010, and 2012. He is an Assistant Professor of Electrical and Computer Engineering at New York University Abu Dhabi, and a Research Assistant Professor at the NYU Tandon School of Engineering. He is the Director of the MoMA Laboratory (nyuad. nyu.edu/momalab). His research interests, funded by industrial partners and the U.S. Government, include robust microprocessor architectures, privacy preserving computation, as well as industrial control systems security. He has authored several publications in IEEE transactions and conferences, and holds patents on privacy-preserving data processing.

He is currently the Co-Chair of the Security track at IEEE International Conference on Computer Design (ICCD) and IEEE International Conference on Very

Large-Scale Integration (VLSI-SoC), and also serves in the technical program committee for various conferences. He has organized several workshops on security, and he currently is the faculty lead for the Embedded Security challenge held annually at Cyber Security Awareness Week (CSAW), Brooklyn, NY, USA.

Robert Czechowski was born in Glogow, Poland, on November 12, 1981. In 2007, he got Rector's award for the best engineering thesis of the specialization: Computer Engineering. He received his M.Sc. degree in Computer Science from the Wroclaw University of Science and Technology in 2009. In 2011, he received his Postgraduate studies diploma – Security management of information systems realized by Wroclaw University of Science and Technology on Computer Science and Management Department. Postgraduate technical studies were dedicated for entrepreneurs and employees only and co-financed by the European Union through the European Social Fund. For five years he worked as ICT specialist in enterprise of an industrial character. Currently, he is a senior programmer on Wroclaw University of Science and Technology. His research interests are in communication and security of smart grid, data flow in smart metering, ICT systems dedicated smart power grids, automatics using AVR microcontrollers and optoelectronics. Presently, he is involved in international grant 'Cyber-Physical Security for Low-Voltage Grids' (NCBiR grant of the ERA-NET No 1/SMARTGRIDS/2014, acronym SALVAGE) and Polish Smart Power Grids Section. Author of several publications on the topics of Digital Security Smart Grids.

Chapter 1

Introduction and motivation

S. M. Muyeen[1] and Saifur Rahman[2]

The energy consumption rate is increasing rapidly at national and international levels, therefore, the energy sustainability, reliability, and carbon footprint are becoming key issues to be addressed in the twenty-first century. The power system is to be re-architected and the changes will require a paradigm shift in the electricity delivery system. The smart grid is the term applied to tomorrow's electricity system. This chapter presents an overview of the smart grid with its general features, functionalities, and characteristics along with its present status and future trend.

1.1 Power system to smart grid

Electric power first saw commercial use in the 1870s. The electrical revolution starts in the late 1880s through the invention of electrical machines and the concept of the power system starts developing. Before reaching 1900s, the power system transmission length exceeded few tens of kilometers which shows light for the development of large power system. The development the power system got a boost in the late 1940s once the electronics revolution is kicked off and reached to a matured stage after entering to the commutation era. In the early twenty-first century, the concerns of resolving the limitations and costs of electrical grid have become apparent. The advanced metering options remove the barrier of averaging peak power prices to all consumers equally, in the 1980s.

In the late 1990s, the environmental issue due to the burning of fossil fuel for mass power production became prominent, so carbon footprint reduction appeared as a major challenge at national and international levels. The depletion of fossil fuel gave a boost to the increase in the renewable power penetration level to the power grid throughout the world. Over the last decade, many renewable energy technologies such as solar, wind, biomass, and wave have advanced significantly, from the viewpoint of conversation efficiency and unit cost production. Among those renewable energy sources, wind, solar, and hydro power stand as true alternatives to conventional technologies for electricity generation. As a result, a distributed power

[1]The Department of Electrical and Computer Engineering, Curtin University, Perth, WA, Australia
[2]Electrical and Computer Engineering Department, Virginia Tech, USA

generation scheme with and without grid connectivity options is becoming popular nowadays, though high penetration of renewable energy has many adverse impact on power system operations. The Renewables Portfolio Standards in the USA, for example, was established in California in 2002 with a goal of increasing the percentage of renewable energy sources to 20% by 2017 and 33% by 2020. In Europe, the target is to raise the penetration from current level of 20% to about 50% by 2050.

On the other hand, electrical power grids are considered as the largest and most complex among all human engineered systems which is becoming more complex every day. Both the high-voltage AC and DC transmission line exists and super-conducting cables are going to be connected with future grid which makes the power grid more complex and vulnerable. The growing concern over terrorist attack in some countries has led to calls for a more robust power grid. The modern power grid is expected to be more robust, stable and reliable, energy efficient, environment friendly, consumer interactive, and threaten free from cyberattack which brought the concept of the smart grid.

The US Energy Independence and Security Act of 2007 (EISA-2007) released the first official definition of the smart grid, which was approved by the US Congress in January 2007, and signed into law by President George W. Bush in December 2007. Ten characteristics of the smart grid are provided in Title XIII of that bill as given below:

1. Increased use of digital information and control technology to improve reliability, security, and efficiency of the electric grid.
2. Dynamic optimization of grid operations and resources, with full cyber-security.
3. Deployment and integration of distributed resources and generation, including renewable resources.
4. Development and incorporation of demand response, demand-side resources, and energy-efficiency resources.
5. Deployment of "smart" technologies (real-time, automated, interactive technologies that optimize the physical operation of appliances and consumer devices) for metering, communications concerning grid operations and status, and distribution automation.
6. Integration of "smart" appliances and consumer devices.
7. Deployment and integration of advanced electricity storage and peak-shaving technologies, including plug-in electric and hybrid electric vehicles, and thermal storage air conditioning.
8. Provision to consumers of timely information and control options.
9. Development of standards for communication and interoperability of appliances and equipment connected to the electric grid, including the infrastructure serving the grid.
10. Identification and lowering of unreasonable or unnecessary barriers to adoption of smart grid technologies, practices, and services.

The power grid from generation to transmission, then to distribution and finally to smart consumption, has undergone a major architectural transformation through novel contributions from power system, computer, and communication engineering disciplines and this re-architected form of power grid is recently being told as smart grid. In a nutshell, the smart grid is simply the smart *Generation-Reticulation-Information & Communication-Decarbonization.*

1.2 Technical aspects—present and future

A smart grid is a modern electric system. It has its own architecture, communications, sensors, automation, and computers to improve the efficiency, reliability, flexibility, and security of the electricity system. This section attempts to draw a very simplified outline of present and future smart grid as presented below.

1.2.1 Smart-grid architecture

It is not easy to define the future smart-grid architecture. The recent smart-grid architecture is close to the form defined by Smart Grid Coordination Group Reference Architecture Working Group (SG-CG/RA) in 2012. In general, the smart grid can be represented by three main layers—the power system component layer, the information and communication layer, and the power and energy management layer as shown in Figure 1.1 which is a modified form of the architecture given in the CEN-CENELEC-ETSI, 2012, report. The power system component layer is composed of power system primary equipment such as generating plants, transmission, and distribution systems. The distributed energy resources (DERs) are also interconnected with a distribution system. The power system component layer and the power and energy management layer are interlinked closely with modern information and communication layer in different aspects.

1.2.2 Smart-grid communications

The smart grid uses advanced and sophisticated information and communication technologies (ICT) to control power delivery in a reliable and efficient way. At present, there exists some ICT infrastructure at the transmission level but it requires bidirectional real-time communication between generation and receiving ends. The smart grid can be regarded as an electricity delivery system that uses information, dual-way, cyber-secure communication technologies with integrated computational intelligence across generation plants, transmission, substations, distribution, and consumption. The present revolution in communication technologies has added much flexibility and control leniency in all electrical components of the power system.

Different communication technologies supported by two main communications media, i.e., wired (power line communication, optical fiber, digital subscriber line, Ethernet, etc.) and wireless (wireless personal area network, wireless LAN, worldwide interoperability for microwave access [WiMax], cellular, satellite, z-wave, etc.), can be used for data transmission between smart meters and electric utilities. Smart-grid communication network can be classified into three subnetworks—home area network (HAN), neighborhood area network (NAN), and wide-area network

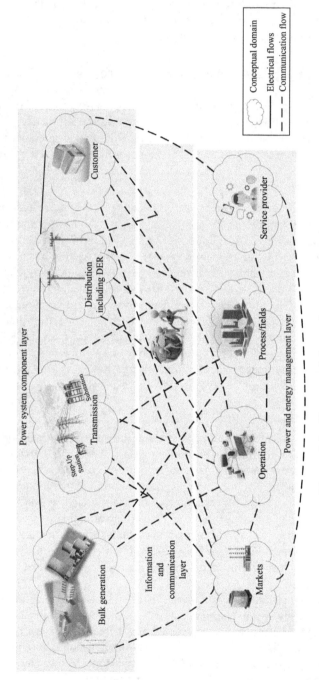

Power system component layer

Customer

Distribution
including DER

Transmission

Subsation

Step-Up
Station

Bulk generation

Information
and
communication
layer

Power and energy management layer

Service provider

Process/fields

Operation

Markets

Conceptual domain

Electrical flows

Communication flow

Figure 1.1 Conceptual form of smart-grid architecture

(WAN). HAN is suitable when the distance is few tens of meters. For a distance in between 100 m and 10 km, NAN is used. WAN is used when the distance is more than 10 km. Different types of protocols/standards are available for these communication technologies such as IEEE P1901.2, IEEE 1901, IEEE 802.3, IEEE 802.15.4, IEEE 802.16m, ADSL, VDSL, 3G, 4G, and Z-wave 400. Data rate also varies among different technologies from few tens of kbps to few tens of Gbps, e.g., the optical fiber WDN protocol supports a data rate of up to 40 Gbps.

1.2.3 Smart measurement and sensing

The power system component and power and energy management layers of the smart grid shown in Figure 1.1 are connected with different kinds of traditional and smart sensors which are required for smooth, reliable, and efficient operations of the smart grid. Some commonly used sensors are voltage sensor, current sensor, speed sensor, temperature sensor, partial discharge sensor, magnetic flux sensor, vibration sensor, pressure sensor, level sensor, medium density sensor, blade pitch angle sensor, wind speed sensor, wind direction sensor, dust-level sensor, axis angle sensor, irradiance sensor, lightning sensor, conductor motion sensor, insulation sensor, energy meter, power meter, power quality meter, phasor measurement unit (PMU), etc.

Smart meters also let market players design and provide new, innovative services. Moreover, smart meters greatly contribute to optimizing network management. They allow better fault identification and localization on MV and LV networks, ensuring faster interventions and reduced outage duration. They also allow detailed monitoring of power quality and increase the capacity to act remotely on the power networks and in particular to manage peak-shaving programs.

1.2.4 Smart transmission and distribution and consumer systems

A key element in the development of smart power transmission systems over the past decade is the tremendous advancement of the wide area measurement system (WAMS) technology, which is also known as the synchrophasor technology. WAMS is superior to the conventional supervisory control and data acquisition (SCADA) system in many senses. WAMS allows monitoring for large electric power grids over wide geographic areas in real time. PMUs are the fundamental of WAMS and are currently being installed at different points in the North American grid, especially under the smart-grid initiatives of the US Department of Energy. PMS can record and communicate at high sampling rate (6–60 samples/s) which is synchronized with GPS. The phasor data concentrator (PDC) unit is responsible for combining data from different PMUs, align data based on the time-stamps, condition and in some cases eliminate the bad data. The IEEE standards related to PMU are C37.238-2011, C37.118.1-2011, C37.118.2-2011, C37.242-2013, C37.244-2013, C37.111, etc.

Supervising and synchronization of WANs were revolutionized in the early 1990s with the expansion of smart-grid research by Bonneville Power Administration (BPA) with prototype sensors capable of quick analysis of abnormalities in electrical quantities over long distances. The outcome of that work was the first operational WAMS in 2000.

Visualizing future form of smart distribution system is not an easy task. The smart grid should be able to provide new abilities such as self-healing, high reliability, energy management, and real-time pricing which may incorporate technologies such as smart metering, advanced communication, automation, distributed generation, and distributed storage system. Advanced digital meters should have the feature of dual-way communication, controllable from remote locations, big memory for recording waveforms, abilities to monitor electrical quantiles such as voltage, current, even harmonics and definitely the structure to support real-time rate and time of use. In recent distribution system, the distribution automation devices operate as an "intelligent node," can interrupt fault current, monitor currents and voltages, communicate with each other, and, in some cases, automatically reconfigure the system based on set or defined objectives. DERs are small sources of generation and/or storage that are connected to the distribution system. If the penetration level of DER is low, the power grid has less possibility to get affected if the protection system is well designed, but the risk increases for high penetration.

The distribution system is becoming more complex after successful commercialization of electric vehicles (EVs) as lot of EVs are going to be connected with present and future grid which requires a bulk amount of electricity or charging. The vehicle-to-grid, the provision of energy or ancillary services from EVs to the grid, may bring new operational changes in the future distribution system. This may also lead to typical power system stability problem and required to address carefully. On the other hand, the definition of energy system is quite large, although for smart electrical grids the inclusion of DERs, EVs creates multi-carrier energy hubs which require optimal energy management system (EMS). Different types of EMS exist for power grid operations. The hierarchical three-level control scheme is found to be effective in microgrid operation; the primary control adopts droop control, voltage and frequency controls are taken care in the secondary control level, where, in the tertiary control level, the optimization and decision-making algorithms are being applied. Centralized and decentralized control and optimal power flow are other popular methods for smart-grid energy management.

The concept of electricity user or consumer has changed drastically. In traditional power systems, the grid operators have largely worked in the paradigm that supply or generation matches with consumer demand. That caused many practical implications in designing and operation of power grid, e.g., power capacity, sizing of system peak, spinning reserve, demand forecasting, etc. On the other hand, the pricing of electricity was based on time-of-use from the very beginning, causing averaging the electricity price. The consumers have no role in demand-side energy management. The situation has changed when consumers start to become empowered consumers after the smart-grid concept is developed. Smart consumers started to take part in energy management and have a greater role in power system operations. To develop a better system, the smart consumers should own and control their detailed consumption data and have direct meter access to their consumption data. The awareness of energy savings at the consumer level should be increased and it is important to bring some changes in the energy consumption behavior of smart consumers.

1.2.5   Energy storage system

Energy storage system (ESS) is going to play a vital role in the present and future smart-grid paradigm. The huge penetration of DERs to the smart-grid requires new grid policies/regulations, additional spinning reserves, and load ramping capabilities to mitigate intermittent and stochastics generation patterns which causes lot of voltage and power fluctuations. The smart way to mitigate those fluctuations is to use ESS at generation, distribution, and consumer ends. The ESS can basically decouple the generation and demand by providing a buffer zone which can be utilized in other purposes, e.g., energy management system where energy can be captured during off-peak hours and delivered during peak hours.

Different types of energy storage technologies are available till to date and researchers and scientists are working on developing more feasible, reliable, and economical storage systems. Modern ESSs are integrated with advanced power electronic converters/inverters which enable fast control of both real and reactive powers through dual-way energy exchange with connection point. The most popular storage device is the battery energy storage system (BESS) in which different types of battery bank are integrated with a bi-directional DC–DC converter for charging and discharging purposes. If it is required to connect the BESS to AC system, then a DC–AC inverter is also integrated with battery bank and DC–DC converter. For large-scale application, NAS battery is getting huge attention throughout the world. Deep-cycle flooded lead acid batteries and deep-cycle valve-regulated lead acid batteries are popular for solar power system. Nickel-metal hydride and lithium-ion batteries find applications in modern electric vehicles.

The superconducting magnetic energy storage (SMES) system is an excellent ESS device which has very fast charging and discharging capabilities. The technology requires SMES coil, two quadrant DC–DC choppers, and DC–AC inverter to connect the unit with the electric grid. The only drawback of this technology is that a cooling system is required for continuous operation which increases the running cost, so this technology could not serve for practical and large applications. On the other hand, a similar technology, energy capacitor system, which uses supercapacitor bank, bi-directional DC–DC converter, and DC–AC inverter, also has the capability to control real and reactive power at a higher speed. This technology finds many applications in renewable energy, electric vehicle, and smart-grid sectors, though it is expensive. A flywheel system has also been identified as future storage device in smart-grid applications after successful commissioning of 20 MW system in New York city in mid-2015 by Beacon Power, USA.

1.2.6   Smart-grid control aspects

The smart grid requires advanced control at component and system levels. Different kinds of industrial controls are applied to generation, transmission, and distribution levels. Recently, different non-linear controls, such as back-stepping control, feedback linearization, model predictive control, and sliding mode control, are applied to control of DERs and interfacing DERs to grid, ESS, and other smart-grid equipment. A non-linear control system can be defined as all those systems that do not follow the principle of homogeneity. The power system is completely non-linear and, therefore, the application of non-linear control is becoming popular in power system applications. Apart from that different intelligent control techniques, such as fuzzy logic and artificial neural network, are being used for the same applications. The main advantages of this control are its simplicity, adaptability, and superior performance. Complex non-linear mapping of system's dynamics through terminal characteristics can be obtained without the need for structured models. These controllers can implement design objectives that are difficult to express mathematically in linguistic or descriptive rules.

The smart grid requires different kinds of optimization at planning and operation stages as given below.

Planning:

- Generation scheduling
- Maintenance scheduling
- Generation mix planning
- Optimal protection and switching device placement
- Prioritizing investments in distribution network

Operative control:

- Constrained load flow
- Power plant operation optimizer
- Unit commitment—economic dispatch
- Optimal power flow
- FACTS (flexible AC transmission system) control
- Voltage/VAr and loss reduction
- Dynamic load modeling
- Short-term load forecast
- Network reconfiguration and load reduction
- Controller tuning

These can be formulated with different kinds of optimization techniques such as constrained and unconstrained optimization problems, linear, non-linear, geometric, and quadratic programming problems, deterministic and stochastic programming problems, and single and multi-objective programming problems.

Apart from that, few different kinds of new control approaches are examined in the smart grid. The application of multi-agent systems (MASs) in smart grid is becoming popular due to their inherent benefits such as increased autonomy, reactivity, proactivity, and social ability. MASs are complex systems composed of several autonomous agents with only local knowledge and limited abilities, but are able to interact in order to achieve a global objective. MAS use a new programming paradigm to implement agents, which is bringing about a new programming paradigm for software engineering called agent-oriented programming. As speedy communication facilities such as fiberoptic, microwave, GSM/GPRS, and 3G, 4G are integrated parts of the smart grid, integration of MAS in smart-grid applications is becoming simple and feasible. In order to implement the MAS, there are several commercial and open-source MAS building toolkits available, e.g., Voyager, Zeus, Tracy, SPRINGS, and JADE. MAS already finds applications in demand/response or energy management of the smart grid.

Compressive sensing is a signal processing technique to acquire and reconstruct a signal effectively or efficiently by obtaining solutions to underdetermined linear systems. It uses the optimization technique, the sparsity of a signal can be exploited to recover it from a very small number of samples compared to the Shannon–Nyquist sampling theorem, through the use of the optimization algorithm. It combines the sampling and compression into one step by measuring minimum samples that contain maximum information about the signal, which eliminates the need to acquire and store a large number of samples only to drop most of them because of their minimal value. Compressive sensing has seen major applications in diverse fields, ranging from image processing to gathering geophysics data. Recently, it finds applications in smart gird such as power line outage and network topology identifications and has promising futures to be applied in many other areas.

1.2.7 Stability and security

Stability is a classical problem in power systems. As the power system is becoming more complex, the voltage, frequency, and small signal stability issue are key challenges to be resolved. The challenges are becoming more prominent due to the grid integration issues of DERs to the existing power grid. FACTS devices are identified as effective tools to enhance the power system stabilities. Many series and shunt-connected FACTS devices are in operations at different places in the world.

In the smart grid concept, SCADA, WAMS plays an important role in control and operation of equipment in generating, transmitting, and distributing plants/ stations. SCADA contain computers and applications that handles prime functions in providing required and important services to utilities such as electricity, natural gas, gasoline, water, transportation systems, and many other critical infrastructures such as oil and gas industries, rigs, and processing plants. By allowing the collection and analysis of data and control of equipment such as pumps and valves from remote locations, SCADA networks provide great efficiency and are widely being used. However, security part of the SCADA system is potentially vulnerable in many cases and often subject to cyberattack. Information and Technology department typically works on the assumption that the SCADA system is beyond the access level of intranet or internet or remote access points, which is not true and invites attackers through different media. The system also lacks real-time data analyzing and bad data detection tools which do not allow real-time protection. The same misconception applies to WAMS as well. This put the smart-grid equipment and infrastructures at high risk, therefore, stability and security aspects should be handled with proper care to make the system robust and reliable.

1.2.8 Policy, regulation, standard

To make an effective smart grid, a lot of policies and regulations are yet to be developed. Policies and regulations are required at the transmission system operator level for successful grid architecture, asset management, power technologies, network operations, and market designs. On the other hand, in the distribution system operator (DSO) level, we need policies and regulations on network planning and assent management, integration of smart customers, integration of DER, and network operations. The DSO has responsibility over the full distribution network and encourages the integration of the consumer in the most effective and economical way to maintain the integrity, safety, and service level of the network while ensuring overall energy efficiency. Smart regulation is required for rewarding and incentivizing capital expenditures and improving the evaluation of operational expenditures for smart grid.

The smart grid is a system of systems from different sources. The systems are not just a collection of components but also communication protocols, information and data models, software implementations of algorithms, etc. Therefore, it is very important that components and subsystems from different suppliers can work together which is known as interoperability. The components should work as a plug-and-play fashion as closely as possible, maintaining security and reliability of the overall system. This requires proper standardization at the international level. The interoperability and standards can give a boost to the development and deployment of control solutions for the smart grid.

1.3   What is inside the book

The key issues addressed in each chapter are highlighted briefly in the following sections.

In Chapter 2, available definitions of the smart grid are presented at the beginning from which the smart-grid concept—in the context of present and future—can be figured out. The prime focus of this chapter is to introduce the architecture of present and future smart grid. The ICT and its impact on smart grid and different functions and roles of the smart grid are also presented in detail.

In Chapter 3, the smart-grid communication technologies, standards, and protocols for the physical layer operations are discussed in detail. Different kinds of existing and future wired and wireless communications technologies, e.g., power line communication, cellular network, IP networks, ZigBee, Wi-Fi, WiMAX, etc., are presented in detail. The smart grid communication network is basically explained under three umbrellas, i.e., access tier or HAN, distribution tier or NAN, and core tier or WAN. Comparative studies among different technologies have also been shown in this chapter.

Different types of measuring and sensing devices used in smart grids are discussed in Chapter 4. The chapter starts with the fundamentals and definition of sensors and sensing technologies followed by a sensing mechanism. The smart-grid sensor and measurement device classifications have been created. Finally, the use of smart sensors in generation, transmission, distribution, and consider ends has been demonstrated.

In Chapter 5, the smart transmission system and the wide-area monitoring system are discussed. The details of PMU, its components and standards are presented. The architecture of WAMS along with its various applications in smart grid is depicted. A nice case study on a pioneering WAMS system is illustrated at the end.

The bad data attack is becoming an emerging threat because of growing dependency of digital measurements in monitoring and control applications of a smart grid. Chapter 6 discusses general estimation techniques and focus is given mainly on bad data detection techniques. A case study on smart-grid application in oscillation monitoring is presented as well.

Energy management is identified as a critical issue for smart grid, especially when a large number of DERs are connected with power network. It is very important to attain minimum cost or power loss along with fulfilling other constraints to make the power and energy flow optimum. In Chapter 7, the optimum energy management issues of a smart grid are discussed in detail. A case study has also been presented based on a two-level centralized energy management approach of the smart grid which is helpful for students and researchers.

In Chapter 8, the key components of smart distribution network such as DERs, advancing metering infrastructure, smart appliances, operating time, and mode based load are covered. The distribution architecture, key challenges for distribution network, protocol, and application layer have also been presented. A nice case study on frequency balancing of the smart grid is demonstrated at the end.

Smart consumer concept, types, and their roles and responsibilities are discussed in Chapter 9. Three types of smart consumers, namely, industrial, commercial, and residential, are described in detail along with demand-side management of the smart grid. The smart-grid consumer awareness, consumer behavior, need for policy changes, etc. are the other prime issues presented in the chapter.

Due to the presence of many renewable energy sources, the ESS is part and parcel of the present and future smart grid. In Chapter 10, different types of ESSs such as battery, supercapacitor, flywheel, and pumped hydroelectric storage have been presented. A brief description of the power electronic converter/inverter required for modern ESSs has also been presented. The chapter contains useful technical data for different storage technologies. Finally, a case study has been illustrated where ESS is used to resolve different power system problems when intermittent renewable sources are connected to the power network.

The plug-in vehicles are becoming very popular and is identified as the future of transportation system. Therefore, a large number of electric vehicles are going to be connected with the future network which will cause different types of energy management, stability, and reliability problems. In Chapter 11, an advanced control and optimization methods have been utilized to integrate plug-in vehicles with smart grid by providing optimal charging and discharging profiles. A case study of unit commitment integrating plug-in vehicles in the form of nonconvex mix-integer optimization problem is formulated and numerical results are demonstrated.

In Chapter 12, an advanced control technique based on multi-agent system (MAS) is presented which has various applications in the smart grid such as power system restoration, electricity trading, and optimization. Two case studies, i.e., MAS in a microgrid and application of MAS in smart-grid transmission/generation, have been presented which is very useful for future researchers.

The applications of a newly born theorem in signal processing and system identification, the compressive sensing-sparse recovery (CS-SR), are demonstrated in smart power grid monitoring, security, and reliability assessment, in Chapter 13. The chapter starts with a short background on compressive sensing-sparse recovery theorems and techniques followed by the state of the art in CS-SR applications in smart-grid technology. At the end, three distinctive monitoring problems are addressed elaborately, i.e., power line outage identification (POI-SRP), power network topology identification (PNTI-SRP), and line parameters dynamic modeling (LPDM-SSRP). A wide variety of case studies are presented as well.

The power system stability enhancement using FACTS devices is the content of Chapter 14. This chapter presents the basic principle and operation of various shunt, series, and series–shunt types of FACTS controllers. The optimum placement of the FACTS devices in the smart grid is the key feature of this chapter. Case studies using IEEE 30-bus and the New England 68-bus test systems are presented to check the effectiveness of the optimization techniques.

The start grid is becoming vulnerable due to cyberattacks as the smart-grid components have computational and communication capabilities. In Chapter 15, first the nature of the smart-grid treats is identified followed by describing different smart-grid architecture layers such as operation, network, software, and hardware, which are vulnerable to cyberattack. The attack methodologies and countermeasures are also discussed in detail. The chapter basically outlines what security measures are to be adopted to minimize the security risks of present and future grids.

In Chapter 16, the smart-grid participants, roles, functions, structure, models, and dynamics are presented. In the next stage, the smart-grid security management for power system operator, distribution system, smart metering network, and consumer levels are discussed. The policies and regulations of future smart grid are also demonstrated.

1.4   Conclusions

Key challenges for the future include, but not limited, to large-scale integration of renewable energy or DERs, stability and reliability improvement, inclusion of bulk energy storage devices, and protecting the grid from cyberattack. It is required to develop suitable policies and regulations at generation, transmission, and distribution levels. Proper coordination is also essential between transmission and distribution system operators. The new regulation should be customer friendly which involve them deeply in successful grid operations by converting them from typical customer to smart and empowered customer. More standards are to be developed and more R&D investments are to be made to make the grid smarter, flexible, reliable, efficient, and environment friendly.

Bibliography

[1]   California Public Utilities Commission. "California Renewables Portfolio Standard (RPS)." [Online]. Available: http://www.cpuc.ca.gov/PUC/energy/Renewables/index.htm.

[2]   European Commission. 2012. "Energy roadmap 2050." [Online]. Available: http://ec.europa.eu/energy/publications/doc/2012_energy_roadmap_2050_en.pdf.

[3]   "Energy Independence and Security ACT of 2007." [Online]. Available: https://www.gpo.gov/fdsys/pkg/PLAW-110publ140/html/PLAW-110publ140.htm [Accessed July 22, 2016].

[4]   IEEE Standards, IEEE Smart Grid Research-IEEE Vision for Smart Grid Controls: 2030 and Beyond, June 2013.

[5]   J. Trefke, L. Nordström, A. Saleem, S. Rohjans, M. Uslar, and S. Lehnhoff, "Smart Grid Architecture Model Use Case Management in a Large European Smart Grid Project." In: *Proceedings of the Fourth IEEE European Innovative Smart Grid Technologies Conference (ISGT'13)*, IEEE Press, Copenhagen, October 2013.

[6]   Smart Grid Coordination Group, "Smart Grid Reference Architecture," *CEN-CENELEC-ETSI*, Tech. Rep., 2012.

[7]   Richard E. Brown, "Impact of Smart Grid on Distribution System Design." In: *Power and Energy Society General Meeting – Conversion and Delivery of Electrical Energy in the 21st Century*, Pittsburgh, PA, 2008.

[8]   P. Mallet, P.-O. Granström, P. Hallberg, G. Lorenz, and P. Mandatova, "Power to the People—European Perspectives on the Future of Electric Distribution," *IEEE Power & Energy Magazine*, March/April 2014.

Chapter 2

Smart grid architecture—key elements and definitions

Sebastian Lehnhoff and Mathias Uslar

Future, intelligent energy systems will have to face the challenge of integrating a large number of active components into existing energy management and operation schemes. For the successful integration of these components into the electric energy system, along with the new functions, market roles and technologies, information and communication technology (ICT) is a key enabler. In parallel to the technical changes, competition will increase significantly and there will be a need for greater direct intervention in the market in order to guarantee security of supply in a system that is operated closer to its stability boundaries. New sales and business models that rely on digitization and the increased use of ICT will create incentives for consumers to modify their energy usage patterns in order to become active participants in the grid operations. The security and commercial viability of such an energy system are vital for industrialized countries on the way to achieving a sustainable energy supply. ICT and the corresponding communication standards contribute to overcoming challenges of integration and interoperability within these highly decentralized structures. The use of ICT is crucial for improving the integration of the distributed energy resources (DERs) and help match generation to supply and achieve a higher level of customer benefit. This chapter provides an overview on how ICT and communication technology will provide meaningful inputs in terms of technological sets to drive this smart-grid transition process.

2.1 Introduction

From the visionary point of view, future, intelligent energy systems will have to face the challenge of integrating a large number of a variety of so-called active components (generation and consumption devices as well as operational equipment) into existing energy management and operation schemes, allowing for an automated on-line optimization of those systems. This way, a stable and reliable management is guaranteed despite the large-scale extension of energy conversion

OFFIS—Institute for Information Technology, Oldenburg, Germany

systems that depend on fluctuating resources, and are thereby prone to prediction uncertainty [1].

The objective and pace of this energy revolution have already been decided. Politicians, the economy, science and the population at large face immense challenges related to this energy turnaround. For the successful integration of renewable fluctuating generation into the electric energy system, along with the new functions, market roles and technologies, information and communication technology (ICT) is a key enabler [2]. The European electricity system is already undergoing drastic changes. For a number of years, it has been adapting to significantly different energy policy and environmental policy conditions defined by the European Commission [3]. Mandatory participation by industry in the emissions certificates trading market and the government's objective of reducing greenhouse gas emissions are contributing to improved energy efficiency. At the same time, alongside these state interventions in the market, there have been attempts to strengthen competition on the energy market. The course has already been set for the transformation from a historically established and existing centralized structure of conventional generation to a decentralized one, in which renewables and end customers play a major role (see Figure 2.1).

The transformation to renewable energy sources brings with it a stochastic pattern of power generation. The supply of power, which will be influenced much more by weather and market-related aspects due to the increasing proportion of wind farms and photovoltaic (PV) systems, must be balanced with the equally fluctuating demand. In addition, there is the problem that most of this supply relies on decentralized feed-in, or plants that feed in directly to the distribution grid. While the flow of electricity has been until very recently only been from the top down, i.e. from high-voltage to low-voltage levels, there are now increasing levels

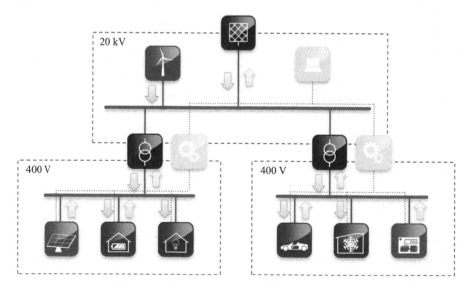

Figure 2.1   From centralized to decentralized generation in future smart grids

of feed-in or feedback from lower voltage levels. The grid infrastructure has to cope with this bidirectional power flow. For example, greater numbers of local smart substations will be constructed so that the distribution grid is equipped to coordinate bidirectional load flows (a major issue being voltage control and adjustment of protection system settings). Throughout the distribution grid, the demand for high-resolution metering, regulation and automation of the electricity flow is growing. In addition to the current developments on the generation side, consumption characteristics will also change in the future. Electric mobility, heat pumps and other consumer appliances will create a new dynamic in the distribution grid and will be integrated into the smart grid. Variable tariffs [4] may also lead to increased grid load. As the switchover to renewable energy sources progresses, it should be anticipated that electricity will be used increasingly to provide heating and to power mobility solutions like electric vehicles (EVs).

To integrate renewable energy sources into the grid, the grid infrastructure must be adapted at all technological levels [5,6]. The electricity system of the future will appear different, not only because of the addition of major lines for long-distance transmission, but also in terms of the paradigms applied in the distribution grid. To balance between supply and demand as well as utilization of storage systems will play an increasingly important role. A large heterogeneous variety of technologies will be used, depending on the time scale (from the sub-second range up to seasonal balancing). In parallel to the technical changes, new developments are also reaching the energy market. While competition will increase significantly, there will also be a need for greater direct intervention in the market in order to guarantee security of supply in a system that is operated closer to its stability boundaries. In conjunction with the technical changes, therefore, the market structures will also have to undergo significant adaptation, and end customers will have a much greater direct impact and influence on market activities. New sales and business models that rely on digitization and the increased use of ICT will create incentives for consumers to modify their energy usage patterns in order to become active participants in the grid operations. The large installed capability of renewable energy sources for generation will regularly lead to situations in which the overall system produces more power than is currently needed. New markets or market rules must be created for this situation, to follow the energy sector's objectives.

The security and commercial viability of the energy supply are vital for industrialized countries on the way to achieving a sustainable energy supply. ICT and the corresponding communication standards contribute to overcoming challenges of crucial systems integration and interoperability within these highly decentralized structures. The use of ICT is crucial for improving the integration of the DERs and help match generation to supply and achieve a higher level of customer benefit [7].

Within this chapter, we will take those aforementioned changes into account when providing an overview on how ICT and communication technology will provide meaningful inputs in terms of technological sets to drive this smart-grid transition process. First, we will provide an introduction to the various existing smart-grid definitions as well as strongly related research objectives in energy informatics from around the globe with a strong focus on the large demo regions the USA, Europe and China. After documenting the global consensus, we focus on providing a vision on how the so-called smart grid will evolve based on the context defined in this section. The section will focus on emerging system properties of smart grids as a system-of-systems (SoS) and discuss relevant research questions arising from this paradigm from Northrop *et al.* [8].

After providing the theory, we will present a way how smart grid-architectures can be documented in order to properly deal with the technological aspects of the SoS challenges [9,10].

We will provide an overview how US and EU methods [11] can be combined using the smart-grid architecture model (SGAM) in order to deal with control, communications and security requirements and their elicitation. In addition, standards which will provide a meaningful technological basis will be briefly presented and put in context. Based on the presented technical basis, this chapter draws conclusions for the most striking issues arising from the operational point of view, defining (future) key functions needed in grid operations based on the proposed technologies.

2.2   Smart grid—a definition

With bidirectional power flows occurring from the feed-in of renewables, operation and monitoring of the infrastructure changes. New sensors with corresponding data, new actors and new control technology have to be deployed and incorporated into the legacy systems.

This new grid operation paradigm is usually called the smart grid. According to the most prominent definition by the US NIST (National Institute of Standards and Technology) [12], the smart grid "is a modernized grid that enables bidirectional flows of energy and uses two-way communication and control capabilities that will lead to an array of new functionalities and applications."

Unlike the existing, old-fashioned grid, which primarily delivers electricity in a one-way flow from generator to outlet, the smart grid will permit the two-way flow of both electricity and information. With this definition in place, the convergence of ICT, communication networks and power grid operations had an impact in the discipline of informatics.

For Europe, the Smart Grid European Technology Platform defined the term as follows: "Smart Grid is an electricity network that can intelligently integrate the actions of all users connected to it—generators, consumers and those that do both— in order to efficiently deliver sustainable, economic and secure electricity supplies." To properly cope with the problem of dealing with this ICT-based control, the discipline of *Energy Informatics* emerged, coined mainly by the definition from Watson *et al.* [13].

Within this contribution, the next section will provide an overview on the existing definitions for the term and will discuss current shortcomings of the existing individual definitions as well as a definition based on the theory of ultra-large-scale system (ULSS) by Northrop *et al.* [8]. Therefore, the scope is defined and the so-called smart grid as a critical infrastructure is analyzed by its inherent attributes.

Related to different definitions of smart grids and related varying R&D objectives—especially in ICT and energy information systems integration—different definitions of the term *Energy Informatics* have emerged since the first work done by Watson *et al.* in 2010 [13]. Watson and Boudreau as IS scholars take the information systems point-of-view as their biased conviction and propose the new subfield of IS (information systems) [14] *Energy Informatics* as follows:

> *Energy Informatics is concerned with analyzing, designing, and imple- menting systems to increase the efficiency of energy demand and supply systems. This requires collection and analysis of energy data sets to sup- port optimization of energy distribution and consumptions networks.*

Kossahl *et al.* [15] argue in 2012 that, although the definition by Watson [13] argues that the concept of IS is broader than IT (information technology), their definition of Green IS will be too broad to properly define a field. In addition, the definition is mainly oriented toward economic goals, focusing on the term effi- ciency. For Green IS-based *Energy Informatics*, Kossahl *et al.* identify and define three new sub-fields, *power supply*, *smart grid* and *power consumption* as focal areas based on their literature analysis.

Vom Brocke *et al.* [16] take in 2013 into account the definition for bioinfor- matics as given per the Oxford English Dictionary stating that the analogous defi- nition of *Energy Informatics* should be:

> *Energy Informatics is [. . .] the branch of science concerned with infor- mation and information flows in energy systems, especially the use of computational methods to increase the efficiency of energy demand and supply systems.*

Vom Brocke *et al.* argue that taking this approach prevents *Energy Informatics* from being labeled as a subfield of an existing, broader domain such as IS (in contrary to the argumentation by Kossahl *et al.* [15]). In 2014, Goebel *et al.* [7] try to outline the current scope of *Energy Informatics* research and define the scope from their point of view for *Energy Informatics* as follows, mainly based on the definition by Watson *et al.* [13]:

> *As ICT is expected to support the transition to sustainable economies by enabling two main developments, a) the increase of energy efficiency beyond what [traditional] engineering can do and b) the efficient inte- gration of renewable sources of energy by making power systems smarter. The Energy Informatics research addresses these high-priority goals by focusing on well-defined research challenges. Economic considerations are part of the evaluation and take into account existing institutional frameworks.*

Based on this definition, Goebel *et al.* define high-level use cases or rather domains for energy informatics (EI) research from their point of view. One of the main issues all the existing definitions have in common is their background in IS research (Business Informatics), which is a subfield of applied computer science.

For this contribution, we take and follow the perspective from the software and systems engineering point of view, mainly focusing on *practical computer science*—which is located between theoretical and applied computer science (informatics) in the sense that it researches and develops concepts and methods for solving concrete practical ICT-driven problems, e.g. the development of data-/ information models or programming languages to efficiently solve specific problems. The next chapter will take the systems engineering view on the smart grid and *Energy Informatics* and motivate the needed scientific research from this point of view, derived from grounded system theory.

The discipline of systems engineering is defined by INCOSE [17] as follows:

> *Systems engineering is an interdisciplinary approach and means to enable the realization of successful systems. It focuses on defining customer needs and required functionality early in the development cycle, documenting requirements, then proceeding with design synthesis and system validation while considering the complete problem. It integrates all the disciplines and specialty groups into a team effort forming a structured development process that proceeds from concept to production to operation. It considers both the business and the technical needs of all customers with the goal of providing a quality product that meets the user needs.*

Building on top of this, the most relevant aspects in the context of *Energy Informatics* is the SoS research. Sommerville [9] defines system-of-systems according to the US Department of Defense (DoD) definition as follows:

> *A System-of-Systems is defined as a set or arrangement of systems that results when independent and useful systems are integrated into a larger system that delivers unique capabilities.*

In addition, Sommerville *et al.* [18] argue that the overall complexity of an SoS stems from both number and type of relationships between individual components and systems and its environment. If only a small number of relationships are present, and in addition, those changes occur only slowly over time, predictions can be made and deterministic models for the properties can be developed. However, SoS seems to be of a different class as there are many different dynamic relations, systems act non-deterministic and overall systems' characteristics cannot be predicted from analysis of the individual parts of the system. For those systems, the so-called *epistemic complexity* emerges—the complexity can be deduced from the lack of knowledge about the SoS rather than from its inherent characteristics [19]. The proper way of dealing with this kind of complexity in SoS; at least two different initiatives have emerged. The UK government has pledged as much as 25 Million Euro to the Large-Scale Complex IT Systems initiative (www.lscits.org) [20] and written entirely independent, but with the same results, the Carnegie Mellon University and its Software Engineering Institute has coined the term of ULSS [21]. Both share the same scope; the following attributes are seen for ULSS. An ULSS has an unprecedented scale in some of the following dimensions:

- Lines of code
- Amount of data stored, accessed, manipulated and refined
- Number of connections and interdependencies
- Number of hardware elements
- Number of computational elements
- Number of system purposes and user perception of these purposes
- Number of routine processes, interactions and "emergent behaviors"
- Number of overlapping policy domains and enforceable mechanisms
- Number of people involved in some way

For obvious reasons, for this type of system, new paradigms and challenges for the Software engineering process emerge, as they are needed [21,22] to cope with the "wicked" problem [23,24]. Northrop *et al.* [8] assert that currently, fundamental gaps in the current understanding of software and its development at ULSS present profound impediments to achieve development objectives. They argue that those gaps have to be considered strategic and not tactical, meaning that incremental research in established categories will most probably not address them. Sommerville *et al.* completely agree and point out the fundamental reasons from their point of view why current software engineering methods are considered to be unsuitable for building twenty-first century ULSS. Sommerville points out that three fundamental reductionist assumptions are not true for systems like smart grids [18]. First, the owners of a system do not control its development. This is obvious for smart grids as regulatory issues like unbundling come into play, but also customers as loads or political stakeholders. In addition, decisions are not made rationally and are not purely driven by technical criteria. Third, just like Rittel [23] argued, there is no definable problem and clear system boundaries. Nam and Pardo argue that for those reasons, obviously the discipline of Smart Cities shall be considered a ULSS and treated this way from a systems engineering perspective [25]. Anvaari *et al.* [21,22] point out the urgent need to take into account the ULSS theory when dealing with the smart grid from an ICT perspective. Therefore, we draw the conclusion that smart grid shall be treated like a ULSS from the research point of view.

Within this section, we have motivated to treat the smart grid as a coalition of systems with the trait of a so-called wicked system [23]. Based on the assumptions for both requirements engineering and governance from the traditional reductionist software engineering point of view, a new way is needed to deal with smart grids from the ICT perspective, mainly focusing on the sub-discipline of systems engineering from the discipline of software engineering. The individual aspects of ULSS will be the basis for a discussion of requirements toward *Energy Informatics* as a discipline. Based on the information gained about the concept of a ULSS as a problem, the existing view on *Energy Informatics* as a discipline, and the wicked systems concept, we motivated a working definition for the discipline from the systems engineering point of view, focusing on tackling the actual problem smart grid as a cyber-physical system [26]. The next section will cope with the results, which yield from these drawn conclusions and define research questions which are addressed by control, communication and security standards.

2.3   Smart-grid future vision

Energy Informatics from the Systems Engineering perspective is a working definition that is strongly related with the future vision of smart grids. Based on the grounded theory of ULSS, software engineering and LSCITS, Sommerville *et al.* [9] as well as Lewis and Collopy [10] motivate the road to pave the way for a smarter way to deal with the systems aspects for smart grids in research. Their Top Ten research agenda can also be applied to smart grids. The authors raise the following questions:

1. How can we model and simulate the interactions between independent systems?
2. How can we monitor coalitions of systems and what are the warning signs of problems?
3. How can systems be designed to recover from failure?
4. How can we integrate socio-technical factors into systems and software engineering methods?
5. To what extent can coalitions of systems be self-managing?
6. How can we manage complex, dynamically changing system configurations?
7. How can we support the agile engineering of coalitions of systems?
8. How should coalitions of systems be regulated and certified?
9. How can we do a probabilistic verification of systems?
10. How should shared knowledge in a coalition of systems be represented?

While Sommerville *et al.* [9] again use the term coalition of systems; one can see ULSS, cyber-physical system, etc. as analogous [10]. We consider *Energy Informatics* to be the discipline (from the view of practical computer science) to address those questions for the critical infrastructure of smart grids. In addition to this, we propose the following definition:

> *Energy Informatics in scope of practical computer science is a discipline dealing with the software engineering aspects of Smart Grid as a Ultra-Large Scale System. Its research shall mostly address aspects of interactions simulation, monitoring, recovery, socio-technical dimensions, self-x properties, dynamic reconfiguration, agile engineering, certification, verification and knowledge sharing. EI researchers treat the critical infrastructure Smart Grid like a so called wicked system with all its inherent attributes.*

Rather universal are the requirements regarding an increase in grid visibility, efficiency and reliability as well as security, which are emphasized in nearly all vision statements and definitions to be presented in this chapter. We will highlight the results from very prominent roadmaps defining control, communication and security standards for the smart grid.

Within the next section, we will focus on the SGAM model [27,28] from the M/490 mandate of the European Commission to the international standardization bodies as it is widely used in EI and based on modeling and analysis principles established practical computer science. The standards form the roadmaps evaluated can be mapped onto this common reference designation system in order to compare the individual efforts.

2.3.1 The German roadmap E-Energy/smart grid

The national smart grid standardization roadmap for Germany considers [29] national and international standards exhaustively, focusing on the smart grid's ICT infrastructure. Basic elements of the roadmap consider an overview on various international standardization efforts as well as specific recommendations for standards relating to different parts of the smart grid. Not only core standards for the future smart grid in terms of ICT have been identified, but also the improvement of existing and the development of new standards has been taken into account.

Furthermore, a second edition was planned to incorporate experiences gained from the realization of the recommendations. The second edition includes a multi-utility aspect (gas, heat and water) as well as communication standards for further focal topics such as market communications, high-voltage direct current and flexible alternating current transmission system.

2.3.2 US NIST IOP roadmap

In the USA, the Energy Independence and Security Act authorized the Department of Commerce in 2007 to coordinate the development of an interoperability framework. This framework is coordinated by NIST and aims at interoperability (IOP) between and among smart-grid systems and equipment with special respects to standards in the fields of ICT protocols and data models. NIST designed an action plan to support and foster the development of important standards. The first version of the framework includes an abstract reference model consisting of about 80 standards, which are related to the smart grid directly or on a meta level. Furthermore, 15 key areas and gaps—where new or improved standards are needed—have been identified. In conclusion, 16 core standards are recommended by NIST. The latter document, NIST Framework and Roadmap for Smart Grid Interoperability Standards, Release 1.0 [12], is the output of the first phase of the three-phase NIST plan. Standards listed are to be regarded as neither exhaustive nor exclusionary. Apart from the 16 core standards, additional standards, specifications, requirements, guidelines and reports for further review are presented and discussed, too.

2.3.3 IEC SMB SG 3

In February 2010, the SG 3 (Strategy Group "Smart Grids") [30] published a draft of a roadmap including their own standards and 11 high-level recommendations. The SG 3 was appointed by the IEC Standardization Management Board (SMB).

The main focus of the IEC roadmap is on improved monitoring and control of all components within the network. Therefore, it is necessary to achieve a higher level of syntactic and semantic interoperability between all involved components and solutions. Increasing energy consumption, further spread of DER, sustainability of generation and distribution, competitive market prices, security of energy supply as well as the maturing infrastructure are seen as main drivers for the changing process. Furthermore, common requirements regarding a smart-grid reference architecture, which should be based on the seamless integration architecture (SIA), are defined by the roadmap.

Altogether, more than 100 standards have been identified, described and prioritized, whereas five standard series are ranked as "core" and nine as "high priority." The roadmap includes 12 main application areas, 6 general topics and 44 recommendations overall.

2.3.4   German BMWi E-Energy program

E-Energy is a public funding program of the Federal Ministry of Economics and Technology (BMWi) [31] and the Federal Ministry of Environment, Nature Conservation and Nuclear Safety (BMU) and consists of six model regions spread across Germany. The current objective is to create the so-called Internet of Energy. In the course of that framework, the BMWi commissioned a study to analyze the standardization environment of the six projects in late 2008. Based on expert interviews, literature research and the consortium's experiences of IEC and Institute of Electrical and Electronics Engineers (IEEE) participation, a set of standards has been identified. Eleven standards for eight topics were recommended as core standards. The recommendations are not absolutely specific for the funded model regions and, thus, applicable to other smart grid projects. Besides IEC standards, the study focuses on converging technologies between the ICT and automation domains. The study was the basic document and overview for the national roadmap and one of the inputs for the IEC roadmap.

2.3.5   Microsoft Smart Energy Reference Architecture

The Microsoft Smart Energy Reference Architecture (SERA) is a comprehensive reference architecture based on NIST work and Microsoft [31] products. It addresses the technology integration throughout the scope of the smart energy ecosystem and the surrounding systems. It should help utilities to create an integrated utility by providing a method of testing the alignment of IT. To meet the necessary requirements of power utilities IT infrastructure, there has been close cooperation between key power industry partners (i.e. Accenture, Alstom Power, ABB, ESRI, Itron Inc. and OSIsoft Inc.) and Microsoft. The main intention of the reference architecture is to accelerate solutions development under the aid of specific Microsoft products. SERA tries to assure to enable developers to provide enhanced, more cost-effective, secure and scalable solutions.

2.3.6   *The State Grid Corporation of China framework*

The State Grid Corporation of China (SGCC) [31] has defined a smart-grid road-map which will have a huge impact on all vendors and markets since China will be one of the largest markets and vendors of products for the upcoming smart grid based on open standards. The first version of the SGCC framework takes into account 8 domains, 26 technical fields and 92 series of standards.

The eight domains include planning, power generation, transmission, substation equipment and communication, distribution, utilization, dispatching and ICT.

For the initial development, SGCC has taken into account several existing standardization roadmaps, e.g. IEC SG 3, NIST Interoperability Roadmap, IEEE P2030, CEN/CENELEC/ETSI Working Groups, German DKE Roadmap and Japanese METI Roadmap.

As motivation for their efforts, several reasons are coined. After the age of information, they see an upcoming age of intelligence where the integration of clean energy requires both a strong and smart grid. The strong and smart grid is considered to be needed to tackle climate change and environment deterioration—the smart grid is essential to optimize the allocation of energy resources. The strong and smart grid is defined as an intelligent power system encompassing power generation, transmission, transformation, distribution, consumption and dispatching. The strong and smart grids will be a shift in terms of function of the grid. According the SGCC definition, the grid itself will no longer be a simple carrier of transmission and distribution of electricity, but more an integrated and intelligent platform for the internet of things (IoT), internet network, communication network, radio and TV networks. The sharp line between generation-side and demand-side will blur.

SGCC has worked out a fast paced three-stage plan. Stage 1 is the planning and trial phase for the years 2009 and 2010 for technical and management specification formulation, key technology R&D and pilot programs. Stage 2 from 2011 till 2015 focusses on speeding up the construction of the UHV grid, and rural distribution network, to establish preliminary smart grids operation. The aim for stage 2 is to achieve technical breakthroughs and extensive application of key technology and equipment. Stage 3 from 2016 to 2020 is the leading and enhancing phase, where basically the construction of the strong and smart grid is completed, therefore enhancing resource allocation abilities, security levels and operational efficiency of the grid. Those stages also have reflecting stages in the needed standardization efforts. In stage 1, they plan for standard formulation and establish a preliminary standards framework. The work focused on developing and amending standards to have pilots finish in due time. Stage 2 wants to renew and amend existing standards, complement necessary standards and complete the SG standards framework. Furthermore, one focus will be the promotion of domestic standards into international standards. In stage 3, those should be promoted to world leading standards providing opportunities for Chinese vendors, making domestic standards all-around international ones.

For the first batch of smart-grid standards, SGCC has identified 22 standards overall, 10 domestic ones and 12 international ones. The following list comprises the international standards:

- Terminology and modeling of smart grid: ISO/IEC 62559
- Standard series on Substation Communication network and System: IEC 61850
- Interface of Power Company Data Exchange Platform—Distribution Management System: IEC 61968
- Specifications on Open Geographical Data Interoperability: OGC Open GIS
- Technology Regulations on Integration of Distributed Generations into Power Grid: IEEE 1547
- Standard Series on Electric Vehicle Charging and Discharging: IEC 61851
- Standard Series on Application Program Interface of Energy Management Systems (EMS): IEC 61970
- IEC 60870 Standard Series on Transmission Control Protocol: IEC 60870
- IEC 62351 Power System Management and Associated Information Exchange—Data and Communications Security: IEC 62351
- IEC 62357 Power System Control and Associated Communications—Object Model, Service Facilities and Protocol Architecture with Reference: IEC 62357
- ISO/IEC 27000 Standard Series on Information Security Management System: ISO/IEC 27000 series
- ISO/IEC 15408 Information Technology Security Evaluation Criteria: ISO/IEC 15408

Those standards have also been in the scope with the IEC SG 3, containing their five core standards. Of particular interest for the IEC TC 57 might be the Chinese initiative for so-called simplified common information model (CIM) series standards. CIM/E will be a data description specification, CIM/G a power grid description specification, CIM/S a simple service description specification and CIM/M a dynamic message encoding specification. Those items will be proposed to the IEC through the national Chinese committee as new work item proposals.

The next section will take the recommendations from those studies into account and provide a consolidated view on the core standards which have been identified for communications, control and security in smart grids and put them into the SGAM model to show, where they can be applied.

2.4 Smart-grid architecture model

In order to derive what kind of functions to implement and what kind of systems to build or set-up for executing a future smart-grid use case a number of questions need to be addressed—this covers the aspect of the formulation of a wicked problem. In general, these may be categorized as functional and non-functional issues [5,6]. Functional issues may address the system's flexibility for, e.g. providing some kind of power reserve for load balancing, or the system's capability to detect certain phenomena that may be characteristic of a stability issue (if these are part of the given use case). However, for the development of a proper supporting ICT-system non-functional requirements may be more relevant and address issues like:

- Is the Quality of Service of the ICT- and automation system sufficient?
- Is the system safe/secure?
- Are all the relevant systems connected and able to communicate with each other?
- Are certain measurements sufficiently accurate/precise?

In order to manage these kinds of issues and questions (especially the latter ones), the use-case-based requirements engineering method utilizes a reference architecture to identify participating system components and their interfaces to each other (both ICT-to-ICT interfaces and ICT-to-power system connections).

A properly defined use case should then be able to identify relevant actors and components consistently from within this reference architecture and specify their interactions with each other. To this end, the European Union has issued a mandate (M/490) for the standardization of smart-grid functionalities and use cases [3] to three European standards organizations (CEN, CENELEC and ETSI) to develop a common framework for representing functional information data flows between power system domains and their relevant subdomains as well as their architectures (the SGAM) [2]. Together with a set of consistent standards for the proper integration of all actors through interoperable communication protocols and data models, this framework supports the implementation of ICT systems in that domain. Additionally, a sustainable process and method is developed for use-case specification and formalization allowing stakeholder interactions, gap analysis and (for the scope of this paper most important) the derivation of non-functional requirements necessary for the successful execution of a given task.

The ISO/IEC 42010 defines a reference architecture as a description of a system's structure in terms of interactions between its element types and their environment [32]. Thus, a reference architecture defines restrictions for an instantiation of a concrete architecture and specifies the syntax as well as the semantics for describing a (technical) system.

The SGAM is the reference architecture model as well as a reference designation system for describing the smart grid and especially ICT- and automation systems within this domain.

Starting from a contextual "component layer" spanning the power system domains in terms of the energy conversion chain and its equipment against the hierarchical levels of power system management and automation (process, field, station, operation, enterprise, market) the so-called zones (see Figure 2.2). The purpose of this layer is to emphasize the physical distribution of all participating components including actors, applications, power system equipment (at process and field levels), protection and remote controlled devices, network infrastructure (communication devices and connections) as well as any kind of computing systems.

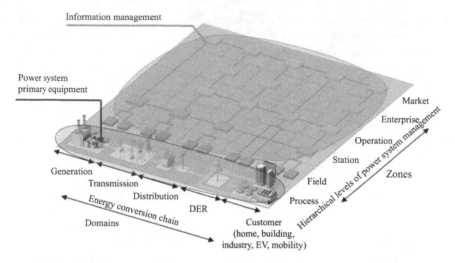

Figure 2.2   Smart grid component layer (CEN–CENELEC–ETSI, 2012)

In order to use the framework for the development of interoperable smart-grid architectures, the component layer is extended by additional interoperability layers (see Figure 2.3) [33] derived from the high-level categorization approach developed by the GridWise Architecture Council (GWAC) [34] (see Figure 2.4). Often referred to as "GWAC stack," its layers comprise a vertical cross-section of the degrees of interoperation necessary to enable various interactions and transactions within a smart grid. Basic functionality (e.g. interaction with field equipment, transcoding and transmitting data) is confined to the lower component and communication layers. Standards for data and information modeling and exchange are defined on the information layer while the top functional and business layer deal with business functionality. As the functions, capabilities and participating actors increase in terms of complexity, sophistication and number, respectively, more layers of the GWAC stack (Figure 2.4) are utilized in order to achieve the desired interoperable results. Thus, each layer typically depends upon and is enabled (through the definition of well-known interfaces) by the layers below it.

Within this framework for a sustainable development process, which is based on the IEC PAS 62559 (which was initially developed by the EPRI in the US as part of the IntelliGrid project and adopted by the IEC as a Publicly Available Specification in 2008), use cases can be formally described detailing functional and performance requirements in order to assess its applicability and (if so) appropriate standards and technologies [35]. A use case is a class specification of a sequence of actions that a system can perform interacting with other actors of the system [36]. Thus, the relevant contents of use case are actor specifications, assumptions about the use-case scenario, contracts and preconditions that exist between the actors and are necessary to set the use case into motion, triggering events (leading to the described scenario or "activating" the use case) and a numbered list of events detailing the underlying control and communication processes. Unified modeling language-based class and

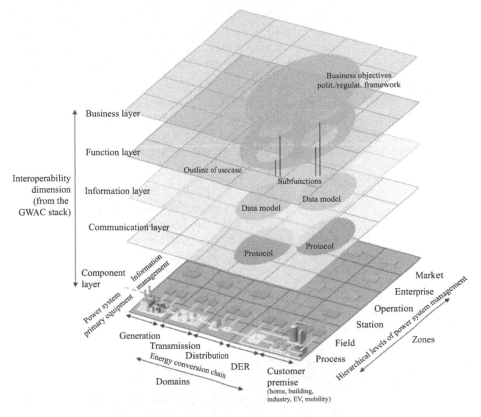

Figure 2.3   The smart-grid architecture model (CEN-CENELEC-ETSI, 2012)

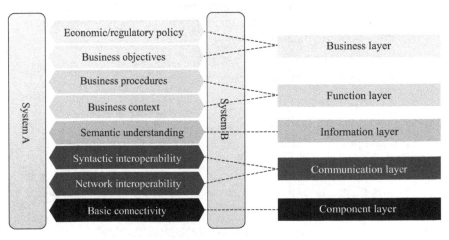

Figure 2.4   Mapping of the GWAC stack onto the SGAM

sequence diagrams are used to describe the actor setup and how processes operate with one another and in what order. For the mapping of the use case onto the SGAM, the following steps are recommended [27] starting from the SGAM-function layer and "drilling down" to the components, communication and information layers:

1. Identify those domains and zones, which are affected by the use case
2. Outline the coverage of the use case in the smart grid plane
3. Distribute sub-functions or services of the use case to appropriate locations in the smart grid plane
4. Continue down to the next layers while verifying that all layers are consistent, there are no gaps and all entities are connected.

This process model allows the detailed specification of non-functional requirements for each step within the use-case sequence diagram, regarding, e.g. the performance of the communication between two entities or the availability of system components. In order to provide a consistent classification and to enable a mapping of requirements on suitable standards and technologies, the PAS 62559 provides comprehensive checklists for non-functional requirements ([37], p. 57ff.) within the areas of:

•   Configuration
•   Quality of service (QoS)
•   Security requirements
•   Data management issues
•   Constraints and other issues

This detailed specification provides all the necessary information for designing and implementing the actual (interoperable) smart grid ICT- and automation system. The QoS-requirements do not only allow specifying what standards and protocols to use but also support the decisions regarding technology or (organization, management) algorithms in order to achieve the necessary performance, availability of the system (acceptable downtime, recovery, backup, rollback, etc.), the frequency of data exchanges and the necessary flexibility for changes in a future smart grid interconnecting a vast number of sensors and actuators.

2.5   Communication and standards

As mentioned in the sections before, currently several studies regarding smart grids proposing the use of standards are developed. To enable comparison of the different approaches and stress out their differences, we choose to identify the standards proposed by the different studies.

Through this analysis, the following standards can be regarded as the consensus on core IT standards for the future transition of the electric distribution grid toward a smart grid as identified at the international level:

•   IEC 61970/61968: Common Information Model (CIM) [35]
•   IEC 61850: Substation Automation Systems and DER

- IEC 62351: Security for the smart grid [38]
- IEC 62357: TC 57 Seamless Integration Architecture
- IEC 60870: Communication and Transport Protocols
- IEC 61400-25: Communication and Monitoring for Wind Power Plants
- IEC 61334: DLMS
- IEC 62056: COSEM
- IEC 62325: Market Communications using CIM

The mentioned standards are referenced by most of the approaches and play an important role within the IEC SIA. All of these standards are developed within the IEC TC 57, but unfortunately by different working groups within TC 57.

The IEC TC 57 SIA 62357 provides an important framework for standardization in the electric energy domain. Most of the roadmaps and studies introduced in this chapter identified the SIA as the core framework for standardization in smart grids in terms of automation and power system management. Figure 2.5 shows the different layers of the SIA in SGAM and both combines the important standards from IEC TC 57 and IEC TC 13 and depicts the relations between them in SGAM. Furthermore, two major interfaces are included, on the one hand an interface for

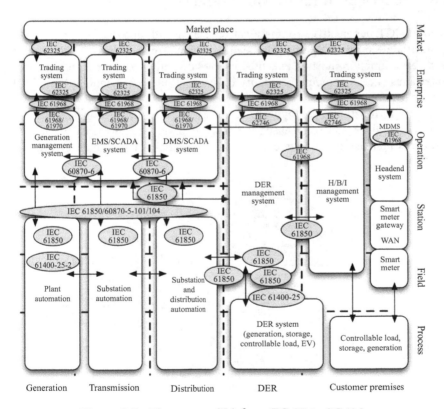

Figure 2.5   The current SIA from TC 57 in SGAM

market communications and on the other hand an interface for smart metering. The layers can be summarized as follows:

- Application and business integration
- Power system integration
- Security and data management

The SIA includes object and data models, services and protocols as well as interfaces between systems, communication architectures (e.g. service-oriented architecture), processes and data formats. The basic data model and domain ontology for .the future smart grid is the CIM (IEC 61970/61968), which provides interfaces for the primary and secondary IT in terms of energy management system (EMS) and distribution management system (DMS). Leading communication protocols are standardized in IEC 60870 (transport protocols) and IEC 61850-7-4xx (substation automation and DER communication). In the field of smart metering, the SIA tries to standardize applications and functions not too much to leave enough space for innovations and vendor-specific implementations—and of course, work done by other TCs outside TC 57.

2.6   Functions of future smart-grid components

From an ICT perspective, communication and control functions together with their appropriate security measures can be classified in terms of the time constraints (i.e. how fast a function must be executed or a valid result is expected) and information constraints (i.e. the amount or quality of data that has to be acquired for a function to be executed successfully).

This is illustrated in Figure 2.6. In smart grids, there is a general shift to functions and applications based on forecasted data or telemetry from volatile resources increasing the time constraints. Additionally, the amount of data that is

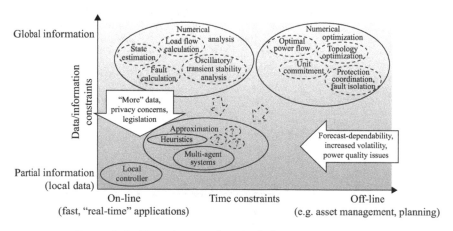

Figure 2.6   Functions in electrical, future power systems

acquired at the process or field level (and potentially aggregated at the station level), is increasing by several orders of magnitude. This necessitates the utilization of on-line mechanisms that are continuously executed for approximating optimized and stable system configurations. Appropriate mechanisms and control schemes are available and rely heavily on ICT-based communication (e.g. multi-agent systems for heuristic optimal control [39]).

In the following, we briefly discuss these aspects for key control and communication functions and components of future smart grids.

2.6.1 Smart device interface components

Smart devices are electronic devices that are generally connected to other (smart) devices or networks via different (wired or wireless) communication protocols such as local area network, power line communication (PLC), Bluetooth, Wi-Fi, 3G, etc. Such devices are deemed "smart" as they can operate interactively and autonomously to some extent. It is often related to ubiquitous computing—a concept from software engineering and computer science where computing tasks are distributed across many devices, locations and data/information formats in order to appear anytime and everywhere. Smart devices are typically characterized by (minimum) set of system hardware and software ICT resources [40], remote external service access and execution, local (somewhat) autonomous service execution as well as access to specific external environments, i.e. human interaction, process (physical world) interaction and distributed ICT interaction with the ICT environment. Smart devices are the main components to interface decentralized resources and processes as well as users in future smart grids and make up the runtime environment for the hitherto discussed SoS. They also serve as the platforms dealing with aforementioned timing and data/information constraints and issues for the functions discussed in the following.

2.6.2 Advanced forecasting

Forecasting systems are used to predict future developments and events. In smart grids, these systems are required in various areas including the planning of systems, optimization of market participation of consumers and producers and the scheduling of timely maintenance processes. With the increasing proportions of stochastic generation from wind power and PV installations, weather forecasts and the subsequent generation forecasts (characterized by local wind speeds and solar irradiation) are playing an increased role. Advanced forecasting has to deal with increased amount of data that may be correlated for increasing the quality of forecast together with a decrease in forecasting horizons. The latter supports the notion that short-term weather fluctuations may be more accurately predicted minutes-in-advance rather than days or weeks ahead. Real-time short-term prediction is typically coupled with more reactive systems that are capable of adjusting to the underlying stochastic behavior.

Advanced forecasting is a major function in smart grids that is significantly further developed together with the increasing penetration of fluctuating feed-in.

Forecasting systems are required for industrial "energy logistics," in order to mitigate the pricing and volume risks. This is of relevance both for generation and for consumption. As the investments and operational costs for storage systems decrease, the availability—i.e. the current loading and capacity of these storages will also have to be integrated into advanced forecasting regimes. Storage may comprise bulk or distributed plants (and even infrastructures when integrating systems beyond electric power, e.g. gas or district heating environments) or even mobile storage facilities. The mobile storage provided by EVs requires advanced forecasting systems to also take into account the change in dis-/charging locations and therefore different grid connection points.

Today, utilities buy energy volumes for domestic customers on the basis of statistical standard load profiles. With the introduction of smart meters or—more generally—advanced metering infrastructures (AMI), the customers' behavior will be forecasted more accurately (individually as well as at aggregated levels). Forecasts of consumption both for domestic households and industrial customers will support generation scheduling and planning of energy procurement. New customers (operators of decentralized generation and consumption units of varying degrees of flexibility and control as well as owners of EVs) necessitate the inclusion of forecasts from outside the domain, e.g. mobility forecasts to predict the dis-/charging patterns and grid connection points of EVs.

At the grid level, advanced forecasting systems are required to estimate the loading and utilization of specific grid operating equipment in smart grids. This is relevant for grid expansion and maintenance, but also for short-term demand response (DR) applications. DR is gaining in importance, since the higher levels of uncontrolled feed-in to the distribution grid are already causing voltage control problems and these will further increase. They can either be dealt with curtailing the feed-in or more economical methods, e.g. price-driven DR applications. In neighboring energy sectors, e.g. the gas sector, forecasting systems are equally important since there is a strong correlation between gas-driven heating processes and the utilization of electric energy.

2.6.3   *Control of demand, generation and storage units*

In smart grids, electrical load shifting and load reduction flexibilities together with flexible production processes in the industrial context are the basis for industrial demand-side management (DSM) activities. In the scope of the paradigm shift from load-led to generation-led operations, control of generation and storage flexibilities becomes equally if not dominantly of importance. Analogous to DSM, two basic options for generation and storage control can be identified. First is a direct influence from outside on the industrial requirement for electricity.

Already today, transmission system operators have the option of disconnecting industrial consumers from the grid using telecontrol systems. Moreover, direct intervention is possible via virtual power plants [41]. The flexibilities of generation and consumption systems operated by industry are scheduled to optimize market returns. The power capacities are made available on the energy trading markets and the balancing power markets. However, the more complex a manufacturing process, the more complex and costly it is to integrate into industrial DSM.

In the case of industrial DR, there is no direct control and the primacy of (the incentive-based) control over the demand, generation and storage systems lies with the systems' operators. In the context of DR, these incentives may be timely differentiated tariffs to shift their electricity requirements to periods in which the electricity is cheaper to buy or feed-in reimbursement is higher. Time-based tariffs can be applied to both the energy procurement side and the grid operation side.

Alongside energy procurement, tariff-based incentives for grid usage and operation are also relevant to industrial energy management. These allow grid operators and industrial customers to agree on control interaction if the actions of the industrial customer are beneficial to the grid utilization (voltage levels and power flows). To leverage the savings potential for industrial companies, it is necessary to safely and securely integrate resource (demand, generation and storage) management into the manufacturing organizations' enterprise resource planning programs.

2.6.4 Data transmission and monitoring

Data transmission and monitoring as a service of an automated infrastructure is necessary for implementing metering processes and for high-performance transmission and processing of mass data from AMI installed at the premises of end-customers. Alongside smart meters themselves, a communication gateway to the field level is also required [42,43]. This may be incorporated into the smart metering device, as a plug-in communications module or as a separate smart device (performing additional forecasting, aggregation or even security—anonymizing—services). Alternatively, the data from digital meters for various units or appliances in a household are encapsulated and passed through a single gateway for outside distribution. Such a multi-utility communication (MUC) controller standardizes the functionality and interfaces of the electricity, gas, water and heating meters for the end customer, the meter operators and the metering service providers.

Within households, it is possible to combine MUCs with interfaces for the end users, functioning as feedback systems to provide transparency for energy usage or presenting price-based incentives for the adjustment of energy usage. A range of wide area network (WAN) transmission technologies and protocols are available to transmit the data from communications-enabled MUCs to a meter operator's/ service provider's back-end system. In the case of an indirect WAN connection, the communication unit may be connected via PLC to the data aggregators typically built into local substations, from which a WAN connection provides access to the back-end systems. This is usually connected with the aggregation/anonymization of individual load profiles. Hence, the ruling on security requirements is particularly important, since customer acceptance will depend significantly on security. However, rigid security rules in Europe have already proven to act as a massive obstacle to AMI/MUC roll-out.

In terms of the data transmission and monitoring infrastructure and data processing, there are various approaches: in the context of automated meter reading

(AMR) infrastructures, the energy consumption data is recorded via a one-way remote reading process while it is typically not possible for switching commands, tariff information, software updates, etc., to be transmitted. An AMI expands the capabilities of the AMR by adding the option of bidirectional communications between the metering operators/service providers and the field level. The main challenge for data transmission and monitoring in smart grids and in the context of digitization in general is the huge increase in data volumes that have to be processed (in real-time by so-called complex event processors) and stored to be used in advanced forecasting systems (big data is an active research perspective in data mining and artificial intelligence-related fields, e.g. machine learning).

2.6.5   *Power flow and energy management*

Smart grid (transmission and) distribution substations will be managed by using integrated and interconnected smart (following the aforementioned notion of smart devices/environments) grid control systems. Following the established hierarchical control process associated with the individual voltage levels, integrated power flow and energy management systems monitor the grid condition on-line and may automatically implement grid stabilization measures such as distributed provision of ancillary services, feed-in curtailment, load shedding or topology switching actions.

Today's control systems communicate via telecontrol systems with the associated substations (e.g. transformer substations) [44], in which the switch panels and switchgear are controlled and their condition is reported back to the grid control system. The station control system assumes the local control and monitoring tasks in the transformer substations, while the telecontrol systems automate on-site control and monitoring. Additionally, the energy management systems coordinate the load situation in collaboration with the major power plants and industrial consumers, to the extent that is necessary for ensuring the availability of the power supply.

Aforementioned control systems manage the supply situation in the extra-high, high- (HV) and medium-voltage (MV) levels. These individual levels may also be linked together horizontally in order to increase reliability of supply. Currently, the MV and LV distribution grids are generally operated without detailed (i.e. information beyond the aggregated load situation at the feeder) knowledge of the current grid condition. Due to the top-down design of the supply system, it has been possible to ensure high-quality supply in the face of the anticipated grid load.

In smart grids, the expanded installation of sensors and measuring equipment that register power flows and the grid conditions over a wide area together with actuators to help control the load at the LV level will become vital. Smart (metering) devices and additional sensors within the grid already provide a range of current measurements at the end-points of the LV grid. By equipping smart devices at the LV level with autonomous agents, it will be possible to evaluate acquired data, using real-time algorithms to determine scheduling or switching commands for controllable consumer devices and storage facilities. The objective is to ensure the most efficient usage of the LV grid (that is as close to its operational constraints without jeopardizing security of supply) downstream of the transformer, maintaining permissible voltage ranges and power flows. These sections of the grid typically feed around a hundred connection points (e.g. households).

In smart grids, grid control systems can be deployed at lower remote levels. Critical grid conditions can then be mitigated automatically by the locally controlled system with no or only little need for interaction with an upstream instance. The management of the individual agents can be restricted to the assignment of parameters and rules, and dealing with a selection of potential problems (or the dynamic prioritization of control objectives). A (possibly reduced) model of the physical grid, with the grid topology and the installed cables, serves as a basis replicating the corresponding energy feed-in and demand positions. Hence, grid operators are provided with a view over their grid condition, together with the demand that occurs in the grid, and creates a dataset that can be used to plan future grid construction. Moreover, together with the corresponding knowledge of the generation and consumption profiles, grid operators may (safely and securely) enable connected customers to participate in the balancing power markets (via an aggregator) or manage decentralized resources to maintain operational feasibility.

In terms of organizing power flow and energy management under these timely as well as with respect to data volumes challenging conditions, self-organizing mechanisms [34,45] appear more attractive than hierarchical ones, albeit with the additional challenge of providing sufficient support for workflow and decision-making, as well as ensure that the increasing complexity stays manageable for human operators. For this reason, the software architecture of control systems is moving away from monolithic structures and becoming more modular.

2.7 Conclusions

Distributed energy systems have become more and more popular over the time, especially in the context of the energy turnaround across the globe. Important lessons learned and best practices to treat the smart grid as ULSS have been gathered. One particular aspect from the functional grid operations point of view is the integration of the existing power distribution and transmission grids and the ICT-induced communication grids as control overlay networks. This aspect is mainly driven by a focus of better utilization of the existing assets owned by the utilities and third parties. Overall, the main driver of the integration is a proper interoperability between the various new components to operate the grid [46]. For the overall topic of interoperability, different system views such as technical interoperability, organizational interoperability or functional interoperability have to be taken into consideration when integrating components in the smart grid domain.

In this chapter, the aspects of interoperability for communications, control and security from different viewpoints have been highlighted [47], which have been evaluated in several research projects over the years.

The aspect of policy-making and standards for technical integration in the utility sector is of high importance [48]. Over the years, standardization has put together a meaningful and mostly consistent set to deal with technical and governance issues and provide a sustainable way of designing and analyzing systems across legislative periods. In parallel to the NIST, Germany started working on this aspect as early as 2009, outlining the need for standards for large federal funding schemes. Another aspect of standardization is the level of so-called semantic integration to achieve interoperability [12]. For the electric utility domain, the common information model (CIM) has proven to be a successful candidate to achieve this level of interoperability. This technology has to be put in context with everyday utility IT/operational technology practice in automation. The CIM [49] as well as other control and measurement standards has to work seamlessly together to achieve a proper overall interoperability. With the CIM focusing mainly on the supervisory control and data acquisition aspects of a utility's control architecture, the need for integration with automation standards has been clearly outlined. Relevant work on mediator-based integration of the IEC 61850 and CIM standards, addressing this integration at various levels [50,51] exists.

As discussed in this chapter, architectural aspects [52,53] and domain-specific models [54] have to be taken into consideration to properly assess the solutions proposed not only for interoperability, but also for (technical) maturity as well as to quantify non-functional requirements like IT security. One prerequisite to answer those questions is a proper documentation of the requirements from a domain-specific reference model [55,56], use case and user story method. Based on the outlined need for a standardized use-case documentation with the focus on non-functional requirements, the IntelliGrid methodology has been enhanced to cover a standardized template, a standardized serialization and a collaborative use-case management repository, which is used in the context of the EU M/490 mandate [57,58]. Addressing the aspect of interoperability at the (system) architectural level, the SGAM has been developed during the work of the M/490 mandate. It revisits the five individual levels of the GridWise interoperability stack but puts them in context with standardization and IntelliGrid compliant use cases. After the creation of models and use cases, one particular aspect is to properly assess key performance indicators or maturity levels to the newly developed systems [59]. As the domain has to be considered an emerging one, previous work on this topic from Healthcare, Banking or Software Engineering in general has to be re-shaped to make it work.

Finally, the application of the methods and governance processes derived from the smart grid already developed can be put in context with a multi-domain view, not only taking electricity distribution and transmission into account but extending its view to closely related domains where the single artifacts can be properly re-used in a different but close context. This aspect will deal with the topic of both multi-utility systems (gas, water, waste water and heat) [60] as well as non-utility aspects like a shared infrastructure at home for the IoT or services (e.g. ambient-assisted living gateways for home communication, home automation buses). The individual aspects cover lessons learned for methods, technology, policy-making and technology assessment.

References

[1] S. Lehnhoff, S. Rohjans, H.-J. Appelrath, "ICT-Challenges in Load Balancing across Multi-Domain Hybrid Energy Infrastructures". *IT – Information Technology*, Vol. 55, No. 2 (2013), http://dx.doi.org/10.1524/itit.2013.0009, Oldenbourg Wissenschaftsverlag.

[2] J. Bruinenberg, L. Colton, E. Darmois, *et al.*, "Smart grid coordination group technical report reference architecture for the smart grid version 1.0 (draft) 2012-03-02". Tech. Rep., 2012.

[3] European Commission, Smart Grid Mandate—Standardization Mandate to European Standardisation Organisations (ESOs) to support European Smart Grid deployment, 2011.

[4] S. Ruthe, C. Rehtanz, S. Lehnhoff, "On the Problem of Controlling Shiftable Prosumer Devices with Price Signals". *International Journal of Electric Power & Energy Systems – IJEPES*, Vol. 72 (2015), pp. 83–90, ISSN: 0142-0615, http://dx.doi.org/10.1016/j.ijepes.2015.02.014.

[5] S. Rohjans, C. Daenekas, M. Uslar, "Requirements for Smart Grid ICT architectures". In *Third IEEE PES Innovative Smart Grid Technologies (ISGT) Europe Conference*, 2012.

[6] C. Daenekas, "Deriving Business Requirements from Technology Roadmaps to Support ICT-Architecture Management", In: *First International Conference on Smart Grid Technology, Economics and Policies (SG-TEP 2012)*, 2012.

[7] C. Goebel, H.-A. Jacobsen, V. del Razo, *et al.*, "Energy Informatics". *Business and Information Systems Engineering,* Vol. 12 (2013), DOI: 10.1007/s12599-013-0304-2.

[8] Northrop *et al.*, "Ultra-Large-Scale Systems: The Software Challenge of the Future". CMU SEI Report, 2006.

[9] Sommerville, Cliff, Calinescu, *et al.*, *Large-Scale Complex IT Systems*, 2011.

[10] Lewis, Collopy, "The Role of Engineering Design in Large-Scale Complex Systems". *American Institute of Aeronautics and Astronautics*, 2012.

[11] Trefke, Rohjans, Uslar, Lehnhoff, Nordstrom, Saleem, "Smart Grid Architecture Model Use Case Management in a Large European Smart Grid Project". In: *Innovative Smart Grid Technologies Europe (ISGT EUROPE), 2013 Fourth IEEE/PES* (pp. 1–5). DOI:10.1109/ISGTEurope.2013.6695266, IEEE Publishing, 2013.

[12] US DoE: NIST Framework and Roadmap for Smart Grid Interoperability Standards, Release 3.0 (Draft), May 2014.

[13] Watson, Boudreau, "Information Systems and Environmentally Sustainable Development: Energy Informatics and New Directions for the IS Community", *MIS Quarterly* 34(1), 2010.

[14] Winter, "Design Science Research in Europe", EJIS (17), 2008.

[15] Kossahl, Busse, Kolbe, "The Evolvement of Energy Informatics in the Information Systems Community – A Literature Analysis and Research Agenda", *ECIS Proceedings 2012*, 2012.

[16] Vom Brocke, Fridgen, Hasan, Ketter, Watson, "Energy Informatics: Designing a Discipline (and Possible Lessons for the IS Community)", ICIS 2013.

[17] Goth, "Ultralarge Systems: Redesigning Software Engineering?", *IEEE Software 2008*, p. 91ff, 2008.

[18] Baxter, Sommerville, "Socio-Technical Systems: From Design Methods to Systems Engineering", *Interacting with Computers*, 2008.

[19] Huynh, Tran, Osmundson, "Architecting of Systems of Systems for Delivery of Sustainable Value", *Second Symposium on Engineering Systems*, MIT Cambridge, 2009.

[20] Heering, Mernik, "Domain-Specific Languages as Key Tools for ULSSIS Engineering", *ULSSIS 2008*, Leipzig, ACM, 2008.

[21] Anvaari, Cruzes, Conradi, "Smart Grid Software Applications as an Ultra-Large Scale System: Challenges for Evolution".

[22] Anvaari, Cruzes, Conradi, "Challenges on Software Defect Analysis in Smart Grid Applications", *SE-Smart Grids Zurich 2012*, IEEE Publishing, 2012.

[23] Rittel, Webber, "Dilemmas in a General Theory of Planning", *Policy Sciences* 4, 1973.

[24] Ritchey, "Wicked Problems: Modelling Social Messes with Morphological Analysis", *Acta Morphologica Generalis* 2(1), 2013.

[25] Nam, Pardo, "Smart City as Urban Innovation: Focusing on Management, Policy, and Context", *ICEGOV2011*, Estonia, 2011.

[26] Wedde, Lehnhoff, Rehtanz, Krause, "Von eingebetteten Systemen zu Cyber-Physical Systems", 2009.

[27] CEN, CENELEC, ETSI, "SGCP Report on Reference Architecture for the Smart Grid v0.5 (for sg-cs sanity check) 2012-01-25", Tech. Rep., 2012.

[28] Englert, Uslar, "Europäisches Architekturmodell für Smart Grids—Methodik und Anwendung der Ergebnisse der Arbeitsgruppe Referenzarchitektur des EU Normungsmandats M/490". In: *VDE-Kongress 2012—Intelligente Energieversorgung der Zukunft, Stuttgart*, VDE-Verlag, 2012.

[29] Uslar, Rohjans, Specht, González, Trefke, "Das Standardisierungsumfeld im Smart Grid—Roadmap und Outlook". *Elektrotechnik und Informationstechnik*, 128(4), 135–140, 2011.

[30] Uslar, Rohjans, Bleiker, *et al.*, "Survey of Smart Grid standardization studies and recommendations—Part 2". In: *Innovative Smart Grid Technologies Conference Europe (ISGT Europe), 2010 IEEE PES* (pp. 1–6). DOI:10.1109/ISGTEUROPE.2010.5638886, IEEE Publishing, 2010.

[31] Rohjans, Uslar, Bleiker, *et al.*, "Survey of Smart Grid Standardization Studies and Recommendations". In: *First IEEE International Conference on Smart Grid Communications*, IEEE Publishing, 2010.

[32] ISO/IEC/IEEE, ISO/IEC/IEEE 42010 ed1.0, "Systems and Software Engineering—Architecture Description", 2011.

[33] Uslar, Grüning, "Zur semantischen Interoperabilität in der Energiebranche: CIM IEC 61970". *Wirtschaftsinformatik*, 49(4), 295–303. Vieweg, 2007.

[34] L. F. Grillo, T. K. Datta, C. Hartner, "Dynamic Late Lane Merge System at Freeway Construction Work Zones". In: *Transportation Research Record*, Vol. 2055, pp. 3–10, 2008.

[35] M. Uslar, M. Specht, S. Rohjans, J. Trefke, J. M. Gonzalez, *The Common Information Model CIM: IEC 61968/61970 and 62325 P Practical Introduction to the CIM*. Springer, 2012.

[36] A. Wegmann and G. Genilloud, "The role of 'roles' in use case diagrams", EPFL-DSC DSC/2000/024, 2000.

[37] International Electrotechnical Commission (IEC), Publicly Available Specification (PAS) 62559 IntelliGrid Methodology for Developing Requirements for Energy Systems, 2008.

[38] Uslar, Rosinger, Schlegel, "Security by Design for the Smart Grid: Combining the SGAM and NISTIR 7628". In: *Proceedings of the IEEE COMPSAC* 2014, IEEE Publishing, 2014.

[39] F. Schloegl, S. Rohjans, S. Lehnhoff, J. Velasquez, C. Steinbrink, P. Palensky, "Towards a Classification Scheme for Co-Simulation Approaches in Energy Systems". In: *Proceedings of the 2015 International Symposium on Smart Electric Distribution Systems and Technologies (EDST)*, IEEE Press, 2015.

[40] C. Steinbrink, S. Lehnhoff, "Challenges and Necessity of Systematic Uncertainty Quantification in Smart Grid Co-Simulation". In: *Proceedings of the IEEE EUROCON 2015-International Conference on Computer as a Tool (EUROCON)*, IEEE Press, 2015.

[41] Uslar, Rohjans, Specht, "Technical Requirements for DER Integration Architectures". *Energy Procedia*, 20, 281–290. DOI:10.1016/j.egypro. 2012.03.028, 2012.

[42] S. Lehnhoff, W. Mahnke, S. Rohjans, M. Uslar, "IEC 61850 based OPC UA Communication—The Future of Smart Grid Automation". In: *Proceedings of the 17th international Power Systems Computation Conference (PSCC'12)*, Stockholm, Sweden, IEEE Press, 2011.

[43] Rohjans, Uslar, Specht, Tröschel, Niesse, "Towards Semantic Service Integration for Automation in Smart Grids". *International Journal of Distributed Energy Resources* 8(2), 2012.

[44] M. Büscher, M. Kube, S. Lehnhoff, K. Piech, S. Rohjans, J. Trefke, "Towards a Process for Integrated IEC 61850 and OPC UA Communication: Using the Example of Smart Grid Protection Equipment". In: *Proceedings of the 40th Annual Conference of the IEEE Industrial Electronics Society (IECON'13)*, IEEE Press, Dallas, November 2014.

[45] S. Lehnhoff, M. Blank, T. Klingenberg, M. Calabria, W. Schumacher, "Distributed Coalitions for Reliable and Stable Provision of Frequency Response Reserve – An Agent-based Approach for Smart Distribution Grids". In: *Proceedings of the IEEE International Workshop on Intelligent Energy Systems (IWIES) Collocated with IECON 2013—The 39th Annual Conference of the IEEE Industrial Electronics Society*, IEEE Press, 2013, invited paper.

[46]   Broy, Fettweis, Herzog, *et al.*, "Interoperabilität und offene Standards im IT-Bereich; Definitionen, Voraussetzungen, Bedeutung und Herausforderungen". Acatech.

[47]   M. Faschang, S. Rohjans, F. Kupzog, E. Widl, S. Lehnhoff, "Requirements for Real-Time Hardware Integration into Cyber-Physical Energy System Simulation". In: *Proceedings of the 2014 Workshop on Modeling and Simulation of Cyber-Physical Energy Systems (MSCPES)*, IEEE Press, 2015.

[48]   S. Rohjans, S. Lehnhoff, S. Schütte, F. Andrén, T. Strasser, "Requirements for Smart Grid Simulation Tools". In: *Proceedings of the 23rd IEEE International Symposium on Industrial Electronics (ISIE'14)*, Istanbul, Turkey, IEEE Press, June 2014.

[49]   Uslar, Rohjans, Specht, Gonzales, "What Is the CIM lacking?" In: *IEEE SmartGridComm 2010*, IEEE Publishing, 2010.

[50]   Santodomingo, Rohjans, Uslar, Rodriguez-Mondejar, M. Sanz-Bobi, "Ontology Matching System for Future Energy Smart Grids". *Engineering Applications of Artificial Intelligence* 32, 242–257. DOI:10.1016/j.engappai. 2014.02.005, 2014.

[51]   Santodomingo, Rohjans, Uslar, Rodriguez-Mondejar, Sanz-Bobi, "Facilitating the Automatic Mapping of IEC 61850 Signals and CIM Measurements". *IEEE Transactions on Power Systems* 28(4), 4348–4355. DOI:10.1109/TPWRS.2013.2267657, 2013.

[52]   J. Trefke, L. Nordström, A. Saleem, S. Rohjans, M. Uslar, S. Lehnhoff, "Smart Grid Architecture Model Use Case Management in a large European Smart Grid Project". In: *Proceedings of the Fourth IEEE European Innovative Smart Grid Technologies Conference (ISGT'13)*, IEEE Press, Copenhagen, October 2013.

[53]   S. Lehnhoff, S. Rohjans, M. Uslar, W. Mahnke, "OPC Unified Architecture: A Service-Oriented Architecture for Smart Grids". In: *Proceedings of the ICSE 2012 International Workshop on Software Engineering Challenges for the Smart Grid*, IEEE Press, Zurich, Schweiz, 2012.

[54]   S. Lehnhoff, S. Rohjans, R. Holzer, F. Niedermeier, H. de Meer, "Mapping of Self-Organization Properties and Non-Functional Requirements in Smart Grids". In: *Proceedings of the Seventh International Workshop on Self-Organizing Systems (IWSOS 2013)*, Lecture Notes in Computer Science (LNCS), Springer, Berlin, 2013.

[55]   Gonzalez, Dänekas, Trefke, Uslar, "Supporting Interoperability in Smart Grids". In *Enterprise Interoperability* (pp. 401–407). DOI:10.1002/ 9781118561942.ch57, Wiley, New York, 2012.

[56]   Gonzalez, Uslar, "An Ontology-Based Method to Construct a Reference Model Catalogue for the Energy Sector". In: S. Smolnik, F. Teuteberg, and O. Thomas (Eds.) *Semantic Technologies for Business and Information Systems Engineering: Concepts and Applications* (pp. 16–39). IGI Global. DOI:10.4018/978-1-60960-126-3.ch002, 2010.

[57]   Trefke, Gonzalez, Uslar, "Smart Grid Standardisation Management with Use Cases". In: *Energy Conference and Exhibition (ENERGYCON), 2012*

IEEE International (pp. 903–908). DOI:10.1109/EnergyCon.2012.6348279. IEEE Publishing, 2012.

[58] Santodomingo, Uslar, Göring, *et al.*, "SGAM-Based Methodology to Analyse Smart Grid Solutions in DISCERN European Research Project". In: *Proceedings of the IEEE EnergyCon 2014*, IEEE Publishing, 2014.

[59] Uslar, Rohjans, Cleven, Wortmann, Winter, "Towards an Adaptive Maturity Model for Smart Grids". In: *Proceedings of the 17th International Power Systems Computation Conference PSCC Stockholm 2011*, 2011.

[60] Uslar, Andrén, Mahnke, Rohjans, Stifter, Strasser, "Hybrid Grids: ICT-Based integration of Electric Power and Gas Grids—A Standards Perspective". In: *Third IEEE PES Innovative Smart Grid Technologies (ISGT) Europe Conference*. IEEE Publishing, 2012.

Chapter 3

Smart-grid communications and standards

Noman Bashir[1], Naveed Ul Hassan[1], Chau Yuen[2]
and Wayes Tushar[3]

A reliable and efficient communication and networking infrastructure will connect the functional elements within the smart grid. Different physical data communication technologies for the smart grid will empower the legacy power grid with the capability to support two-way energy and information flow. The smart grid will rely on several existing and future wired and wireless communication technologies (e.g., power line communication, cellular network, internet protocol networks, ZigBee, Wi-Fi, Worldwide Interoperability for Microwave Access, etc.). Also, advanced techniques for accessing the network and routing the information to the different nodes of the network will be required. In this chapter, we discuss smart-grid communication network and divide it into three tiers, i.e., home area network, neighborhood area network, and wide-area network. We present smart-grid communication technologies, standards, and protocols for the physical layer operations. We next discuss the medium access control (MAC) and network layer protocols of the smart-grid communication technologies. In the end, we present a case study for the implementation and performance evaluation of various smart-grid algorithms and communication infrastructure.

3.1 Introduction

In this chapter, we provide an overview of smart-grid communication technologies and standards. Traditional electricity grid is commonly divided into transmission and distribution networks. Transmission networks transport electricity from power plants over high-voltage lines to various substations, while distribution networks supply low-voltage electricity to residential, commercial, and industrial consumers. In a traditional grid, distribution network, including consumer premises, generally

[1]Electrical Engineering Department, SBASSE, LUMS, Lahore, Pakistan
[2]Engineering Product Development, Singapore University of Technology and Design (SUTD), Singapore, Singapore
[3]SUTD-MIT International Design Center, Singapore University of Technology and Design (SUTD), Singapore, Singapore

lacked communication capabilities. However, communication technologies lie at the heart of smart-grid vision. Communication infrastructure is being deployed on top of the traditional power grid in order to enable bi-directional data communication between various nodes, e.g., user appliances, smart meters, homes, grid stations, etc. Increased communication capabilities of smart grids can be effectively utilized for efficient management, control and monitoring of nodes, better integration of renewable energy sources and enhancement of grid stability and resilience.

Smart gird communication network can be divided into three sub-networks/ tiers, namely, access tier or home area network (HAN), distribution tier or neighborhood area network (NAN), and core tier or wide-area network (WAN). HAN collects information about the energy consumption and state of various appliances that are located inside the consumer premises, where the actual energy consumption occurs. HAN allows the end customers to more actively participate in grid management. It also facilitates regulatory authorities and grid operators to implement dynamic pricing, demand-side management algorithms and the integration of small distributed renewable energy sources, e.g., solar panels installed on rooftops, with the main grid. NAN, on the other hand, is an intermediate network between HAN and WAN, which primarily connects multiple HANs to the core network. The core tier of the smart-grid communication network consists of a WAN that creates a communication path and connects multiple NANs with the core utility systems.

Smart-grid communication networks transmit data from source to destination. In data communication networks, a layered architecture is usually considered for information delivery from source to destination. Each layer performs a specific set of tasks according to well-defined protocols. In the ISO/OSI model, there are seven layers [1]. The three lower layers in this model, i.e., physical layer, data link layer (DLL) and the network layer, are actually responsible for data communication, while the upper layers provide an interface with the smart-grid applications and end users. The physical layer is responsible for the transmission of bits over the physical medium (channel). The DLL is responsible for controlling the shared access to the channel, synchronization, framing, and error control issues. The network layer routes packets across any tier of the communication network for their successful delivery at the destination. As the most important functionality of DLL is to control the shared medium access when there are multiple nodes in the network, this layer is also popularly known as MAC layer, particularly in wireless sensor networks (WSNs) literature.[1] Numerous protocols exist for different layers depending on the application demands and several other requirements. The service providers offer services defined in the three lower layers. The discussion in this chapter is therefore mainly concerned with the communication technologies and protocols for the physical, MAC and network layers. In Section 3.2, we discuss smart-grid communication network and divide it into three tiers, i.e., HAN, NAN, and WAN.

---

[1]It is important to note that medium access control is one functionality of data link layer. However, throughout this chapter, the terms DLL and MAC layer will be used interchangeably.

In Section 3.3, we present smart-grid communication technologies, standards and protocols for the physical layer operations. In Sections 3.4 and 3.5, we, respectively, discuss the MAC and network layer protocols of the smart-grid communication technologies. We present the testbed, developed at Singapore University of Technology and Design (SUTD), Singapore, for the implementation and performance evaluation of various smart-grid algorithms and communication infrastructure in Section 3.6. The chapter is concluded in Section 3.7.

3.2   Smart-grid communication network

Smart-grid communication network can be divided into three sub-networks as shown in Figure 3.1. HAN creates communication paths for smart meters, home appliances, plug-in electric vehicles, and various sensors to monitor and control home environment (rightmost section). NAN establishes the communication links among multiple smart meters, multiple HANs, and data collectors deployed at the distribution network level of the power system (center section). Finally, WAN creates communication paths between the data concentrators at various NANs, data centers, and other metering and monitoring infrastructure deployed at main power generation plants or at the transmission network level (leftmost section). Both wired and wireless communication technologies can be used to build HANs, NANs, and WANs. Wired technologies offer simple, low cost, and reliable solutions. Wireless technologies, on the other hand, can be easily deployed and are highly scalable.

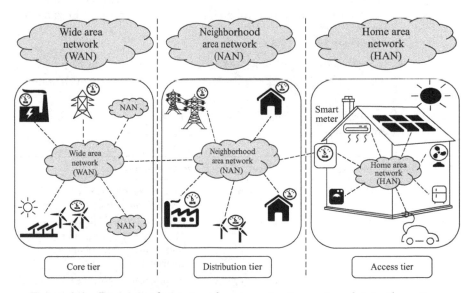

Figure 3.1   Division of smart-grid communication network into three tiers

3.2.1   Access tier HAN

Communication capabilities of HANs enable the application development for efficient home energy management, user comfort, and demand response applications. For instance, an efficient air-conditioning control application may provide the user a web/android/iOS interface to control the appliances, turn them ON/OFF based on occupancy, automatically conserve energy by pre-cooling or heating based on historical usage patterns and engage the appliances in demand response programs or automatically respond to dynamic pricing signals. Similar networks, are respectively, referred to as building area networks (BANs) and industrial area networks, when applied in buildings (often commercial spaces) and industrial spaces. Home appliances, smart meters, and environmental sensors generally send the collected data (power readings, state, temperature, occupancy, humidity, etc.) to a home gateway controller that further transmits the gathered information over NAN for energy management or other related services. The home gateway controller also receives control commands from the consumer or the grid operator and communicates this information over HAN to appropriate actuators that might be connected with the appliances in order to alter their operational state or power consumption levels [2]. As the communication requirements of each appliance may differ, the home gateway controller should also be equipped with appropriate interfaces for multiple communication technologies [3].

The coverage area of a HAN is generally small and is expected to be less than 100 m. Smart-grid applications in HANs are also not expected to be data-rate intensive. The typical data rates will be in the range of few Kbps. Different protocols based on IEEE 802.15.4, IEEE 1901, and IEEE P1901.2 and other proprietary standards generally suit the coverage area and data-rate requirements of HANs for smart-grid applications. The data rates, coverage area, and the type of devices that will require connectivity through HANs are summarized in Table 3.1.

3.2.2   Distribution tier—NAN

NAN, also known as the distribution tier of the smart-grid communication network, is responsible for enabling the state estimation and real-time control of multiple

Table 3.1   *Comparison of communication requirements of HAN, NAN and WAN*

Network/tier	Data rate	Coverage area	Devices/nodes
HAN	1–100 Kbps	1–100 m	Home appliances, smart meters, sensors, plug-in electric vehicles
NAN	100 Kbps–10 Mbps	100 m–10 km	Multiple smart meters, multiple HANs, DERs, protection circuits and control equipment on distribution network
WAN	10 Mbps–1 Gbps	10–100 km	Multiple NANs, power plants, PMUs, protection circuits and control equipment on transmission network

HANs, protection circuits, and control equipment on the distribution network and small-scale distributed energy sources (DERs), e.g., solar and wind generation plants. NAN enables the interconnection between the core tier of smart-grid communication infrastructure, i.e., WAN, with access tier infrastructure, i.e., HAN. NAN also provides the communication support to facilitate the management of data collected from multiple HANs or smart meters. NAN is sometimes referred to as a field area network (FAN), when it gathers data for smart-grid monitoring from the power system infrastructure, such as power lines and towers.

Coverage area of a NAN is generally less than 10 km, while the data-rate requirements might be several Mbps. Physical layer technologies to build NANs are generally based on both wired and wireless technologies. The data rates, coverage area, and the type of devices that will require connectivity through NANs are given in Table 3.1.

3.2.3 Core tier—WAN

The core tier of the smart-grid communication network, also called WAN, connects nodes spread over a large geographical area. WANs terminate at the core utility systems of the grid operator. WANs connect multiple NANs/FANs, power generation plants, protection circuits and control equipment on the transmission network and phasor measurement units (PMUs) to core utility systems. WAN comprises high capacity, high bandwidth, and robust communication links that form the backbone of smart-grid communication infrastructure. The coverage area of a WAN is much larger than that of a NAN. The data-rate requirements are also large. The data rates, coverage area, and the type of devices that will require connectivity through WANs are given in Table 3.1.

3.3 Communication technologies, standards and protocols for physical layer operations

In this section, we discuss the communication technologies for each tier of the smart-grid communication network. These technologies enable the actual transmission of bits over the channel and enable physical layer operations. There are numerous wired and wireless communication protocols that can satisfy the data-rate and coverage area requirements of the HAN, NAN, and WAN. Wired communication technologies are superior to wireless technologies in terms of reliability, bandwidth, security, and lower maintenance costs. On the other hand, wireless communication technologies offer flexible deployments and high scalability. These characteristics of wireless technologies often prove to be crucial for the rapid deployment of smart-grid communication network in the areas without any existing communication infrastructure. The recent efforts to improve the energy efficiency of mobile-connected devices have made wireless technologies even a better option [4]. Some of the more advanced wireless broadband technologies also have the capability to provide bandwidth and data rates comparable to those of popular wired technologies [5,6].

Table 3.2   Important communication technologies for physical layer operations of HAN, NAN, and WAN

Sub-network/tier	Medium/channel	Communication technologies
HAN	Wired	PLC, Ethernet
	Wireless	Bluetooth, ZigBee, WLAN, Z-Wave
NAN	Wired	PLC, DSL, Ethernet, fiber-optics
	Wireless	ZigBee, WLAN, WiMAX, cellular, CR networks (TVWS)
WAN	Wired	PLC, Metro Ethernet, fiber-optics (SONET)
	Wireless	WiMAX, cellular, satellite

In Table 3.2, we list down some important communication technologies, divided into wired and wireless categories, that can be used for HAN, NAN, and WAN operations. A communication technology is either suited for a specific tier or for multiple tiers. To build a HAN, popular wired technologies include narrowband and broadband power line communication (PLC) and Ethernet, while wireless options include Bluetooth, ZigBee, wireless LAN (WLAN), and Z-Wave. For NAN, narrow-band and broadband PLC (BB-PLC), digital subscriber lines (DSL), Ethernet, and fiber-optics are the possible wired technology options, while ZigBee, WLAN, Worldwide Interoperability for Microwave Access (WiMAX), cellular standards, e.g., GSM, 3G, 4G and LTE, and cognitive radio (CR) networks, including TV White spaces (TVWS), are some of the stronger candidates on the wireless side. Finally for WAN, wired options include narrow-band PLC (NB-PLC), passive optical networks (PONs), Metro Ethernet for the core network with some wired technologies such as Internet protocol (IP)/multi-protocol label switching ,and fiber-optics (SONET [synchronous optical networking]). Wireless technologies for WANs and NANs are generally the same. It is also important to note that sometimes, multiple communication technologies can be simultaneously used for different purposes inside the same network tier. For example, both Bluetooth and ZigBee may be used inside HAN or both PLC and cellular communication technologies may be used inside NAN.

In the following subsections, we provide an overview of important wired and wireless communication technologies. A summary of the standardization activities for all the technologies is also presented.

3.3.1 Wired communication technologies

The most important wired communication technologies for smart grids are PLC, optical communication, DSL, and Ethernet.

3.3.1.1 Power line communications

PLC technologies leverage the existing power lines for communication, which offers service providers an advantage of utilizing already available infrastructure for both power and data transmission [7]. The cost-effectiveness, ease of

Table 3.3 Powerline communication families and standards summary

PLC family	Standards	Data rate	Frequency band	Applications
Narrow-band PLC	IEEE P1901.2 G3-PLC PRIME	Up to 500 Kbps	3–500 KHz	HAN, NAN, WAN
Broadband PLC	IEEE 1901 HomePlug 1.0 HomePlug Turbo HomePlug AV HomePlug CC HomePlug Green PHY HomePlug BPL	Up to 100 s of Mbps	1.8–250 MHz	HAN, BAN, NAN

deployment, ubiquitous nature, and advanced standardization efforts make PLC a popular choice to build HAN, NAN, as well as WAN [8,9]. In a typical NAN for large buildings or a residential community comprising several households, the data from smart meters can be collected at the data concentrators using power lines and then further transmitted to the grid operator or any third-party energy management service provider via wireless communication links [10]. In France's Linky meter project that aims to upgrade the 35 million traditional meters to Linky smart meters, BB-PLC is selected as the communication technology to establish NAN and then GPRS for WAN [11]. A similar project has also been carried out by the Italian electric utility, ENEL. In this project, 32 million smart meters were deployed. Again, PLC technology was adopted to collect smart meter data from several households to the nearest data concentrator, while GPRS was used for WAN [12]. However, the use of power lines for reliable data transmission is not straightforward and faces significant technical challenges. The key challenges are noise, high signal attenuation, and interference from other sources, such as nearby external electromagnetic sources, power cables, and electrical appliances [13].

PLC families
The PLC technologies can be divided into two families, i.e., NB-PLC and BB-PLC. This classification is based on the frequency band of operation and the achievable data rates. Table 3.3 provides a summary of standards, data rates, frequency bands, and potential applications of NB-PLC and BB-PLC technologies.

- *Narrow-band PLC (NB-PLC)*: NB-PLC systems operate in 3–500 kHz frequency bands. The data rates provided by NB-PLC systems range from 1 bps to 500 Kbps. NB-PLC systems can be deployed on both low- and high-voltage power lines. The pilot deployments demonstrate the ability of NB-PLC systems to cover very large areas up to 150 km [14], which makes it suitable for HAN, NAN, and WAN.
- *Broadband PLC (BB-PLC)*: BB-PLC systems operate in high-frequency bands (2–30 MHz). These systems utilize more bandwidth and can provide data rates

up to 200 Mbps. However, the higher data rates come at the cost of reduced coverage area and lower communication reliability. Due to these issues, BB-PLC is considered to be more suitable for HAN and NAN. Recent developments in the BB-PLC technologies have resulted in modems that can provide data rates of around 10 Mbps over the distances of 8 km [7].

PLC standards

Many parallel standardization efforts have also been carried out for the PLC technologies. These efforts have resulted in a number of competing standards as shown in Table 3.3.

- **IEEE P1901:** The IEEE P1901 standard was developed by the IEEE P1901 Working Group for BB-PLC. The aim was to meet the requirements of in-home multimedia and smart-grid applications. This standard includes physical and MAC layer protocols [15]. It also meets the safety requirements set by the regulators of several countries and also ensures the compatibility with other wired and wireless communication systems. IEEE P1901.2, on the other hand, is a worldwide standard for NB-PLC.
- **G3-PLC:** G3-PLC is a PLC standard introduced by European Regional Development Fund and Maxim Integrated. It aims to support interoperability, robustness, cost-effectiveness, and cyber security in smart-grid communication networks. It can be used to transmit smart meter readings and to build in-home real-time applications.
- **PRIME:** PRIME is a global open-source PLC standard developed by PRIME Alliance that provides multi-vendor interoperability for NB-PLC. It specifies the physical layer, MAC layer, convergence layer, and management plane. It is being used for AMI applications in smart-grid deployments [16].
- **HomePLUG:** HomePlug Power line Alliance is the most important industrial association that develops widely accepted standards for BB-PLC systems. The alliance has released multiple standards over the last couple of years, which have progressively increased the supported data rates for PLC systems. HomePlug 1.0 and HomePlug Turbo have the capability to support data rates of 4 and 85 Mbps, respectively, while HomePlug AV supports 200 Mbps. HomePlug AV is a very advanced technology that provides high-quality and multi-stream networking. It is convenient to use and also backward compatible with HomePlug 1.0 and HomePlug Turbo. HomePlug Command and Control (HPCC) is designed for low-cost applications that is suitable for creating a reliable HAN between a smart meter and electric appliances. HomePlug Green PHY specification is a standard developed by the Smart Energy Technical Working Group within the HomePlug Power line Alliance that supports lower data rates, lower power consumption, IP networking support, and interoperability within HomePlug. This standard is particularly suitable for the smart-grid applications used in HAN [17]. HomePlug broadband over powerline (BPL) allows inter-networking with devices using the BPL technology (IEEE 1901 standard).

3.3.1.2 Optical-fiber communication

Electric utilities have widely used optical-fiber communication technologies to build communication backbone for creating a network of control centers and substations [18]. Optical-fiber has been traditionally preferred due to its ability to provide high bandwidth (tens of Gbps) over very large distances (several kilometers). It is also advantageous for the high-voltage environment of power system due to its robustness and resistance against radio frequency and electromagnetic interference (RFI/EMI). Optical-fiber is regarded as one of the most prominent communication technologies in the future smart grid [19,20]. However, high installation and maintenance cost of the optical-fiber technology remains a major bottleneck.

PON, wavelength division multiplexing (WDM), and SONET/synchronous digital hierarchy (SDH) are the various forms of optical-fiber communication. The choice of particular technology will depend upon the application requirements in terms of response time, reliability, and quality of service (QoS).

- **PON:** It is a technology that enables the deployment of various point-to-multipoint network topologies by using a single fiber [21]. It supports a data rate of up to 2.5 Gbps over the range of 60 km. PON has made fiber-to-the-home possible, which also opens ways to form access networks. Ethernet PON is a variant of PON technology that utilizes the standard Ethernet protocol for communication over the optical network. It also supports the interoperability with existing IP-based networks.
- **WDM:** It is an effective technology to exploit bandwidth capacity of optical-fiber networks. It supports data rates up to 40 Gbps with a coverage area of up to 100 km. It enables the use of the same fiber to support multiple streams of data simultaneously, such as supporting uplink and downlink traffic using the same fiber.
- **SONET/SDH:** SONET and SDH are essentially the same standards. The SONET is used in the USA and Canada, while SDH is used in rest of the world. These are time-division multiplexing (TDM)-based architectures designed to carry high capacity traffic. With a coverage area of around 100 km, they can support data rates up to 10 Gbps.

3.3.1.3 Digital subscriber lines

DSL is the set of high-speed digital data communication technologies that enable the transmission of data over the telephone lines. It is preferred by the electric utilities as it avoids the cost of deployment for a dedicated communication infrastructure. However, because of its low reliability and down time issues, DSL is not a suitable technology for time-critical smart-grid applications.

DSL standards

There are many open and proprietary standards for DSL technologies. However, the three most widely used standards are VDSL, ADSL and HDSL. The data rates,

Table 3.4   DSL standards and their versions

Standard	Version	Coverage	Data rate	Applications
ADSL	ADSL	Up to 4 km	8 Mbps	NAN/FAN
	ADSL2	Up to 7 km	12	
	ADSL2+	Up to 7 km	24	
VDSL	VDSL	Up to 1.5 km	52–85 Mbps	NAN/FAN
	VDSL2	Up to 300 m	Up to 200 Mbps	
HDSL	HDSL	Up to 3.6 km	2 Mbps	NAN/FAN

coverage area and utilization of these standards to build smart-grid communication architecture are given in Table 3.4.

- **ADSL:** Asymmetric DSL (ADSL), ADSL2, and ADSL2+ are the three variants of ADSL technology. ADSL can support data rates of 8 Mbps in downstream and 1.3 Mbps in upstream with a coverage range of up to 4 km. ADSL2 provides the capability to support 12 Mbps in downstream and up to 3.5 Mbps in upstream with a coverage area of up to 7 km. ADSL2+ supports 24 Mbps in downstream and 3.3 Mbps in upstream with the coverage range similar to that of ADSL2 [22]. The typical latency for the unidirectional flow of data using ADSL is approximately 20 ms. This value fulfills the medium latency requirements for the exchange of non-critical information among protection units, control functions, and synchrophasors [22].
- **VDSL:** Very high bit rate DSL (VDSL or VHDSL) is another version of DSL for shorter distances (around 1.2 km). It can support 52 Mbps downstream and 16 Mbps upstream data rates using copper cables, while 85 Mbps downstream and upstream data rates on coaxial cables. Second-generation systems (VDSL2) can support data rates of up to 200 Mbps in both downstream and upstream directions with a coverage area of 300 m [23].
- **HDSL:** High-speed DSL (HDSL) can support data rates up to 2.048 Mbps over a distance of 3.6 km. HDSL is not rate adaptive and the line rate is always 1.544 Mbps or 2.048 Mbps.

3.3.1.4   Ethernet

Ethernet is among the most widely used networking technologies for wired local-area networks (LANs) and metropolitan area networks (MANs). Its widespread use is ascribed to its simplicity, low installation cost, ease of installation and maintenance, reliability, and interoperability. Ethernet is a frame-based technology, where the stream of data is divided into shorter pieces called frames. Each frame contains source address, destination address, and error-checking data to make the system reliable.

Ethernet uses a shared medium approach where more than one devices use the same medium to communicate with each other. As multiple hosts use the same medium, the simultaneous transmissions give rise to data collisions. Ethernet technology uses the carrier sense multiple access/collision detection (CSMA/CD) protocol to handle this problem. However, the collision is inherently limited by the network devices, such as switches and routers [24].

Ethernet standards

IEEE has led Ethernet standardization efforts and released IEEE 802.3 in 1983, followed by many subsequent versions. IEC 61850 is another Ethernet standard that is used in substation automation systems of power grids.

- **IEEE 802.3:** This standard has improved significantly over time in order to support higher data rates and provide larger network coverage. Ethernet baseband is defined in standards, such as 10BASE5, where the first numerical digit indicates the data rate (in Mbps) and the last number or letter defines the type and length of the cable. For example, 10BASE5 standard can support a data rate of 10 Mbps over the distance of 500 m using thick coaxial cable. Some other Ethernet standards include 10BASE2, 100BASE-TX, 1000BASE-SX, and 10GBASE-LX4 [24].
- **IEC 61850:** It is an Ethernet standard for communication within substations for automation and protection applications [25]. It brings advantages to sub-station automation systems by reducing the complexity and maintenance cost. It defines communication protocols and ensures interoperability between the substation equipment. IEC 61850 also defines a detailed data structure for each of the physical device inside the substation, consisting of logical devices, logical nodes, and data relevant to each node.

3.3.2 Wireless technologies

There are numerous wireless communication technologies and standards. The three prominent families of IEEE wireless standards, include IEEE 802.15, IEEE 802.11, and IEEE 802.16. Wireless communication technologies that are based on IEEE 802.15 standards, e.g., Bluetooth and ZigBee, are suitable for personal area networks (PAN) that include HANs and sometimes NANs. Wireless communication technologies that are based on IEEE 802.11 standards, e.g., Wi-Fi, are suitable for LAN that also include HANs and NANs. Wireless communication technologies that are based on IEEE 802.16 standards, e.g., WiMAX, are suitable for metropolitan area networks (MAN) that include NANs and WANs. Cellular communication standards, GSM, GPRS, EDGE, 3G, 4G, LTE, and A-LTE, can also be used in smart-grid applications. CR networks, which allow devices to establish wireless communication in freely available spectrum bands, e.g., TVWS or un-utilized frequency bands of licensed cellular network operators, can also be exploited for smart-grid communication needs. Satellite communication can also be used to provide wireless connectivity and to build WANs. Additionally, some proprietary wireless communication technologies, such as Z-wave, are also used in smart-grid applications. In the following, we provide an overview of these wireless communication technologies.

3.3.2.1   IEEE 802.15-based wireless networks

IEEE 802.15 is a family of wireless communication standards for wireless PANs (WPANs). These technologies are suitable for HANs and sometimes NANs. There are two major standards within this family, i.e., IEEE 802.15.1 and IEEE 802.15.4. IEEE 802.15.1 defines the physical and MAC layer specifications for wireless connectivity with fixed, portable, and moving devices in personal operating space. On the other hand, IEEE 802.15.4 is the reference standard for low-power, low-cost, and low data-rate WPAN that specifies the physical and MAC layers [26]. The IEEE 802.15.4-based physical layer is capable of providing data rates up to 250 Kbps, with a coverage area of about 10 m. Recent advancements in physical layer specifications support relatively higher data rates over a longer range [27]. This standard also supports star, tree, and mesh networking topologies for WPANs. In each of the topology, PAN coordinator has the responsibility of managing the entire network. The tree and mesh topologies also need routers, which are special nodes with a task of relaying messages to establish connections between end devices and PAN coordinator. The standard also defines a hybrid MAC protocol for channel sharing and power management. It combines the reservation-based channel access schemes (TDMA [time-division multiple access]) and contention-based channel access schemes (CSMA) to exploit the benefits of both schemes. A new task group, 802.15.4g, is working on the physical layer enhancements to the legacy 802.15.4 standard in order to make it more suitable for smart utility networks, including smart grids [28]. Bluetooth is a popular technology that is based on IEEE 802.15.1, while ZigBee is based on the IEEE 802.15.4 standard.

- **Bluetooth:** It is a low-power, short-range WPAN network technology for connecting mobile and fixed devices based on the IEEE 802.15.1 standard. The operating frequency range of Bluetooth is 2.4–2.4835 GHz and it offers a data rate of up to 721 Kbps [29]. It supports both point-to-point and point-to-multipoint communication and offers a coverage area anywhere between 1 and 100 m depending upon network configuration, which makes it more suitable for use in HANs. Bluetooth has two class versions: (i) legacy Bluetooth 3.0+HS and (ii) newly introduced low-energy Bluetooth 4.0. A brief overview and comparison of both the Bluetooth versions is presented in Table 3.5.

Bluetooth 3.0+HS defines two network architectures, namely Piconet and Scatternet. In a Piconet network, seven slave devices can simultaneously communicate with one master node. In Scatternet topology, more than one Piconets are interconnected through a shared node. Bluetooth 4.0 uses the traditional star and point-to-point topologies. Bluetooth is a suitable technology for local monitoring

Table 3.5   Bluetooth specifications

Bluetooth version	Coverage area	Date rate	Power consumption	Network topology
Bluetooth 3.0+HS	Up to 100 m	1–3 Mbps	100 mW	Piconet, Scatternet
Bluetooth 4.0	150 m	1 Mbps	10 mW	Star, Point-to-point

applications in smart-grid-based substation automation systems [30]. Bluetooth 4.0 is particularly designed for low-power applications for transmitting a small amount of data with low latency. The recent improvements in the form of Bluetooth v4.2 can also support several Internet of Things applications, including connected home deployments. However, Bluetooth has significant drawbacks in terms of security and its performance is also severely influenced by the surrounding communication links.

- **ZigBee:** It is a set of application profiles and network specifications for low data rate, short-range, and low duty cycle applications. It is developed, promoted, and maintained by the ZigBee Alliance. ZigBee, based on the IEEE 802.15.4 standard, is the most widely used technology for WPANs in residential, industrial, as well as commercial environments. It is considered as an ideal technology for a variety of applications, such as home automation, automatic meter reading, energy monitoring, and smart lighting. U.S. National Institute for Standards and Technology regards ZigBee and ZigBee Smart Energy Profile (SEP) as the most appropriate standard for smart-grid HANs [31]. ZigBee SEP provides an interface for managing appliances that monitor, control, and automate the delivery and use of energy [32]. The standardization efforts for ZigBee in the smart-grid domain have also resulted in two SEP versions, called SEP 1.x and SEP 2.0. SEP 2.0 was developed in cooperation with other standardization groups, such as HomePlug and IP for Smart Objects industrial alliances [33,34]. SEP 2.0 is a comprehensive profile that supports the control of apartment buildings and plug-in hybrid electric vehicles. It also ensures interoperability between ZigBee and IPv6-based networks.

The universal popularity of ZigBee is mostly due to its simplicity, low complexity, and low cost. The ZigBee protocol stack consists of four layers, physical, MAC, network, and the application layers. ZigBee implements IEEE 802.15.4-based physical and MAC layers but it also specifies additional layers to support network security, network management, and various other applications. ZigBee has interoperability features and the application profiles allow the specification of the supported devices, data formats, communication models, and message types. ZigBee can provide data rates up to 250 Kbps in all of the 868 MHz, 915 MHz, and 2.4 GHz bands.

3.3.2.2 IEEE 802.11-based Wireless Networks

IEEE 802.11 is a suite of wireless communication standards for WLAN. WLAN technologies, based on IEEE 802.11 standards, provide robust, high-speed, point-to-point, and point-to-multipoint wireless communication support and are suitable for use in HAN and NAN of many smart-grid applications [35]. The prominent examples of such applications include home automation solutions, monitoring and control of DERs, and automation, protection, and control of substation equipment. These technologies use simple and flexible CSMA/CA-based channel access schemes. WLAN technologies are less energy efficient. These technologies might also suffer from signal availability and reliability issues in harsh environments, e.g., in high-voltage environment with lot of EMI [36]. However, applying proper path engineering and system design techniques, some of these problems can be avoided [37].

Using error correction algorithms, message acknowledgment and data buffering can also be helpful. The recent advancements in antenna design also make WLAN-based systems more immune to electromagnetic and RFI [38].

WLAN standards

The legacy IEEE 802.11 standard, first released in 1997, proposed the standard for WLANs. It included three non-interoperable technologies, frequency hopping spread spectrum, direct sequence spread spectrum (DSSS), and infrared (IR). The data rates were limited to 1–2 Mbps. IEEE 802.11 adopted the spread spectrum technology to allow more users to occupy the same frequency band, while trying to minimize interference with other users [37]. This standard has been frequently revised to increase the data rates, coverage area, and to support new features. The transmission ranges of WLAN technologies depend upon various factors, such as antenna type, modulation schemes, transmission powers, and wireless environment. Some prominent versions of IEEE 802.11 standards for WLANs are listed below and their comparison is given in Table 3.6.

- **IEEE 802.11b:** IEEE 802.11b, also known as Wi-Fi, offers a maximum data rate of 11 Mbps. Wi-Fi operates in the unlicensed frequency band of 2.4 and 5 GHz. It uses the DSSS modulation technique.
- **IEEE 802.11s:** IEEE 802.11s builds wireless mesh networks (WMNs) by supporting multi-hop transmissions, while using the 802.11 physical layer.
- **IEEE 802.11p:** IEEE 802.11p is a dedicated standard to support wireless access for vehicular environment. It is expected to be the key enabling technologies for vehicle-to-grid (V2G) systems.
- **IEEE 802.11e:** IEEE 802.11e standard offers QoS features, such as traffic prioritization, scheduling, and admission control that are suitable for delay sensitive applications.
- **IEEE 802.11n:** IEEE 802.11n standard defines the orthogonal frequency division multiplexing-based physical layer and provides multiple input multiple output (MIMO) antenna support. This latest Wi-Fi standard can provide very high data rates, up to 600 Mbps. The experimental results demonstrate the range of 802.11n network to be around 300 m.

3.3.2.3   IEEE 802.16-based wireless networks

IEEE 802.16 series of wireless communication standards are developed for Wireless MANs [39]. WiMAX technology is based on these standards. WiMAX was

Table 3.6   WLAN standards

Standard	Coverage	Data rate	Applications
IEEE 802.11b	Up to 40 m	Up to 11 Mbps	HAN, AMI
IEEE 802.11s	Up to 300 m	Up to 54 Mbps	HAN, mesh networks
IEEE 802.11p	Up to 1 km	Up to 54 Mbps	Vehicular environment
IEEE 802.11e	Up to 300 m	Up to 54 Mbps	V2G, QoS
IEEE 802.11n	Up to 300 m	Up to 600 Mbps	High throughput

released in 2001 and it targeted wireless connectivity for rural and sub-urban areas, with an objective to support interoperability. It complements IEEE 802.11 (WLAN) as it can support thousands of simultaneous users over larger areas, controls the channel bandwidth based on the number of connections, and applies more advanced QoS mechanisms. IEEE 802.16-based networks also have some disadvantages in terms of network management complexity and operation in licensed frequency bands.

The original version of IEEE 802.16 standard defined the operating frequency range to be 10–66 GHz for large coverage areas, i.e., between 7 and 10 km with data rates of up to 100 Mbps [39]. In order to support interoperability, the WiMAX forum later defined standards to operate in lower frequency bands. The 3.5 and 5.8 GHz frequency bands are assigned for fixed communication. The 3.5 GHz spectrum band is for licensed operators, while 5.8 GHz spectrum band is for unlicensed operators. The later band provides data rates of up to 70 Mbps at distances of up to 48 km [40]. Similarly, 2.3, 2.5, and some portion of 3.5 GHz have also been assigned for licensed mobile communications. The licensed spectrum is advantageous as they allow operators to use high transmission power to provide more coverage over longer distances. The different versions of WiMAX technologies exist that support several advanced features, such as multicast and broadcast services, MIMO, orthogonal frequency division multiple access (OFDMA), and various types of adaptive coding and modulation schemes. IEEE 802.16j standard [41] specifies various types of multi-hop relaying techniques to enable flexible deployments for large area coverage. IEEE 802.16m [42] advancement to IEEE 802.16 standard aims to provide a very high throughput of 1 Gbps at low mobility and at least 100 Mbps at high mobility.

The high data rates and large coverage areas make WiMAX technology well suited for WANs, e.g., to interconnect large facilities (power plants, offshore renewable generation plants, etc.) with the centralized control centers. WiMAX networks can also be helpful in implementing real-time pricing schemes to the consumers based on real-time energy consumption. The fast communication capabilities of WiMAX also enable the grid operators to employ fast outage detection and restoration plans.

3.3.2.4 Cellular communication

Cellular communication technologies are broadly classified into generations that are labeled, 1G, 2G (GSM), 3G (UMTS), 4G (LTE-A), and 5G. There are also intermediate generations labeled as 2.5G (GPRS and EDGE) and 3.5G (HSPA). Each generation of cellular communication technology outperforms the previous generation in terms of data rates, energy efficiency, and latency. Cellular communication systems typically operate in licensed frequency bands, e.g., 900, 1,800, 1,900 MHz, etc. The data rates provided by various generations typically vary from few Kbps to several hundred Mbps [40]. It is also important to note that the cellular communication technologies evolve rapidly and the latest generations support very high data rates and can provide sophisticated communication services. Interested readers are referred to References 43 and 44 to find out more details of cellular communication technologies.

Cellular communication networks are universally available for voice and data communication. As these networks operate in licensed spectrum bands, they are fully maintained by their network operators. The use of cellular technologies for smart-grid applications has certain advantages, such as nearly ubiquitous coverage, no installation costs, and reliability [45]. The main disadvantages, however, are the high operational costs and the involvement of third-party cellular network operators into the system. Moreover, the performance of cellular data services is highly dependent on the number of users in the same base station, which can result in variable throughput and poor latency. In several countries, electric utilities have traditionally used GSM, GPRS and EDGE for data communication in supervisory control and data acquisition (SCADA) systems [46–48].

In smart grids, the major application of cellular communication so far has been the transfer of data from smart meters to the grid operator. Often, cellular communication modules are embedded into the smart meters to enable automatic communication. Itron, for example, has developed an electricity meter, called SEN-ITEL, which has an integrated GPRS module that establishes communication with a server-running Smart Synch's transaction management system [49]. Linky smart meter project in France and another similar project by ENEL in Italy also used cellular communication to transmit the aggregated data from several smart meters to utility operator. Another example is the Echelon's Networked Energy Services system, which again uses a GSM network operated by T-Mobile [49].

3.3.2.5   CR networks

The wireless spectrum is an expensive resource and the license price for exclusive usage of only a few MHz bandwidth is often several billion US dollars [50]. Measurements performed at various geographical locations, worldwide, however, reveal that the actual licensed spectrum utilization is only around 5%–15%. CR technology exploits this fact and is aimed at the opportunistic utilization of unoccupied RF bands without disturbing the operations of licensed spectrum owners [51]. CR technology also encourages more operators to establish their own networks, comprising only a few base stations, as there are no license fees. These CR network operators can potentially offer valuable communication services for smart-grid applications [52,53].

Some unlicensed frequency bands, e.g., TVWS, are also made freely available in several countries for novel applications. These frequency bands can also be opportunistically exploited by CR network operators. TVWS are the frequency bands in the very high-frequency and ultra-high-frequency bands that are recently vacated due to the transition from analog to digital TV broadcasts. In September 2010, the Federal Communications Commission unanimously approved new rules for the use of unlicensed TVWS spectrum [54]. Office of Communications in the UK also allows free usage of TVWS [55]. In Singapore, TVWS is also made freely available and its use is currently under trial [56]. TVWS comprises relatively large frequency blocks with attractive wireless propagation characteristics and their free availability has the potential to unleash a myriad of new applications and services. However, due to the dynamic nature of spectrum holes and competition among several CR network operators, the freely available spectrum may not be sufficient to cater for the QoS requirements of many smart-grid applications. Novel algorithms and system designs are therefore required in order to overcome these challenges [57–59].

3.3.2.6 Microwave communication

Microwave communication is a wireless communication technology using the electromagnetic radiation with frequencies between 500 MHz and 40 GHz [60]. The range of a microwave communication link is limited to around 50 km due to curvature of the Earth. One way to increase the range of the microwave communication is to use the repeaters at periodical intervals. This form of microwave communication is called terrestrial communication. The range of microwave can also be enhanced by increasing the height of microwave antennas or by using satellites orbiting the Earth [60]. The later arrangement is referred to as satellite communication systems and is described in detail next.

Satellite communication

Satellite communication is also a wireless communication technology. Satellite communication is enabled by low Earth orbit, medium Earth orbit and geostationary Earth orbit satellites. These satellites orbit at different altitudes around the Earth [61]. They are characterized with variable bandwidth and latency. Satellite communication is usually expensive. However, it is particularly useful in areas where wired or wireless communication technologies cannot be easily deployed. Satellite communication technology has long been used by the electric utilities for SCADA systems for locations that are either beyond the reach of traditional communication networks or pose challenges in deploying new networks. The recent enhancements in satellite communication technologies suggest that they can also be used for a variety of smart-grid applications, such as forming backup backhaul transport services or providing backup communication services at critical substations [23].

3.3.2.7 Z-Wave

Z-wave is a short-range, low-power, low-cost and reliable, proprietary wireless technology developed by ZenSys. It specifies a five-layered protocol stack, consisting of physical, MAC, transfer, routing and application layers. It operates at 900 MHz and supports data rates of up to 40 Kbps. Its coverage area is limited to only 30 m for indoor applications, while it can extend up to 100 m for line of sight outdoor applications. Z-Wave 400 series operates in 2.4 GHz frequency band and can provide up to 200 Kbps in data rates. This technology was specifically designed for short-range remote control applications in residential and small commercial environments. It has been proposed as a candidate technology for HANs due to its support for mesh networks. Power plugs with integrated Z-wave communication capabilities are available for commercial and experimental use in the market.

3.3.3 Summary of communication technologies for smart-grid applications

Table 3.7 provides a summary of wired and wireless communication technologies for smart-grid applications discussed in this section. The table presents an overview of the standardization activities, transmission range, maximum data rate and applicability scope of each technology in smart-grid communication systems.

Table 3.7 Summary of communication technologies

Communication technology	Standards/protocols	Coverage	Data rate	Applications
		Wired communication technologies		
PLC	**NB-PLC:** IEEE P1901.2 PRIME, G3-PLC	Up to 150 km	Up to 500 Kbps	HAN, NAN, WAN
	BB-PLC: IEEE 1901 HomePlug 1.0, Turbo HomePlug AV, CC, BPL HomePlug Green PHY	Up to 8 km	Up to 100 s of Mbps	HAN/BAN, NAN
Optical-fiber	PON WDN SONET/SDH	Up to 60 km Up to 100 km Up to 100 km	Up to 2.5 Gbps 40 Gbps 10 Gbps	WAN WAN WAN
DSL	ADSL VDSL HDSL	Up to 7 km Up to 1.5 km Up to 3.6 km	8–24 Mbps 52–200 Mbps 2 Mbps	NAN/FAN NAN/FAN NAN/FAN
Ethernet	IEEE 802.3 IEEE 61850	Up to 100 m –	Up to 10 Gbps –	HAN, BAN, NAN SAS
WPAN	**IEEE 802.15.1:** Bluetooth **IEEE 802.15.4:** ZigBee	Up to 100 m Up to 100 m	721 Kbps 250 Kbps	V2G, HAN V2G, HAN
WLAN	IEEE 802.11x	Up to 100 m	2–600 Mbps	HAN, NAN, V2G

Technology	Standard	Range	Data rate	Application
WiMAX	IEEE 802.16	0–10 km	128 Mbps Up / 28 Mbps down	NAN, WAN / WAN
	IEEE 802.16m	0–100 km	100 Mbps mobile / 1 Gbps fixed	NAN, WAN
Cellular	2G	Several km	14.4 Kbps	V2G
	2.5G	Several km	144 Kbps	NAN, WAN, smart meters
	3G	Several km	2 Mbps	NAN, WAN, smart meters
	3.5G	Several km	14 Mbps	NAN, WAN, smart meters
	4G	Several km	100 Mbps	NAN, WAN, smart meters
Microwave communication	Direct Microwave Link	Up to 50 km	Up to few hundred Mbps	WAN
	Satellite Internet	100–6,000 km	1 Mbps	WAN
CR (TVWS)	—	Several km	Up to few Mbps	NAN, WAN
Z-Wave	Z-wave	Up to 30 m	40 Kbps	HAN
	Z-wave 400	Up to 100 m	40 Kbps	

3.4   MAC layer communication protocols

Any communication network in smart grids, e.g., HAN, NAN, and WAN, generally comprises multiple nodes (sensors, appliances, etc.). These nodes often transmit the collected data or receive control commands from a higher level using a shared medium (channel). Simultaneous data transmissions by multiple nodes on the same channel can create interference and data collisions, which can result in packet loss. In communication networks, MAC layer communication protocols are responsible for controlling the access to the shared channel by individual nodes in order to avoid collisions and packet loss [62]. MAC layer communication protocols also play a decisive role in defining the energy efficiency, communication efficiency, and latency of the network. Simultaneous consideration of all these issues, while designing MAC layer communication protocols for HAN, NAN, and WAN, is a challenging task [62,63].

In some applications, MAC layer communication protocols also control the wake up and sleep schedules of the sensor nodes for energy conservation. Error control for reliable and efficient communication is also performed at the MAC layer. Various approaches, such as forward error correction (FEC) and automatic repeat request (ARQ), are generally used. The FEC scheme achieves reliable communication by adding some redundant bits to the original packet. The ARQ scheme controls the retransmission of lost packets, efficiently utilizes the communication bandwidth but adds some additional latency. The simplistic ARQ schemes are not suitable for smart-grid applications but their carefully designed advanced versions may prove to be feasible in certain environments. The reliability objective can also be achieved by using a hybrid approach that benefits from the advantages of both ARQ and FEC [64]. A hybrid approach incrementally increases the error resiliency of a packet through retransmissions and might be suitable for some smart-grid applications in harsh physical environments, requiring real-time delivery (extremely low latency) [64,65]. Smart-grid communication networks might also consist of heterogeneous nodes with variable latency requirements. For example, some sensor in the network might be used to collect critical information at aperiodic intervals, while another node in the same network might be tasked to periodically gather and transmit a large amount of non-critical information. MAC layer communication protocols are also responsible for dealing with such heterogeneity in the latency requirements of various sensor nodes in the network.

There are numerous MAC layer protocols that are specifically designed for WSNs. Depending on the application, communication networks in smart grids, particularly HANs, generally comprise several wireless sensors sharing the same communication channel. WSN-based MAC protocols can be used for smart-grid applications. However, it is also important to note that WSN and smart grids sometimes do not share the same MAC layer objectives. For example, many MAC layer protocols in WSN are designed for short-distance and low data-rate applications with a prime focus on energy conservation. These protocols, based on the resource constrained assumptions, can result in significant communication delays, which might not be suitable for real-time communication requirements of smart-grid applications. However, recent research efforts have focused on developing new MAC layer protocols for smart grids as well as modifying the existing WSN-based MAC layer protocols to suit smart-grid application needs [66,67].

MAC layer protocols for wired and wireless communication technologies that are mostly used for NAN and WAN are well known in the literature. For example, MAC protocols for fiber-optics, cellular communication, DSL, etc. are not much interesting in the context of smart grids. Our focus in this section will primarily be on the MAC layer protocols for WSN-based smart-grid communication networks that are mostly used to build HANs and sometimes NANs. These protocols can be divided into three classes (i) contention-based MAC layer protocols, (ii) reservation-based MAC layer protocols, and (iii) hybrid MAC layer protocols. These protocols primarily stem from two fundamental multiple access schemes, i.e., TDMA and CSMA. Other multiple access schemes, such as code division multiple access, frequency division multiple access (FDMA), and OFDMA, are not suitable for smart-grid application because of their complexity, memory intensiveness, and high processing requirements [68].

In the contention-based protocols, sensor nodes compete with each other to gain channel access, which may result in unpredictable communication delays. Contention-based random access protocols are generally considered to be more scalable as they do not need timing synchronization between the nodes. In reservation-based protocols, transmission schedules are planned for each of the sensor nodes, which results in deterministic communication delays for each node. Reservation-based protocols, however, often waste precious resources (transmission time, bandwidth, etc.), when a sensor node has no packets for transmission in some scheduled interval. Hybrid MAC protocols combine the multiple MAC schemes with an aim to overcome the drawbacks of the individual schemes.

In the following subsections, we discuss some important MAC layer protocols for wired and wireless communication technologies that are used to build HANs, NANs, and WANs.

3.4.1 MAC layer communication protocols for wired technologies

Various wired technologies are used at different levels of the smart-grid infrastructure. Optical-fiber, DSL, and Ethernet are the widely used wired technologies. The typically used communication protocols for wired communication networks are Modbus, distributed network protocol (DNP), and point-to-point protocol (PPP). It is important to note that DNP and PPP are in fact protocol stacks, i.e., a set of protocols for multiple layers, including the MAC layer.

3.4.1.1 Modbus protocol

Modbus is a communication protocol that is based on the master/slave architecture. The master node, which is capable of sending broadcasts or individual messages to the slave nodes, can initiate communication. The slave nodes respond by performing the actions requested by the master node. Modbus is an open source and also the most widely used communication standard in industrial applications.

The Electronics Industries Association (EIA) standards, e.g., EIA 232 (also called RS 232), EIA 422 and EIA 485, were developed for asynchronous serial transmissions for reliable communication over wired networks. Modbus can be implemented over EIA standards as well as Ethernet and optical-fibers. In substation automation systems, Modbus is commonly implemented using EIA 485. This implementation can support 1 master and 31 slave nodes and can achieve data rates of up to 10 Mbps at short distances of 12 m. It may also be used to connect and aggregate smart meter data in HANs and NANs. On the other hand, implementations of Modbus over Ethernet and optical-fibers can, respectively, support 100 Mbps and 1 Gbps data rates over much longer distances and are thus suitable for NAN and WAN.

3.4.1.2   Distributed network protocol 3.0

Distributed network protocol 3.0 (DNP3), recently adopted as IEEE 1815-2010 standard, is a protocol stack that also includes a MAC layer (or DLL) protocol. DNP3 can be implemented over EIA 232 or EIA 485 standards and Ethernet [69]. Like Modbus, master/slave architecture is employed in DNP3. However, in DNP3, slave nodes also have the capability to initiate a request. In traditional grids, DNP3 is mostly used for substation automation and in SCADA systems [24]. This protocol is not suitable for time-critical smart-grid applications because it cannot provide performance guarantees [69].

3.4.1.3   Point-to-point protocol

PPP, again, is a set of protocols, defined by Internet Engineering Task Force (IETF), in order to establish a direct connection between two nodes. The three components of PPP are the encapsulation component, Link Control Protocol (LCP), and Network Control Protocol (NCP). The encapsulation component performs the transmission task over the specified physical layer. The LCP is MAC protocol as it establishes, configures and tests the link and also negotiates the capabilities of each link. NCP is used to negotiate optional configuration parameters and it facilities network layer routing. This protocol can be implemented over Ethernet and DSL.

3.4.2   MAC layer communication protocols for wireless technologies

In this subsection, we will primarily discuss MAC layer communication protocols for WSN-based smart-grid communication networks (e.g., Bluetooth, ZigBee, etc.). These protocols are broadly categorized as contention-based, reservation-based and hybrid protocols and may be used to build HANs and NANs.

3.4.2.1   Contention-based MAC layer protocols

Contention-based protocols are considered to be the most suitable candidates for dynamic requirements of the smart grid. The most significant contention-based MAC layer protocols are carrier sense multiple access with collision avoidance

(CSMA/CA), RT-MAC, MaxMAC, delay responsive cross-layer (DRX) and fair and delay aware cross-layer (FDRX) and QoS-MAC.

- **CSMA/CA:** CSMA/CA is a contention-based protocol that improves upon the traditional CSMA techniques by avoiding collisions. In CSMA/CA, the nodes only transmit when the channel is sensed to be "idle." In this protocol, a node that has packets to transmit, listens to the shared medium to determine channel availability. If the channel is unavailable, e.g., if some other node is already transmitting, then this node waits for a certain period of time before listening again. The node only transmits its packet when the channel becomes available. In this scheme, collisions can only occur with nodes that might be hidden. This problem can be partially resolved if the sender node that wants to gain access to the channel first transmit a 'Request to Send' (RTS) command to its destination (receiver) node and then wait for the 'Clear to Send' (CTS) command. IEEE 802.11 uses this protocol and asks the sender/receiver nodes to exchange RTS/CTS commands.
- **RT-MAC:** Traditional contention-based MAC layer protocols may suffer from long communication delays due to network contention. RT-MAC introduces a feedback control packet called clear channel (CC) control packet. The CC control packet enables a sensor node to transmit data by setting the clear channel flag to 1. The value of flag is reset on the basis of a clear channel counter of CC control packet [70]. In order to further reduce the end-to-end delay for data transmission, RT-MAC decreases the contention duration for the control packets resulting in faster traveling of data packet. This mechanism enables RT-MAC to support lower end-to-end delays and also improves reliability without any added cost of energy consumption. RT-MAC can be used for real-time smart-grid applications. RT-MAC also improves the reliability of communication by fast reporting of an alarming event. Despite reducing the latency, RT-MAC cannot overcome the RF interference issues that arise in multi-stream communications [62].
- **MaxMAC:** MaxMAC, also called $(E^2 - MAC)$, is an energy-efficient MAC layer protocol. In WSNs, energy efficiency is a major concern. Several researchers have proposed MAC layer protocols that can trade-off energy efficiency with QoS parameters, such as throughput, latency and reliability. However, most of the energy-efficient MAC layer protocols exhibit severe limitations in terms of latency and throughput. These protocols are also not adaptable to the varying traffic load. MaxMAC protocol, however, can adaptively tune the essential parameters at run time to achieve good latency and throughput performance. During high traffic periods, it introduces additional wake up calls in order to achieve high throughput and low latency, while during sparse traffic periods energy efficiency remains the main concern [71]. Additional wake up calls result in extra power consumption for sensor nodes, which makes this protocol suitable for some energy unconstrained smart-grid applications [62].
- **DRX and FDRX:** DRX data transmission and FDRX data transmission are the MAC protocols proposed for smart-grid applications that require low latency and data prioritization [72]. These protocols achieve the delay requirements by

fine tuning MAC layer parameters and fulfilling the service requirements using application layer prioritization. DRX uses end-to-end delay estimation for data prioritization. If the packet is from a time-critical smart-grid application and the estimated delay is greater than the application requirements, DRX will reduce the clear channel assessment duration, which prioritizes the time-critical packets to access the medium before other low priority packets. On the other hand, FDRX promotes fairness in the DRX scheme by preventing few nodes from dominating the use of shared communication channel [72]. FDRX also reduces the collision rate. Both these protocols can also be suitable for use in smart-grid control operations where sudden changes in loads or the generation cycle take place [72].

- **QoS-MAC:** QoS-MAC protocol aims to incorporate QoS for different priority traffic. To provide differentiated services, the MAC layer of communication technologies based on the IEEE 802.15.4 standard, e.g., ZigBee and Bluetooth, is modified with QoS-MAC [73]. This approach results in a much better network performance for high-priority data traffic, which may be well suited to certain smart-grid applications. The QoS-MAC protocol also reduces the latency and improves the throughput of the network. Although this protocol incorporates QoS, still it cannot support the stringent QoS requirements that might be needed for some smart-grid applications.

3.4.2.2   Reservation-based MAC layer protocols

Contention-based protocols suffer from high latency [66]. Reservation-based protocols, on the other hand, have deterministic delays but they are difficult to scale. Important reservation-based MAC layer protocols include tree-based TDMA-MAC and rate-allocation-based MAC.

- **Tree-based TDMA-MAC:** The Tree-based TDMA protocol aims at improving the traditional TDMA scheme by employing the tree topology. The TDMA scheme lacks the flexibility to cater for the fluctuating data traffic and is also hard to adjust to topology changes. In the tree-based scheme, the sink node forms a tree topology and then uses a weight vector to decide the parent node of each device. Depending upon the constructed topology, sink node assigns time slots for each device with their acknowledgment (ACK) information. The devices send the data in an allocated time slot that is assigned by the sink node. The tree-based MAC layer protocol facilitates the aggregation of sensor nodes data and also better energy performance as a fixed time slot is assigned to each node in the network [74]. Topology changes can be easily accommodated because of tree-based topology. This approach also supports high packet transmission rate and low network latency. The tree-based MAC protocol is suitable for use in the HANs of smart grids.
- **Rate-allocation-based MAC:** This protocol assigns different data rates to the nodes based on their delay requirements [75]. This protocol targets the scheduling problem but considers data bits rather than data packets. In every scheduling slot, bits are transmitted based on queue lengths at each node in order to reduce the average delay of the system. The rate-allocation mechanism becomes optimal

if all the sensor nodes have equal queue lengths in every time slot. The rate-allocation scheme demonstrates better energy efficiency. It also has the capability to make an intelligent trade-off between the energy consumption and latency, which is helpful in building smart-grid applications [75].

3.4.2.3 Hybrid protocols for MAC layer

Hybrid protocols combine the good features of contention and reservation-based MAC layer protocols. IEEE 802.15.4 standard, which is widely used to build HANs, also uses a hybrid protocol for its MAC layer. Other important hybrid protocols include EQ-MAC, Z-MAC, and WirelessHART.

- **IEEE 802.15.4 hybrid protocol for MAC layer:** IEEE 802.15.4 standard uses a hybrid protocol that combines CSMA/CA and TDMA schemes to ensure a guaranteed time slot for each node. This protocol facilitates the deployment of numerous WSN-based smart-grid applications, such as remote monitoring, automation, and supervision. This protocol may not be suitable for time-critical and high-performance smart-grid applications due to the limited number of available time slots and channel congestion issues if extensive data exchanges are required [76].
- **EQ-MAC:** EQ-MAC is a hybrid protocol for the MAC layer that provides QoS support by enabling service differentiation for cluster-based sensor networks [77]. EQ-MAC uses the CSMA/CA scheme for the control messages and the TDMA scheme for data traffic. This protocol is sub-divided into two sub-protocols: classifier MAC (C-MAC) and channel access MAC (CA-MAC). The C-MAC sub-protocol classifies the data at the sensor nodes based on its type and importance. CA-MAC is the heart of the EQ-MAC protocol and it actually employs the hybrid channel access scheme for the control and data traffic. Based on the C-MAC classifier, CA-MAC distinguishes between the control traffic and data traffic. As EQ-MAC uses CSMA/CA for sending control messages, it potentially can suffer from high latency, making it unsuitable for some delay sensitive smart-grid applications [62].
- **Z-MAC:** Z-MAC is another hybrid protocol for the MAC layer that combines contention and reservation-based protocols to achieve low communication delay and high channel utilization [78]. It combines the CSMA and TDMA protocols in an intelligent manner to benefit from the advantages of both schemes. The Z-MAC protocol generally uses the TDMA scheme and reverts to CSMA if time synchronization is not possible in the network. It attempts to achieve time synchronization by sending periodic packets to the nodes. This protocol has two basic modes of operation: low-level contention (LCL) and high-level contention. If the protocol is operating in the LCL mode, any of the sensor nodes can make the transmission [79]. However, in a high contention scenario, explicit contention notification is used to inform the nodes and only the owner of the time slot and its immediate neighbor can make the transmissions to reduce contention.
- **WirelessHART:** WirelessHART is a widely used technology for industrial applications, which require real-time communication and high security. It is considered as the most suitable candidate for electric power system applications

inside a substation or a power generation plant. It is also being proposed as a candidate technology for smart-grid HANs. WirelessHART specifies physical, MAC, network, transport and application layers. Unlike ZigBee, it only defines its physical layer based on the IEEE 802.15.4 standard. At the MAC layer, WirelessHART also combines CSMA/CA and TDMA. Other features include multi-hoping, support for mesh networking, and good time synchronization.

3.4.3   Summary of MAC layer communication protocols

Table 3.8 summarizes various MAC layer communication protocols for smart-grid applications. This table also highlights the key objectives of the protocol, its type and energy consumption characteristics.

Table 3.8   Summary of MAC layer communication protocols

Wired communication: MAC layer protocols	
Protocol	**Objectives**
Modbus	Substation automation and control, HAN, NAN Can be implemented over EIA standards, Ethernet and optical-fibers
DNP3	Substation automation and control, SCADA Not suitable for time-critical applications Can be implemented over EIA standards and Ethernet
PPP	WAN, utility backbone networks Can be implemented over Ethernet and DSL

Wireless communication: MAC layer protocols			
Protocol	**Objectives**	**Category**	**Energy**
CSMA/CA	High-throughput support	Contention based	High
RT-MAC	Real-time data streaming support	Contention based	Low
DRX and FDRX	Supports delay and QoS requirements	Contention based	High
MaxMAC	Provides high throughput and low latency	Contention based	High
QoS-MAC	Provides QoS support based on IEEE 802.15.4	Contention based	High
Tree-based MAC	Supports home area network (HAN) applications	Reservation based	Low
Rate-allocation MAC	Reduces average delay	Reservation based	Low
IEEE 802.15.4	Provides QoS support based on IEEE 802.15.4	Hybrid	Low
EQ-MAC	Provides QoS support for single hop sensor networks	Hybrid	Low
Z-MAC	Provides QoS support	Hybrid	High
WirelessHART	Provides interoperability and scalability	Hybrid	High

3.5 Network layer communication protocols

The network layer of a communication network is responsible for the delivery of data packets across the network all the way from the source to the destination. The important functionalists of this layer include unique addressing scheme for the identification of various network nodes, traffic control to avoid congestion and bottlenecks on the network and routing scheme to reach all the destinations and end points of the network. In data communication networks, these protocols are essential to provide connectivity between different nodes of the network. The network layer protocols are also commonly known as the routing protocols.

There are numerous network layer protocols depending on the communication technology, type of devices and network topology. It is also important to note that sometimes a node in a certain network tier has access to multiple communication technologies. For example, INSTEON is a propriety standard that combines RF and PLC technologies to create a mesh network for HAN. Accordingly, appropriate protocols are designed for such hybrid networks to improve their communication reliability. In this section, we categorize these protocols according to the smart-grid infrastructure they fit into, i.e., HAN, NAN, and WAN. Routing protocols for HAN, NAN, and WAN are further classified on the basis of wired, wireless, and hybrid categories.

3.5.1 Routing protocols for HANs

A key wired communication technology, that finds many applications inside smart-grid HANs, is PLC. The leading PLC standards, include IEEE 1901 and HomePlug. Wireless communication technologies are generally preferred to build HAN. Most of the routing protocols for HAN are based on the IEEE 802.15.4 standard, e.g., ZigBee, WirelessHART, and 6LowPAN and ISA100.11a. Some proprietary routing protocols are also used inside HANs.

3.5.1.1 Routing protocols for wired HANs

We only discuss the routing protocol for PLC communication technology. Other wired technologies used in HANs either do not require routing or have well-established routing protocols. The most significant routing protocol that is used in PLC-based HANs is the adaptive channel state routing (ACSR). This algorithm measures the channel stability and node reachability based on the channel state indicator (CSI) and distance metric. The state of the channel is estimated by implementing a periodic routing information packet from each node to the neighboring nodes. If a node receives this packet, the CSI value is incremented and in case of a packet loss it is decremented. During the actual routing phase, the channel is declared to be stable, if the CSI value is greater than a given threshold and vice versa. A probabilistic flooding method is used to determine the optimal path if all of the paths available are declared unstable. Distance metric is established based on the number of nodes between the source and the destination. The value of distance metric is set to 1, if the two nodes can exchange information directly. Otherwise, the repeater nodes are required and distance metric is updated based on their number. The distance metric can also assume an infinite value if the path is unstable. PLC network topology may vary depending upon the changes in the states of different components of the power system. In view of changing network topology, ACSR always attempts to find a reliable and shortest path for communication.

3.5.1.2   Routing protocols for wireless HANs

Most wireless communication technologies that are used at the HAN level are based on the IEEE 802.15.4 standard. ZigBee, 6LowPAN, WirelessHART, and ISA100.11a all have at least IEEE 802.15.4-based physical layer. Z-wave is another important proprietary wireless technology that is used to build HANs. Below, we discuss ZigBee, 6LowPAN, WirelessHART, ISA100.11a, and Z-wave routing protocols.

IEEE 802.15.4-based routing protocols

- **ZigBee routing protocol:** A ZigBee network is formed by three type of devices, namely, ZigBee coordinator, router, and the end device. ZigBee coordinator and router, both having the ability to route packets, are termed as full function devices. On the other hand, the end device is a reduced function device. It has limited functionality of only receiving a packet and then generating a corresponding response. ZigBee standard can support star, tree and mesh network topologies. Star topology is the simplest, one-hop topology, where the end devices are directly connected to the ZigBee coordinator or router. On the other hand, tree and mesh topologies support multi-hop communications. ZigBee network topologies are shown in Figure 3.2.

    The routing protocol in a ZigBee network depends upon the network topology. There are three routing methodologies supported by the ZigBee network: (i) tree routing, (ii) on-demand mesh routing, and (iii) source routing.

    (i)   **Tree routing** is applied at the ZigBee coordinator to support data collection. It is a proactive approach, based on the parent–child relationships, established during the network formation phase. At the start, the ZigBee coordinator defines the maximum number of routers connected to it and the number

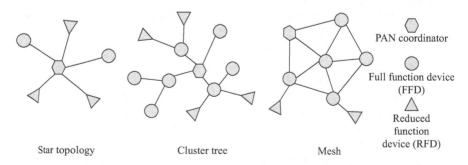

Figure 3.2   ZigBee network topologies

of end devices connected to each of the router. It also defines the maximum depth of the network tree. Based on this knowledge, it assigns an address to each router node or end node when it joins the network. These addresses are provided within an address range that is based on the number of devices connected to the ZigBee coordinator. During the operation phase, these addresses are used to locate any device in the network for routing packets.

(ii) **On-demand mesh routing** is based on the route discovery mechanism from ad hoc on-demand distance vector (AODV). This mechanism does not use well-defined and well-developed routing tables. Instead, it relies on the path request and path reply messages to establish a route from the source to the node. The originating or source device broadcasts path request message, while the destination or end device replies with a path reply message. The network resorts back to the tree routing if it is unable to initiate route discovery or fails to establish a path between the source and end device.

(iii) In **source routing**, the sink node sends a broadcast signal to each of the router in the network. The routers save the sink node as the destination address for all the future routing. During the operation phase, when end devices aim to send data to the sink node, the destination is found by using the reverse route method. In this way, the end devices avoid sending the route request to the same sink node. The sink node is also able to reply immediately to the end device after saving or without saving the route to the routing table.

Tree routing and on-demand mesh routing present significant scalability issues. Development of routing tables at each node allows easier and simplistic routing schemes. However, the number of addresses that can be supported has pre-defined limits. Moreover, any changes in the network topology might also require updating the address tables at most of the nodes inside the network. These schemes also present problems when the number of end devices is large. If there is only one ZigBee coordinator in the network, termed as many-to-one communication, the routers near the sink node will have many route entries that may overflow due to the limited memory capacity. ZigBee Alliance has developed a new specification called ZigBee-Pro to tackle these issues. ZigBee-Pro provides a route aggregation and stochastic addressing mechanism where new devices choose an address in a stochastic manner at the time of joining the network. An address conflict resolution mechanism based on MAC addresses is utilized if two or more devices pick the same address.

- **6LowPAN routing protocol:** IPv6 over Low-power Wireless Personal Area Networks (6LowPAN) is a protocol developed by the IETF 6LowPAN Working Group. The key features of this protocol are neighborhood discovery, auto-configuration, header compression and fragmentation. To enable the transmission of IPv6 packets over IEEE 802.15.4, 6LowPAN builds an adaptation layer between the MAC and network layers. This adaptation layer is responsible for the fragmentation of IPv6 packets (consisting of 1,280 bytes), into IEEE 802.15.4 frame size, which consists of only 127 bytes. It also compresses the 40-byte header of IPv6 into a 2-byte header. It further supports the IPv6 address auto-configuration and neighbor discovery for LowPANs.

6LowPAN utilizes two schemes for the mesh topology of the network, i.e., mesh-under routing and route-over. The mesh-under scheme performs the routing task at the adaptation layer using the IEEE 802.15.4 addresses. As the mesh-under scheme performs the routing and forwarding task at the link layer, a single IP hop will consist of multiple link-layer hops. On the other hand, the route-over scheme takes the routing decision at the network layer, which means that each link-layer hop is an IP hop.

IETF has also developed a routing protocol for low-power and lossy networks, abbreviated as, RPL. It is a good protocol for the route-over routing and can be used in HANs as well as NANs. It is a distance vector routing protocol that uses a directed acyclic graph (DAG) to sort topologies and can support multiple sink nodes. DAG has multiple destination-oriented DAGs (DODAGs), one for each sink node. Each node has an associated rank property to maintain its location in the DODAG. When more than one sink nodes are present, the roots are integrated. In RPL, the sink node of DODAG broadcasts a DODAG Information Object (DIO), filled with information, including objective function and rank. Using this information, the node that joins the DODAG checks the cost of directly reaching the node. The sender of the DIO is then considered a parent and the rank is updated and broadcast. Eventually, each client node possesses information of its predecessor and is able to forward any inward traffic to the sink node by using its parent as the next hop node. Furthermore, a DODAG Destination Advertisement Object (DAO) allows an outward path to be built between the sink node and the node that follows an inward path, inclusive of the originator node.

6LowPAN provides great advantages in terms of interoperability and QoS routing. However, 6LowPAN also has some issues that need to be addressed, such as secure neighborhood discovery, local discovery, and application of IP security to the small home devices.

- **WirelessHART routing protocol:** WirelessHART provides static routing and communication schedules using a central network manager. It uses the graph routing to route messages between the source and the destination. It also uses the source routing for network diagnostics. WirelessHART is a strictly managed protocol in which the network manager has complete knowledge of the network topology and the address of all the devices connected to it. The network manager pulls the neighborhood address tables from each of the network devices. The table contains a list of all the devices with which the particular device can connect. It then utilizes this information to generate graph routes that are not unique and might overlap. A unique graph ID is assigned to each graph, which are distributed to each of the network devices. During the routing process, the source device only writes the specific graph ID in the network header based on the destination.

Enhanced least-hop first routing (ELHFR) implements the graph routing mechanism of WirelessHART. The key improvement over the traditional WirelessHART is that it routes the message to the destination in the least possible hops.

It uses the breadth first search algorithm to build the spanning tree of the graph topology at the gateway. For every network device, the network manager constructs the graphs with least hops to each node in the spanning tree. This ensures that every node in the spanning tree has a single path route with the least hops to the gateway. It also provides multi-path option by generating a redundant path with the same graph ID as the main path for each node in the network. To maintain an updated information, the network manager periodically regenerates the leaf nodes. However, the periodic maintenance does not take into account the changes in network, such as new node entering or leaving the network. The ELHFR addresses this problem by providing a single path for the node when it joins and then updating its graphs during the maintenance cycle. During the operation phase when the nodes receive data, they examine the nature of the destination address. Source routing is used if the destination is not a gateway and the next hop neighbor is selected from the graph table.

- **ISA 100.11a routing protocol:** ISA 100.11a was developed for low data-rate monitoring and automation applications. It builds DLL, network, transport and application layers on top of the IEEE 802.15.4-2006 physical layer. Its operating band is 2.4 GHz. This protocol utilizes channel hoping to enhance reliability and reduce interference. It supports mesh and star network topologies. ISA100.11a has a routing mechanism implemented at the sub-net level and backbone level. The sub-net level mesh routing is performed at DLL where graph routing and source routing are used, while the backbone level routing is performed at the network layer.

Proprietary Z-wave routing protocol
- **Z-wave routing protocol:** Z-wave network consists of controllers and slave devices. Controller is a device that initiates the commands for the slave devices, while the slave devices act upon the controller commands. There are two types of controllers in a Z-wave network depending upon the functions in the network. The primary controller, usually a portable battery operated devices, can be considered as a master with the capability to add or remove devices in the network. These Z-wave devices have the updated network topology and routing table. On the other hand, the secondary controllers are generally static and also have the ability to generate commands. Secondary controllers obtain the routing table from the primary controller upon entering the network. Slave devices can also perform the task of message forwarding if they are asked to do so. However, certain slave devices, known as routing slaves, also have the capability to route packets on their own to the pre-defined static routes. Z-Wave uses the source routing scheme. While initiating a message, the controller embeds the complete route into the frame. In case of the primary controller, it first tries to reach the destination directly. It will only use the default source routing if it is unable to directly find the destination node. It also employs the shortest path mechanism, while selecting a path from source to destination.

3.5.1.3   Routing protocols for hybrid HANs

There are some hybrid approaches that combine PLC with other communication technologies to improve the reliability of communication networks. Below, we discuss the routing protocols of these hybrid approaches.

- **INSTEON:** INSTEON is a proprietary standard that combines RF and PLC technologies to create a hybrid mesh network for home automation systems. An INSTEON device can act as an independent RF device as well as an independent PLC device. It has the complete capability of initiating, receiving, and forwarding a message. INSTEON employs a "simulcast" scheme to deliver the message from the source to the destination. In simulcast, the signal transmitted by the sender is retransmitted by every receiving device. This retransmission is done within a certain time in order to enhance signal strength. In order to achieve path diversity, simulcast employs another mechanism of retransmission using alternate medium. If a device receives the signal using PLC, it will retransmit it using the RF module and vice versa. In order to avoid continuous retransmissions, the protocols only allow four retransmissions.

- **Routing protocol for IPv6-based hybrid RF-PLC network:** Hybrid IPv6-based RF–PLC network architecture has also been proposed by the researchers for BANs. The hybrid network consists of PLC-only nodes, battery-operated RF-only nodes and RF–PLC gateways. The route-over scheme of the 6LowPAN protocol is used at the gateways to ensure interoperability between RF and PLC. The underlying protocol in the route-over scheme is RPL. The routing metric is node's energy estimation, where the routing path is selected on the basis of its energy consumption. In case of a tie-breaker, expected transmission count routing metric is used. RF–PLC gateways work as data collectors and are placed in a way that the RF nodes can reach them in only one hop. The gateways collect the signal and send it to the base station using the backbone PLC network. The improved network lifetime, reduced packet loss and decreased latency can be achieved by increasing the number of RF–PLC gateways.

- **Routing protocol for hybrid PLC-ZigBee network:** Hybrid PLC-ZigBee network has also been proposed for home automation systems. This scheme combines ZigBee communication technology with HPCC PLC technology. The scheme employs the AODV and flooding schemes with some modifications to the forwarding mechanism. The flooding approach is modified in a way that each node forwards the data with a particular sequence number only once. In the modified AODV mechanism, instead of broadcasting, the nodes sequentially request the destination nodes for path. This scheme employs three routing strategies, namely, joint-path, backbone-based, and dual-path. The basic routing strategy is the joint-path strategy, which establishes paths that may consist of both ZigBee and PLC networks. The second routing strategy is called backbone-based and it prefers PLC network over ZigBee network to send the packet. The dual-path strategy uses both the previous schemes. This hybrid network demonstrates better performance in terms of packet delivery, average latency, and network overhead.

3.5.1.4 Summary of routing protocols for HANs

Table 3.9 provides a summary of the routing protocols for HANs. The information about the physical medium, underlying standards, and routing schemes is given.

3.5.2 Routing protocols for NANs

In the following subsection, an overview of the routing protocols for NAN, classified only as wired and wireless are presented. Note that there are no hybrid communication protocols that simultaneously use two different communication technologies for NANs.

3.5.2.1 Routing protocols for wired NANs

In this section, routing protocols for PLC-based NANs are discussed. There are three significant PLC routing protocols, i.e., improved on-demand distance vector (IPODV), geographic routing, and powerline multi-path routing (PMR). Other wired technologies used at the NAN have well established routing protocols.

- **Improved On-demand Distance Vector (IPODV) routing protocol:** IPODV is an improved version of the standard IPODV routing protocol. It enhances the performance of on-demand distance vector (ODV) protocol by improving the neighborhood table management in a way that stable neighbors are preferred during the route selection. In the standard AODV mechanism, the nodes that do not reply even once are deleted from the neighbor table. This leads to poor performance as that node may have been only temporarily lost. As the physical topology of the system hardly changes at the NAN level, the improved mechanism counts the number of recent messages received by neighbor nodes. The link quality is determined by the number of messages received by the node and a threshold value is used to determine if a certain node is to be deleted from the table or not. It also reduces transmission overhead by efficiently managing the route mechanism. IPODV uses the data packets to send hello messages, thus removing the need for separate transmissions. IPODV is much more robust than traditional AODV scheme.

Table 3.9 Summary of routing protocols for HANs

Standard	Version	Physical medium	Routing protocol	Routing methodology
IEEE 802.15	IEEE 802.15.4	Wireless	ZigBee	Tree, OADV, source
	IEEE 802.15.4	Wireless	6LowPAN	Mesh-under, route-over
	IEEE 802.15.4	Wireless	Wireless HART	Graph and source
	IEEE 802.15.4	Wireless	ISA 100.11a	backbone
Proprietary	Z-Wave	Wireless	Z-Wave	Source
IEEE 1901	IEEE 1901.2	Wired	ACSR	Shortest path
	INSTEON	Hybrid	INSTEON	Simulcast
	6LowPAN	Hybrid	IPv6 RF–PLC	Route-over
	ZigBee, HPCC	Hybrid	PLC-ZigBee	Flooding, AODV

- **Geographic routing protocol:** Geographic routing protocols from the WSNs have been proposed for the PLC networks as the nodes in PLC-based NAN are static and their location is pre-determined. Implicit geographic forwarding, beaconless routing, and beacon-based routing are the examples of such protocols. These protocols deploy the greedy perimeter stateless routing if the path is found to be broken. Also, the two routing schemes of the shortest path first (SPF) and flooding mechanism are used. SPF has knowledge about the link quality between the nodes and uses a cost function minimization approach to select the optimal path. In the flooding mechanism, every node that receives a message not intended to it, retransmits it. The evaluations shows that geographic routing protocols are delay and energy efficient for unicast messaging. Simple flooding mechanism, in comparison to SPF, provides lowest latency at the cost of higher energy consumption.

- **Power line multi-path routing (PMR) protocol:** PMR is a routing protocol for NB-PLC that uses an on-demand source routing scheme. When the master node intends to send a signal to the destination slave, it sends route request (RREQ) messages to the network. These messages are received by the destination node, which selects a path among multiple paths and replies back to the master node. It also employs a broadcast control mechanism to control broadcast overhead. It also ensures the fairness in the route discovery process by employing extra back-off time and random packet forwarding. The first node receiving the RREQ, delays its retransmission so that these nodes do not occupy most of the network resources preventing other RREQs from reaching the destination. The messages to be retransmitted are chosen based on random at each node instead of the first-in-first-out strategy. PMR shows better performance in terms of less overhead and finding disjoint routes.

3.5.2.2   Routing protocols for wireless NANs

In this subsection, we discuss routing protocols for wireless NANs. Wireless communication technologies that are used to build NANs are generally based on IEEE 802.11 and IEEE 802.15 standards. Below, we discuss few important routing protocols for communication technologies that are based on these standards.

IEEE 802.15-based routing protocols
There are two important routing protocols known as distributed autonomous depth-first routing (DADR) and hybrid routing protocol (HYDRO).

- **Distributed autonomous depth-first routing (DADR) routing protocol:** DADR is a proactive, scalable and reliable, distance vector routing protocol that quickly adapts to changing link conditions [80]. Depth-first search, routing table, and backtracking mechanisms are used if a link fails and this information is propagated along with the packet. It also avoids a loop by employing a unique Frame ID with the message. It readily learns a new route so when a node is removed, its reliability declines for a while but then restores to 100%. DADR is also more adaptive than other protocols. An alternate route is used, while carrying failed link information, when topology varies and thus a lower overhead is required. In case of link failure, it returns the packet back to its previous sender so that the previous sender can try an alternate route. The routing table in DADR is updated during the periodic HELLO messages among neighbors, during data forwarding and also during positioning messages. DADR has been implemented in several indoor and outdoor experiments and simulations. For example, it has been used to build a real life large scale network of up to 1,500 nodes in Japan and also in the large-scale software simulation study of 2,107 smart meters and 500 relay nodes in a flat mesh network topology [80].

- **Hybrid routing protocol (HYDRO):** Hybrid routing protocol (HYDRO) is a link-state routing protocol that uses a distributed algorithm for DAG [81]. It is mainly used for low-power and lossy networks. In this protocol, each node in the network synthesizes a default route table that includes statistics of the link-layer packet success rate. The nodes also periodically make topology reports and share it with the border router. The border router then analyzes all the topology reports and creates a complete network topology. The source node sends packets to the border router using its default route. As the border router now has a complete view of the network topology, it further forwards these packets to the destination node. Further details about this protocol can be found in Reference 81. The performance of HYDRO has been evaluated on different testbeds [81]. For example, it has been implemented on a testbed consisting of 48 nodes located at the same floor but with a large enough network diameter that required 3–5 hops. The packet delivery ratio (PDR) was high in all the experiments. This protocol was also used in a real application, consisting of 57 nodes that also included nodes placed in remote areas. Border routers were employed and the results in all the experiments showed high PDR.

IEEE 802.11-based routing protocols
IEEE 802.11s is an annex to the single hop IEEE 802.11 WLAN standards, which defines routing protocols to build multi-hop WMNs. IEEE 802.11s is characterized by the binding coordination function, known as enhance distributed channel access (EDCA) and routing and frame forwarding at the DLL, called path selection. EDCA enables different medium access priorities, while path selection enables routing across the network.

- **Hybrid wireless mesh protocol (HWMP):** HWMP is the default routing protocol for IEEE 802.11s based wireless NANs [82]. It is a combination of on-demand reactive routing and tree-based proactive routing. On-demand reactive routing, which is an alteration of AODV protocol, is only employed when a root node is not available. In this case, source node initiates a route discovery mechanism by initiating a Path Request (PREQ) message. The destination or any intermediate node, having destination path information, replies back with unicast Path Reply (PREP) message. This routing scheme enables peer-to-peer communication between the nodes and also reduces the impact of network topology changes. Tree-based proactive routing is used when a root node is available. In this situation, the root node either broadcasts a root announcement (RANN) message or it circulates a PREQ message to every other node in the network. If a node receives a RANN message, it replies with a unicast PREQ and the root node responds back with a PREP message. On the other hand, when a node receives a PREQ message of a desired sequence number or metric from the root node, it replies with a PREQ message and adds this path to the root.

- **Improved hybrid wireless mesh protocol (IHWMP):** This protocol improves the HWMP routing protocol to make it more suitable for smart-grid mesh networks. IHWMP considers the air-time link cost variation, while choosing paths [83]. It proposes a route fluctuation prevention algorithm that stores multiple routing paths along with the most optimal routing path in its routing table. The routing table also contains route information from current and preceding RANN messages. In this manner, the optimal path in the table does not fluctuate and will vary only if the change in the air-time cost in the current path is more than that of the reserve path. Compared to HWMP, this improved algorithm demonstrates a higher PDR, lower end-to-end delay and lesser retransmissions at the MAC layer.

- **Timer-based multi-path diversity routing protocol:** Timer-based multi-path diversity routing protocol is also developed from the HWMP [84]. This protocols aims to meet the smart-grid requirements in terms of reliability, self-healing, and throughput performance. It uses a multi-gateway structure and proactive routing. It periodically broadcasts RANN message from each data aggregator point at random to form multiple paths. If a new RANN message is received, a new path information is added to the routing table. The default path in the routing table is modified only if a better path is found. Furthermore, a backup HWMP buffer periodically saves a data packet from the upper layer when a packet is sent. In case of a link failure, the node can send packets using the backup route. When a back-up route is also not available then on-demand routing is utilized.

- **Secure hybrid wireless mesh protocol:** Secure hybrid wireless mesh protocol (SHWMP) includes protection from attacks [85]. It employs hop-by-hop authentication using a Merkle (hash) tree approach, which is a popular approach for secure verification of nodes in the network. SHWMP provides better PDR as compared to HWMP at the cost of small computational and storage overheads. It is also more robust against routing loops, flooding and routing message modification attacks. Further details of this protocols and another similar approach can be found in References 85 and 86.

3.5.2.3   Summary of routing protocols for NANs

Table 3.10 provides a summary of routing protocols for NANs. In this table information about the physical medium, underlying standards, and routing schemes are given.

Table 3.10 Summary of routing protocols for NANs

Standard	Version	Physical medium	Routing protocol	Routing methodology
IEEE 802.15	IEEE 802.15.4	Wireless	DADR	Distance vector routing
	IEEE 802.15.4	Wireless	Hybrid routing protocol	On-demand source routing
IEEE 802.11	IEEE 802.11s	Wireless	HWMP	On-demand mesh routing, tree
	IEEE 802.11s	Wireless	IHWMP	On-demand mesh routing, tree
	IEEE 802.11s	Wireless	SHWMP	On-demand mesh routing, tree
	IEEE 802.11s	Wireless	Timer-based multi-path discovery	On-demand mesh routing, tree
PLC standards		Wired	IPODV	Distance vector routing
		Wired	Graph routing	SPF, simple flooding
		Wired	PMR	On-demand source routing

3.5.3 Routing protocols for WANs

WAN is the backbone of the whole smart-grid communication infrastructure. Primary wired technologies that are used to build WAN are optical-fibers or metro Ethernet. Routing mechanisms for these technologies are well studied and the smart-grid environment does not give rise to any significant issues. The wireless technologies used to construct WAN are WiMAX, 3G/4G, or GPRS. These technologies achieve communication in one hop, i.e., directly from the source to the destination and do not need routing. However, there are some multi-hop mesh wireless networks that may be deployed at the WAN. The routing protocols discussed in the NAN section can be used for such deployments.

3.6 Case study

In this section, we present the smart-grid testbed, particularly its communication infrastructure that is implemented at the Singapore University of Technology and Design (SUTD) campus. The testbed performs intelligent energy management in multiple residential and commercial units [10]. Residential units are three bedroom apartments, in a student dormitory, with 6–9 resident students in each apartment, while commercial units are faculty offices and shared meeting rooms. In the testbed, an intelligent energy system (IES), which also represents the grid operator end and implemented as a cloud server, provides energy management services to the end users in student dormitory and offices. In the testbed, we have the three tiers of

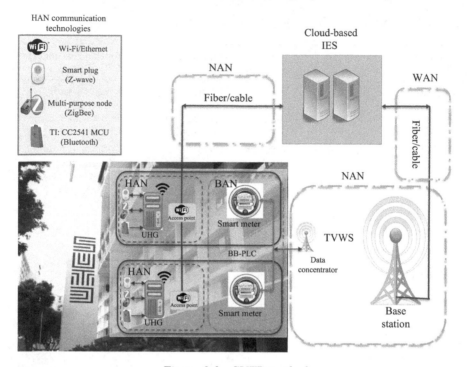

Figure 3.3   SUTD testbed

the smart-grid communication network, i.e., HAN, NAN, and WAN. The aims of the testbed are primarily to demonstrate and validate various smart-grid concepts as well as to identify and test appropriate low latency, efficient and reliable communication infrastructure for each tier. Figure 3.3 describes our communication infrastructure for the residential units. In the next subsections, we give the details of our HAN and NAN. The WAN is based on the traditional fiber/cable internet.

3.6.1   SUTD testbed: HAN

We build an individual HAN for each unit in our testbed. The HAN is equipped with a gateway controller, also called universal home gateway (UHG). The devices, sensors, and actuators, collectively called HAN nodes, communicate with their UHG. We adopt ZigBee, Z-wave, and Bluetooth, as the physical layer technologies for bi-directional communication between various HAN nodes and the UHG. In our testbed, IES can only communicate with the UHGs of the HANs, while it cannot communicate directly with the individual HAN nodes. The communication between UHG and HAN is established through a NAN, which will be described later.

3.6.1.1   UHG

In the SUTD testbed, UHG is built on a Raspberry Pi computer (Model-B Rev 1), which is capable of handling a number of communication protocols. HAN nodes operate on three different communication protocols, i.e., ZigBee, Z-wave, and Bluetooth. These nodes sense and measure different readings, e.g., power consumption, temperature, occupancy, etc., and then transmit the collected information to UHG. As different protocols are used, the delay between the commands from the IES to the UHG and from the UHG to the HAN nodes could pose several challenges depending on the smart-grid application requirements.

In the testbed setup, for the direct communication of UHG with IES or any other cloud-based third-party energy management and also to cater for the asynchronous communication requirements, we have implemented the extensible messaging and presence protocol (XMPP) and RESTful HTTP protocols [87]. XMPP is an application profile of the extensible markup language (XML), which enables the near-real-time exchange of structured yet extensible data between multiple network entities. XMPP is not only capable of sending asynchronous requests but also of supporting a massive number of users. In addition, RESTful HTTP is lightweight, has a simple HTTP request format, and is very easy to implement. Moreover, RESTful HTTP is best suited for applications that require periodic communication. Hence, RESTful HTTP is selected for the periodic uploading of sensor data from the UHG of HANs to the database of the cloud server.

3.6.1.2 ZigBee

We designed a multi-purpose node (MPN), driven by an Arduino Fio microcontroller, consisting of several low-power sensors and actuators. The sensors include a motion detector and temperature, humidity, noise, and Lux sensors. Actuators on MPN include IR Blaster to control the air-conditioning system and potentiometer to control the LED light power supply. MPN communicates with the UHG through a ZigBee communication module. In every hostel unit, each HAN consists of 4 MPNs, one for each bedroom, and one for the living room. While in the office, there are 20 MPNs (one for each faculty office) in one section of an office block connected through multiple ZigBee relays to a single UHG.

3.6.1.3 Z-wave

Z-wave smart-plugs are used to monitor the real-time power consumption of electrical appliances. These plugs communicate with UHG through RaZberry module. These plugs also have built-in relays that respond to demand response signals initiated by IES to delay/deny the operation of electrical appliances.

3.6.1.4 Bluetooth

In order to monitor the environmental parameters, e.g., temperature, humidity, and pressure level of the surroundings, we also distributed several Bluetooth low-energy (BLE) devices based on Texas Instrument's CC2541 MCU among students. The BLE devices are also equipped with an accelerometer, a gyroscope, and a magnetometer to provide information about acceleration, orientation, and magnetization of an object, respectively. The idea is to enable students to build new green energy services. The BLE devices can connect to the UHG of HANs using Bluetooth technology.

3.6.1.5   BB-PLC

In Singapore, student dormitories are high-rise, multi-story, residential buildings, containing several individual apartments. Each apartment has a smart meter that collects the aggregated power consumption values at regular intervals and is required to communicate these readings to IES or the grid operator. The data collected by a single smart meter is generally small. However, the combined data of several hundred smart meters is large. In the testbed, we use BB-PLC to create BAN, which aggregates the data from several smart meters to the data concentrator located at the top of the residential building. This choice of communication to build BAN has certain advantages, including the reuse of existing power line infrastructure, superior performance across thick walls, and easier management of the entire network of smart meters. It is also important to note that smart meter network could be managed by a different service provider than the one managing HANs inside the apartments or offices.

3.6.2   NAN

The HANs created in our case study are required to transmit their data to IES or any other third-party energy management service provider using NAN. In our testbed, we have considered two technologies to build a NAN: (1) TVWS and (2) Traditional fiber/cable internet. We further explain the use of these technologies in our testbed.

3.6.2.1   TVWS

The aggregated data from several smart meters is further transmitted using TVWS to the base station of the service provider. The base station further transmits the received data to the IES cloud using a traditional fiber or cable internet, which forms our WAN. We implement bi-directional TVWS communication. In Table 3.11 we provide the detailed specifications of TVWS spectrum that we have

Table 3.11   Technical specifications of TVWS network deployed at SUTD campus

Attributes	Specification
Frequency bands	630–742 MHz
Occupied channel bandwidth	5 MHz
Channel spacing	8 MHz (configurable to 5, 6, or 7 MHz)
Data rates	13.5, 12, 9, 6, 4.5, 3, 2.25, 1.5 Mbps
Step size	1 MHz
Modulation	BPSK, QPSK, 16-QAM, and 64-QAM
Multiple access	TDMA type
Range	1 km circular radius (3.14 square km)
Number of nodes	Up to 2,000
Transmit power	+12 to +30 dBm (Data concentrator)
Receiver sensitivity	−98 dBm at 1.5 Mbps (typical)
System gain (without antenna)	128 dB (Data concentrator)

used in our testbed at SUTD. Building a NAN using TVWS spectrum, which is a freely available resource (currently under trial in Singapore), provides more operational flexibility, scalability, wide coverage area, and less cost.

3.6.2.2 Fiber/cable Internet

Each UHG of the HAN connects to IES through the traditional fiber or cable internet, which is also used for web surfing or video streaming. For UHGs, fiber/cable internet forms the NAN. On the other hand, the fiber/cable internet link between the TVWS base station and IES can be considered as part of WAN. It should be noted that BAN (using BB-PLC), TVWS, and fiber/cable Internet forms a dedicated and separated network to connect smart meters with utility provider or IES.

3.6.3 Challenges

We faced a number of challenges, including communication delay, finding volunteers for case studies, modeling non-homogeneous scenarios and investigating the mismatch between the theoretical results and the outcomes from the experiments. We only discuss some communication-related challenges that can occur in practical smart-grid deployments. Communication delays can be reduced by doing higher than required sampling, i.e., by taking more readings than the actual requirements of the smart-grid application. However, communication delays can also be caused by the time required for data savings and retrieval at the cloud server as well as the difference between the multiple communication technologies that are often used to build HAN (ZigBee, Z-wave, Bluetooth, etc.). A well thought out efficient and customized script and database design are crucial to reduce the data saving and retrieval delays. Use of multiple communication protocols can lead to timing synchronization issues.

In power grids, during unexpected outage of a scheduled energy resource (e.g., a power plant) or during unexpectedly high peak demand periods, reserve power is required to come online within a very short period of time. Energy reserves are further split into three types, primary (delay: less than 30 s), secondary (delay: less than 15 min) and tertiary (delay: 15 min), depending on the required response times to monitor and control. In smart grids, there is an additional option of demand response management, where instead of engaging energy reserves, some loads are switched OFF to decrease load demand on the grid. Therefore, communication delays should not exceed the requirements of engaging energy reserves. The centralized monitoring and control scheme implemented in our testbed can comfortably respond with time delays similar to the secondary and tertiary reserve requirements. However, it might not be suitable as primary reserve, which requires very fast response time. Implementing distributed control algorithms, instead of centralized algorithms, might enable each node under observation to respond very fast, if necessary, could be a potential way to resolve this issue. Further details on the algorithms and experimental results can be found in Reference 10.

3.7 Conclusion

This chapter provided a detailed overview of various communication technologies, standards, and protocols for smart-grid applications. The three tiers of the smart-grid communication infrastructure were identified as HAN, NAN, and WAN. Different wired and wireless communications technologies were discussed that can be used to build HAN, NAN, and WAN. A detailed overview of MAC and network layer protocols that are relevant to smart-grid communication infrastructure were discussed. Finally, a case study detailing the smart-grid testbed development at SUTD was presented in order to demonstrate the actual implementation of smart-grid communication technologies and communication protocols.

References

[1] N. Benvenuto and M. Zorzi, *Principles of Communications Networks and Systems*. Wiley Online Library, 2011.

[2] C. Gomez and J. Paradells, "Wireless home automation networks: A survey of architectures and technologies," *IEEE Communications Magazine*, vol. 48, pp. 92–101, June 2010.

[3] V. Aravinthan, V. Namboodiri, S. Sunku, and W. Jewell, "Wireless AMI application and security for controlled home area networks," in *Power and Energy Society General Meeting, 2011 IEEE*, pp. 1–8, July 2011.

[4] P. Serrano, A. De La Oliva, P. Patras, V. Mancuso, and A. Banchs, "Greening wireless communications: Status and future directions," *Computers Communications*, vol. 35, pp. 1651–1661, August 2012.

[5] P. P. Parikh, M. G. Kanabar, and T. S. Sidhu, "Opportunities and challenges of wireless communication technologies for smart grid applications," in *Power and Energy Society General Meeting, 2010 IEEE*, pp. 1–7, July 2010.

[6] B. Bennet, M. Boddy, F. Doyle, M. Jamshidi, and T. Ogunnaike, "Assessment study on sensors and automation in the industries of the future: Reports on industrial controls, information processing, automation, and robotics," October 2004.

[7] S. Galli, A. Scaglione, and Z. Wang, "For the grid and through the grid: The role of power line communications in the smart grid," *CoRR*, vol. abs/1010.1973, 2010.

[8] R. P. Lewis, P. Igict, and Z. Zhou, "Assessment of communication methods for smart electricity metering in the U.K.," in *2009 IEEE PES/IAS Conference on Sustainable Alternative Energy (SAE)*, pp. 1–4, September 2009.

[9] V. Paruchuri, A. Durresi, and M. Ramesh, "Securing powerline communications," in *IEEE International Symposium on Power Line Communications and Its Applications, 2008 (ISPLC 2008)*, pp. 64–69, April 2008.

[10] W. Tushar, C. Yuen, B. Chai, *et al.*, "Smart grid testbed for demand focused energy management in end user environment," *IEEE Wireless Communication Magazine*, 2016 (accepted on Jan 07, 2016).

[11] V. C. Gungor, D. Sahin, T. Kocak, *et al.*, "Smart grid technologies: Communication technologies and standards," *IEEE Transactions on Industrial Informatics*, vol. 7, pp. 529–539, November 2011.

[12] R. van Gerwen, S. Jaarsma, and R. Wilhite, "Smart metering," *Leonardo-energy. Org*, vol. 9, 2006.

[13] S. Galli and O. Logvinov, "Recent developments in the standardization of power line communications within the IEEE," *IEEE Communications Magazine*, vol. 46, pp. 64–71, July 2008.

[14] Y. Yang, D. Divan, R. G. Harley, and T. G. Habetler, "Design and implementation of power line sensornet for overhead transmission lines," in *Power Energy Society General Meeting, 2009 (PES '09. IEEE)*, pp. 1–8, July 2009.

[15] S. Goldfisher and S. Tanabe, "IEEE 1901 access system: An overview of its uniqueness and motivation," *IEEE Communications Magazine*, vol. 48, pp. 150–157, October 2010.

[16] "Prime: Powerline intelligent metering evolution," 2008.

[17] V. C. Gungor, D. Sahin, T. Kocak, *et al.*, "Smart grid technologies: Communication technologies and standards," *IEEE Transactions on Industrial Informatics*, vol. 7, pp. 529–539, November 2011.

[18] W. Wang, Y. Xu, and M. Khanna, "Survey paper: A survey on the communication architectures in smart grid," *Computer Networks*, vol. 55, pp. 3604–3629, October 2011.

[19] M. Maier, "Fiber-wireless sensor networks (FI-WSNs) for smart grids," in *2011 13th International Conference on Transparent Optical Networks (ICTON)*, pp. 1–4, June 2011.

[20] L. Jianming, Z. Bingzhen, and Z. Zichao, "The smart grid multi-utility services platform based on power fiber to the home," in *2011 IEEE International Conference on Cloud Computing and Intelligence Systems (CCIS)*, pp. 17–22, September 2011.

[21] M. P. McGarry, M. Reisslein, and M. Maier, "WDM Ethernet passive optical networks," *IEEE Communications Magazine*, vol. 44, pp. 15–22, February 2006.

[22] D. M. Laverty, D. J. Morrow, R. Best, and P. A. Crossley, "Telecommunications for smart grid: Backhaul solutions for the distribution network," in *Power and Energy Society General Meeting, 2010 IEEE*, pp. 1–6, July 2010.

[23] E. Ancillotti, R. Bruno, and M. Conti, "The role of communication systems in smart grids: Architectures, technical solutions and research challenges," *Computer Communications*, vol. 36, no. 1718, pp. 1665–1697, 2013.

[24] J. Ekanayake, N. Jenkins, K. Liyanage, J. Wu, and A. Yokoyama, *Smart Grid: Technology and Applications*. New York: Wiley, 2012.

[25] "IEC 61850: Power utility automation," 2003.

[26] WPAN-WG, "802.15.4-2015 – IEEE approved draft standard for low-rate wireless personal area networks (WPANs)," 2015.

[27]   C. Yoon and H. Cha, "Experimental analysis of IEEE 802.15.4a CSS ranging and its implications," *Computer Communications*, vol. 34, pp. 1361–1374, July 2011.

[28]   K.-H. Chang and B. Mason, "The IEEE 802.15.4g standard for smart metering utility networks," in *2012 IEEE Third International Conference on Smart Grid Communications (SmartGridComm)*, pp. 476–480, November 2012.

[29]   "IEEE standard for information technology – local and metropolitan area networks – specific requirements – Part 15.1a: Wireless medium access control (MAC) and physical layer (PHY) specifications for wireless personal area networks (WPAN)," *IEEE Std 802.15.1-2005 (Revision of IEEE Std 802.15.1-2002)*, pp. 1–700, June 2005.

[30]   H. Zhang, G. Guan, and X. Zang, "The design of insulation online monitoring system based on Bluetooth technology and IEEE1451.5", in *Power Engineering Conference, 2007. IPEC 2007*, pp. 1287–1291, December 2007.

[31]   P. Yi, A. Iwayemi, and C. Zhou, "Developing ZigBee deployment guideline under WiFi interference for smart grid applications," *IEEE Transactions on Smart Grid*, vol. 2, pp. 110–120, March 2011.

[32]   ZigBee Alliance, "ZigBee Smart Energy Profile Specifications". Available from http://www.zigbee.org/ [Accessed 1 May 2016].

[33]   PSO Alliance, "IP for Smart Object (IPSO) Alliance". Available from http://www.ipso-alliance.org/ [Accessed 1 May 2016].

[34]   HomePlug, "HomePlug Powerline Alliance". Available from http://www.homeplug.org/ [Accessed 1 May 2016].

[35]   "IEEE standard for information technology – telecommunications and information exchange between systems-local and metropolitan area networks-specific requirements – part 11: Wireless LAN medium access control (MAC) and physical layer (PHY) specifications," *IEEE Std 802.11-1997*, pp. i–445, 1997.

[36]   P. P. Parikh, M. G. Kanabar, and T. S. Sidhu, "Opportunities and challenges of wireless communication technologies for smart grid applications," in *Power and Energy Society General Meeting, 2010 IEEE*, pp. 1–7, July 2010.

[37]   "Using spread spectrum radio communication for power system protective relaying applications," *Tech report*, July 2005.

[38]   G. Thonet and B. Deck, "A new wireless communication platform for medium-voltage protection and control," in *2004 IEEE International Workshop on Factory Communication Systems*, pp. 335–338, September 2004.

[39]   "IEEE standard for local and metropolitan area networks part 16: Air interface for fixed and mobile broadband wireless access systems amendment 2: Physical and medium access control layers for combined fixed and mobile operation in licensed bands and corrigendum 1," *IEEE Std 802.16e-2005 and IEEE Std 802.16-2004/Cor 1-2005 (Amendment and Corrigendum to IEEE Std 802.16-2004)*, pp. 1–822, 2006.

[40]   "Wireless connectivity for electric substations," *Tech report*, February 2008.

[41] "IEEE standard for local and metropolitan area networks part 16: Air interface for broadband wireless access systems amendment 1: Multihop relay specification," *IEEE Std 802.16j-2009 (Amendment to IEEE Std 802.16-2009)*, pp. 1–290, June 2009.

[42] "IEEE standard for local and metropolitan area networks part 16: Air interface for broadband wireless access systems amendment 3: Advanced air interface," *IEEE Std 802.16m-2011(Amendment to IEEE Std 802.16-2009)*, pp. 1–1112, May 2011.

[43] D. Seo, *Evolution and Standardization of Mobile Communications Technology*. IGI Global, 2013.

[44] V. Garg, *Wireless Communications and Networking*. Morgan Kaufmann, 2007.

[45] P. K. Lee and L. L. Lai, "A practical approach to wireless GPRS on-line power quality monitoring system," in *Power Engineering Society General Meeting, 2007. IEEE*, pp. 1–7, June 2007.

[46] X. Zhang, Y. Gao, G. Zhang, and G. Bi, "CDMA2000 cellular network based SCADA system," in *Power System Technology, 2002. International Conference on PowerCon 2002*, vol. 2, pp. 1301–1306, 2002.

[47] H. G. R. Tan, C. H. Lee, and V. H. Mok, "Automatic power meter reading system using GSM network," in *International Power Engineering Conference, 2007 (IPEC 2007)*, pp. 465–469, December 2007.

[48] A. Mahmood, M. Aamir, and M. I. Anis, "Design and implementation of AMR smart grid system," in *Electric Power Conference, 2008 (EPEC 2008). IEEE Canada*, pp. 1–6, October 2008.

[49] T. K. Reddy and T. Devaraju, "A review of smart grid communication technologies," 2014.

[50] Q. Zhao and B. Sadler, "A survey of dynamic spectrum access: signal processing, networking, and regulatory policy," *IEEE Signal Processing*, vol. 24, no. 3, pp. 78–89, 2007.

[51] S. Haykin, "Cognitive radio: Brain-empowered wireless communications," *IEEE Journal on Selected Areas in Communications*, vol. 23, pp. 201–220, February 2005.

[52] R. Yu, Y. Zhang, S. Gjessing, C. Yuen, S. Xie, and M. Guizani, "Cognitive radio based hierarchical communications infrastructure for smart grid," *IEEE Network Special Issue on Communication Infrastructures for Smart Grid*, pp. 6–14, October 2011.

[53] Y. Liu, C. Yuen, S. Huang, N. U. Hassan, X. Wang, and S. Xie, "Peak-to-average ratio constrained demand-side management with consumers preference in residential smart grid," *IEEE Journal of Selected Topics in Signal Processing*, vol. 8, pp. 1084–1097, June 2014.

[54] Report and O. M. O. F. 10-174, "Unlicensed operation in the TV broadcast bands," 2010.

[55] OFCOM, "Digital dividend: Cognitive access, statement license-exempting cognitive devices using interleaved spectrum," 2009.

[56] I.-C.D.A. (iDA) of Singapore, "Regulatory framework for TV white space operations in the VHF/UHF bands," 2014.

[57] N. U. Hassan, S. Hussain, C. Yuen, and L. Duan, "Tradeoff between spectrum cost and quality of service in a cognitive radio network," in *Global Communications Conference (GLOBECOM), 2013 IEEE*, pp. 1179–1184, December 2013.

[58] N. U. Hassan, C. Yuen, and M. B. Attique, "Tradeoff in delay, cost, and quality in data transmission over TV white spaces," in *International Conference on Communications (ICC), 2016 IEEE*, May 2016.

[59] L. Duan, J. Huang, and B. Shou, "Investment and pricing with spectrum uncertainty: A cognitive operators perspective," *IEEE Transactions on Mobile Computing*, vol. 10, pp. 1590–1604, November 2011.

[60] D. K. Misra, *Radio-Frequency and Microwave Communication Circuits: Analysis and Design*, 2nd ed., Wiley Online Library, 2004.

[61] Y. Hu and V. O. K. Li, "Satellite-based internet: A tutorial," *IEEE Communications Magazine*, vol. 39, pp. 154–162, March 2001.

[62] E. Fadel, V. Gungor, L. Nassef, *et al.*, "A survey on wireless sensor networks for smart grid," *Computer Communications*, vol. 71, pp. 22–33, 2015.

[63] F. A. Silva, "Industrial wireless sensor networks: Applications, protocols, and standards [book news]," *IEEE Industrial Electronics Magazine*, vol. 8, pp. 67–68, December 2014.

[64] M. C. Vuran and I. F. Akyildiz, "Cross-layer analysis of error control in wireless sensor networks," in *2006 Third Annual IEEE Communications Society on Sensor and Ad Hoc Communications and Networks, 2006. SECON'06*, vol. 2, pp. 585–594, September 2006.

[65] B. E. Bilgin and V. C. Gungor, "Adaptive error control in wireless sensor networks under harsh smart grid environments," *Sensor Review*, vol. 32, no. 3, pp. 203–211, 2012.

[66] O. D. Incel, "Survey paper: A survey on multi-channel communication in wireless sensor networks," *Computer Networks*, vol. 55, pp. 3081–3099, September 2011.

[67] Y. Liu, C. Yuen, X. Cao, N. U. Hassan, and J. Chen, "Design of a scalable hybrid MAC protocol for heterogeneous M2M networks," *Internet of Things Journal, IEEE*, vol. 1, pp. 99–111, Feb 2014.

[68] I. Akyildiz and M. C. Vuran, *Wireless Sensor Networks*. New York, NY: Wiley, 2010.

[69] X. Lu, Z. Lu, W. Wang, and J. Ma, "On network performance evaluation toward the smart grid: A case study of DNP3 over TCP/IP," in *Global Telecommunications Conference (GLOBECOM 2011), 2011 IEEE*, pp. 1–6, December 2011.

[70] B. K. Singh and K. E. Tepe, "Feedback based real-time MAC (RT-MAC) protocol for wireless sensor networks," in *Global Telecommunications Conference, 2009. GLOBECOM 2009. IEEE*, pp. 1–6, November 2009.

[71] P. Hurni and T. Braun, "MaxMAC: A maximally traffic-adaptive MAC protocol for wireless sensor networks," in *Proceedings of the Seventh*

European Conference on Wireless Sensor Networks, EWSN'10, pp. 289–305. Berlin, Heidelberg: Springer-Verlag, 2010.

[72] I. Al-Anbagi, M. Erol-Kantarci, and H. T. Mouftah, "Priority- and delay-aware medium access for wireless sensor networks in the smart grid," *IEEE Systems Journal*, vol. 8, pp. 608–618, June 2014.

[73] W. Sun, X. Yuan, J. Wang, D. Han, and C. Zhang, "Quality of service networking for smart grid distribution monitoring," in *2010 First IEEE International Conference on Smart Grid Communications (SmartGridComm)*, pp. 373–378, October 2010.

[74] M. S. Kim, S. R. Kim, J. Kim, and Y. Yoo, "Design and implementation of MAC protocol for smartgrid HAN environment," in *2011 IEEE 11th International Conference on Computer and Information Technology (CIT)*, pp. 212–217, August 2011.

[75] J. Yang and S. Ulukus, "Delay minimization in multiple access channels," in *Proceedings of the 2009 IEEE International Conference on Symposium on Information Theory—Volume 4*, ISIT'09 (Piscataway, NJ, USA), pp. 2366–2370, IEEE Press, 2009.

[76] S. Ullo, A. Vaccaro, and G. Velotto, "Performance analysis of IEEE 802.15. 4 based sensor networks for smart grids communications," *Journal of Electrical Engineering: Theory and Application*, vol. 1, no. 3, pp. 129–134, 2010.

[77] B. Yahya and J. Ben-Othman, "Energy efficient and QoS aware medium access control for wireless sensor networks," *Concurrency Computation: Practice Experience*, vol. 22, pp. 1252–1266, July 2010.

[78] I. Rhee, A. Warrier, M. Aia, J. Min, and M. L. Sichitiu, "Z-MAC: A hybrid MAC for wireless sensor networks," *IEEE/ACM Transactions on Networking*, vol. 16, pp. 511–524, June 2008.

[79] M. Yigit, E. A. Yoney, and V. C. Gungor, "Performance of MAC protocols for wireless sensor networks in harsh smart grid environment," in *2013 First International Black Sea Conference on Communications and Networking (BlackSeaCom)*, pp. 50–53, IEEE, 2013.

[80] T. Iwao, K. Yamada, M. Yura, *et al.*, "Dynamic data forwarding in wireless mesh networks," in *2010 First IEEE International Conference on Smart Grid Communications (SmartGridComm)*, pp. 385–390, October 2010.

[81] S. Dawson-Haggerty, A. Tavakoli, and D. Culler, "Hydro: A hybrid routing protocol for low-power and lossy networks," in *2010 First IEEE International Conference on Smart Grid Communications (SmartGridComm)*, pp. 268–273, October 2010.

[82] S. M. S. Bari, F. Anwar, and M. H. Masud, "Performance study of hybrid wireless mesh protocol (HWMP) for IEEE 802.11s WLAN mesh networks," in *2012 International Conference on Computer and Communication Engineering (ICCCE)*, pp. 712–716, July 2012.

[83] J. S. Jung, K. W. Lim, J. B. Kim, Y. B. Ko, Y. Kim, and S. Y. Lee, "Improving IEEE 802.11s wireless mesh networks for reliable routing in the smart grid infrastructure," in *2011 IEEE International Conference on Communications Workshops (ICC)*, pp. 1–5, June 2011.

[84]   H. Gharavi and B. Hu, "Multigate communication network for smart grid," *Proceedings of the IEEE*, vol. 99, pp. 1028–1045, June 2011.

[85]   M. S. Islam, M. A. Hamid, and C. S. Hong, "Transactions on computational science vi," *Ch. SHWMP: A Secure Hybrid Wireless Mesh Protocol for IEEE 802.11s Wireless Mesh Networks*, pp. 95–114, Berlin: Springer-Verlag, 2009.

[86]   J. Ben-Othman and Y. I. S. Benitez, "On securing HWMP using IBC," in *2011 IEEE International Conference on Communications (ICC)*, pp. 1–5, June 2011.

[87]   S. K. Viswanath, C. Yuen, W. Tushar, *et al.*, "System design of internet-of-things for residential smart grid," *IEEE Wireless Communications Magazine – Special Issue on Enabling Wireless Communication and Networking Technologies for the Internet of Things*, March 2016.

Chapter 4

Measurement and sensing devices for the smart grid

*Ahmed S. Musleh, Ahmed Al-Durra
and Mohammed A. Abou-Khousa*

This chapter discusses the measurement and sensing devices used in smart grids. The fundamentals, the background of sensors, and the basic definitions of sensing technologies are presented where the architecture and sensor elements have been debated. A sensing mechanism with the sensing element modes and the measurements error is concisely defined. Furthermore, the classifications of sensors and measurement devices used in smart grids are summarized starting from the classic devices and ending with the most advanced ones. The basic measurements, smart meters, and miscellaneous are presented with illustrations. The use of these devices in the different smart grid sections—generation, transmission, distribution, and end consumer or customer—is succinctly illustrated.

4.1 Introduction

Modern power-grid networks consist of numerous power substations, generators, and lines which are connected across hundreds to thousands of kilometers. The large scale, complexity, and connectivity of power systems is reflected in the complex behavior that the power grids exhibit. A blackout, which could be defined as either a short- or long-term loss of the electric power in an area, is a rare incident. Nevertheless, the increasing public dependence on electricity makes this infrequent blackout extremely harmful for the economy and society. Due to the connected and complex nature of the grid, there is a possibility that any small failure in the grid can cascade through the system to result in a complete blackout [1]. Failure reasons could vary from basic demand–supply mismatch to harsh weather conditions affecting some parts of the system. These reasons are the main derive toward the development and use of measurement and sensing devices in power grids.

Department of Electrical Engineering, The Petroleum Institute, Abu Dhabi, UAE

Sensors in power grids have evolved significantly in the past decades. However, the main jump in the sensors development has been during the technological revolution realized in the last decade. The introduction of the computerized sensors and digitalized measurements, combined with the new challenges and motivations toward advanced smart grids, has boosted the applications of sensing devices in modern power grids. Sensing devices in a smart grid have a huge variety in the technology used, the process sensed, and the level of importance and reliability maintained. In smart grids, some sensors are used to maintain the operations of the grid, while others are used just to collect data for analysis. Some sensors have an extremely high reporting time (e.g., microseconds), while others have tremendously relaxed reporting time (e.g., days). In this chapter, the fundamentals and the backgrounds of the sensing devices are discussed. Furthermore, the main sensors and metering devices used in smart grids are briefly presented.

4.2 Sensor development background

A *sensor* is often defined as a "device that receives and responds to a signal or stimulus" [2]. According to the definition of the International Bureau of Weights and Measures, the sensor is defined as the "element of a measuring system that is directly affected by a phenomenon, body, or substance carrying a quantity to be measured" [3]. Such definitions are indeed broad and almost describe all practical sensing elements. Examples of common sensors include the human eye, a liquid-in-glass thermometer, a thermocouple, strain gauge, rotor of a turbine flow meter, photocell of a spectrometer, to name a few. In some fields, the term "detector" is often used in lieu of "sensor" to describe the same functional element [3].

Sensors are typically used in measurement systems that are designed for a specific task; make a measurement of certain process quantity. A measurement is an activity whereby the value of a quantity (such as force, voltage, and temperature) is estimated experimentally or empirically. Measurement is an essential activity in branch of technology and science. For instance, it is only through measurements we know the time, the temperature of our bodies, pressure in a pipeline, speed of our car, our electrical energy consumption, etc. Measurements are important aspects of business and engineering decision-making (e.g., measuring how much energy we consume on monthly basis, compliance to standard, etc.). Measurements are the basis for establishing scientific facts (e.g., Ohm's Law), monitoring manufacturing processes, and maintaining human and environmental health and safety (e.g., measuring the amount of H_2S). Governments and industries spend billions of dollars yearly to purchase, commission and maintain measurement, and test equipment (imagine how many utility meters are installed in residential areas only).

We begin by defining a process as a system that generates information (e.g., chemical reactor, a jet fighter, a gas platform, a generator, etc.). To perform a measurement, a measuring system or instrument is needed (e.g., a measuring tape is an instrument to measure length). A measurement system is an information system that presents an observer with a numerical value (an estimate) corresponding to the

Figure 4.1 Basics of making a measurement

variable being measured. The variable being measured is called measurand or commonly quantity under measurements (e.g., length, resistance, current, voltage, power, pressure, temperature, etc.). For instance, a multimeter is a measuring system devised to estimate the value of resistance, voltage, or current. The true value of the measurand is unknown to the observer (otherwise, there will be no need to make measurement). It is the objective of the measuring system to provide and accurately estimate this unknown quantity.

It is the fact of life that every measurement—even the most careful—always has an error as defined in the following equation:

$$\text{Measurement error: } E \triangleq measured\ value - true\ value = O - I \qquad (4.1)$$

This error arises from factors ranging from hardware imperfection, variability in the environment and variable being measured to systematic as well as random errors and noise. Since the true value of the measurand is unknown to us, it is practically impossible to know the measurement error for sure. Our lack of knowledge about the sign and magnitude of measurement error is called *measurement uncertainty*. A measurement uncertainty estimate is the characterization of what we know statistically about the measurement error. It is statistically described by the worst-case scenario of combination of the standard deviations of all measurement errors (i.e., assuming that all error sources are independent). Therefore, a measurement result is only complete when accompanied by a statement of the uncertainty in that result. In other words, it is not sufficient to produce the numeric value of a measurand without specifying how much we can trust this value.

4.2.1 Fundamentals of measurements

There are five major fundamental aspects, concerning any measurement:

- Measurement principle
- Measurement method
- Measurement model
- Measurement procedure
- Measurement system (or instrument) [referred to as "measuring system" in many standards]

Each measurement we make is based on a principle realized in a method according to a model while following a certain procedure and utilizing a measuring system.

The **measurement principle** is a phenomenon serving as a basis of a measurement. The phenomenon can be of a physical, chemical, or biological nature.

Specific principles of measurement are applied in order to determine measured variables/quantities. For instance,

- Thermoelectric effect applied to the measurement of temperature.
- Energy absorption applied to the measurement of the amount of substance concentration.
- Induce EMF as per Faraday's Law applied to measure the flow rate of conductive fluids.

The practical application of a measurement principle gives rise to a **measurement method**.

The measurement method is a generic description of a logical organization of operations used in a measurement. There are three major measurement methods:

- **Substitution measurement**: a quantity is determined by substituting for it a known quantity which produces the same effect.
- **Differential measurement**: the difference between a known quantity and an unknown quantity is determined.
- **Null measurement**: the measurement is balanced to bring the indicator to zero. The Wheatstone bridge is one of the most common null measurement methods.

Depending on how the measured valued is obtained, measurements methods are categorized as **direct** or **indirect**.

- In direct methods of measurement (also known as *comparative* or *relative* methods of measurement), the desired value of the measured variable is determined by comparison with a reference value of the same measured variable. Measurements where the measured value is received directly, without supplementary calculations and usually from the reading of a single measuring device, can also be classified as direct methods of measurement (e.g., resistance thermometer: resistance = temperature).
- In the case of indirect measurements (sometimes referred to as *multivariate measurements*), the desired value of a measured variable is derived from different physical variables. The measured value is determined from these variables using the given physical relationship.

A **measurement model** is mathematical relation among all quantities known to be involved in a measurement. A general form of a measurement model is the equation: $O = f(I_1, I_2, \ldots, I_n)$, where O, the output quantity in the measurement model, is the **measurand**, the quantity value of which is to be inferred from information about input quantities in the measurement model; $I_1, I_2, \ldots, I_n$.

Each input quantity in a measurement model must be measured, unless its value can be otherwise obtained. The measurement model contains all important parameters which affect the measurements. Often times, *calibration factors* are introduced to account for the measurement sensitivity to parameters such as material type and temperature. The calibration factors are determined in a process called *calibration*. The calibration is done with known inputs and under controlled environment.

The **measurement procedure** is a detailed description of a measurement according to one or more measurement principles and to a given measurement method, based on a measurement model and including any calculation to obtain a measurement result. The measurement procedure is usually documented in sufficient detail to enable an operator to perform a measurement. A measurement procedure can include a statement, concerning a target measurement uncertainty. A measurement procedure is sometimes called a standard operating procedure.

Finally, the **measuring instrument** is a device used for making measurements in principle or in association with one or more secondary devices. A measuring system could be a set of one or more measuring instruments. In many instances, it integrates other supporting devices as a supply, assembled and formed to give information used to generate estimates of the measurand. A measuring system may as well be composed of only one measuring instrument.

4.2.2 Measurement system architecture

Measurement systems perform a complete measuring task, from the initial sensing (detection) to the final presentation (indication and communication). The estimate provided by the measurement system should be as accurate as possible. The measurement system consists of several elements or blocks that tailored to achieve that end. It is possible to identify four types of elements as shown in Figure 4.2, although in a given system one type of element may be missing or may occur more than once. Each element affects the measurement in a certain way.

4.2.2.1 Sensing element

The sensing element is directly exposed to the process and gives an output which depends in some way on the variable to be measured. The sensing element interacts with the process in a "certain way" leading to an output which is a consequence of this interaction. This interaction can be physical, chemical, biological, or the combination of those. In general, the sensing element can be:

- **Sensor**: *element of a measuring system that is directly affected by a phenomenon, body, or substance carrying a quantity to be measured* [3] (e.g., resistive strain gauge).
- **Transducer**: *device, used in measurement, which provides an output quantity having a specified relation to the input quantity* [3]. This relation is due to "conversion" from one type of energy to another (e.g., microphone).

Nowadays, there are a variety of sensors with a vast range of sensing mechanisms. These sensing mechanisms can be based on physical, chemical, biological, or the combination of those interactions. The physical sensing mechanisms can be based on electricity, magnetism, electromagnetism, mechanical, or electromechanical

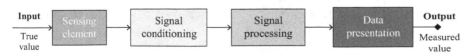

Figure 4.2   Basic elements of measurement systems

Table 4.1   Types of sensing elements

Element type	Example principles	Example elements
Resistive	Resistance of a conductive material depends on many parameters including temperature and geometry	Resistive temperature detector
Capacitive	Capacitance between two conductors depends on the geometry and permittivity	Level sensor, displacement sensor
Inductive	Inductance of a conductor is function geometry and permeability	Proximity sensors
Electromagnetic	Time-varying magnetic fields produce EMF (Faraday Law of induction)	Faraday accelerometer
Thermoelectric	The junction potential between two different metals is a function of temperature	Thermocouple
Piezoelectric	Voltage induced across piezoelectric material block is a function of force and geometry	Acceleration, force, pressure sensors
Piezoresistive	Resistance of a semiconductor material depends on force and geometry	Strain gauge

interactions. Physical sensing mechanisms that are tightly linked to electricity, magnetism, and electromagnetism are used in the majority of industrial measurement systems. In general, sensing elements based on such mechanisms can be grouped into the types described in Table 4.1.

4.2.2.2   Signal conditioning element

The output of the sensing elements can be a resistance, capacitance, and inductance, voltage in mV range that is related directly or indirectly related to the measurand. The second element of the measurement chain is the signal conditioning element. The output of the sensing element is fed into the signal conditioning element. The primary function of the signal conditioning element is to convert the output of sensing element into a form suitable for further processing. This form is usually a DC voltage, a DC current, or an AC voltage with variable frequency.

Often times, the sensor and observer (display) of the measurement system are not physically close, and hence, the output of the sensor needs to be converted to a form which can be transported reliably over physical media such as cables. Even if the sensor and display are close, the output of the sensor needs to be converted into format that is compatible to the subsequent elements in the measurement chain leading to the display. This is the basic function of the signal conditioning element. The signal conditioning element performs its main function while aiming to:

- enhance the signal-to-noise ratio and measurement range
- combat interference or reduce its effect
- reduce measurement error (e.g., offset correction/compensation, non-linearity, etc.)
- accomplishing all of the above at competitive cost!

In general, there are four common analog circuits that are used for signal conditioning in measurement systems:

- Deflection bridges
- Amplifiers (instrumentation amplifiers and lock-in amplifier)
- Filters
- Current transmitters

It is not uncommon that the above elements can be combined in one measurement chain (e.g., deflection bridge followed by an amplifier, filter, and current transmitter).

4.2.2.3   Signal processing element

The output signal from the conditioning elements is usually in the form of a DC voltage, DC current, or variable frequency AC voltage. In many cases, calculations must be performed on the conditioning element output signal in order to establish the value of the variable being measured. For instance:

- calculation of temperature from a thermocouple EMF signal
- calculation of total mass of product gas from flow rate and density signals
- offset subtraction and compensation

These calculations are referred to as signal processing and are usually performed digitally using a general-purpose computer or microcontroller (MCU). The output of the most signal conditioning elements are analogue. Hence, there should be processing element which would convert this output into a digital representation first. The signal processing element takes the output of the conditioning element and converts it into a form more suitable for further processing and presentation. Examples include

- Analog-to-digital converter which converts a voltage into a digital form for input to a computer, MCU or digital signal processor (DSP)
- Computer/MCU/DSP which calculates the measured value of the variable from the incoming digital data

Typical calculations performed by the signal processing elements include

- Data conversion
- Auto-zeroing and executing self-calibration routines
- Correction for sensing element non-linearity and dynamic compensation

4.2.2.4   Data presentation element

This element presents the measured value in a form which can be easily recognized by the observer. This element could simply be an indicator. In modern sensors, this element often encompasses connectivity subsystems to execute certain communication protocols and coordinate measurement transmission over telemetry links.

4.3   Fundamentals of sensing mechanism

In every measuring system, its elements respond to physical quantities. The term "respond" refers to the description of a reaction. The response describes, for example, how the output value (voltage) of a temperature sensor behaves in relation to the input signal (temperature). For measurement system design and analysis, the effect of each element as well as when all elements are cascaded in a measurement chain should be modeled. Basically, capturing how they collectively affect the input. One of the best ways to do that is through the development of the so-called measurement model defined earlier. To develop a measurement model for the whole system, each element is modeled independently and then combined into the model of the chain.

4.3.1   Element model

Each element (e.g., the thermocouple), say the ith element, can be modeled by finding the relation between its input (I_i) and output (O_i) as depicted in Figure 4.3. The obtained relation enables us to predict the *element's response* to any given input. This response depends on the *element's characteristics (attributes)*. Depending on the observed effect of these characteristics with respect to time, these characteristics can be categorized as follows:

1. **Static (or steady state) characteristics**: these are the relationships which may occur between the output O and the input I of an element when I is either at a constant value or changing slowly. With the thermocouple example, the knowledge of these characteristics allows us to answer questions like: What are the maximum and minimum temperatures we can measure?
2. **Dynamic characteristics**: these determine the dynamic response of the element as the input is varying with time. With the thermocouple example, the knowledge of these characteristics allows us to answer questions like: if the temperature changes from 10 to 11 °C within 0.5 s, would the output voltage change fast enough to reflect the new temperature value?

The overall element's response depends on both characteristics. There are two general types of static characteristics: **systematic** and **statistical characteristics**. Systematic characteristics are those that can be exactly quantified by mathematical expression or graphical means. These are distinct from statistical characteristics which cannot be exactly quantified using deterministic expressions.

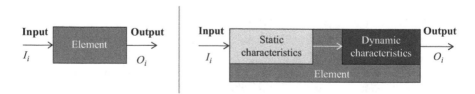

Figure 4.3   Element's model

4.3.1.1 Systematic characteristics

Range: The input range of an element is specified by the minimum and maximum values of I, i.e., I_{MIN} to I_{MAX}. The output range is specified by the minimum and maximum values of O, i.e., O_{MIN} to O_{MAX}.

Span: It is the maximum variation in input or output, i.e., input span is $I_{MAX} - I_{MIN}$, and output span is $O_{MAX} - O_{MIN}$.

Linearity: An element is said to be linear if corresponding values of I and O lie on a straight line. The **ideal straight line** connects the minimum point $A(I_{MIN}, O_{MIN})$ to maximum point $B(I_{MAX}, O_{MAX})$. The Ideal line equation is given by

$$O_{Ideal} = KI + a \qquad\qquad\qquad (4.2)$$

where the slope $K = (O_{MAX} - O_{MIN})/(I_{MAX} - I_{MIN})$ defines the *sensitivity* of the element and the intercept $a = O_{MIN} - KI_{MIN}$ defines its *zero-offset* or *bias*. Random environmental disturbances such as temperature fluctuations usually modify the sensitivity of the element and/or is zero bias.

The ideal straight line defines the ideal characteristics of an element. In many cases, the straight-line relationship is not obeyed and the element is said to be **non-linear**. Non-linearity can be defined in terms of a function $N(I)$ which is the difference between actual and ideal straight-line behavior.

Sensitivity: This is the change, ΔO, in the output, O, for unit change, ΔI, in the input, I, i.e., it is the ratio $\Delta O/\Delta I$. In the limit that ΔI tends to zero, the ratio $\Delta O/\Delta I$ tends to the *derivative*, dO/dI (i.e., the rate of change of O with respect to I). The sensitivity quantifies the element's ability to respond to small changes in the input. For a linear element, dO/dI is equal to the slope or gradient K of the straight line.

Resolution: It is defined as the smallest change, ΔI_R, in the input, I, that causes perceptible (observable) change in the output, O. For an input change below this ΔI_R, the output will stay constant.

Hysteresis: For a given value of I, the output O may be different depending on whether I is increasing or decreasing. Hysteresis is the difference between these two values of the output O.

4.3.1.2 Statistical characteristics

While the systematic characteristics, such as linearity, are deterministic in nature, there are some static characteristics that are not deterministic; we describe those as being statistical characteristics. Statistical characteristics describe the random variations in the element's behavior. These variations cannot be exactly quantified using deterministic expressions. However, they can be described using statistical models.

Repeatability: It is the ability of an element to give the same output for the same input, when repeatedly applied to it. Usually quantified by the standard deviation of the measurement.

Tolerance: It is the maximum error that is to be expected in a given value.

Accuracy: The accuracy of measurement is the agreement between a measured quantity value and a true quantity value of a measurand. A measurement is said to be more accurate when it offers a smaller measurement error.

Precision: It defines the closeness of agreement between indications or measured quantity values obtained by replicate measurements on the same input. Measurement precision is commonly expressed numerically by measures of imprecision such as the error standard deviation, σ_E. Measurement precision is often used to define measurement repeatability.

4.3.1.3   Dynamic characteristics

If the input, I, to an element is changed suddenly, from one value to another, then the output, O, will not instantaneously change to its new value. For example, if the temperature input to a thermocouple is suddenly changed from 25 to 100 °C, sometime will elapse before the EMF output completes the change from 1 to 4 mV. The characteristics of the element's response to input variations is referred to as dynamic characteristics, and these are most conveniently summarized using a *transfer function*, $G(s)$. First-order and second-order elements are two of the most frequently used element types.

4.3.2   Measurements errors

It is extremely important in any measurement system to reduce errors to the minimum possible level and then to quantify the maximum remaining error that may exist in any instrument output reading. The starting point in the quest to reduce the incidence of errors arising during the measurement process is to carry out a detailed analysis of all error sources in the system. Each of these error sources can then be considered in turn, looking for ways of eliminating or at least reducing the magnitude of errors.

In general, we can consider two broad types of errors: **static** and **dynamic errors**. Static errors in measurement systems can be further divided into those that arise during the measurement process and those that arise due to later corruption of the measurement signal by induced noise during transfer of the signal from the point of measurement to some other point. Errors arising during the measurement process can be divided into two groups, known as **systematic errors** and **random errors**.

4.3.2.1   Systematic errors

Systematic errors describe *errors in the output readings of a measurement system that are consistently on one side of the correct reading, i.e., either all the errors are positive or they are all negative* [4]. Sources of systematic errors include

- Hardware imperfection (e.g., mutual coupling or cross-talk, directivity, switch isolation, etc.)
- System disturbance due to measurement
- Systematic variations in the environmental disturbances (sensitivity and zero bias drift)
- Wear in instrument components
- Connecting leads and cables

Most of the systematic error can be reduced or completely eliminated by careful instrument design and/or *calibration*. During calibration, the measurement system is used to measure a known input (i.e., using a calibration standard). Depending on the number of error sources, multiple calibration standards can be used. Other systematic error reduction techniques include opposing input (introducing an equal and opposite environmental input that cancels it out) and high-gain feedback systems. Intelligent/smart systems are often used to auto-compensate the drifts.

4.3.2.2 Random errors

Random errors in sensor readings are caused by unpredictable variations in the measurement system. *They are usually observed as small perturbations of the measurement either side of the correct value*, i.e., positive errors and negative errors occur in approximately equal numbers for a series of measurements made of the same constant quantity [4].

Therefore, random errors can largely be eliminated by calculating the average of a number of repeated measurements, provided that the measured quantity remains constant during the process of taking the repeated measurements. The random errors are usually quantified using the static response of the system.

4.3.2.3 Dynamic errors

The dynamic error of the measurement system is the difference between the measured signal and the true signal as a function of time. This type of error is encountered when reading the instrument before it has reached its steady state. This type of premature reading produces a dynamic error.

4.4 Classifications of smart grid sensors and meters

The power grid is a very complicated system that has many types of measurements devices and sensors which vary in purpose, importance, and the technologies that they use. The following are the basic sensors breakdown divided into three sections as basic measurements, smart meters, and miscellaneous.

4.4.1 Basic measurement devices

Basic measurement devices are the ones that have been used since decades. They are the devices that are essential to the operation and control of the power system. They do not have advanced technologies and output like the smart meters do, yet power systems cannot operate without them. Basic measurement devices in smart power grid include voltage sensor, current sensor, power meter, and power quality meter which is responsible for detecting abnormalities in the system such as harmonic, flicker, and frequency and phase oscillation. Each of these devices is summarized briefly below.

4.4.1.1 Voltage sensor

Voltage sensor is a device that measures the voltage potential at a specific point or node in the power grid. This sensor is considered to be as one of the two basic sensors in electrical engineering which are the voltage and current sensors. The main purpose of this sensor is to give a feedback on the voltage readings. This feedback is of a great importance to the operation, control, and protection of the power grid. High voltages in the power grid are considered to be an obstacle for voltage sensing; thus, voltage sensor has many types and equipment that have been developed through decades of research and developments. Each of these types has different rating and usages. The basic types include

- **Resistive voltage divider:**

  This is the most simple voltage sensor. It is based on voltage division rule over known resistors. Figure 4.4 shows the schematic of the circuit.

  The voltage in here is reduced using the simple voltage divider low. From this low, the original voltage to the measured one can be related as follows:

$$V_{out} = V_{in}\left(\frac{R_2}{R_1 + R_2}\right) \qquad (4.3)$$

  In this voltage sensor, the voltage V_{out} is measured using a normal voltage sensor device, then using the above equation we can find the original voltage to be measured which is V_{in}. This type of voltage sensors is considered to be the simplest type. Furthermore, it can be used for measuring AC and DC voltages. However, it is not used extensively nowadays for its low reliability and losses.

- **Potential transformer (PT):**

  Here, the voltage is dropped using a PT. Just as any other transformer, the turn ratio defines how much the voltage drop is. Figure 4.5 shows the schematic of the PT. PT is based on the ferro-magnetic characteristics of the metal. Depending on the turn ratio of the coils, the voltage can be decreased to a level that is suitable to be measured by a voltage sensing apparatus. This device is being used extensively by the industry for its reliability and a wide range of rating that it can be used in. Figure 4.6 shows two different types and shape of PTs. Figure 4.6(a) shows the PT manufactured by ABB Company. It is available

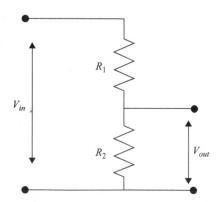

Figure 4.4.   Resistive voltage divider

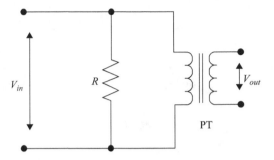

Figure 4.5 Potential transformer

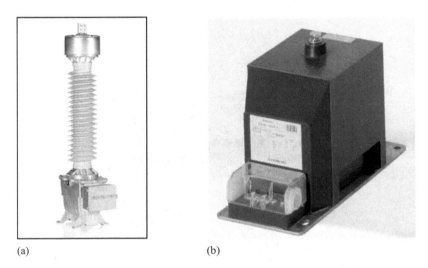

(a)   (b)

Figure 4.6 Potential transformer (a) ABB [5] and (b) Siemens [6]

from 72 up to 800 kV. It can withstand a wide range of shifting conditions including polar and desert climates [5]. Another type of the PT is shown in Figure 4.6(b). Here, a block-type design is adopted by Siemens Company [6].

- **Hall effect voltage sensor:**

  This sensor measures direct and alternating voltages with good insulation between the primary and secondary circuits. The primary voltage V_{in} to be measured is applied directly to the sensor terminals. An input resistance R_E is necessary to limit the input current to the sensor. The primary current flowing across the primary winding via this resistance generates a primary magnetic flux. The Hall probe placed in the air gap of the magnetic circuit provides a voltage relative to this flux. The electronic circuit amplifies this voltage and translates it into a secondary current. This secondary current multiplied by the number of turns of secondary winding cancels out the primary magnetic flux that created it.

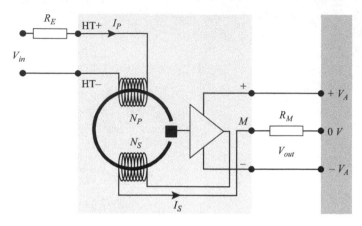

Figure 4.7   *Hall effect voltage sensor [7]*

Figure 4.8   *Hall effect voltage sensor ABB [7]*

The voltage sensor measures instantaneous values. The secondary output current is consequently exactly proportional to the primary voltage at any instant. It is an exact replica of the primary voltage. This secondary current is passed through a measuring resistor. The measuring voltage V_{out} at the terminals of this measuring resistor is therefore also exactly proportional to the primary voltage V_{in} [7]. Figure 4.7 shows a typical diagram of the Hall effect voltage sensor.

A Hall effect voltage sensor manufactured by ABB is shown in Figure 4.8. It measures AC and DC with a measuring range of 600–5,000 V.

4.4.1.2 Current sensor

As the voltage sensor, current sensor is the basic measurement device in electrical engineering. It is considered to be more difficult to build a current sensor than building a voltage sensor. This difficulty comes from the high current values that are required to be measured. In most of the cases, current is measured by transforming it into voltage quantity that is proportional to it. The following are some of the basic types of current sensors that the industry uses nowadays.

- **Resistive shunt current sensor:**

  As the voltage divider voltage sensor, this sensor is the most basic form of current sensing in electrical engineering. Figure 4.9 illustrates the basic schematic of this sensor technology. The output voltage from the known resistor can be used to measure the current that is passing through the resistor. This type of current sensors is considered to be the simplest type. Furthermore, it can be used for measuring AC and DC currents. However, it is not used extensively nowadays for its low reliability and losses.

- **Current transformer (CT):**

  This sensor is used to measure alternating current. It has the same working principle as the PT. Like any other transformer, a CT has a primary winding, a magnetic core, and a secondary winding. The turn ratio of these windings determines the relation between the output current readings and the input current. The following figure shows an example of the CT. The main electric line goes through the hall of it as shown in Figure 4.10.

  Figure 4.11 shows two types of CTs manufactured by Schneider Electric Company (a) [8] and Siemens Company (b) [2]. In (a), the basic shape is presented, while (b) shows a block-type CT.

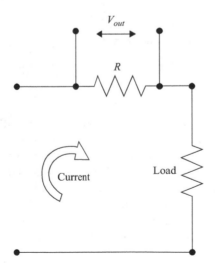

Figure 4.9   Resistive shunt current sensor

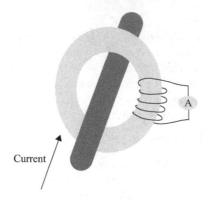

Figure 4.10   Current transformer [7]

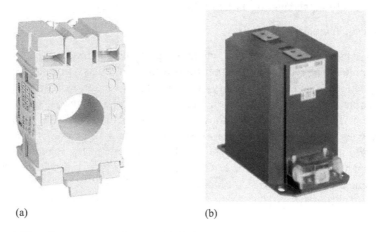

(a)   (b)

Figure 4.11   Current transformer (a) Schneider Electric [8] and (b) Siemens [2]

- **Hall effect current sensor:**

  As the Hall effect voltage sensor, the primary current flowing crossways the sensor generates a primary magnetic flux. The magnetic circuit stations this magnetic flux. The Hall probe placed in the air gap of the magnetic circuit provides a voltage relative to this flux. The electronic circuit amplifies this voltage and puts it into a secondary current. This secondary current multiplied by the number of turns of secondary winding cancels out the primary magnetic flux that created it. The current sensor measures instantaneous values, so it can measure both DC and AC currents [7]. Figure 4.12 shows the basic Hall effect current sensor.

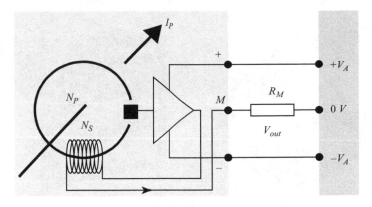

Figure 4.12   Hall effect current sensor

Figure 4.13   Hall effect current sensor ABB

Figure 4.13 shows an example of the Hall effect current sensor from ABB which is very similar to the CT. However, the inside components are different, as this one has more components than the usual CT which is basically windings.

4.4.1.3   Power meter

Power meters are very important devices that are used at each level of the power grid, starting from the generation side to the customer side. These devices give us the knowledge needed for operating and controlling the grid in the best possible scenario with respect to the reduction of losses and increasing system capacity. Basically, a power meter is a combination of voltage and current sensors. It uses data obtained from these sensors in order to get more meaningful data on another

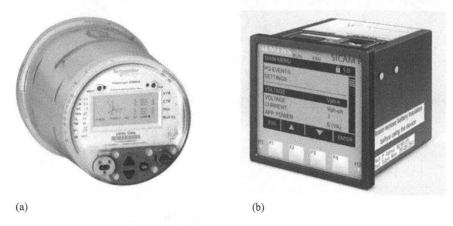

(a)   (b)

Figure 4.14   Power meter (a) Schneider Electric [9] and (b) Siemens [10]

level. Power meters thus could be seen as a higher level of measurement devices in the power grid. A power meter gives the following data:

- **Apparent power**: this is the power that is the combination of the active and reactive power. It is derived using the following equation:

$$S = V_{rms} * I_{rms} \tag{4.4}$$

- **Active (true) power**: this is the actual power that is being dissipated using resistive components. It is calculated from the average of the instantaneous power shown in the following equation:

$$P(t) = V(t) * I(t) \tag{4.5}$$

- **Power factor**: this is the parameter that gives an indication of the apparent power to the true power ratio. Thus, it is derived from the following equation:

$$\cos(\phi) = \frac{P}{S} \tag{4.6}$$

Other measurements that may be present are the reactive power and the energy which are derived from the above data. Figure 4.14 shows two examples of power meter used in power grids [9,10]. Both these meters are capable of measuring voltages, currents, and the different types of the power. Some power meters may have the attribute of measuring some power quality aspects such as harmonics; Figure 4.14(b) is one of those [10].

4.4.1.4   Power quality meter

This meter is used to find out any problems that may distort the waveform of the voltage or the current at any point in the grid. It is very important for the control and protection of the power grid. It is used to identify what are the problems that

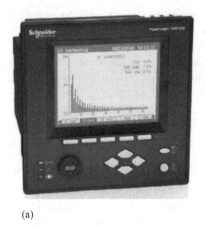 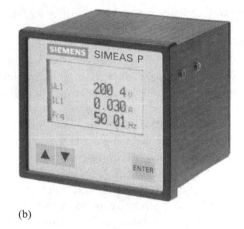

(a)   (b)

Figure 4.15   Power quality meter (a) Schneider Electric [11] and (b) Siemens [12]

are present so that counter solutions shall be carried. It measures all power quality issues such as voltage sag, voltage swell, flicker, oscillations, over voltage, under voltage, distortion, and harmonics. In Figure 4.15, two examples of the power quality meter manufactured by Schneider Electric Company and Siemens Company are presented.

Meter in (a) offers many advantages such as advanced power quality analysis coupled with revenue accuracy, multiple communications options, web compatibility, and control capabilities. Also, it can be used in low-voltage (LV) to high-voltage systems [13]. Meter in (b) can get more than 100 values including voltages (phase-to-phase and/or phase-to-ground), currents, power types and energy, power factor, phase angle, harmonics, total harmonic distortion, frequency and symmetry factor, and energy output [7].

4.4.2   Smart measurement devices

Smart measurement devices are new devices characterized by the advanced use of technology and communication advances in order to facilitate the measurement system and to get more accurate and precise insight of the power grid at its different levels from the generation to the customer side. In the power grid, they could be summarized in two devices, phasor measurement unit (PMU) and smart meters. The two types are introduced briefly next.

4.4.2.1   Phasor measurement unit

PMUs are power system devices that provide synchronized measurements of real-time phasors of voltages and currents [14–16]. The synchronization is achieved by the same time sampling of voltage and current waveforms using timing signals from the global positioning system (GPS). The global reference time is helpful in capturing the wide-area snapshot of the power system. PMU takes voltage and current signals and output voltage phasor, current phasor, frequency, and the rate of change of frequency as stated in Reference 16. The synchronized phasor measurements elevate the standards of power system monitoring and control to a new level [14]. One of the most important issues that need to be addressed in the emerging technology of PMUs is location selection. The planned system application influences the required number of installations. The cost of PMUs limits the number that will be installed although an increased demand in the future is expected to bring the cost down [15].

The first PMU prototype was built in Virginia Polytechnic Institute (Virginia Tech) in 1992; after that, many standards have followed in order to specify the specification of PMUs along with the accepted output [17]. The updated synchrophasor standard IEEE C37.118.1-2011 defines the requirements for the PMU measurements in terms of the steady-state performance evaluation quantities such as total vector error, frequency error, and rate of change of frequency error [18].

Any sinusoidal waveform can be represented by a unique complex number known as a phasor. Consider a sinusoidal signal represented by the following equation:

$$x(t) = X_m \cos(wt + \phi) \tag{4.7}$$

The phasor representation of this sinusoidal waveform is given by

$$X(t) = \frac{X_m}{\sqrt{2}} \angle \phi \tag{4.8}$$

Figure 4.16 shows how the waveform is represented by its phasor. Here, the magnitude (Mag.) of the signal refers to the root-mean-square value of the measured waveform. Additionally, the phase angle (ϕ) is the angle between the peak of the sine wave and the point at which the time equals to zero. In other words, the phase angle is a relative value that is computed by simultaneously comparing it to a reference phasor that is synchronized to a common time [14].

Synchrophasor is the term used to describe a phasor which has been calculated at an instant known as the time tag of the synchrophasor [14]. In order to obtain the

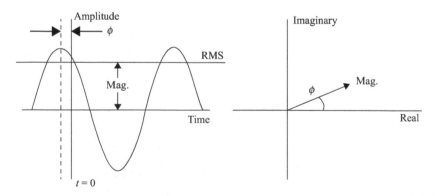

Figure 4.16   Phasor representation of sinusoidal signal

simultaneous measurement of phasors across a wide area of the power system, it is necessary to synchronize these time tags, so that all phasor measurements belonging to the same time tag are truly simultaneous [15]. This advantage differentiates the PMU technology from the now used monitoring system which is basically a supervisory control and data acquisition (SCADA) system. Furthermore, PMU can provide more than 60 samples per cycle, which makes it incomparable to SCADA which can give up to 4 samples per cycle [15]. Figure 4.17(a) shows an example of PMU manufactured by ABB Company. In Figure 4.17(b), another PMU type is presented from Siemens Company. This type is basically a disturbance recorder with an integrated PMU. With a high sampling rate and exceptional frequency response, it is capable of a detailed analysis of grid disturbances. With the incorporated PMU, vector quantities of voltages and currents with high precision with respect to amplitude, phase angle, and time synchronization are measured [19].

Figure 4.18 shows the typical block diagram of the PMU where it clearly shows the main parts of any PMU.

PMU must calculate the phasor given by (4.8) using the sampled data of the input signal given by (4.7). The anti-aliasing filters present in the input to the PMU is used immediate before a signal sampler to restrict the bandwidth of a signal to approximately or completely satisfy the sampling theorem over the band of interest since the theorem states that unmistakable reconstruction of the signal from its

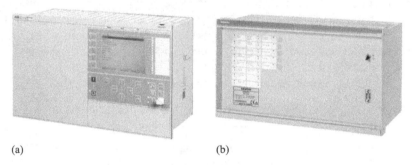

(a) (b)

Figure 4.17   PMU (a) ABB [20] and (b) Siemens [19]

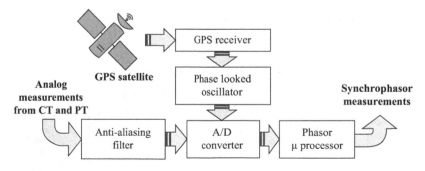

Figure 4.18   Block diagram of the PMU

samples is possible when the power of frequencies above the Nyquist frequency is zero. The synchronization is achieved by using a sampling clock which is phase locked to the one pulse per second synchronization signal provided by a GPS receiver [16]. The signals get processed in the phasor microprocessor and the output is sent via communication lines.

The synchronized phasor measurement technology is relatively new; consequently, a numerous number of research centers on the world are actively testing and developing new applications of this technology. Applications can be classified in these three areas [15]:

- Power system real-time monitoring
- Advanced network protection
- Advanced control schemes

4.4.2.2   Smart meter

Smart meters are advanced copy of the traditional power meters. They are characterized with the advanced use of communication and information technology that will make them a very sophisticated and elegant device. Smart meters are able to measure normal power quantities as well as advanced calculations such as prediction and forecast of power consumptions, and analysis of a consumption pattern. Furthermore, smart meters give the user the ability to control its consumption habits which in return would lead the user toward reducing its consumption significantly.

Nowadays, smart meters could be seen as separate devices that are distributed across the house (sockets and appliances), or what is known as smart sockets. Figure 4.19 shows a smart metering system that is based on smart sockets which have been developed in Reference 21.

Figure 4.20 Examples of smart power meter used in power grids.

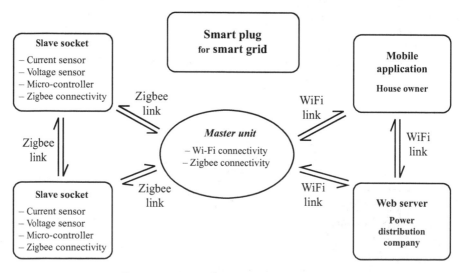

Figure 4.19   Smart metering system [21]

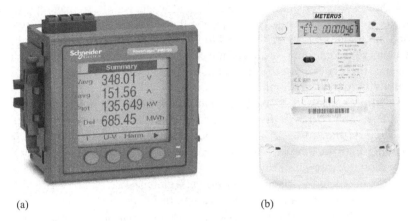

(a)   (b)

*Figure 4.20   Smart power meters (a) Schneider [11] and (b) OSGP-based smart
meter [22]*

Figure 4.20(a) shows an advanced power meter manufactured by Schneider
Company. It measures all power-related data as well as some power quality info
such as frequency and up to the 15th harmonic. It is used on LV and medium-
voltage (MV) systems. Also, it can predict demand for load management purposes
[11]. Figure 4.20(b) shows an example of a smart meter based on the open smart
grid protocol (OSGP) [22]. It is used in Europe which has the ability to decrease
loads, disconnect–reconnect remotely, and interface to gas and water meters [23].

4.4.3   Miscellaneous sensing devices

Miscellaneous sensors are the ones that have no electrical meaning. They might be
considered as extra sensors. However, they are very important to the operation,
control, and the protection of the smart power grid. These sensors vary dramatically. The following list summarizes some of the basic miscellaneous sensors used
in the smart grid.

- **Speed sensor**: it is used to quantify the speed of any moving apparatus or
  component that is used in the power grid. Mainly, it is used to measure the
  speed of the rotor shaft of the generator, fuel turbine, and wind turbine.
- **Temperature sensor**: it measures the temperature of the components as well
  as the medium surrounding them. It is used to measure the temperature of the
  generator's components and windings, turbines' components, transformer's
  windings, solar panels' components, converter's components, etc.
- **Partial discharge sensor**: it is used to enable the operator to detect any
  abnormal conditions, manufacturing problems, or deteriorated parts in the
  electrical insulation system of the electrical apparatus. It is basically used in
  generators, transformers, and transmission lines. It is a very complicated sensor
  that may vary in its types and technologies that are used as well as the frequency band adopted [24]. In Reference 25, a special ultraviolet sensor has
  been introduced in order to detect partial discharge in transmission lines.

- **Magnetic flux sensor**: this sensor is used whenever electromagnetic components are present. Basically, it measures the magnetic field inside the generator and the transformer which is used to identify turn-to-turn or coil-to-coil shorts which are the common results of the electrical and thermal aging. Shorted turns reduce the generator efficiency, and may lead to thermal asymmetry which results in increased vibration [5].
- **Vibration sensor**: it is used in identification and detection of any vibration that any components in the power grid exhibit. These components may include generators, transformers, and turbines. High levels of vibration may lead to rubbing and fracture of affected components [26]. A new method for online vibration monitoring of rotating electric generators without a need for installed sensors dedicated is discussed briefly in Reference 27.
- **Pressure sensor**: it measures the pressure inside the fuel turbine to ensure the safe operation of the turbine. Also, it is used to control the combustion of the fuel.
- **Level sensor**: this sensor is used for sensing the level of any liquid used in the power-grid component, starting from the fuel to the oil used in transformers and gearboxes.
- **Fuel quality sensor**: it is used for monitoring the quality of the fuel and its various characteristics which may be used for the economic operation of the power plant. Its importance varies depending on the fuel type (gas, diesel, nuclear fuel).
- **Medium density sensor**: it measures the density of the medium within the electrical components. The major use of it is measuring the density of the gas within the breakers.
- **Continuous emission monitoring systems**: it is an integrated system of gas analyzers, gas sampling system, temperature, flow and opacity monitors that are packaged with a data acquisition system to demonstrate environmental regulatory compliance of various industrial sources of air pollutants [28]. Basically, this system monitors gases such as CO, NO_2, SO_2, and O_3. In Reference 29, an environmental air pollution monitoring system for monitoring the concentrations of major air pollutant gases has been developed, complying with the IEEE 1451.2 standard.
- **Blade pitch angle sensor**: it is used for the position feedback in the closed loop control for the pitch angle of the rotor blades. This is very important for the speed control of the wind turbine.
- **Blade speed sensor**: it is used for measuring the speed of the rotation of the blades which is very important for the gearbox operation and control.
- **Torque sensor**: it measures the torque exhibited by the wind turbine blades to ensure the safe operation of the turbine. Also, it is used to control the operation of the wind turbine [30].
- **Wind speed sensor**: it measures the speed of the wind which is of a very importance for the operation and control of the wind turbine, gearbox, and generator. Furthermore, wind speed data is used for power management on grid level.

- **Wind direction sensor**: it is used for identifying the direction of the wind with respect to the blades. This data is important for the operation and control of the wind turbine, gearbox, and generator.
- **Dust-level sensor**: it measures the level of the dust layer on the solar panels. This area of research has attracted much focus for its great significance intended for the development of the self-cleaning solar panels.
- **Axis angle sensor**: it is used to measure the axis angle in the solar panels tracking systems. These measurements are used to control the rotation of the solar panels in the single- and dual-axis tracking systems so that the angle of incidence between the incoming sunlight and a solar panel is minimized.
- **Irradiance sensor**: it measures the sunlight radiation and its angle of incident which is used in the tracking system as well as the operation and control of the solar system. This is very important for analyzing and boosting the efficiency of the solar system.
- **Lightning sensor**: these sensors measure the peak magnitude and time of lightning currents flowing in the shield wires. These sensors are being researched to understand the distribution of lightning currents on transmission lines and to validate lightning location systems [31].
- **Conductor motion sensor**: it measures the motion of the conductor which is used for the data analysis and protection of the grid [31].
- **Insulation sensor**: it measures the condition of the insulation materials used in the underground cables and other electrical components. This is extremely important for the reliability and protection of the grid. A smart cable failure prediction system based on a novel sensing device for underground cable condition assessment has been introduced in Reference 32.

4.5 Use of sensors and measuring devices in smart grid

Power grids are generally divided into four main parts: generation, transmission, distribution, and end consumer or customer. Each part has specific function, gears, and ratings. For each part, several monitoring and measurements devices have been developed and implemented. These sensors and measurement devices are generated from the need of acquiring specific parameters' data throughout the grid which are of a very importance to the operation, reliability, and security of power grids. The emergence of smart grid has pushed forward the development of sensors and measuring devices in grids through the advanced technologies, and facilities that it offers. The following are the main types of sensors and measuring devices used in power grids divided by the grid's four main parts.

4.5.1 Generation side

Since generation units are a key part of power systems, it is essential to maintain and ensure the safe and stable operation of them. In the generation part of the power system, sensors are mainly used for monitoring the power generation devices and its surrounding environment. Generation devices' sensors are used to monitor the condition of the devices, to measure the important parameters that will ensure the stable behavior, and to set a control action that will produce the desired output from the devices. Although monitoring the condition of the generation units might be the same for all types of units when it come to the output (electrical power), monitoring the process of generation differs from one type to the other. For instance, machine-based generation units such as wind and gas turbines have different measurement devices from those used in semiconductor-based units such as solar plants. Similarly, sensing surrounding environment differs from one generation type to another. For example, renewable generators might have a huge dependency on the environment sensors. Solar panel-based generation units need to have a continuous data on the weather conditions, regarding the irradiance of the sun, and wind turbine-based generation units need data regarding the wind speed and direction.

The generation side of power grids could be seen as two basic components. A device that generate electrical energy (generator, solar panels, etc.), and a device that can adjust the electrical energy generated so that it suits the ratings and needs of the power grid (transformer, converter, etc.).

Figure 4.21 shows the basic components in the generation side of a smart grid (fuel, wind, and solar) and summarizes the basic use of sensors in it. The figure shows only three types of power generation plants. Although the industry is not limited to these three types, but they represent the basics in power generation as most of the rest types have similarities with at least one of them. Furthermore, the figure lists the basic measurement devices and sensors used in smart grid's power generation side.

4.5.2   Transmission side

The transmission side of power grids is the main player in power delivery. Transmission lines are the ones that connect the distant power-grid stations. They are spreaded over long distances that may reach to thousands of kilometers. Transmission systems are among the largest and most diverse, remotely located investments. Although these systems have been standing for decades, still many problems and challenges are being faced and anticipated for the near future that may cause problems for the operation of the transmission systems. Challenges may include

● Present transmission lines and substations are aging while the required reliability is increasing.
● The need to maximize the utilization of the system, and thus operate closer to the edge of transmission system limits.
● The need to increase the available capacity of the present transmission system.
● An increasing penetration of distributed generation systems and advanced power electronics devices.
● The need to integrate increasing amounts of renewable energy, which are highly variable, intermittent, and unpredictable.

These challenges have pushed researchers worldwide to investigate and come up with a new more reliable and secure transmission system which could be seen basically as the transmission side of the smart grid. For this to happen, an ongoing effort to research and develop sensor technologies is being carried to aid utilities in

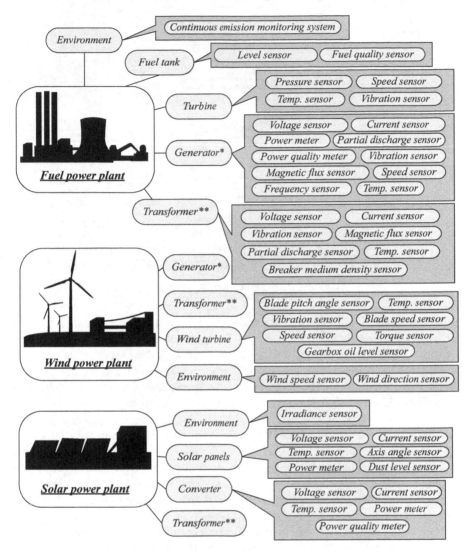

Figure 4.21   Generation side sensors

addressing the aforementioned challenges. Figure 4.22 shows the basic components in the transmission side of a smart grid and summarizes the basic use of sensors in it.

The figure shows only the basic component in transmission systems. Although the industry is not limited to these four components, they represent the basics in power transmission as most of the rest components have similarities with at least one of them.

4.5.3   Distribution side

The distribution side is one of the most complicated parts of the smart grids. It is the side that receives the electric power from the transmission system and delivers

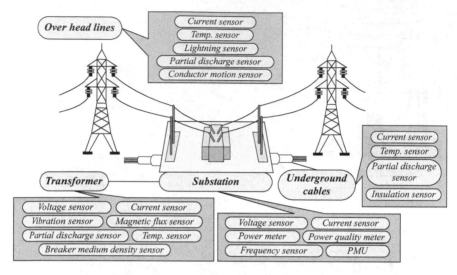

Figure 4.22   Transmission side sensors

it to the customers. Unlike the transmission system, distribution system is a highly fluctuating system for its continuous changing loads. From this perspective, the transmission system could be seen as a more stable and stronger system that the distribution system even though is much larger in size. Similar to the transmission system, the distribution system is also facing an increasing amount of challenges that will harden its control and operation. The emergence of a new type of loads such as electric vehicles, as well as the distributed generation systems, could be seen as the most important challenge to be faced currently and in the near future. Alongside the development of these technologies, advanced sensor technologies are being investigated and developed to have a clear vision and sight on the operation and control of the distribution system. Figure 4.23 shows the basic components in the distribution side of the smart grid and summarizes the basic use of its sensors.

The figure shows only the basic components in distribution systems. Although the industry is not limited to these components, they represent the basics in power distribution as most of the rest components have similarities with at least one of them.

4.5.4   Customer side

The end consumer in the smart grid is considered to be the customer side, which is the side that supposed to consume the electrical energy produced in the generation side. Traditionally, this part of the power grid shall be the most relaxed and stable part of the grid. However, the emergence of new loads and micro-distributed generation systems has made this part an equal complicated part as the other parts in the power grid. Moreover, the complexity of this part has been enlarged with the introduction of smart homes, smart sockets, and smart appliances. Figure 4.24

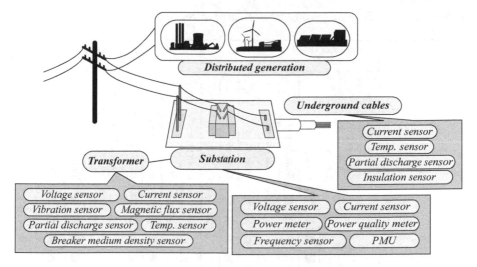

Figure 4.23   Distribution side sensors

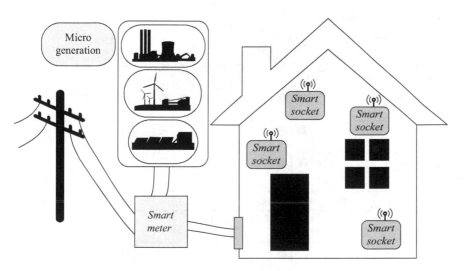

Figure 4.24   Generation side sensors

shows the basic components in the customer side of smart grid and summarizes the basic use of sensors in it.

4.5.5   Summary of sensing devices used in the smart grid

Table 4.2 summarizes the use of sensors and meters used in the different sides and components in the smart grid. It shows only the basic components of the power grid.

Table 4.2   Summary of sensing devices used in the smart grid

	Generation side								Transmission side				Distribution side				Customer side	
	Environment	Fuel tank	Turbine	Generator	Transformer	Wind turbine	Solar panels	Converter	Over-head lines	Underground cables	Transformer	Substation	Distributed generation	Underground cables	Transformer	Substation	Micro generation	House
Voltage sensor				■	■		■	■			■	■	■		■	■	■	
Current sensor				■	■		■	■	■	■	■	■	■	■	■	■	■	
Power meter				■			■	■			■	■			■	■	■	■
Power quality meter				■				■			■	■			■	■		
PMU												■				■		
Smart meter																		■
Smart socket																		■
Speed sensor			■	■		■							■				■	
Temperature sensor				■	■	■	■	■	■	■	■		■		■	■	■	
Partial discharge sensor				■	■				■	■	■		■	■	■		■	
Magnetic flux sensor				■	■						■		■		■		■	
Vibration sensor			■	■	■	■					■		■		■		■	
Pressure sensor			■										■				■	
Level sensor		■				■							■				■	
Fuel quality sensor		■											■				■	
Medium density sensor					■						■		■		■		■	
Continuous Emission Monitoring Systems (CEMS)	■												■				■	
Blade pitch angle sensor						■							■				■	
Blade speed sensor						■							■				■	

Table 4.2   (*Continued*)

	Generation side								Transmission side				Distribution side				Customer side	
	Environment	*Fuel tank*	*Turbine*	*Generator*	*Transformer*	*Wind turbine*	*Solar panels*	*Converter*	*Over-head lines*	*Underground cables*	*Transformer*	*Substation*	*Distributed generation*	*Underground cables*	*Transformer*	*Substation*	*Micro generation*	*House*
Torque sensor						■							■				■	
Wind speed sensor	■												■				■	
Wind direction sensor	■												■				■	
Dust level sensor							■						■				■	
Axis angle sensor							■						■				■	
Irradiance sensor	■												■				■	
Lightning sensor									■									
Conductor motion sensor									■									
Insulation sensor										■				■				

4.6   Conclusions

In this chapter, sensing devices and meters used in smart grids were discussed briefly. Starting from the fundamentals and the background of sensors, the basic definitions of sensing technology were presented where the architecture and sensor elements have been debated. Furthermore, the sensing mechanism with the sensing element modes and the measurements error were concisely defined. On the other hand, the classifications of sensors and measurement devices used in smart grids were summarized starting from the classic devices and ending with the most advanced ones. The use of these devices in the different smart grid sections was also deliberated. All in all, the development of the new monitoring systems in the smart grids needs to be resilient and flexible due to the wide range of applications which need to be addressed. The development need to address the overall integration and architecture of the measurement devices used in the smart grid rather than focusing on the component level, and the cooperation of the technology researchers, industry, and the standards organization is a must.

References

[1] A. Singh, J. Bapat and D. Das, "Distributed Health Monitoring System for Control in Smart Grid Network," in *Innovative Smart Grid Technologies – Asia (ISGT Asia), 2013 IEEE*, Bangalore, 2013.

[2] F. Jacob, *Handbook of Modern Sensors, Physics, Designs, and Applications*, New York: Springer, 2010.

[3] I. B. o. W. a. M. (BIPM), "International Vocabulary of Metrology Basic and General Concepts and Associated Terms (VIM-3)," 2012, [Online]. Available: http://www.bipm.org/en/publications/guides/vim.html.

[4] J. Bentley, *Principle of Measurement Systems*, London: Pearson Ed. Ltd., 2005.

[5] "Capacitor Voltage Transformer CPB (72–800 kV)," ABB, 2016, [Online]. Available: http://new.abb.com/high-voltage/instrument-transformers/voltage/cpb. [Accessed 31 January 2016].

[6] Siemens, "4M Protective and Measuring Transformers," Siemens, Berlin, 2009.

[7] ABB, "Current Sensors Voltage Sensors, Technical Catalog," ABB, France, 2014.

[8] "CT: Current Transformers," Schneider Company, [Online]. Available: http://www.schneider-electric.ae/en/product-range/950-ct/. [Accessed 20 January 2016].

[9] "PowerLogic ION8650: Power Meter," Schneider Company, [Online]. Available: http://www.schneider-electric.com/en/product-range/61053-powerlogic-ion8650. [Accessed 27 January 2016].

[10] "SICAM P850 Power Meter," Siemens, [Online]. Available: http://w3.siemens.com/smartgrid/global/en/products-systems-solutions/power-quality-measurements/power-meter/Pages/sicam-p850.aspx. [Accessed 18 May 2016].

[11] "Schneider Electric Advanced Energy Meter," Schneider Company, [Online]. Available: http://panelmeters.weschler.com/item/egories-product-by-manufacturer-schneider-electric/gy-meters-schneider-electric-advanced-energy-meter/pm5110? [Accessed 31 January 2016].

[12] "SICAM P50 Power Meter," Siemens, [Online]. Available: http://w3.siemens.com/smartgrid/global/en/products-systems-solutions/power-quality-measurements/power-meter/pages/sicam-p50.aspx. [Accessed 18 May 2016].

[13] S. Company, "Gain Energy Insight and Control with PowerLogic," [Online]. Available: http://www.optimizar.com.ar/documentacion/Productos-Medidores/Datos%20T%C3%A9cnicos%20ION%207550-7650-IME%20(Optimizar.com.ar).pdf. [Accessed 29 January 2016].

[14] J. D. L. Ree, V. Centeno and J. S. Thorp, "Synchronized Phasor Measurement Applications in Power Systems," *IEEE Transactions on Smart Grid*, vol. 1, no. 1, pp. 20–27, 2010.

[15] B. Singh, N. Sharma, A. Tiwari, K. Verma and S. Singh, "Applications of Phasor Measurement Units (PMUs) in Electric Power System Networks Incorporated with FACTS Controllers," *International Journal of Engineering, Science and Technology*, vol. 3, no. 3, pp. 64–82, 2011.

[16] K. E. Martin, D. Hamai, M. G. Adamiak, *et al.*, "Exploring the IEEE Standard C37.118-2005 Synchrophasors for Power Systems," *IEEE Transactions on Power Delivery*, vol. 23, no. 4, pp. 1805–1812, 2008.

[17] K. Martin, "Phasor Measurement Systems in the WECC," in *2006 IEEE PES Power Systems Conference and Exposition*, Atlanta, GA, pp. 132–138.

[18] "IEEE Standard for Synchrophasor Measurements for Power Systems," in IEEE Std C37.118.1-2011 (Revision of IEEE Std C37.118-2005), pp.1–61, 28 Dec 2011.

[19] "Digital Fault Recorder with Integrated Phasor Measurement Unit (PMU)," Siemens, [Online]. Available: http://w3.siemens.com/smartgrid/global/en/products-systems-solutions/power-quality-measurements/recorder-measure-ment-unit/pages/simeas-r-pmu.aspx. [Accessed 13 May 2016].

[20] ABB, "Phasor Measurement Unit RES670 2.0 IEC Application Manual," 2014, [Online]. Available: http://www.mena.abb.com/product/db0003db 004281/4384fabdebb3c7c4c12579d5002543cf.aspx. [Accessed 30 January 2016].

[21] A. Musleh, M. Debbouza and M. Farook, "Smart Plug for Smart Grid: Power Monitoring and Management System," BSc Thesis, Department of Electrical Engineering, Abu Dhabi University, Abu Dhabi, 2014.

[22] ETSI, "Group Specification GS OSG 001: Open Smart Grid Protocol," 2012, [Online]. Available: https://commons.wikimedia.org/wiki/File:Intelli-genter_zaehler-_Smart_meter.jpg.

[23] "Smart Meter," Wikipedia, 16 May 2016, [Online]. Available: https://en.wikipedia.org/wiki/Smart_meter. [Accessed 19 May 2016].

[24] Y. J. Kim, S. H. Hong, T. S. Kong and H. D. Kim, "On-Site Application of Novel Partial Discharge Monitoring Scheme for Rotating Machine," *in 2015 Electrical Insulation Conference (EIC)*, Seattle, 2015.

[25] J. Liu, Y. Yang, J. Wang and Q. Wang, "Novel Sensor System for Online Partial Discharge (PD) Detecting on Insulator of Transmission Lines," in *Second IEEE Conference on Industrial Electronics and Applications, 2007 (ICIEA 2007)*, Harbin, 2007.

[26] F. Ewert, "Online Monitoring – Early Detection and Diagnostics of Initiating Damage in Turbogenerators," in *Power Gen Europe 2014*, Cologne, 2014.

[27] P. V. Junior, M. A. S. Bobi, C. R. Gomes, H. S. Gomes and M. P. d. Nascimento, "Vibration Monitoring of Electric Generators Without Sensor Dedicated," in *2010 IEEE International Conference on Industrial Technology (ICIT)*, Vi a del Mar, 2010.

[28] F. Xiaoliang and Z. Haiming, "Design CEMS for Flue Gas from Thermal Power Plant," in *Power and Energy Engineering Conference, 2009 (APPEEC 2009)*. Asia-Pacific, Wuhan, 2009.

[29] N. Kularatna and B. H. Sudantha, "An Environmental Air Pollution Monitoring System Based on the IEEE 1451 Standard for Low Cost Requirements," *IEEE Sensors Journal*, vol. 8, no. 4, pp. 415–422, 2008.

[30] M. R. Wilkinson, F. Spinato and P. J. Tavner, "Condition Monitoring of Generators and Other Subassemblies in Wind Turbine Drive Trains," in *IEEE International Symposium on Diagnostics for Electric Machines, Power Electronics and Drives, 2007 (SDEMPED 2007)*, Cracow, 2007.

[31] "Sensor Technologies for a Smart Transmission System," *Electric Power Research Institute (EPRI)*, 2009.

[32] J. L. Lauletta, Y. Sozer and J. A. D. Abreu-Garcia, "A Novel Sensing Device for Underground Cable Condition Assessment," in *Electrical Insulation Conference (EIC), 2015 IEEE*, Seattle, 2015.

Chapter 5

Smart transmission and wide-area monitoring system

Yong Liu[1], Shutang You[1] and Yilu Liu[1,2]

Global positioning system (GPS) time-synchronized phasor measurement units (PMUs) were introduced in the 1980s and have been gradually deployed throughout the electric power system. At present, almost all the major countries in the world have ambitious PMU deployment plans. The applications of a PMU-based wide-area monitoring system (WAMS) in electric transmission grid management have attracted continuous interest as a result. In this chapter, after a brief introduction of its principles and development history, the architecture of WAMS is introduced and its many applications in smart transmission are summarized. Furthermore, FNET/GridEye, which is a pioneering WAMS system, is also introduced in this chapter as an example of WAMS.

5.1 Introduction

As a complementary system for the conventional supervisory control and data acquisition (SCADA) system, wide-area monitoring system (WAMS) allows monitoring increasingly complex behaviors exhibited by large electric power grids over wide geographic areas in real time. It substantially enhances operators' situational awareness and facilitates the operation of a more reliable, efficient, and secure power grid.

Phasor measurement unit (PMU), as the fundamental measurement device in WAMS, was invented in 1988 at Virginia Tech. Early in the 1980s, some attempts [1–3] were made to measure phase angle differences directly. These studies used the satellite and radio transmission to synchronize phase angle measurements (no magnitude information was included). As the adopted approaches to estimate phase angle were not generic phasor calculation methods, technology developed in these pioneer studies were no longer used in most modern WAMS systems. The work that led to the invention of modern PMU originated from the research on computer-based

[1]Department of Electrical Engineering and Computer Science, The University of Tennessee, Knoxville, TN, USA
[2]Oak Ridge National Laboratory, Oak Ridge, TN, USA

relaying of transmission lines in 1977. Due to the computational capability constraint of solving six fault loop equations at that time, Dr Arun G. Phadke and others at Virginia Tech proposed the symmetrical component algorithm for transmission line relaying [4]. Along with this algorithm, this work also presented an efficient method to calculate the positive sequence components of voltage and current. These positive sequence components, which constitute power system phasors, serve as fundamental power system state attributes and are essential in most power system analysis and calculations. The paper [5] authored by Dr Phadke *et al.* in 1983 identified the wide application potential of positive sequence phasor measurement in power systems, marking the start of the modern PMU technology. At that time, the start operation of the global positioning system (GPS) system provided an effective approach to synchronize measurements over a large geographical area. Under these conditions, the first PMU was invented in 1988 and the first commercial PMU was built in 1992.

The development and improvement of PMU has been driven by the needs from the power industry. The development of first PMU was funded by American Electric Power Service Corporation (AEP), Department of Energy, Bonneville Power Administration (BPA), and later New York Power Authority (NYPA). The first batch of PMUs assembled at Virginia Tech was installed in the BPA, AEP, and NYPA systems. Afterwards, PMU commercialization and standardization has been reducing its cost and increasing its deployment. Some major blackouts, such as the 2003 blackout in the USA also boosted PMU deployment. In addition, various applications based on synchrophasor measurement technology for the control room have been or are being developed.

The primary function of PMU is to obtain the high-precision time-stamped phasor estimation of voltage and current waveforms. Figure 5.1 shows an example of two sinusoid voltage waveforms and their phasor representations. Calculating the phasor of a waveform is often realized by fast Fourier transform or discrete

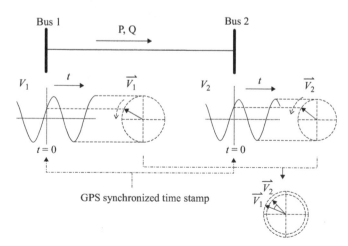

Figure 5.1   Synchronized phasor representations of sinusoid waveforms

Fourier transform (DFT) that can obtain the phasor representation at the desired frequency. Aided by GPS synchronization, the two phasors can be time stamped so that they can be put together under a common reference. The factors that may influence the calculation of phasors include sampling rates, filtering, window length, phase calculation algorithm, frequency estimation methods, output filtering, and reporting rates. In real measurement environments, signals may be also mixed with harmonics and distortions. To achieve the best performance, various algorithms have been developed and discussed in the literature [6].

5.2 Architecture of WAMS

5.2.1 PMU installation

PMUs are usually installed in high-voltage substations or power plants. Each phasor (either voltage or current) will require three connections that measure all three phases. A scheme diagram of the PMU installation on one phase at a substation is shown in Figure 5.2. The potential transformer (PT) and current transformer (CT) scale down the voltage and current at the second windings, in order to match the inputs of A/D converters in the PMU. The burden represents the V A rating of the electronic instruments presented to the PT/CT secondary circuits, whereas the attenuator is used for adjusting the output amplitude. Figure 5.3 shows the hardware architecture and functionality components of a generic PMU. The analog input will first go through a low-pass anti-aliasing filter. There are two typical strategies for anti-aliasing filtering. One is to use an analog filter that has a cut-off frequency higher than the Nyquist criterion. The other is to use a much higher sampling rate and add a digital "decimation filter" in addition to the analog filter. In this case, the cut-off frequency of the analog filter will be much higher than the Nyquist criterion. A combination of analog and digital filters makes the filtering performance more robust to aging and temperature changes. In addition, the high sampling rate can increase phase calculation precision, and it provides

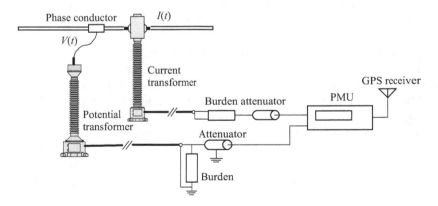

Figure 5.2   PMU installation and connection scheme diagram

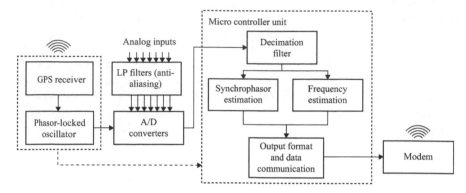

Figure 5.3   PMU hardware architecture

PMUs the potential to act as digital fault recorders if data recording is possible for analog signal samples. The main procedures in the micro control unit are calculations of voltage and current phasors, as well as frequency and rate of change of frequency (ROCOF). These data are time stamped using GPS signals and are transferred to the control room via the communication links.

5.2.2   Synchrophasor data management

A direct upper level of PMUs in WAMS is a phasor data concentrator (PDC). The main and basic functions of PDC are to combine data from many PMUs, align data based on the time stamps, condition and reject bad data for further processing. As PDC is the only device that communicates multiple PMUs, it also monitors the overall status of the WAMS system. Many PDCs have storage in which some local applications can be implemented with less data latency. Typically, PDCs can be categorized into three levels based on their installation location, capacity, and functionality, namely substation PDC, regional control center PDC, and ISO (independent system operator) or operation center PDC. A PDC that concentrates data from several PDCs is also called a super PDC. Together with PMUs installed at substations, these three levels of PDCs can form a hierarchy structure (Figure 5.4) or a distributed network structure (Figure 5.5). The hierarchy structure is simpler in configuration and management as the data stream is mostly going upward. The distributed structure provides better flexibilities as each PDC can request and access real-time or off-line data from any PMUs connected to the network.

The hierarchy structure is more common for the deployment of WAMS within one region or ISO, while the distributed network structure is more suitable for an interconnection-level power grid that consists of multiple utilities and ISOs. These ISOs, which adopt the hierarchy structure within its control area, require the PMU data sharing as they are in one interconnected power grid, thus a hybrid structure may be a candidate for an interconnection-level power grid. An example of such a hybrid structure is the North American Synchro-Phasor Initiative Network (NASPInet) proposed for the US Eastern Interconnection (EI) [7], as shown in Figure 5.6.

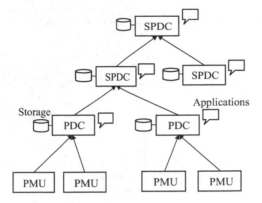

Figure 5.4 The hierarchy structure

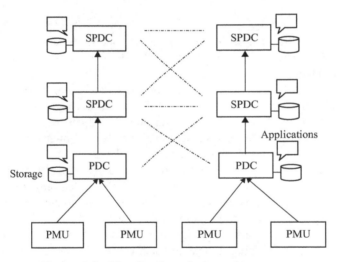

Figure 5.5 The distributed network structure

The sharing of PMU data between utilities and ISOs is realized by the commutation infrastructure with a distributed structure through a secure firewall called phasor gateways.

The communication between PMUs and PDCs can adopt various protocols, such as the commonly used Internet or serial connection protocols. For Internet-based PMU data communication, the transmission control protocol (TCP) and the user datagram protocol (UDP) are the two commonly used transportation protocols. TCP requires handshaking to achieve a robust communication link, while UDP uses a simpler transmission model, requiring less bandwidth and has better real-time performance. Compared with Internet protocols, serial communication further minimizes the bandwidth but limitation exists on its maximum data throughput.

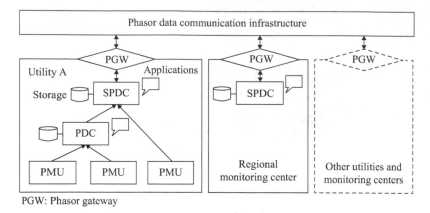

PGW: Phasor gateway

Figure 5.6   North American Synchro-Phasor Initiative Network (NASPInet) architecture

Some local PMU-based control functions can be directly implemented by hardware PDCs. In control centers, however, handling a large quantity of PMU data and the heterogeneity of various functions are beyond the computational capability of a commercial hardware PDC. As synchrophasor data are more valuable if it can be processed in real time, control centers often require a high-performance PMU data management platform realized by running software PDC on more powerful hardware. These PMU data management platforms have similar functions as PDCs so they can be categorized as one type of super PDCs.

There are some proprietary and open-source software available for synchrophasor data management. Software PDC is usually installed on generic or special computing platforms to process PMU data in real time and support real-time and off-line applications. For example, OpenPDC, developed by Grid Protection Alliance, is an open-source platform designed to accommodate various standard PMU input protocols, provide phasor data transformation and replication, and support user-configurable output streams [8]. These functions are realized by three layers of adapters, each of which is a collection of functional classes that can manage, process, and respond to dynamic changes in fast moving streaming time-series data in real time. A diagram showing the OpenPDC input/action/output interface adapters is shown in Figure 5.7.

5.2.3   Synchrophasor standards

To improve the compatibility and technology transfer, a variety of standards related to PMU have been published. Current versions of PMU-related standards are shown in Figure 5.8. These standards cover various areas including the technical specifications of phasor measurement and data transfer/storage; PMU calibration, testing, and installation; PDC requirements; and the GPS time protocol [9,10].

Similar to other IEEE standards, PMU standards keep updating with technology improvements. For example, the IEEE 1344 standard was published in 1995 as

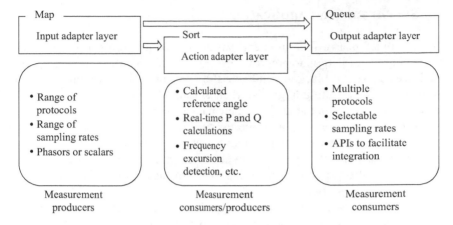

Figure 5.7   OpenPDC software architecture

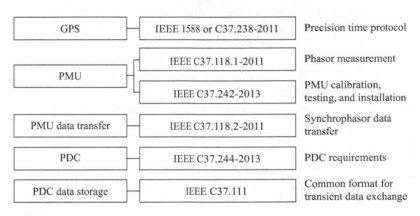

Figure 5.8   IEEE standards related to phasor measurement systems

the first PMU-related standard. It specified technical requirements on time syn-chronization and data sampling. The defined data transmission format for a single PMU followed the COMTRADE syntax. This standard was replaced by IEEE C37.118 standard series published in 2005. The updated standard added PMU test methods and error limit specifications. The data transmission format was improved by adding status and error indications. In addition, the standard was adapted for network communications and was able to handle multiple PMUs. In 2011, this standard was spited into two standards: C37.118.1 and C37.118.2. Compared with the 2005 version, C37.118.1 added phasor measurement requirements under dynamic conditions and included measurement definitions for frequency and the ROCOF. It also specified the latency test for data output delay. Its latest amend-ment C37.118.1a-2014 removed a few un-achievable requirements using currently available hardware.

5.3   Synchrophasor applications

As synchronized phasor measurement was introduced to electrical power grids, it has become an ideal tool to improve protection and control of power systems. Experiences on recent blackouts indicate that conventional protection and control system aiming at protecting large disturbances cannot prevent cascading failures as more components are operating near their full capacities. The fast and GPS time-stamped PMU measurements improve existing applications and opens the possibilities of various application by providing much faster, more accurate system data. With PMU measurements, operators are able to operate the power system in a more efficient and controlled way by applying more advanced monitoring, control, and remedy action applications.

Table 5.1 shows a list of WAMS applications and their data quantity and communication requirements. Figure 5.9 shows the matrix of industry needs—dependency on synchrophasor measurements for these applications. It can be seen that some applications are not demanding in terms of data quantity or transmission speed, but are still considered as high-value applications by the industry. These applications include angle and frequency monitoring, congestion management, event and oscillation detection, and dynamic parameter estimation. Some other applications that are critically needed by industry have very high requirements on PMU data quantity and legacy, such as automatic wide-area control and stabilization, power system restoration, and reliability action schemes. Realizing these advanced functions need more deployment of PMUs and an upgrade of communication infrastructures.

Table 5.1   WAMS applications and requirements on the number of PMUs and communication

No.	Applications based on WAMS	PMU quantity requirements	Communication requirements
1	Angle/frequency monitoring	Low	Medium
2	Adaptive protection	Medium	High
3	Automatic wide-area control	High	High
4	Congestion management	Medium	Medium
5	Dynamic state estimation	Medium	High
6	Event detection	Low	Medium
7	Oscillation detection	Low	Medium
8	Parameter estimation (dynamic)	Medium	Low
9	Parameter estimation (steady state)	Low	Low
10	Post-event analysis	Medium	Low
11	Power system restoration	High	Medium
12	Reliability actions schemes	High	High
13	Renewables integration	Medium	Medium
14	Thermal line rating	Low	Low
15	Transient stability prediction	High	High
16	Voltage stability monitoring	Medium	Medium
17	Wide-area stabilization	High	High

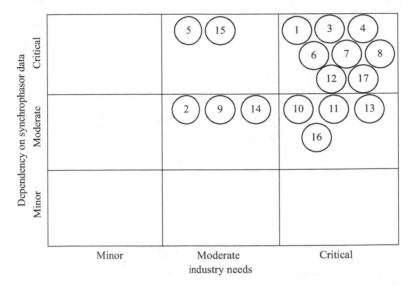

*Figure 5.9   The industry needs and synchrophasor data dependency matrix
(numbers correspond to the applications in Table 5.1)*

5.4   Case study—FNET/GridEye

As a pilot WAMS system, FNET/GridEye was originally deployed in 2004 and has
been operated by the University of Tennessee (UT) and Oak Ridge National Laboratory (ORNL) since 2009. It has served utilities, academics, and policy-makers with
valuable synchrophasor data from the North America power grid, even worldwide.
The FNET/GridEye system consists of a large number of GPS time-synchronized
frequency disturbance recorders (FDRs) as well as a data collection and processing
center. The measurement data obtained from FNET/GridEye make it possible for
various situational awareness applications [11]. Therefore, the FNET/GridEye system
will be introduced in this section as an example of the WAMS system.

5.4.1   Sensor design

The measurement sensors used by FNET/GridEye are referred to as FDRs. Each
FDR in the network measures the voltage magnitude, phasor angle, and frequency
from single-phase signal input. So, each FDR is actually a single-phase PMU. Three
generations of FDRs have been developed so far. Despite the hardware component
difference, all three generations share the same principle. Figure 5.10 shows a photo
of the currently deployed Generation-II FDR, whereas Figure 5.11 demonstrates its
hardware layout. As shown in Figure 5.11, the analog–digital conversion (ADC)
component periodically samples the conditioned voltage signal after the voltage
transducer and anti-aliasing filter with the help of digital signal processor (DSP)
oscillator pulses. To achieve high sampling precision, these oscillator pulses are

Figure 5.10   Photo of Generation-II FDR

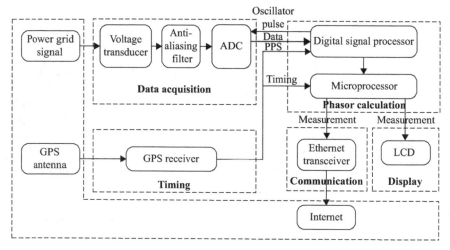

Figure 5.11   Generation-II FDR hardware block diagram

regulated by the one pulse per second signal provided by the GPS receiver. Phasor values are calculated in the DSP by an improved DFT algorithm [12,13], time stamped in the microprocessor, packaged by the Ethernet transceiver, and then transmitted over Ethernet.

Both hardware and firmware have been upgraded recently to further improve the phasor measurement accuracy for Generation-III FDR [14–16]. Specifically, a 16-bit ADC with ultra-high-precision bandgap voltage reference is used to replace the 14-bit ADC with internal voltage reference in Gen-II FDRs. The signal-to-noise ratio of the sampling circuit can be improved as a result. Additionally, an adaptive synchronous sampling algorithm is implemented in Gen-III FDRs, which enables the maximum timing error between two consecutive samples to be less than 10 ns [14]. Due to these upgrades, the steady-state phase angle and frequency measurement error of the Gen-III FDRs are less than 0.005° and 0.00006 Hz, respectively. The phasor measurement accuracy under dynamic conditions has also been improved by utilizing an improved DFT-based dynamic phasor measurement algorithm in Gen-III FDRs, which relies on multiple-step digital filters and a dynamic error compensation module [15]. The testing results show that the accuracy of Gen-III FDRs far exceeds the PMU Standard C37.118.1-2011 and C37.118.1a-2014.

5.4.2 Server framework

The measurements collected by FDRs are transmitted to data centers at UT and ORNL where all the FDR measurement data are concentrated and processed. Physically, a FNET/GridEye data center operates on several dedicated server machines, e.g., data server, application server, web server, backup server, etc.

Functionally, a data center can be treated as a multi-layer data management system as shown in Figure 5.12. The first and most important component of a data center is the data concentrator, where real-time measurements are extracted from the network TCP/Internet protocol data package, interpreted, error checked, time aligned, and then streamed into different layers for application or storage. The second layer of a data center hierarchy is composed of the real-time application agent and the data storage agent. The real-time application agent consists of various real-time application modules (e.g., disturbance detection, oscillation detection).

The third layer is the non-real-time application agent. Applications implemented on this layer (e.g., event replay applications, frequency statistical analysis, and other user-requested web services) are operated on various saved data formats (e.g., txt, mysql data) instead of real-time streaming data. This multi-layer data management system successfully deals with the various time requirements of different functionalities and accomplishes the efficient collection, storage, and utilization of real-time data. In the following two sections, various real-time and non-real-time applications run on the data center will be introduced.

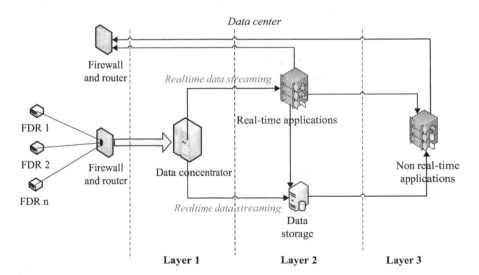

Figure 5.12   FNET/GridEye data center structure

Note that as FNET/GridEye relies on the public Internet for data communication right now, the latency and other communication issues are understandably more severe than traditional WAMS. Only real-time monitoring and diagnostic applications have been implemented online so far because they have relatively lower requirements for communication. If the measurement data are intended to be used for real-time control purposes, internal or dedicated communication networks will be needed.

5.4.3   Real-time applications

As mentioned above, FNET/GridEye applications can be divided into real-time and non-real-time applications roughly by their response time frame. Real-time applications require response within seconds or even sub-seconds after receiving the measurement data, while non-real-time applications have more flexible timing requirements or are upon request [17]. In this subsection, some of the important real-time applications are presented.

5.4.3.1   Real-time visualization of measurement data

Real-time visualization of wide-area measurement data is one of the FNET/GridEye system's most important applications. Correlating streaming wide-area frequency and voltage angle measurements with corresponding FDR geographical location information, the FNET/GridEye system creates an intuitive real-time visualization tool that helps operators better interpret what happens in the power grid in real time. More importantly, an electric utility usually has access to information about its own system but very limited access outside its control area. FNET/GridEye provides full coverage and thus presents a whole picture of the entire North American power system. Figure 5.13 shows a snapshot of the real-time frequency contour map of the North American power grids. These maps can be accessed online through the FNET/GridEye web services.

5.4.3.2   Disturbance recognition and location

Disturbances such as losing generators or transmission lines occur in the power grid frequently—often on an hourly basis. To prevent a single disturbance from escalating into large-area blackouts, the first step is to detect and locate the disturbance in the fastest manner.

Continuously screening the streaming frequency measurement data, the FNET/GridEye disturbance recognition module uses the rate of average frequency change df/dt as an indicator of power system disturbance. If df/dt exceeds a pre-defined threshold, the disturbance detection module will be triggered. Note that the thresholds vary for different power grids and different seasons. Once a frequency disturbance is detected, it will be categorized into generation trip, load shedding, line trip, etc. based on its characteristics using an artificial neural network algorithm [18] and then a geometrical tri-angulation algorithm making use of the time difference of arrival (TDOA) will be utilized to locate this disturbance [19,20]. If it is a generation trip, the net active power change ΔP can be estimated by multiplying the average frequency deviation Δf and the empirical coefficient beta value β. ABC points of the

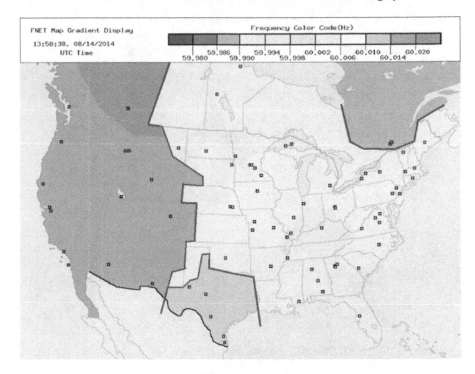

Figure 5.13   FNET/GridEye real-time frequency contour map

frequency curve will also be calculated. All this information as well as a frequency plot will be presented in a FNET/GridEye event report (as shown in Figure 5.14) and sent out automatically to service subscribers, such as utility operators, in seconds.

5.4.3.3   Inter-area oscillation detection and modal analysis

Small-signal stability is a key concern of power system operators. Utilizing a frequency or angle-based oscillation detection algorithm, FNET/GridEye is effective in monitoring the low-frequency inter-area oscillations that occur in a power grid [21]. Employing a multi-channel matrix pencil algorithm, a modal analysis of each oscillation event can also be performed. As shown in Figure 5.15, once an oscillation is detected, the frequency and relative angle of involved FDRs will be plotted and the FDRs that record the largest oscillation amplitude will be listed. The frequency and damping ratio information of dominant modes will also be computed and given. Note that this is also an automatic service that can be sent to service subscribers in real time.

5.4.3.4   Ambient data-based oscillation mode frequency and damping ratio estimation

Power system ambient data are the natural response of the system due to small-magnitude disturbances, random load switching, etc. Despite the embedded higher noise, rich oscillation modal information can still be abstracted. This FNET/GridEye

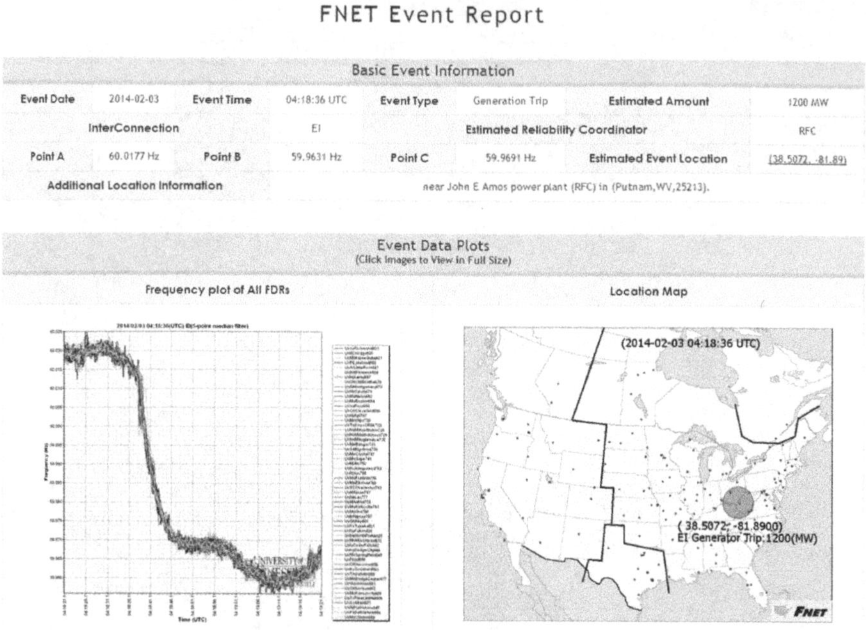

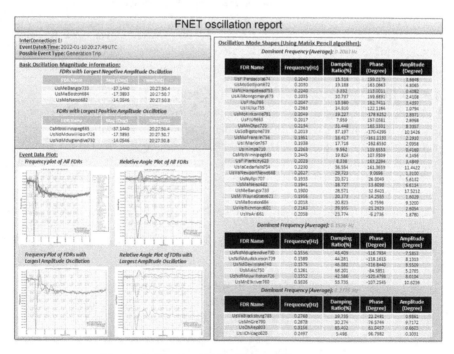

Figure 5.14   FNET/GridEye disturbance recognition and location report

Figure 5.15   FNET/GridEye oscillation report

application module employs an empirical mode decomposition filter [22] to de-trend the ambient signal first and then utilizes an auto-regressing moving-average model method [23] to abstract the oscillation mode frequency and damping ratio information. Most importantly, by use of a multi-channel parallel processing design, this function module has the capability to process hundreds of streaming FDR measurements at the same time, which gives a whole picture of the entire North American power grid oscillation information in real time for the first time ever. Figure 5.16 shows the oscillation mode frequency and damping ratio calculation results of one FDR. In addition, FNET/GridEye allows incorporating other signal processing modules for various ambient oscillation monitoring and analysis purposes [24].

5.4.3.5 Islanding and off-grid detection

Islanding is an extremely dangerous phenomenon that occurs when one or more generators are no longer working synchronously with the rest of the power system.

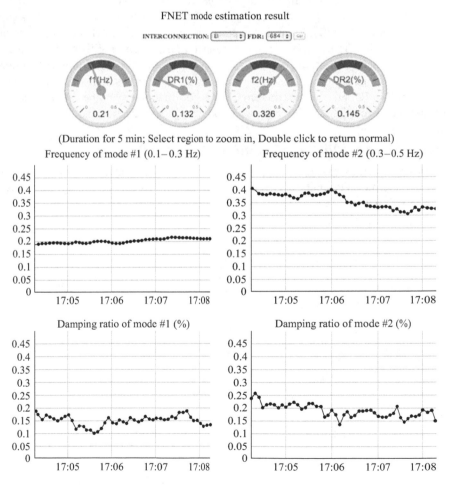

Figure 5.16   Online oscillation mode frequency and damping ratio display

Islanding can escalate to blackouts if no control action is taken in time. The FNET/ GridEye islanding detection module calculates the integration of frequency difference between each FDR and the system average and sends out islanding warnings to grid operators if one or more FDR detect abnormally large frequency differences, which indicates that certain sub-systems (or generators) have become islanded from the others [25,26]. Figure 5.17 shows an islanding event detected by this module.

5.4.3.6   Measurement-based model construction

Phasor measurement data provide first-hand knowledge of power system dynamic behaviors. On the other hand, circuit-based power system models always have limitations of the amount of details that can be completely and accurately included. Taking advantage of the real-time wide-area measurements, one of phasor measurements' novel applications is to develop data-driven models that can be updated online to estimate or predict system responses [27].

Figure 5.18 is the estimated voltage angle response of an event from such a data-driven model. It shows a good match with the actual measurement. This process can be summarized as: firstly, identify the transfer function model between the "output FDR" (cross in Figure 5.19) and the "input FDR" (star in Figure 5.19) using measurement data of certain contingencies, and then use the trained transfer function model and the measurement data from the "input FDR" to estimate the response of the "output FDR" of other contingencies. This estimation technique can be used for missing measurement data interpolation and wrong data identification.

5.4.4   Non-real-time applications

Besides the various real-time applications, FNET/GridEye also developed a series of non-real-time data analytics applications, some of which are briefly introduced in this section.

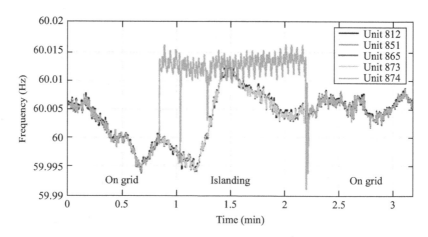

Figure 5.17   Islanding event detected by FNET/GridEye

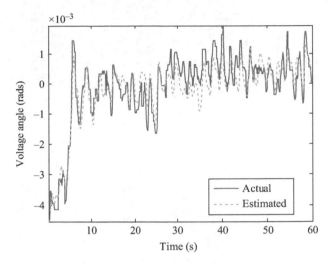

Figure 5.18 Estimated angle response from a measurement data-driven model

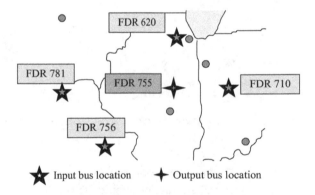

Figure 5.19 Illustration of the measurement data-driven model development

5.4.4.1 Event replay and post-event analysis

Blackouts cause disastrous losses in large areas. For instance, the 2008 Florida blackout led to the loss of 22 transmission lines, 4,300 MW of generation, and 3,650 MW of customer service or load in two-thirds of the Florida area. FNET/GridEye and PMU measurements of this blackout were used to replay this event, one screenshot of which is shown in Figure 5.20. It can be clearly noticed that this event originated from the Florida area, and then propagated to the entire EI. This event replay movie facilitates the blackout's post-event analysis and helps avoid similar blackouts in the future.

5.4.4.2 Measurement-aided model validation

As one of the most important non-real-time applications, FNET/GridEye frequency measurements have been utilized to validate the dynamic model of US power grids,

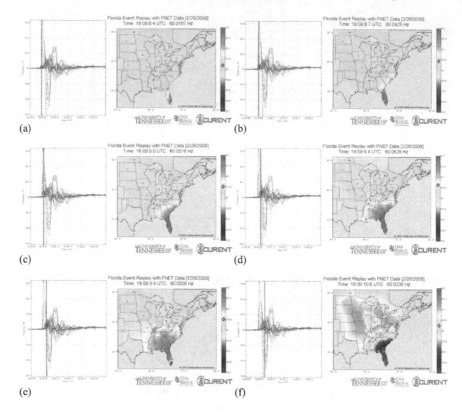

Figure 5.20   Florida blackout replay enabled by FNET/GridEye frequency measurement

such as the Eastern system. By comparing the real frequency response recorded by FNET/GridEye to the dynamic simulation results from the EI multi-regional modeling working-group model, it is revealed that the current EI dynamic model is far from accurate enough to give a creditable frequency response. Our validation experience tells us that the parameters of machine inertias, governor settings, loads, etc. need to be carefully tuned to match the real system response recorded by FNET/GridEye [28]. In Figure 5.21, incorporating the previously un-modeled governor dead-bands into the EI model, its frequency response simulation accuracy was significantly improved [29]. This example reveals the great potential of FNET/ GridEye measurement in large-scale power system model validation.

5.4.4.3   Electromechanical speed map development

Electromechanical waves propagate at different speeds in the power grids. This phenomenon has long been observed in the US power grids and can be explained by different generator and load densities in different regions [30]. Accurate calculation of the electromechanical wave propagation speed is important to understand the power grid dynamic characteristics. FNET/GridEye provides a measurement-based

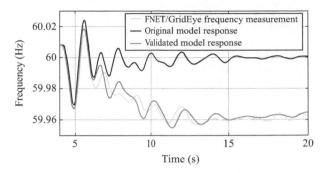

Figure 5.21 FNET/GridEye frequency measurement model validation

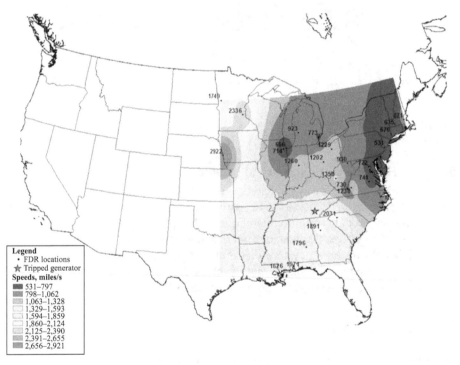

Figure 5.22 Electromechanical speed map calculated by FNET/GridEye

solution to this problem: the propagation speed of an electromechanical wave traveling from one FDR to another can be calculated easily by dividing the distance between those two FDRs by the TDOA. With as many as hundreds of almost evenly deployed FDRs and electromechanical waves detected in the EI system, the average propagation speed between any two FDRs can be obtained, based on which EI system propagation speed contour map is available (as shown in Figure 5.22) [18].

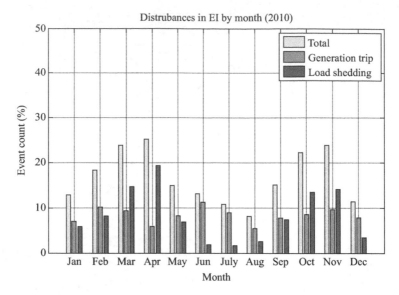

Figure 5.23   EI system disturbance distribution over a year in 2012

In this way, a trustworthy estimation of the propagation speeds over the EI territory can be obtained making use of FNET/GridEye frequency measurements.

5.4.4.4   Historic data statistical analysis

Since the FNET/GridEye system went online in 2004, a large amount of data have been collected from FDRs located within the USA and around the world. By employing some data mining techniques, FNET/GridEye historic data can be extremely informative. For instance, statistical analysis is able to demonstrate how the disturbances in the EI system distribute over a year (as shown in Figure 5.23) [31] and the impact of social events such as the FIFA World Cup or the NFL SuperBowl on the power grids can be analyzed [32].

5.5   Conclusions

The uniqueness of WAMS is that it uses a common time reference to synchronize all the measurements over a wide area. Compared with SCADA systems, advantages of WAMS are primarily due to the time-stamped phasor measurements, as well as its high resolution. The time resolution of WAMS (10–60 samples per second, much higher than SCADA systems' 1 sample per 2–4 s) allows situation awareness at the sub-second level and system dynamics can be captured more accurately. WAMS will continue to benefit a wide variety of power system monitoring, analysis and control functions and contribute to the reliable and economic power transmission system operation.

References

[1] G. Missout and P. Girard, "Measurement of bus voltage angle between Montreal and Sept-Îles," *IEEE Transactions on Power Apparatus and Systems*, vol. PAS-99, pp. 536–539, 1980.

[2] G. Missout, J. Beland, G. Bedard, and Y. Lafleur, "Dynamic measurement of the absolute voltage angle on long transmission Lines," *IEEE Transactions on Power Apparatus and Systems*, vol. PAS-100, pp. 4428–4434, 1981.

[3] P. Bonanomi, "Phase angle measurements with synchronized clocks-principle and applications," *IEEE Transactions on Power Apparatus and Systems*, vol. PAS-100, pp. 5036–5043, 1981.

[4] A. G. Phadke, M. Ibrahim, and T. Hlibka, "Fundamental basis for distance relaying with symmetrical components," *IEEE Transactions on Power Apparatus and Systems*, vol. 96, pp. 635–646, 1977.

[5] A. G. Phadke, J. Thorp, and M. G. Adamiak, "A new measurement technique for tracking voltage phasors, local system frequency, and rate of change of frequency," *IEEE Transactions on Power Apparatus and Systems*, vol. PAS-102, pp. 1025–1038, 1983.

[6] A. G. Phadke and J. S. Thorp, *Synchronized Phasor Measurements and Their Applications*. New York City: Springer Science & Business Media, 2008.

[7] R. B. Bobba, J. Dagle, E. Heine, *et al.*, "Enhancing grid measurements: Wide area measurement systems, NASPInet, and security," *IEEE Power and Energy Magazine*, vol. 10, pp. 67–73, 2012.

[8] Grid Protection Alliance, "OpenPDC–Open Source Phasor Data Concentrator Software Package," Version 2.1 (SP1). 6 Mar 2015. Available from http://openpdc.codeplex.com/.

[9] K. Martin, D. Hamai, M. Adamiak, *et al.*, "Exploring the IEEE standard C37. 118–2005 synchrophasors for power systems," *IEEE Transactions on Power Delivery*, vol. 23, pp. 1805–1811, 2008.

[10] K. E. Martin, G. Benmouyal, M. Adamiak, *et al.*, "IEEE standard for synchrophasors for power systems," *IEEE Transactions on Power Delivery*, vol. 13, pp. 73–77, 1998.

[11] Z. Zhong, C. Xu, B. J. Billian, *et al.*, "Power system frequency monitoring network (FNET) implementation," *IEEE Transactions on Power Systems*, vol. 20, pp. 1914–1921, 2005.

[12] L. Wang, J. Burgett, J. Zuo, *et al.*, "Frequency disturbance recorder design and developments," in *Power Engineering Society General Meeting, 2007*. IEEE, 2007, pp. 1–7.

[13] T. Xia and Y. Liu, "Single-phase phase angle measurements in electric power systems," *IEEE Transactions on Power Systems*, vol. 25, pp. 844–852, 2010.

[14] L. Zhan, J. Zhao, J. Culliss, Y. Liu, Y. Liu, and S. Gao, "Universal grid analyzer design and development," in *2015 IEEE Power & Energy Society General Meeting*, 2015, pp. 1–5.

[15] L. Zhan, Y. Liu, J. Culliss, J. Zhao, and Y. Liu, "Dynamic single-phase synchronized phase and frequency estimation at the distribution level," *IEEE Transactions on Smart Grid*, vol. 6, pp. 2013–2022, 2015.

[16] L. Zhan and Y. Liu, "Improved WLS-TF algorithm for dynamic synchronized angle and frequency estimation," in *2014 IEEE PES General Meeting/Conference & Exposition*, 2014, pp. 1–5.

[17] Y. Zhang, P. Markham, T. Xia, *et al.*, "Wide-area frequency monitoring network (FNET) architecture and applications," *IEEE Transactions on Smart Grid*, vol. 1, pp. 159–167, 2010.

[18] P. N. Markham, "Data mining and machine learning applications of wide-area measurement data in electric power systems," PhD diss., University of Tennessee, 2012.

[19] R. M. Gardner, J. N. Bank, J. K. Wang, A. J. Arana, and Y. Liu, "Non-parametric power system event location using wide-area measurements," in *2006 IEEE PES Power Systems Conference and Exposition*, 2006, pp. 1668–1675.

[20] R. M. Gardner, J. K. Wang, and Y. Liu, "Power system event location analysis using wide-area measurements," in *2006 IEEE Power Engineering Society General Meeting*, 2006, p. 7.

[21] T. Xia, Y. Zhang, L. Chen, *et al.*, "Phase angle-based power system inter-area oscillation detection and modal analysis," *European Transactions on Electrical Power*, vol. 21, pp. 1629–1639, 2011.

[22] N. E. Huang, Z. Shen, S. R. Long, *et al.*, "The empirical mode decomposition and the Hilbert spectrum for nonlinear and non-stationary time series analysis," *Proceedings of the Royal Society of London A: Mathematical, Physical and Engineering Sciences*, pp. 903–995, 1998.

[23] D. J. Trudnowski, J. W. Pierre, N. Zhou, J. F. Hauer, and M. Parashar, "Performance of three mode-meter block-processing algorithms for automated dynamic stability assessment," *IEEE Transactions on Power Systems*, vol. 23, pp. 680–690, 2008.

[24] S. You, J. Guo, G. Kou, Y. Liu, and Y. Liu, "Oscillation mode identification based on wide-area ambient measurements using multivariate empirical mode decomposition," *Electric Power Systems Research*, vol. 134, pp. 158–166, 2016.

[25] Z. Lin, T. Xia, Y. Ye, *et al.*, "Application of wide area measurement systems to islanding detection of bulk power systems," *IEEE Transactions on Power Systems*, vol. 28, pp. 2006–2015, 2013.

[26] J. Guo, Y. Zhang, M. A. Young, *et al.*, "Design and implementation of a real-time off-grid operation detection tool from a wide-area measurements perspective," *IEEE Transactions on Smart Grid*, vol. 6, pp. 2080–2087, 2015.

[27] Y. Liu, K. Sun, and Y. Liu, "A measurement-based power system model for dynamic response estimation and instability warning," *Electric Power Systems Research*, vol. 124, pp. 1–9, 2015.

[28] L. Chen, "Wide-area measurement application and power system dynamics," PhD diss., University of Tennessee, 2011.

[29] G. Kou, S. Hadley, and Y. Liu, "Dynamic model validation with governor deadband on the eastern interconnection," *Oak Ridge Nat. Lab., Power and Energy Syst. Group, Oak Ridge, TN, USA, Tech. Rep. ORNL/TM-2014/40*, 2014.

[30] J. S. Thorp, C. E. Seyler, and A. G. Phadke, "Electromechanical wave propagation in large electric power systems," *IEEE Transactions on Circuits and Systems I: Fundamental Theory and Applications*, vol. 45, pp. 614–622, 1998.

[31] Y. Ye, "Wide-area situational awareness application developments," PhD diss., University of Tennessee, 2011.

[32] Y. Lei, Y. Zhang, J. Guo, *et al.*, "The impact of synchronized human activities on power system frequency," in *2014 IEEE PES General Meeting/Conference & Exposition*, 2014, pp. 1–5.

Chapter 6

Bad data detection in the smart grid

Haris M. Khalid[1] and Ahmed Al-Durra[2]

This chapter will discuss bad data detection techniques and their application in oscillation monitoring. Utilization of synchrophasor measurements for wide-area monitoring applications enables system operators to acquire real-time grid information. However, intentional injections of false synchrophasor measurements can potentially lead to inappropriate control actions, jeopardizing the security, and reliability of power transmission networks. An attacker can compromise the integrity of the monitoring algorithms by hijacking a subset of sensor measurements and sending manipulated readings. Such an approach can result to wide-area blackouts in power grids. This chapter considers bad data detection techniques with special focus on oscillation monitoring. To achieve an accurate supervision, a Bayesian inference technique has been discussed for each monitoring node using a distributed architecture.

6.1 Introduction

Due to the increasing dependency of digital measurements for monitoring and control applications, bad data attack is an emerging threat. If a sensor is successfully attacked, its stored information can be compromised [1–4]. Given the criticality of power systems in the context of the national security, wide-area monitoring system (WAMS) applications such as oscillation detection is an attractive attack target [5]. If a phasor measurement unit (PMU) is successfully attacked, its stored information can be unnoticeably compromised. This can result in a significant impact on public safety and economic losses [6–10]. Currently, it is a time-consuming task to detect and identify data attacks as an adversary can choose the site of attack judiciously and design the attack vector carefully [11].

Many methods have been proposed to identify abnormal data segments and isolate attacked sensors in recent years. Most of them are published to enhance

[1]Department of Electrical and Electronics Engineering, Sharjah Higher Colleges of Technology (SHCT), UAE.
[2]Department of Electrical Engineering, The Petroleum Institute, Abu Dhabi, UAE

static application such as state estimation [6,12,13], power flow analysis [8,9,14], and electricity market [7]. Static monitoring applications primarily focus on monitoring the operating point of the system and address slow dynamics in the range of minutes to hours [15]. In contrast, few have been proposed for dynamics monitoring applications, which tracks transient dynamics in the order of seconds or less.

6.2 Possible approaches

6.2.1 *State estimation*

Several methodologies in the areas of state estimations were developed over the past decades. Literature on the types of state estimation algorithms were presented in References 16–19. State estimation also has different approaches based on the application of the algorithms such as conventional state estimation [16], distributed state estimation, or multi-area state estimation [20]. Depending on the timing and evolution of the estimates, state estimation schemes may be broadly classified into two basic distinct paradigms: static state estimation and dynamic state estimation [21]. Another extension of static state estimation also included the sequential state estimation which has the advantage of being able to perform updates with partial measurement set [22]. This enabled the method to address the problem of data loss and bad data. A static state estimation algorithm based on linear programming known as least absolute value was also developed in [23,24]. Generally, under normal operating conditions, the power system is regarded as a quasi-static system that changes steadily but slowly [17]. Therefore, in order to continuously monitor the power system, state estimators must be executed at short intervals of time. But with the inherent expansion of power systems, with the increase of generations and loads, the system becomes extremely large for state estimation to be executed at short intervals of time since it requires heavy computation resources. Therefore, a technique known as tracking state estimation [25,26] was developed. Once state estimates were calculated, the method simply update the next instant of time using a new measurement set obtained for that instant, instead of again running the entire static state estimation algorithm. Tracking estimators help energy management systems to keep track of the continuously changing power system without actually having to execute the entire state estimation algorithm. This allows continuous monitoring with reasonable utilization of computing resources.

6.2.2 *Weighted least squares*

One of the most commonly used types of static state estimation in utilities is the weighted least squares methodology [18]. It was formulated as an optimization problem with a notion of minimizing the squares of the differences between the measured and estimated values calculated using the corresponding power flow equations. The weighted least square uses the Newton–Raphson algorithm to obtain the state estimates. There have been numerous findings on different variations of weighted least square further to improve specific aspects of the algorithm. A fast decoupled state estimator [27,28] is an example in which voltage magnitudes and phase angles are processed separately. The voltage magnitude values are concerned with the reactive power measurements while angles were related to active power measurements. Regularized least square for power systems in Reference 29 proposed a type of weighted least square that was able to function in cases of partial observability.

6.2.3 Dynamic state estimation

The extended Kalman filter (EKF) is the most widely used algorithm to perform dynamic state estimation [30]. Other forms of Kalman filters such as unscented Kalman filter [31] and iterative EKF [32] were also proposed in the literature. Other algorithms used to perform dynamic state estimations include artificial neural networks [33] and fuzzy logic [34] which are also computationally complex. Generally, dynamic state estimations are well suited when the dynamics of the power systems are smooth and follow the historical value. In other words, they could fail to accurately estimate when there exists a bigger change in operating points.

6.2.4 Bad data analysis using chi-squared test and normalized residual test

One of the essential functions of a state estimator is to detect bad measurements and to identify and eliminate them accordingly [35]. Bad data analysis could be performed during the estimation process or post-estimation. When using the weighted least squares estimation algorithm for state estimation, detection, and identification of bad data is done after the estimation process by processing the measurement residuals. The analysis is essentially based on the properties of the residuals, including their expected probability distribution.

Chi-squared test for bad data detection was presented in References 36 and 37. It uses the properties of the chi-squares probability density function to compare with the objective function of weighted least squares. Chi-squared test was able to detect bad data but does not identify locations.

Alternatively, the normalized residual test was able to detect as well as identify the locations of occurrences [35,36,38]. The normalized residual test was developed based on the statistical characteristics of the measurement residuals. Detection and identification could also be accomplished by further processing of the residuals as in the hypothesis testing identification methods [35,36,39]. Although both methods used the residual sensitivity matrix to represent the sensitivity of the measurement residuals to the measurement errors, hypothesis testing identification was more complex and computationally costly due to the further processing of the residuals. Hence, hypothesis testing identification was used to detect and identify bad data in this thesis. However, hypothesis testing identification was observed to exhibit some limitations as noted in Reference 36. The primary limitation was the inability to track bad data if it occurred at critical locations. To resolve this issue, utilizations of PMU measurements were proposed in recent literature.

6.3 Case study: oscillation monitoring

One mature dynamic monitoring application is oscillation detection [40–42]. Such low-frequency dynamics were only observable by analyzing measurements from PMUs. Today, various types of oscillation detection schemes have been installed in many transmission utilities to monitor the inter-area oscillations within critical tie-lines. Hence, oscillation detection is more likely to be subjected to intentional data-injection attacks than other dynamic monitoring applications [5]. Moreover, these data attacks are assumed to take place in PMUs installed in substations.

6.3.1 Consequences of an attack

In an event of an attack, the following two negative consequences can occur due to inaccurate monitoring and time complexity of the algorithms. (1) If the PMU data is altered in a way that is not detectable as false dynamics by oscillation monitoring schemes, the perceived observable state of the system will be wrong. This may lead to improper control actions endangering the security of the system. (2) The malicious intent might not be to hide the attack. An example is the denial of service, where the system operator loses the observability in a critical region.

6.3.2 Distinction between a fault and a cyber-attack

Although both types of perturbations can lead to abnormal operations, the notion of a fault and cyber-attack is distinctly different. A fault is considered as physical events that affect the power grid behavior, where the inherent transient dynamics are observable in neighboring substations and can be correlated in a time scale [43–46]. In contrast, a cyber-attack decouples from the physical world [4,6,11,12,47]. Thus, the false dynamics embedded inside attacked measurements may not correlate with other locations in time. The key is to establish a link with neighboring metering devices, and perform correlation studies.

6.3.3 Difference from static monitoring applications

In contrast to static monitoring application, impacts of acting on incorrect information are experienced relatively faster leading to more destabilizing issues. The static monitoring approach of analyzing a state-space model derived from linear differential algebraic equations is not suitable for tracking transient dynamics such as inter-area oscillations [5,6,13]. The fundamental issue is the linearized equations that are restrictive representation of the nonlinear dynamic transients caused by system perturbations. Another concern is that the reaction time for addressing transient dynamics is much less in comparison with static applications such as state estimation. Despite significant efforts have been invested in preventing cyber-attacks for static monitoring applications [6–9,13,47], researchers have not fully investigated the impact of cyber-attacks for monitoring inter-area oscillations. The reason is WAMS applications for monitoring transient dynamics is still an emerging research where main focus to date has been towards application development [16–18]. Hence, novel methodologies are needed to prevent cyber-attacks in oscillation detection schemes.

6.4   Modeling an attack in oscillation detection schemes

In this section, the three critical research tasks to establish immunity towards cyber-attacks for oscillation detection schemes are described.

6.4.1   State-space representation of a power grid

In the field of real-time dynamic monitoring, especially for WAMS applications, the notion is to become less dependent on classic models and adopt real-time system identification techniques. The reason is that classic differential equations are less representative of continuous random load variations, line temperature variations, and other operational uncertainties. Although using differential equation-based models are suitable for some steady-state or static applications such as state estimation or automatic generation control, it is not suitable for monitoring electromechanical interactions of synchronous generators. Therefore, system parameters are not extracted from offline predetermined power system models. Instead, the proposed method extracts desired parameters from PMU measurements.

A power grid prone to data-injection attacks can be modeled as a nonlinear dynamical system. Continuous load perturbations are part of noise-induced transitions, which can be expressed as follows:

$$ax_{t+1} = f(x_t, w_t), \quad t = 0, 1, \dots, T \tag{6.1}$$

where α is the constant matrix, $f(\cdot)$ is the nonlinear function representing the state transition model, $x \in R^r$ is the state variable, and the superscript r is the number of monitored dynamic modes. For the case of oscillation detection, the state variable represents the electromechanical oscillation, and r refers to the number of oscillations in the subspace R. The process noise of the recursive scheme is $w \in R^r$ at the time t over a monitoring window of T time instances. Suppose that a power grid described in (6.1) is monitored by N number of PMUs. We propose a distributed scheme that processes the estimated oscillatory dynamics of each PMU nodes at a centralized track fusion center. The observation vector z^i for extracting electromechanical oscillations at the ith node is

$$z_t^i = h_t^i(x_t) + v_t^i, \quad i = 1, 2, \dots, N \tag{6.2}$$

Note that the ith node may be subjected to an attack. The terms $z^i \in R^{p^i}$ and p^i are the number of measurements made by the ith PMU. The nonlinear function representing the local observation matrix of the ith sensor is $h_t^i(\cdot)$, and $v^i \in R^{p^i}$ refers to the observation noise.

6.4.2   Constraints of a power grid

From (6.1) and (6.2), a power grid can be governed by the following constraints:

$$x_t \in X_t, w_t \in W_t \quad \text{and} \quad v_t \in V_t \tag{6.3}$$

where X_t, W_t, and V_t are assumed to have Gaussian probability distribution function. Once the observation model is constructed from synchrophasor measurements corrected from the affected location, the corresponding state representation of electromechanical oscillations can be formulated in the frequency domain.

6.4.3 Electromechanical oscillation model formulation

According to Hauer [48], a measured noise-induced signal containing K number of electromechanical oscillations can be modeled in the frequency domain. As a result, (6.2) can be expressed as follows:

$$z_t^i = \sum_{k=1}^{K} A_k e^{(-\sigma_k + j2\pi f_k)tT_s} + v_t^i, \quad t = 1, 2, \ldots, T \tag{6.4}$$

where A_k is the complex amplitude of the kth mode, σ_k and f_k are the corresponding damping factor and oscillatory frequency, respectively. The sampling time is represented as T_s.

However, estimating oscillatory parameters for an accurate WAMS will require the complete observability of the observation matrix. This is quiet challenging in the presence of data-injection attacks. Locational awareness for each node is required, considering the fact that installed PMUs may also malfunction during an attack. This requires classification of the attack, followed by its characterization and modeling.

6.4.4 Characterization of an attack: an example

An initial characterization about the unobservable attacks can be possibly induced by Bayesian inference. Assume that a bus of power grid is attacked as shown in Figure 6.1. Considering the Bayesian inference, the probabilities on *a-prior* distribution over the oscillatory states at ith node are $p(x_t^i)$, and the observation matrix is $p(H_t^i|x_t^i)$. The resultant posterior distribution over the observations can be represented by the Bayesian inference as follows:

$$p(x_t^i|z_t^i) = \frac{p(x_t^i)p(z_t^i|x_t^i)}{p(z_t^i)} \tag{6.5}$$

To quantify the uncertainty of possible data-injection attacks, the density of the predicted synchrophasor observations is required to be computed. This can be obtained by averaging over the uncertainty of data-injection attacks on the oscillatory states and the observation matrix. Let z_{pr}^i represent the predicted synchrophasor observations at ith node, then $z_{pr,t}^i$ can be presented in the form of predictive distribution as follows:

$$p(z_{pr,t}^i|z_t^i) = \sum_{x_t^i} \int dH_t^i p(z_{pr}^i|H_t^i, x_t^i, z_t^i) p(H_t^i|x_t^i, z_t^i) p(x_t^i|z_t^i) \tag{6.6}$$

H_t^i is a hypothesis extracted from the observation signal about the presence or absence of the attack signal.

This distribution will later assist in the development of the probability of attack vectors.

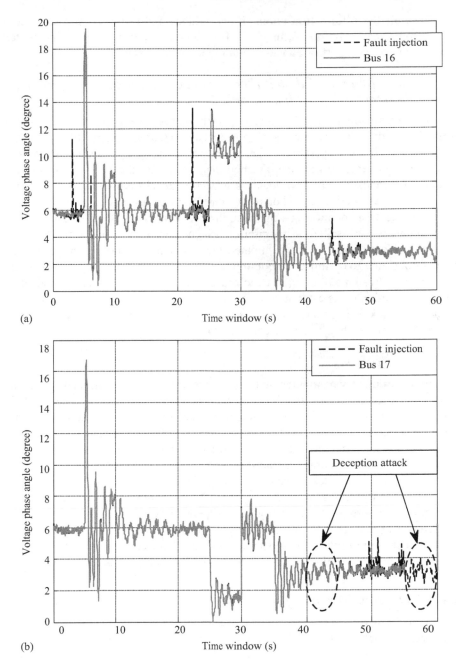

Figure 6.1   Profile of (a) Bus 16 and (b) Bus 17 with random fault injections

Once all the information about the covariance and estimated states are collected from local PMU nodes, they will be treated at the distributed fusion center which is an integral part of attack-tolerant monitoring system.

6.4.5   Significance of distributed architecture towards information of cyber-attack

To create a cascading failure caused by lightly damped electromechanical oscillations, the injected data can be assumed to imitate regular small-amplitude load variations as seen in daily operations. Therefore, characterizing a plausible attack or loss of information needs to be done. Furthermore, from a practical viewpoint, we can assume that attacked nodes are local as wide-area attacks are less feasible from a geographical perspective. Based on these considerations, it is not possible to identify abnormalities if a monitoring scheme utilizes one or local PMU measurements. Conventionally, an oscillation monitoring scheme can be installed in the PMU or local phasor data concentrator. Estimated oscillatory parameters are transmitted to the control center to optimize the communication bandwidth. Building on top of this configuration, we intend to request each recursive monitoring scheme to send additional information of its estimated covariance matrix and state vector. As a result, a centralized track fusion center is proposed. Similar to (6.5), the observation model of the track fusion center z_t^{TF} can be expressed as follows:

$$z_t^{TF} = H_t^{TF} x_t + \omega_t^{TF} \tag{6.7}$$

Estimating oscillatory parameters in the presence of abnormal or attacked nodes will require the complete observability of the oscillation observation matrix. This requires the calculation of correlation information from the initial estimates of the observation model. Treated correlation information will then assist in the removal of faulty oscillatory parameters to ensure an attack-free update of oscillatory parameters to the system operators. Based on this concept, a refined covariance matrix generated is then sent back to each monitoring node to improve the observability of the wide-area dynamics of inter-area oscillations. However, an accurate power oscillation monitoring scheme is required to operate near real time. This can be very challenging in the presence of data-injection attacks. To reduce the computational effort of determining the initial estimates and the error covariance matrix at each PMU node, diagonalization of the system model into subsystems can be proposed.

6.4.6   Diagonalization of a system into subsystems

Note that the attacked system at node i can be diagonalized up to N number of subsystems. Considering the diagonalization of $N = 2$ subsystems, using the theory of robust eigenvalue placement, the system (6.4) and (6.5) can be decomposed into L and R non-singular matrices

$$La R = \begin{bmatrix} \alpha_1 & 0 \\ \alpha_2 & 0 \end{bmatrix}, \quad L\kappa R = \begin{bmatrix} \kappa_1 & 0 \\ \kappa_2 & \kappa_3 \end{bmatrix}, \quad L\psi R = \begin{bmatrix} \psi_1 \\ \psi_2 \end{bmatrix}, \quad H^i R = \begin{bmatrix} H_1^i \\ H_2^i \end{bmatrix}^* \tag{6.8}$$

where $\alpha_1 \in R^{n_1 \times n_2}$ is a non-singular lower triangular, $\kappa_1 \in R^{n_1 \times n_1}$ is quasi-lower triangular, $\kappa_3 \in R^{n_2 \times n_2}$ is non-singular lower triangular. Transforming $x_t = R[x_{1,t}^* \ \ x_{2,t}^*]^*$, where $x_{1,t} \in R^{n_1}, x_{2,t} \in R^{n_2}$. The system can be transformed into the following two diagonalizable subsystems by taking the inverse of high-dimensional matrices of (6.1) and (6.2) using linear minimum variance:

$$x_{1,t+1} = \kappa_0 x_{1,t} + \psi_0 w_t \tag{6.9}$$

$$x_{2,t} = \overline{\kappa} x_{1,t} + \overline{\psi} w_t \tag{6.10}$$

$$z_t^i = \overline{H}_t^i x_{1,t} + \overline{v}_t^i \tag{6.11}$$

where $x_{1,t}$ and $x_{2,t}$ are the states of subsystem 1 and subsystem 2, respectively. $\kappa_0, \psi_0, \overline{\kappa}, \overline{\psi}, \overline{H}$ and $\overline{v}$ are diagonalized variables, which are computed from the inverse of weighted matrices α_1 and κ_3. Note in the subsystem transformation, only first subsystem will have the prediction and filtering stage, whereas the rest of $N-1$ subsystems will only have the filtering stage. Each subsystem is a smaller matrix than the original model, which would then improve the computing speed required to update the covariance matrix at each monitoring instance. Once the subsystems are constructed from the system affected by the data-injection attacks, the interactions between them shall be evaluated. Moreover, by handling the noise and state constraints of (6.3), the immunity of the estimation results during data-injection can be increased. Referring to (6.9)–(6.11), the resultant noises w_t and v_t will have the diagonalizable expected value:

$$E\left[\begin{bmatrix} \overline{w}_t \\ \overline{v}_t^i \end{bmatrix}, \left[\overline{w}_t^* \ \overline{v}_t^{2*}\right]\right] = Q_t^{1,2} \delta_t^{1,2}$$

where $Q_t^{1,2}$ is the process noise correlation factor between subsystems (6.1) and (6.2), $\delta_t^{1,2}$ is the Kronecker delta function used for shifting the integer variable after the presence or absence of noise.

In this step, we will integrate a centralized filter to remove estimated parameters from attacked sensor nodes while providing accurate covariance matrix for the individual monitoring nodes. This established a closed loop monitoring system. Thus, the resilience of inter-area oscillation detection against data-injection attacks can be improved.

6.4.7 Detection bad data using initial observation analysis

Once the probability of attack vectors is developed, the attack can be detected by doing an initial observation analysis of the measurements. This can be achieved by taking the difference between the given and predicted observation of the oscillation state:

$$Z_{t+1}^i = \left[z_{t+1}^i - z_{pr,t+1}^i\right] = \sum_{t=1}^T \psi_{t-1}^* \theta_t^{H(1)} \Delta H_t^i + v_t^i \tag{6.12}$$

where the vector Z_{t+1}^i is the innovation calculated for the ith node. z_{t+1}^i and $z_{pr,t+1}^i$ are the data-injection-free (nominal) and predicted-affected observation outputs, respectively. $\Delta H_t^i = \Delta H_{d,t}^i - H_t^i$ is the perturbation in H_t^i. $\theta_t^i = \delta z_t^i / \delta H_t^{i*}$ is the gradient used to identify the perturbation due to data-injection attacks. ψ_t is the data vector formed from past outputs and reference inputs at each node.

References

[1] A. G. Tartakovsky, B. L. Rozovskii, R. B. Blažek, and H. Kim, "A novel approach to detection of intrusions in computer networks via adaptive sequential and batch-sequential change-point detection methods," *IEEE Transactions on Signal Processing*, vol. 54, pp. 3372–3382, 2006.

[2] S. Zhu, S. Setia, S. Jajodia, and P. Ning, "An interleaved hop-by-hop authentication scheme for filtering of injected false data in sensor networks," in *IEEE Symposium on Security and Privacy*, 2004, pp. 259–271.

[3] V. Shukla and D. Qiao, "Distinguishing data transience from false injection in sensor networks," in *Fourth Annual IEEE Communications Society Conference on Sensor, Mesh and Ad Hoc Communications and Networks*, San Diego, CA, 2007, pp. 41–50.

[4] F. Pasqualetti, F. Dorfler, and F. Bullo, "Control-theoretic methods for cyberphysical security: Geometric principles for optimal cross-layer resilient control systems," *IEEE Control Systems*, vol. 35, pp. 110–127, 2015.

[5] H. M. Khalid and J. C.-H. Peng, "Immunity towards data-injection attacks using track fusion-based model prediction," *IEEE Transactions on Smart Grid,* DOI: 10.1109/TSG.2015.2487280, in press October 2015.

[6] F. Pasqualetti, F. Dorfler, and F. Bullo, "Attack detection and identification in cyber-physical systems," *IEEE Transactions on Automatic Control*, vol. 58, pp. 2715–2729, 2013.

[7] X. Le, M. Yilin, and B. Sinopoli, "Integrity data attacks in power market operations," *IEEE Transactions on Smart Grid,* vol. 2, pp. 659–666, 2011.

[8] A. H. Mohsenian-Rad and A. Leon-Garcia, "Distributed internet-based load altering attacks against smart power grids," *IEEE Transactions on Smart Grid*, vol. 2, pp. 667–674, 2011.

[9] H. Yi, M. Esmalifalak, N. Huy, *et al.*, "Bad data injection in smart grid: Attack and defense mechanisms," *IEEE Communications Magazine*, vol. 51, pp. 27–33, 2013.

[10] T. T. Kim and H. V. Poor, "Strategic protection against data injection attacks on power grids," *IEEE Transactions on Smart Grid*, vol. 2, pp. 326–333, 2011.

[11] S. Cui, Z. Han, S. Kar, T. T. Kim, H. V. Poor, and A. Tajer, "Coordinated data-injection attack and detection in the smart grid: A detailed look at enriching detection solutions," *IEEE Signal Processing Magazine,* vol. 29, pp. 106–115, 2012.

[12]   M. Ozay, I. Esnaola, F. Vural, S. R. Kulkarni, and H. V. Poor, "Sparse attack construction and state estimation in the smart grid: Centralized and distributed models," *IEEE Journal on Selected Areas in Communications*, vol. 31, pp. 1306–1318, 2013.

[13]   H. Fawzi, P. Tabuada, and S. Diggavi, "Secure estimation and control for cyber-physical systems under adversarial attacks," *IEEE Transactions on Automatic Control*, vol. 59, pp. 1454–1467, 2014.

[14]   Y. Yuan, Z. Li, and K. Ren, "Modeling load redistribution attacks in power systems," *IEEE Transactions on Smart Grid*, vol. 2, pp. 382–390, 2011.

[15]   C. Rehtanz, J. Béland, G. Benmouyal, *et al.*, "Wide area monitoring and control for transmission capability enhancement," *CIGRE Technical Brochure*, 2007.

[16]   A. Monticelli, "Electric power system state estimation," *Proceedings of the IEEE*, vol. 88, pp. 262–282, 2000.

[17]   N. Shivakumar and A. Jain, "A review of power system dynamic state estimation techniques," in *Joint International Conference on Power System Technology and IEEE Power India Conference, 2008 (POWERCON 2008)*, 2008, pp. 1–6.

[18]   W.-g. Li, J. Li, A. Gao, and J.-h. Yang, "Review and research trends on state estimation of electrical power systems," in *Power and Energy Engineering Conference (APPEEC), 2011 Asia-Pacific*, 2011, pp. 1–4.

[19]   A. Leite da Silva and D. Falcao, "Bibliography on power system state estimation (1968–1989)," *IEEE Transactions on Power Systems*, vol. 5, pp. 950–961, 1990.

[20]   A. Gómez-Expósito, A. de la Villa Jaén, C. Gómez-Quiles, P. Rousseaux, and T. Van Cutsem, "A taxonomy of multi-area state estimation methods," *Electric Power Systems Research*, vol. 81, pp. 1060–1069, 2011.

[21]   Y.-F. Huang, S. Werner, J. Huang, N. Kashyap, and V. Gupta, "State estimation in electric power grids: Meeting new challenges presented by the requirements of the future grid," *Signal Processing Magazine, IEEE*, vol. 29, pp. 33–43, 2012.

[22]   A. Simoes-Costa and V. Quintana, "An orthogonal row processing algorithm for power system sequential state estimation," *IEEE Transactions on Power Apparatus and Systems*, vol. PAS-100, pp. 3791–3800, 1981.

[23]   A. Abur and M. K. Celik, "Least absolute value state estimation with equality and inequality constraints," *IEEE Transactions on Power Systems*, vol. 8, pp. 680–686, 1993.

[24]   M. K. Celik and A. Abur, "A robust WLAV state estimator using transformations," *IEEE Transactions on Power Systems*, vol. 7, pp. 106–113, 1992.

[25]   A. S. Debs and R. Larson, "A dynamic estimator for tracking the state of a power system," *IEEE Transactions on Power Apparatus and Systems*, vol. PAS-89, pp. 1670–1678, 1970.

[26]   D. Falcao, P. Cooke, and A. Brameller, "Power system tracking state estimation and bad data processing," *IEEE Transactions on Power Apparatus and Systems*, pp. 325–333, 1982.

[27] A. Garcia, A. Monticelli, and P. Abreu, "Fast decoupled state estimation and bad data processing," *IEEE Transactions on Power Apparatus and Systems*, vol. PAS-98, pp. 1645–1652, 1979.

[28] A. Monticelli, "Fast decoupled state estimator," in *State Estimation in Electric Power Systems*, Springer, 1999, pp. 313–342.

[29] M. C. de Almeida, A. V. Garcia, and E. N. Asada, "Regularized least squares power system state estimation," *IEEE Transactions on Power Systems*, vol. 27, pp. 290–297, 2012.

[30] T. Zhai, H. Ruan, and E. E. Yaz, "Performance evaluation of extended Kalman filter based state estimation for first order nonlinear dynamic systems," in *42nd IEEE Conference on Decision and Control, 2003*, 2003, pp. 1386–1391.

[31] G. Valverde and V. Terzija, "Unscented Kalman filter for power system dynamic state estimation," *IET Generation, Transmission & Distribution*, vol. 5, pp. 29–37, 2011.

[32] M. Brown Do Coutto Filho, and J. S. de Souza, "Forecasting-aided state estimation—Part I: Panorama," *IEEE Transactions on Power Systems*, vol. 24, pp. 1667–1677, 2009.

[33] A. Sinha and J. Mondal, "Dynamic state estimator using ANN based bus load prediction," *IEEE Transactions on Power Systems*, vol. 14, pp. 1219–1225, 1999.

[34] J.-M. Lin, S.-J. Huang, and K.-R. Shih, "Application of sliding surface-enhanced fuzzy control for dynamic state estimation of a power system," *IEEE Transactions on Power Systems*, vol. 18, pp. 570–577, 2003.

[35] E. Handschin, F. Schweppe, J. Kohlas, and A. Fiechter, "Bad data analysis for power system state estimation," *IEEE Transactions on Power Apparatus and Systems*, vol. 94, pp. 329–337, 1975.

[36] A. Abur and A. G. Exposito, *Power System State Estimation: Theory and Implementation*, CRC Press, 2004.

[37] H. M. Merrill and F. C. Schweppe, "Bad data suppression in power system static state estimation," *IEEE Transactions on Power Apparatus and Systems*, pp. 2718–2725, 1971.

[38] J. Chen and A. Abur, "Placement of PMUs to enable bad data detection in state estimation," *IEEE Transactions on Power Systems*, vol. 21, pp. 1608–1615, 2006.

[39] T. Van Cutsem, M. Ribbens-Pavella, and L. Mili, "Hypothesis testing identification: A new method for bad data analysis in power system state estimation," *IEEE Transactions on Power Apparatus and Systems*, vol. PAS-103, pp. 3239–3252, 1984.

[40] C. Lu, B. Shi, X. Wu, and H. Sun, "Advancing China's smart grid: Phasor measurement units in a wide-area management system," *IEEE Power and Energy Magazine*, vol. 13, pp. 60–71, 2015.

[41] W. Sattinger and G. Giannuzzi, "Monitoring continental Europe: An overview of WAM systems used in Italy and Switzerland," *IEEE Power and Energy Magazine*, vol. 13, pp. 41–48, 2015.

[42] V. Madani, J. Giri, D. Kosterev, D. Novosel, and D. Brancaccio, "Challenging changing landscapes: Implementing synchrophasor technology in grid operations in the WECC region," *IEEE Power and Energy Magazine*, vol. 13, pp. 18–28, 2015.

[43] Y. Zhang and J. Jiang, "Bibliographical review on reconfigurable fault-tolerant control systems," *Annual Reviews in Control*, vol. 32, pp. 229–252, 2008.

[44] M. Blanke and J. Schröder, *Diagnosis and Fault-Tolerant Control*, vol. 2, Springer, 2006.

[45] C. Hajiyev and F. Caliskan, *Fault Diagnosis and Reconfiguration in Flight Control Systems*, vol. 2, Springer, 2013.

[46] F. García-Nocetti, *Reconfigurable Distributed Control*, Springer, 2005.

[47] F. Skopik and P. D. Smith, *Smart Grid Security: Innovative Solutions for a Modernized Grid*, Elsevier Science, 2015.

[48] J. F. Hauer, "Application of Prony analysis to the determination of modal content and equivalent models for measured power system response," *IEEE Transactions on Power Systems*, vol. 6, pp. 1062–1068, 1991.

Chapter 7

Optimal energy management in the smart grid

Eleonora Riva Sanseverino[1], Maria Luisa Di Silvestre[1], Gaetano Zizzo[1], Ninh Nguyen Quang[2], Adriana Carolina Luna Hernandez[3] and Josep M. Guerrero[3]

In this chapter, the problem of energy management in smart grids is outlined. Optimized energy management is considered here as the operation of energy and power flow control in the aim of attaining minimum cost or minimum power losses while meeting technical constraints. Of course, according to the type of energy system in which such operation is carried out, the meaningful variables and objectives in the problem may largely change. As the extension of the system increases, the influence of the physical behaviour of the electrical power lines takes a more important role. Power electronics takes instead an increasing influence, as the dimension of the power system decreases although Kirchhoff's laws must always be verified. Energy management is also an interesting issue in multi-carrier energy hubs that implement a recent tendency in energy systems interfacing smart grids. They include different energy resources that need to be managed differently, but in an integrated fashion. This chapter offers an overview of energy management in a small system using a two-level centralized approach.

7.1 Introduction

Environmental impact, economic operation, and homegrown are the main pillars over which sustainability in energy systems is founded. These three factors affect differently various energy systems based on size and final use of energy. For this reason, sustainable operation and management may radically change looking at the same problem in different contexts.

It is well known that the word 'optimization' refers to the possibility to attain the maximum benefit at the minimum cost. For this reason, it can be well referred to the concept of sustainability. As recently claimed by Jeffrey Sachs [1], '... we will need all energy sources that meet three conditions: homegrown (for national

[1]Department DEIM, University of Palermo, Bldg. 9 Viale delle Scienze, 90128 Palermo, Italy
[2]Institute of Energy Science, Hanoi, Vietnam
[3]Department of Energy Technology, Aalborg University, Aalborg, Denmark

security), low-cost (for competitiveness) and environmentally safe. With improved technologies, there is a place for fossil fuels, renewables, and nuclear in the mix...'. Therefore, sustainability for an energy system refers to the possibility to optimally (minimizing expenses, environmental impact and maximizing energy security) mix different energy sources to cover energy needs. On the other hand, the definition of energy system is quite large, although for smart electrical grids it usually refers to the mixture of generation technologies and energy resources interfacing the electrical power grid. Such interfacing can take place at a transmission or distribution level. At the transmission level, the inception of *storage technology* [2–5] and *industrial multi-carrier energy hubs* [6,7] poses interesting questions on their optimal management.

Storage technology is typically used at the transmission level to provide regulation services or energy cost mitigation, although, in general, it is considered as an enabling technology, such as Information and Communication Technology (ICT), for smart-grid implementation. Based on the purpose of the energy storage management, it can be carried out in different time frames ranging from milliseconds to one hour.

1. Short-time applications (from milliseconds to seconds) are devoted indeed to rapid spinning reserve implementation, primary control of frequency, ride through capability and power quality. These applications require high power density batteries and delivering short bursts of high power [6].
2. Applications delivered in few seconds to minutes such as secondary and tertiary frequency regulation ancillary services, smoothing of output wind power, demand-side applications, active and reactive power control, harmonic compensation and black start capability. In these cases, limited power and energy density batteries are needed. Energy needs to be stored for a small amount of time.
3. Applications delivered in minutes up to few hours. In these cases, utility side applications can be implemented, such as market imbalance, arbitrage, load balancing and peak load shaving (from 2 to 4 h duration). In this case, the reliability of systems with a large amount of renewable energy, isolated systems and micro-grids can be improved and support for transport and automobile feeding can be provided and new generation and transmission infrastructures construction can be delayed. High energy density batteries are needed for these applications [5].
4. Long-term/multi-MWh applications (days) avoid new generation and transmission construction cost. These applications require very high energy density batteries.

In most of the above cases, strategies rely on forecasts about renewable energy availability, price fluctuations and demands for cost–benefit optimization [8].

The concept of multi-carrier energy hub is relatively recent. The first study about this new technology dates back to 2007 by Geidl [7]. These have given a formal definition of a multi-carrier energy hub as follows [9]: 'unit where multiple energy carriers can be converted, conditioned and stored; such a system represents

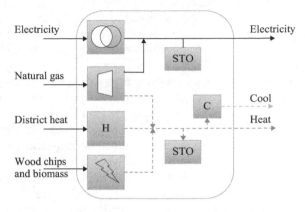

Figure 7.1   Multicarrier energy hub structure

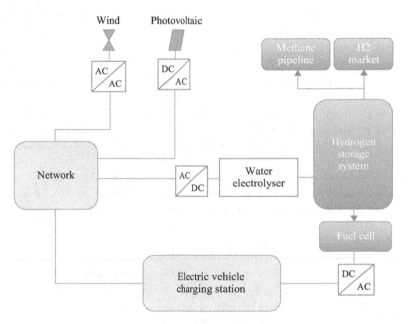

Figure 7.2   Multicarrier industrial energy hub structure

an interface between different energy infrastructures and/or loads'. In the same work, a typical structure of multi-carrier hub is shown, along with conversion, storage and transmission devices (Figure 7.1).

Industrial energy hubs, such as those considered in [10,11] developed in the frame of the FP7 EU Ingrid project, employ also hydrogen and fuel cells as conversion units. Electrical energy is used for electrolysis. Figure 7.2 shows the particular industrial energy hub considered in these papers. Minimum cost is achieved by storing, selling hydrogen or using it again to create electricity.

At the *distribution* level, energy management optimization is devoted to managing *energy storage systems and other resources as distributed generation units and flexible loads* to achieve minimum operating cost, minimum CO_2 emissions or maximum efficiency of distribution systems.

Storage is thus again a crucial enabling technology because most of the distributed generation employs renewable energy generation whose production peak does not typically match the peak load. At the distribution level, however, storage also allows frequency and voltage regulation especially under islanded operation of portions of distribution grid. Such portions of distribution systems, when hosting a large amount of renewable energy and provide flexible operation, are also referred to as *micro-grids* (MGs). The latter are defined from the Department of Energy (US) as 'a set of interconnected loads and distributed energy resources within clearly defined boundaries that acts as a single controllable entity with respect to the grid. A grid can connect and disconnect from the grid to operate it both in grid-connected or islanded mode'.

7.2   Micro-grid turns to smart grid

MGs technology is particularly relevant in islanded areas or in modern distribution networks to increase reliability and reduce operating costs by flexible disconnection from the main grid. In this field, the use of energy management systems is quite relevant and aims at managing efficiently loads, generation and storage units at different time scales and for different purposes, similarly as it was described in Section 7.1.

Different from standard power systems, in MGs, the inertia of generation systems is quite limited as most of the generation sources are interfaced by means of conversion units. For this reason, energy storage is an enabling technology as it allows one to support frequency deviation as it is shown in Reference 12, where this criterion is also used for sizing purposes.

In what follows, a summarized review of the state of the art about energy management systems for MGs is reported [13], as these systems are quite challenging in terms of complexity of their architecture (Figure 7.3). The energy management system (EMS), also called micro-grid central controller (MGCC), schedules the energy production of dispatchable generators and the energy delivery or storage for energy storage systems (ESS), as well as ideal switching time of flexible loads (FLEX LOADS). Wind and photovoltaic generation is not dispatchable and will inject all the power produced.

In MGs, hierarchical control architecture is adopted. The EMS occupies the highest layer [14–18]. At the lowest level, the field layer integrates: flexible loads; distributed energy resources (DER), which can be renewable energy sources (RES), storage systems or conventional generators; power converters (PC), which act as an interface between DER and the grid; grid components, such as transformers, switchgears and lines. The local field control layer operates based on the instructions given from higher control levels and includes basic functions such as local

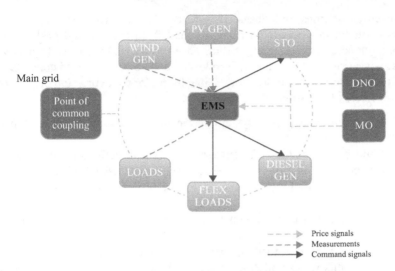

Figure 7.3   Micro-grid structure and energy resources

generation control, storage management and local consumption management. The aim of energy management is also to interface the MG with the energy market main actors (distribution network operator [DNO] and market operator [MO]). Locally, it may provide power quality control functions as well as identify optimized set points and status of the field components. As it is clear from this picture, ICT is of great relevance for enabling all of the EMS functions and MG technology in general.

7.2.1   Hierarchical control for micro-grids

MGs management includes different tasks at different time scales. In order to properly handle these technical issues, a hierarchical three levels control scheme has been recently proposed and accepted as an efficient solution for MGs operation [14,17–19]:

1. *Primary control layer.* The droop-control method is often used in this level to emulate physical behaviours that make the system stable and more damped. It is normally centred around the optimized set points and promptly (within milliseconds) provides new operating points by means of interfacing PCs droop control law.
2. *Secondary control level.* It ensures that voltage and frequency into the MG are restored at the rated values. In addition, such a control level can include a synchronization control loop to seamlessly connect or disconnect the MG to or from the distribution system. Distributed algorithms are quite interesting approaches at this level. It can be deployed in times ranging up to seconds.
3. *Tertiary control level.* The mission of tertiary control is the implementation of an intelligent EMS in the MG. At this level, the EMS tries to optimize its operation mostly based on economic issues, but also accounting on technical

capacities of lines and buses, if required. These objectives can be attained based on information from both MG and from the external grid. Decision-making and optimization algorithms are employed at this stage. Typically, this stage is carried out at a centralized level, but also decentralized approaches are of recent interest in the literature. Time frame for the expression of this control level ranges from minutes to hours.

Also, conventional power systems are managed based on a three-level hierarchical control [20].

It should be noted that different terminology and definitions of the different hierarchical control functions have been recently discussed [19]. Sometimes tertiary control is also employed for the coordination of different MGs, as defined in [14,17,18].

In conventional structure, both tertiary and secondary control can be implemented in the EMS. However, dynamics of the different control levels can be considered decoupled due to their different execution times (at least an order of magnitude). In this way, modelling and stability analysis for MG systems turns to be largely simplified.

7.2.2   Centralized and decentralized management

Functions of the EMS module can, however, be implemented with both a centralized or decentralized architecture. The level of intelligence of local controllers installed at the field level on generation, storage and load units influences the extent to which decentralization can be carried out. Local controllers can either execute commands from upper layers or take local decisions. Both architectures show positive or negative aspects depending on the MG use (supplying residential, commercial or military users), the legal framework and physical characteristics (size, location, ownership and topology).

Centralized EMS [21–25] requires data collection from the market and the distributor as well as from the MG field layer. Such data are the basis for deploying optimized centralized scheduling ensuring real-time observability of the system, data security and straightforward implementation. On the other hand, the EMS module needs to be computationally powerful in order to process a large amount of data while taking proper decisions. Large bandwidth communication is, however, needed to allow timely information exchange although such centralized architecture does not seem to be resilient to failure events. Limited expandability is another critical limit. Centralized EMS architecture is thus more suitable in the following cases: (1) small-scale MGs where gathering centralized information and decision-making can be implemented at low computation cost; (2) when all parties within the MG have a unique goal and the EMS can operate the MG from a single perspective; (3) MGs in which privacy is a priority (i.e. military applications) [21–25].

Decentralized architectures instead seem suitable when (1) the MG is large (tens of nodes) or generation, consumption and storage units are largely dispersed; (2) the energy resources are owned by different parties having different interests and require local decision-making (DM) schemes; (3) fast reconfigurations and frequent modifications of the systems, such as adding or removing existing nodes, are required and (4) higher reliability of the system is needed [14,26,27]. Efficient wireless communication protocols [28] as well as distributed algorithms [29] are enabling technologies for distributed control.

Multi-agent system (MAS) for distributed energy management has recently become an important research field. It can have the same hierarchy and functions as centralized ones, but transferring the DM authority to local side by increasing the intelligence level of load controllers (LCs).

7.2.3   *Energy management system and optimal power flow*

As it was recalled above, the EMS may integrate an optimal power flow (OPF) module to assess technical feasibility of optimally economic solutions. It is indeed well known that while smaller systems may not account for efficiency of power flow transfer from sources to loads, in larger systems, this issue cannot be neglected. Moreover, lines ampacity and voltage drops must be kept within limits and often cost-based energy resources dispatch does not account for such issues. Therefore, in general, power systems will require an OPF module to operate and verify technical feasibility of solutions accounting for lines ampacity and voltage drops. While the cost-based energy management issue can be formulated as a linear problem, OPF is highly nonlinear and its solution, especially under an islanded condition, is not trivial. In this chapter, the technical-economic optimal solution is provided in two steps deployed in two different time frames.

The first, carried out every hour, is the assessment of the economic optimum by means of a mixed integer linear programming approach; the second, carried out every few minutes, is the identification of a technically feasible solution through the OPF solution in the neighbourhood of the economic optimum, see Figure 7.4. As the figure shows, the OPF provides optimized set points to fast ramping units, type II and type III and type I generators with ramping times that are compatible with the elementary time interval chosen for this regulation level (ΔT).

Regarding generation units, a broad classification can be done referring to the dynamic of the generation units in [30,31]:

1. *Type I generators*. It provide the base load; these have slow ramping times, typically tens of minutes (base units). They can be scheduled hourly every 24 h or every ΔT with a time compatible with ramping limits.
2. *Type II generators*. It provide the regulation service to match generation and demand (fluctuations); these have fast ramping times, typically seconds (peaking units). They share the amount of power that can vary within the elementary time interval of 10–15 min.
3. *Type III generators*. It provide both services; these have fast ramping times, typically seconds. They can both supply the base load and the regulating service contributing to regulation.

At this point, it should be evidenced that other technical-economic approaches exist in the literature of power systems and that they provide solutions to the

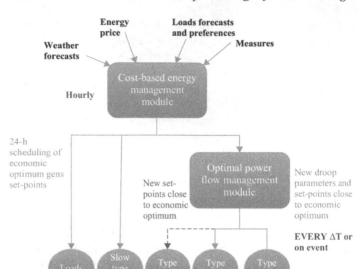

Figure 7.4   Energy management system structure for technical-economic optimization

problem using complex and time-consuming nonlinear optimization approaches. The proposed framework for energy management in smart MGs allows accurate identification of operational set points accounting for possible forecasting error through an execution, monitoring and replanning approach [32] in the energy management module and technical feasibility of proposed solutions by means of the OPF module.

7.3   Energy management module in smart MGs

A growing interest on the optimal management of smart grids has been oriented to be able to efficiently deliver sustainable, economic and secure electricity supplies due to the increasing complexity of medium- and low-voltage distribution even at a family consumption level [14,33].

Particularly, the micro-grid, as part of the smart-grid, includes a variety of DER units (distributed generation and distributed storage units) and different types of end energy users in both grid-connected and islanded mode. These systems require an energy and power management strategy in order to optimally supply the loads in short-term power balancing and long-term energy management requirements of generation and demand-side strategies [14]. The energy management module, however, will deploy scheduling for all energy resources.

The energy management level can provide directly the energy references for type I generators since they cannot be controlled every few minutes. For the other types of generation units, OPF should be performed as well to set feasible references for the MG operation.

Regarding demand-side management, some loads can be managed according to economic criteria. The unmanageable loads are considered as parameters, whereas the manageable loads are considered variables of the problem. In this approach, these are essentially *deferrable loads* that are scheduled according to the preferred time defined by the user.

7.3.1 Problem description and general formulation

The optimization model is defined to schedule reference profiles for both generation and consumption side in an MG. In this way, the optimization problem aims to minimize the operating cost by setting the energy provided by each generator, the energy of the controllable loads and estimating the level of energy in the energy storage systems (ESSs).

The model uses the following indexes:

- $h = \{1,2,\ldots,T\}$ time slots
- $g = \{1,\ldots,ng\}$ generators
- $i = \{1,\ldots,nl\}$ load
- $a = \{1,\ldots,na\}$ manageable appliances in each load

7.3.2 Variables definition

In the optimization problem, the variables are presented in lowercase and are defined in terms of the presented indexes as

$eg^{(g)}(h)$ is the energy provided by generator g at each time slot h.

$e_{sload}^{(i)(a)}(h)$ is the shiftable part of the i-th load at each time slot h.

$soc^{(g)}(h)$ is the state of charge of the storage of active generators at each time slot h.

$p_{ESS}^{(g)}(h)$ is the power that can be charged or discharged at the ESS at each time slot h.

$x_{load}^{(i),(a)}(h)$ are the ON/OFF commands for the shifted energy of the i-th load at each time slot h.

$x_{ON}^{(i),(a)}(h)$ are the ON commands for the shifted energy of the i-th load at each time slot h.

$x_{OFF}^{(i),(a)}(h)$ are the OFF commands for the shifted energy of the i-th load at each time slot h.

The model provides reference set points in each time slot for all the generators. However, for generators of types 2 and 3, those references can be set also with a higher frequency to adjust power flow and provide optimized feasible operational references.

7.3.3   Objective function

The objective function minimizes the operating costs and, at the same time, considers the customer preferences. It can be defined as

$$
cost = \sum_{h=1}^{T}\sum_{g=1}^{n_g} Cg^{(g)} * eg^{(g)}(h) + \sum_{h=1}^{T}\sum_{i}^{nl}\sum_{a=1}^{na} C_{load}{}^{(a)} * e_{sload}^{(i)(a)}(h)
$$

$$
* \left(1 - X_{prefer}{}^{(i)(a)}(h)\right) \tag{7.1}
$$

In the first term, related to the cost associated with the generation, $Cg^{(g)}$ is the unitary cost of energy supplied by the generator g and $eg^{(g)}(h)$ is the energy generated by generator g at each h.

The second term in (7.1) is a penalization for scheduling the appliances outside the hours of customer preference. $e_{load}^{(i)(a)}(h)$ is the scheduled shiftable energy profile for the appliances in the i-th load and $C_{load}{}^{(a)}$ is the penalty coefficient associated with activate the appliances at an inconvenient time. Accordingly, the binary parameter $X_{prefer}{}^{(i)(a)}(h)$ has been included to define the desired time for using the appliance a that belongs to the load i for each time slot h, and is equal to 1 if the users prefer to use the appliances at that time.

7.3.4   Constraints definition

7.3.4.1   Energy balance

The energy balance between generation and demand should be fulfilled, and can written as

$$
\sum_{g=1}^{n_g} e_g^{(g)}(h) = \sum_{i=1}^{n_i}\left\{ E_{nsload}{}^{(i)}(h) + \sum_{a=1}^{na} e_{sload}^{(i)(a)}(h)\right\}, \quad \forall h \tag{7.2}
$$

where the term on the left side is the aggregated energy provided by the n_g generation units at each time slot h. The term on the right side corresponds to the aggregated consumption of the n_i loads. These loads are composed by a non-shiftable parameter defined as $E_{nsload}{}^{(i)}(h)$, and a controllable part defined as the sum of the shifted appliances $e_{sload}^{(i)(a)}(h)$, defined previously.

7.3.4.2   RES-based active generation

Some of the generation units of the micro-grid can be active generators integrated by an RES and an ESS [2]. These generators are type III. Defining the index $\{a_g = \{1,\ldots,n_{ag}\} \in g | n_{ag} < n_g\}$, the energy of these generators can be expressed as

$$
e_g^{(g)}(h) = E_{RES}^{(g)}(h) + p_{ESS}^{(g)}(h) * \Delta h, \qquad \forall g \in a_g, h \tag{7.3}
$$

where $E_{RES}^{(g)}(h)$ is the RES energy profile along the time horizon and $p_{ESS}^{(g)}(h) * \Delta h$ is the energy of the ESS estimated by the optimization model for the n_{ag} active generators.

The available energy in the ESS can be defined by means of its state of charge, which is an estimation of this energy depending on its capacity presented in percentage, and can be written as

$$soc^{(g)}(h) = soc^{(g)}(h-1) - \phi^{(g)} * p_{ESS}^{(g)}(h) * \Delta h, \qquad \forall g \in a_g, h \qquad (7.4)$$

where $\phi^{(g)}$ is a storage coefficient that depends on the technology of the ESS. For battery-based storage, this term is inversely proportional of the nominal capacity in kWh.

Additionally, it is needed to include a constraint on the global balance of the storage system in order to ensure at least the same condition of the current day in the next day. This condition can be defined as

$$\sum_{h=1}^{T} \left(soc^{(g)}(h) - soc^{(g)}(h-1) \right) \geq 0, \qquad \forall g \in a_g, h \qquad (7.5)$$

If the MG includes independent ESSs, which are not part of active generators, the constraints (7.4) and (7.5) should be included in the model as well as the energy contribution of the ESS should be added in energy balance equation (7.2).

7.3.4.3 Constraints related to the shiftable load

The shiftable loads are indeed appliances that are turned on/off at convenient times to reduce the cost. In this way, the energy of the appliances can be written as

$$e_{sload}^{(i),(a)}(h) = P_{app}(a) * \Delta h * x_{load}^{(i),(a)}(h), \qquad \forall i, a, h \qquad (7.6)$$

where $P_{app}(a) * \Delta h$ is the average energy used by the appliance at each time slot and $x_{load}^{(i),(a)}(h)$ is the on/off command of the appliance.

Besides, the appliances should be used during predefined time, which can be written as

$$\sum_{h=1}^{T} x_{load}^{(i),(a)}(h) * \Delta h = duration^{(a)}, \qquad \forall i, a, h \qquad (7.7)$$

where $duration^{(a)}$ is a parameter that includes the operation time of the appliances. But, most of these appliances have continuous operation and it is important to ensure that the devices are not turned off until they finish their operation. Therefore, the binary variables $x_{ON}^{(i),(a)}(h)$ and $x_{OFF}^{(i),(a)}(h)$ are defined to establish when the appliance is started and stopped, respectively, and consequently, they have to fulfil the conditions

$$x_{ON}^{(i),(a)}(h) \geq x_{ON}^{(i),(a)}(h-1), \qquad \forall i, a, h > 1 \qquad (7.8)$$

$$x_{OFF}^{(i),(a)}(h) \geq x_{OFF}^{(i),(a)}(h-1), \qquad \forall i, a, h > 1 \qquad (7.9)$$

The relationship of these variables with $x_{load}^{(i),(a)}(h)$ can be defined as

$$x_{load}^{(i),(a)}(h) = x_{ON}^{(i),(a)}(h) - x_{OFF}^{(i),(a)}(h), \qquad \forall i, a, h$$

$$x_{load}^{(i),(a)}(h) \geq 0, \qquad \forall i, a, h$$

(7.10)

7.3.4.4   Variables boundaries

In general, all the variables of an optimization model should be properly delimitated in order to define a feasible region where the optimal solution can be found. Accordingly, the energy provided by the generation units are bounded in the range

$$P_{min}^{(g)} * \Delta h \leq e_g^{(g)}(h) \leq P_{max}^{(g)} * \Delta h, \qquad \forall g, h$$

(7.11)

where $P_{min}^{(g)}$ and $P_{max}^{(g)}$ are the minimum and maximum power amounts that the generators can be supplied, respectively.

Regarding the ESSs in the micro-grid, the SoC and power are limited to be in the ranges

$$SoC_{min}^{(g)} \leq soc^{(g)}(h) \leq SoC_{max}^{(g)}, \qquad \forall g \in a_g, h$$

(7.12)

$$P_{ESS_{min}}^{(g)} \leq p_{ESS}^{(g)}(h) \leq P_{ESS_{max}}^{(g)}, \qquad \forall g \in a_g, h$$

(7.13)

where $SoC_{min}^{(g)}$ and $SoC_{max}^{(g)}$ are the minimum and maximum values of the SoC, whereas $P_{ESS_{min}}^{(g)}$ and $P_{ESS_{max}}^{(g)}$ are the boundaries of the ESS power, respectively.

Lastly, the sum of the non-shiftable load and the shiftable load should be limited as

$$Ens^{(i)}(h) + \sum_{a=1}^{na} e_{sload}^{(i)(a)}(h) \leq E_{load\,max}^{(i)}, \qquad \forall i, h$$

(7.14)

where $E_{load\,max}^{(i)}$ is a maximum load requirement.

7.3.5   *Global optimization methods for linear integer problems*

A mixed integer linear programming model is based on a linear formulation that can ensure to get a global optimal solution. However, as they include integer or binary variables, it becomes in a non-convex problem that must be solved by some kind of systematic search. In fact, the integer programming problems have many local optima and finding a global optimum to the problem requires a linear relaxation to prove that a particular solution dominates all the feasible points.

To solve these kinds of problems, exact methods have been defined such as cutting plane approaches based on polyhedral combinatory, enumeration techniques and relaxation and decomposition techniques. Additionally, some heuristics and meta-heuristic approaches, as well as to population-based evolutionary algorithms, have been designed to solve such a class of optimization problems [34]. These kinds of methods randomly generate candidate solutions which satisfy the integer constraints. The initial solutions are usually far from optimal, but these methods through recombination provide new candidate solutions, through methods, such as integer- or permutation-preserving mutation and crossover, that continue to satisfy the integer constraints, but may show better objective values. This process is repeated until a sufficiently 'good solution' is found. Generally, these methods are not able to 'prove global optimality' of the solution [35].

The method used in this case for solving the problem is the branch and bound approach. This method starts by finding the optimal solution to the 'relaxed' problem without the integer constraints (via standard linear or nonlinear optimization methods). If in this solution the decision variables with integer constraints have integer values, then no further work is required. If one or more integer variables have non-integral solutions, the branch and bound method chooses one such variable and 'branches', creating two new sub-problems where the value of that variable is more tightly constrained. These sub-problems are solved and the process is repeated, until a solution that satisfies all of the integer constraints is found [36].

In this case, the solver BONMIN has been chosen in GAMS, general algebraic modelling system, and it uses exact methods such as branch and bound to solve mixed integer linear programming problems.

7.4 Optimal power flow module in smart micro-grids

As energy management, OPF is essentially a tertiary-level optimized operation issue in electric power systems and has been for a long time a concern of many researchers. For this purpose, many optimization techniques have been used, both deterministic and heuristics-based, such as 'the steepest descent' method [37], particle swarm optimization method [38], fuzzy rules method [39], dynamic programming [40], global optimization with convex relaxation [41,42] and so forth.

In addition, optimization problems have been solved considering the presence of energy storage systems, which are critical in islanded MG systems [37,42–46].

In Reference 47, a methodology for unbalanced three-phase OPF for decision-making in a smart grid is presented.

In the above-mentioned research works, OPF for three-phase balanced and unbalanced MGs is formulated considering the real powers injected from generators as variables. However, to the best knowledge of the author, there is no study concerning OPF in islanded MGs where generated and consumed powers depend on frequency and voltage levels and operating frequency is constrained as well. Such a level of details is instead required as in islanded MG systems none of the generators can take the role of slack bus and the balance between generated and consumed power should be considered as strictly precise. Moreover, the generated power from inverter interfaced depends on the frequency through a linear dependency. Such linear dependency based on droop parameters can also be optimized.

In Reference 37, particle swarm optimization is used to choose the droop parameters and then perform the load flow analysis using the formulation seen in Reference 48. In the paper, however, the OPF is not dealt with the three-phase load flow formulation in which loads and generators depend on voltage and frequency.

In Reference 49, it is shown that with P/V droop control, the DG units that are located electrically far from the load centres automatically deliver a lower share of the power. This automatic power-sharing modification can lead to decreased line losses; therefore, the system shows an overall improved efficiency as compared to the methods focusing on perfect power sharing. Such a concept of unequal power sharing is developed in this paper, where droops are optimized based on global objectives such as power losses, the latter being an optimization objective that seems concurrent with dynamic stability of the system [50]. Many papers have indeed investigated the issue of voltage and frequency stability with reference to the droop parameters values as well as the set point value [51,52], but no work has been found about a theoretically convincing approach when micro-grids show a higher R/X ratio.

7.4.1   *Problem description and general formulation*

The OPF in this work is solved to minimize power losses and the general model for OPF calculation encompasses power lines, loads, generators, including their control loops such as droop characteristics. The problem is highly nonlinear. The optimization variables with this formulation are new droop parameters for primary regulation; moreover, the operating solution produces an iso-frequency working condition for all units with operating frequency within admissible ranges.

In Reference 50, it has been shown that the power losses term is connected to the droop parameters values and thus such choice influences the steady-state and dynamic operation of the micro-grids.

Let P_i denote the calculated three-phase real power injected into the micro-grid at bus i. The formulation to calculate P_i can be expressed as follows:

$$P_{i(Kg,Kd)} = \sum_{j=1}^{n_{br}} |V_i||V_j||Y_{ij}|cos(\theta_{ij} - \delta_i + \delta_j)$$

(7.15)

where

V_i and V_j are the voltages at bus i and bus j, depending on the linear droop parameters, Kg and Kd, at droop buses where inverter interfaced units are installed.

δ_i and δ_j are the phase angles of the voltages at bus i and bus j, depending on Kg and Kd at droop buses.

Y_{ij} is the admittance of branch ij
θ_{ij} is the phase angle of Y_{ij}
n_{br} is the number of branches connected into bus i.

So, the total real power loss of the system, namely the objective function (OF) of the OPF problem can be calculated as follows:

$$OF_{(Kg,Kd)} = P_{Loss} = \sum_{i=1}^{n_{bus}} \left(P_{i(Kg,Kd)} \right)$$

(7.16)

where n_{bus} is the number of buses in the system. The constraints, as it will be detailed later, concern line ampacities below rated limits, voltage drops below rated values, frequency between admissible bounds and limits on droop parameter variations.

7.4.2 Optimization variables

In this chapter, both for P–f droop generation units and for Q–V droop generation units, the optimization variables are the parameters of inverter interfaced units K_G and K_d

$$K_G = \left(K_{G1}, K_{G2} \dots, K_{Gn_g}\right) \qquad (7.17)$$

$$K_d = \left(K_{d1}, K_{d2} \dots, K_{dn_g}\right) \qquad (7.18)$$

where n_g is the number of generators. Therefore, the generated real and reactive powers P_{Gi} and Q_{Gi} of generator i are, respectively, expressed as linear functions of voltage and frequency displacements according to the terms (7.17) and (7.18).

7.4.3 Constraints

The optimal dispatch issue is that to find the set of droop parameters (K_{Gi}) and (K_{di}) and relevant operating frequency and bus voltages minimizing the function expressed in (7.16), subject to the constraint that generated power should equal total demands plus total power losses (P_{Loss}):

$$\sum_{i=1}^{n_{gr}} P_{Gri} = \sum_{i=1}^{n_d} P_{Li} + P_{Loss}$$
$$\sum_{i=1}^{n_{gr}} Q_{Gri} = \sum_{i=1}^{n_d} Q_{Li} + Q_{Loss} \qquad (7.19)$$

where P_{Gri} is the real power of generator i, P_{Li} is the real power of load bus i and n_d is the number of load buses.

The following inequality constraints, expressed as follows, must be met:

$$K_{Gimin} \leq K_{Gi} \leq K_{Gimax}, \quad i = 1 \text{ to } n_g \qquad (7.20)$$

$$K_{dimin} \leq K_{di} \leq K_{dimax}, \quad i = 1 \text{ to } n_{gr} \qquad (7.21)$$

$$\Delta f = f - f_0 \leq 0.02 \qquad (7.22)$$

$$I_{branchi} \leq I_{maxbranchi}, \quad i = 1 \text{ to } n_{br} \qquad (7.23)$$

where $K_{Gimin}, K_{Gimax}, K_{dimin}, K_{dimax}$, respectively, are the minimum and maximum values of the droop parameters for P–f and Q–V droop generators. Δf is the operating frequency deviation, $I_{branchi}$ is the current in the i-th branch and $I_{maxbranchi}$ is the maximum current in the i-th branch and n_{br} is the number of branches in the system.

7.4.4 Heuristic GSO-based method

The OF (7.16) for the considered OPF is highly nonlinear due to the nonlinear relation between power losses and generated power. The variables $(K_{Gi}$ and $K_{di})$ do not appear explicitly in the equation but they are linearly related to the generated power. For this reason, the use of classical nonlinear optimization methods seems to be inadequate due to the difficulty in including constraints and unbalanced loading conditions.

Moreover, when the OF is highly nonlinear, the search space is typically multimodal. Hence to analyse such a complex model it is required to search for a global optimum. The global optimization capability is important when dealing with complex nonlinear models and heuristics can be a suitable choice.

GSO [53] is a relatively recent heuristic method. Within GSO, agents are at first deployed randomly in the search space. Each agent in the swarm decides the movement direction using the strength of the signal picked up from its neighbouring agents. To some extent, this is similar to the luciferin's induced glow of a glow-worm which is used to attract mates or preys. The brighter the glow, the stronger the attraction.

And the best and brightest will be chosen in the end of iterations as the solution of problem. Therefore, the glow-worm metaphor is employed to represent the underlying principles of this particular optimization approach. In this chapter, this methodology solves the issue including constraints about maximum frequency deviation, line ampacity and voltage drop limits to select the solution in the code. Such constraints are used in the selection phase, giving preference to feasible solution by measuring with a penalty term the constraints violation.

Pseudocode of the GSO algorithm considering frequency, line ampacity and voltage drops constraints is shown in Figure 7.5. When selecting solutions for recombination, those showing frequency, voltage drops and branch currents out of bounds are still kept in the swarm, but are chosen with a lower probability. Fitness is indeed based on ranking of solutions to keep a stable selection pressure in the probabilistic choice of the target vector. Selection probability indeed depends on luciferin a quantity calculated for each agent which, in turn, depends on fitness and constraints. For more details, refer to Reference 53.

Another issue usually faced when applying GSO is the choice of the termination condition. It is difficult to know if the result attained at a given iteration is the best solution. To solve this issue, a number of iterations (n) are previously given as

Initialize Archive A
Repeat Until **Termination Condition**
        Do m times
                Step 1: deterministic **choice (selection)** *of the base vector*
                Step 2: probabilistic **choice (selection)** *of the target vector*
                (Roulette Wheel technique based on l(t)) considering frequency,
                voltage drop and line ampacity constraints
                        Step 3: **recombination**
                END m
                Step 4: create **new population** *(replace A)*
        END
        A = **archive**
        m = **archive size**

Figure 7.5   Pseudocode of GSO

termination condition. The same parameter n can be increased until results with no more improvements or negligible improvements are attained.

7.5   Case study

The proposed technical-economic optimization has been carried out on the 9-bus test system as depicted in Figure 7.6. For the case study, references for generators and shiftable loads are established by considering economic optimization and after that, the OPF is performed to include frequency constrained operational asset.

The case study is composed of one type III generator (DG2) and two type I generators (DG1 and DG3).

The type I generators are microturbines with rated power of 30 and 10 kW. The elementary production cost of each microturbine is 0.01 and 0.013 [€/kWh], respectively. The ramping time of each microturbine is set around 1 s/kW.

The generator DG2 is a photovoltaic (PV)-based active generator integrated by a storage system with a capacity of 40 kWh and a photovoltaic plant with a peak power of 40 kWp. The parameters related to the energy storage system are presented in Table 7.1. In the table, $SoC^{(DG2)}(0)$ is the state of charge of the electrical storage system at bus 2; $SoC_{max}^{(DG2)}, SoC_{min}^{(DG2)}$, respectively, are the maximum and

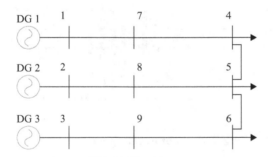

Figure 7.6   9-Bus test system

Table 7.1   Energy storage system electrical features

Parameter	Value
$SoC^{(DG2)}(0)$ (%)	60
$SoC_{max}^{(DG2)}$ (%)	100
$SoC_{min}^{(DG2)}$ (%)	20
$P_{ESS_{max}}^{(DG2)}$ (%)	20
$P_{ESS_{min}}^{(DG2)}$ (%)	−20
ϕ^{DG2} (%)	0.77

minimum percentages of the state of charge of the electrical storage system at bus 2. $P_{ESS_{max}}^{(DG2)}$, and $P_{ESS_{min}}^{(DG2)}$, respectively, are the maximum and minimum power injected and released by the ESS.

The storage coefficient that defined the relationship between energy and SoC can be established as

$$\phi^{DG2} = \frac{100\%}{Nominal\ Capacity\ [kWh]}$$

The PV energy profile along 24 h is reported in Figure 7.7.

Regarding the consumption, Load 1 and Load 2 at buses 4 and 5 are residential buildings partially manageable, with a peak power of 22.5 kW, while Load 3 is an office building with nonmanageable loads. The non-controllable part profile of each load is shown in Figure 7.8.

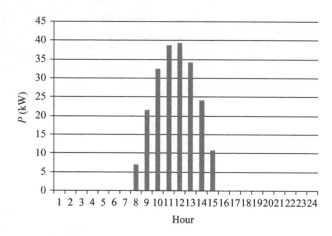

Figure 7.7   Photovoltaic generation profile

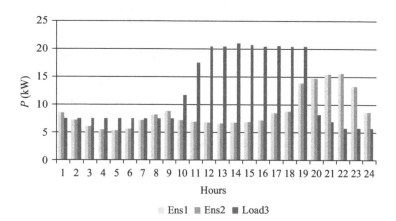

Figure 7.8   Noncontrollable part of loads

Both residential buildings have 30 flats each and the manageable load in each flat is composed of two appliances: a washing machine and a dishwasher. The operational parameters of each appliance and the preferred time of use are shown in Table 7.2.

Those parameters are included in the model implemented in the algebraic modelling language called GAMS and using a solver for mixed integer linear programming. The optimal value of the objective function obtained for a time span of 24 h is 7.36€.

The results of the scheduling for the generation units are presented in Table 7.3 and Figure 7.9.

Table 7.2   Manageable appliances features

Appliance	Duration (h)	Power (kW/h)	Preferred time
Washing machine	2	0.67	6:00–22:00
Dishwasher	2	0.65	6:00–22:00

Table 7.3   Energy management module output: optimal set points of each generator

Hour	Optimal generation references		
	DG1	DG2	DG3
1	10	4.7261	10
2	10	2.1547	10
3	10.2533	0	10
4	10.5647	0	10
5	10	0.3075	10
6	10	0.1519	10
7	10	5.4597	10
8	10	14.9069	10
9	10	20.0325	10
10	10	15.3481	10
11	10	23.09	10
12	10	23.7845	10
13	10	22.1875	10
14	10.12637	26.11643	10
15	30	6.73	10
16	30	3.3831	10
17	30	3.4361	10
18	30	5.1859	10
19	30	11.4605	10
20	30	7.2689	10
21	30	7.645	10
22	30	4.7583	10
23	22.4047	0	10
24	13.0333	0	10

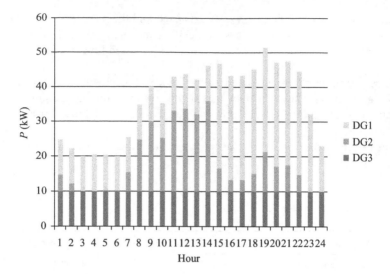

Figure 7.9   Generators set points outputted by the energy management module

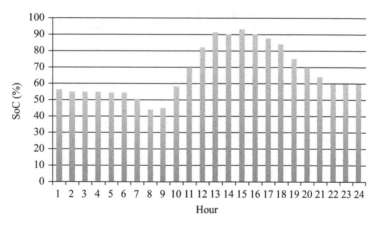

Figure 7.10   State of charge of the battery at bus 2

Generator DG3 is scheduled to provide the minimum predefined energy (10 kW) since it has a slightly higher operating cost per unit compared to Generator DG1. In the case of Generator DG2, most of its energy is provided by the PV generation but also part of this energy is exchanged with the battery. To illustrate the behaviour of the energy in the battery during the time horizon, the state of charge is presented in Figure 7.10.

Regarding the demand side, the obtained shiftable energy profile for each load is shown in Table 7.4. The complete scheduling of the appliances for each load is presented in Appendix A. Most of the manageable load is scheduled to the defined preferred time. This fact can be seen also in Figures 7.11 and 7.12, where the loads at the buses 4 and 5 are presented.

Table 7.4 *Energy management module output: optimal set points of each load*

Hour	Optimal shifted load	
	Load 1	Load 2
1	0	0
2	0	0
3	0	0.67
4	0.67	1.34
5	1.34	0.67
6	1.34	0
7	1.965	1.34
8	9.555	1.34
9	12.1	2.635
10	6.475	2.635
11	10.54	1.25
12	7.905	1.92
13	2.59	5.895
14	4.51	7.19
15	1.92	10.36
16	0	8.395
17	0	5.76
18	0.67	6.385
19	3.84	5.18
20	3.84	5.805
21	4.555	5.09
22	3.885	3.84
23	0	0
24	0	0

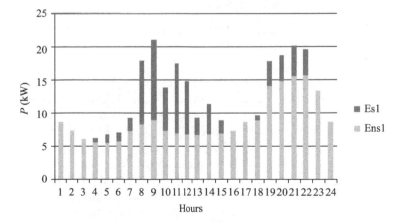

Figure 7.11 *Load at bus 4. Ens1 is the non-controllable load and Es1 is the controllable load*

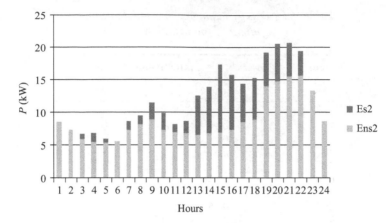

Figure 7.12   Load at bus 5. Ens2 is the non-controllable load and Es2 is the controllable load

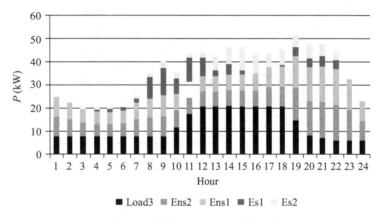

Figure 7.13   Aggregated load

The whole aggregated demand is shown in Figure 7.13. It includes the non-controllable part of Loads 1 and 2 (Ens1, Ens2), Load 3, and the results of the optimization for the shiftable loads Es1 and Es2. Comparing these results with the generation scheduled profiles (Figure 7.9), the majority of demand is shifted to the time when generator DG2 (the cheapest) provides more energy.

Most of the manageable load is scheduled to the defined preferred time.

After the energy management module, the GSO-based OPF has been carried out for the 9-bus system represented in Figure 7.6, the bus data and line data are shown in Tables 7.5 and 7.6. Based on the above results of the energy management module, a lower bound value, P_{min}, and an upper bound value, P_{max}, of the

Table 7.5   Bus data of 9-bus test system

Bus number	Type generator	Generator (pu)		Exponent of loads	
		VG0i	f0i	Alpha	Beta
1	PQ	–	–	0	0
2	Droop	1.07	1.07	0	0
3	PQ	–	–	0	0
4		1.07	1.07	2	2
5		1.07	1.07	2	2
6		1.07	1.07	2	2
7		1.07	1.07	2	2
8		1.07	1.07	2	2
9		1.07	1.07	2	2

Table 7.6   Lines data of 9-bus system (pu)

Bus nl	Bus nr	R	X	I_{max}
6	9	0.01288089	0.00084849	2.8
4	7	0.01288089	0.00084849	2.8
5	8	0.01288089	0.00084849	2.8
5	6	0.01173823	0.00030459	2.1
4	5	0.01173823	0.00030459	2.1
1	7	0.00692521	0.08702493	2.8
2	8	0.00692521	0.08702493	2.8
3	9	0.00692521	0.08702493	2.8

output power of each generator at each hour as generated power constraints have been assumed.

The P_{min} and P_{max} values of each generator at each hour are shown in Table 7.7. The following underlying hypotheses have been made: the base power and base voltage for per unit calculations have been set to $S_B = 10$ kVA and $V_B = 380$ V. Tables 7.5 and 7.6 show the electrical data of the network. The frequency and voltage dependency of the real and reactive power supplied to the loads are represented using an exponential model as described in Reference 54.

Using the proposed GSO method, we get the optimal generated power results of each generator as shown in Table 7.8 and Figure 7.14, P_{loss} is the power losses corresponding to each operating condition.

Figure 7.15 shows a comparison of the set points outputted by the two modules. As it can be observed, the OPF module produces set points that sometimes are above the economic optimum sometimes below. It is indeed unpredictable to what extent the dependency on the frequency of the model used for loads and generators will affect the equilibrium among generated power, loads and losses. In general,

Table 7.7 P_{max} and P_{min} of each generator (pu)

Hour	Max (kW)			Min (kW)		
	DG1	DG2	DG3	DG1	DG2	DG3
1	14.0000	12.0000	14.0000	10.0000	0.0000	10.0000
2	14.0000	12.0000	14.0000	10.0000	0.0000	10.0000
3	14.0000	12.0000	14.0000	10.0000	0.0000	10.0000
4	14.0000	12.0000	14.0000	10.0000	0.0000	10.0000
5	14.0000	12.0000	14.0000	10.0000	0.0000	10.0000
6	14.0000	12.0000	14.0000	10.0000	0.0000	10.0000
7	14.0000	12.0000	14.0000	10.0000	0.0000	10.0000
8	14.0000	20.9069	14.0000	10.0000	8.9069	10.0000
9	14.0000	26.0325	14.0000	10.0000	14.0325	10.0000
10	14.0000	21.3481	14.0000	10.0000	9.3481	10.0000
11	14.0000	29.0900	14.0000	10.0000	17.0900	10.0000
12	14.0000	29.7845	14.0000	10.0000	17.7845	10.0000
13	14.0000	28.1875	14.0000	10.0000	16.1875	10.0000
14	14.0000	30.0000	14.0000	10.0000	14.1164	10.0000
15	30.0000	12.7300	14.0000	26.0000	0.7300	10.0000
16	30.0000	12.0000	14.0000	26.0000	0.0000	10.0000
17	30.0000	12.0000	14.0000	26.0000	0.0000	10.0000
18	30.0000	12.0000	14.0000	26.0000	0.0000	10.0000
19	30.0000	17.4605	14.0000	26.0000	5.4605	10.0000
20	30.0000	13.2689	14.0000	26.0000	1.2689	10.0000
21	30.0000	13.6450	14.0000	26.0000	1.6450	10.0000
22	30.0000	12.0000	14.0000	26.0000	0.0000	10.0000
23	24.4047	12.0000	14.0000	20.4047	0.0000	10.0000
24	15.0333	12.0000	14.0000	11.0333	0.0000	10.0000

*Table 7.8   OPF module output: optimal output power of
each generator*

Hour	Optimal output power (kW)			P_{loss} (kW)
	DG1	DG2	DG3	
1	10.0000	5.6400	10.0000	0.4460
2	10.0000	3.5120	10.0000	0.4010
3	10.0000	1.9160	10.0000	0.3840
4	10.0000	2.1850	10.0000	0.3860
5	10.0000	1.8890	10.0000	0.3800
6	10.0000	1.7930	10.0000	0.3770
7	10.0000	6.2440	10.0000	0.4630
8	11.1410	12.2380	10.0000	0.8350
9	12.7550	14.5620	10.0000	1.1160

(Continues)

Table 7.8　(Continued)

Hour	Optimal output power (kW)			P_{loss} (kW)
	DG1	DG2	DG3	
10	10.5550	13.0480	10.0000	0.8130
11	12.0240	17.1320	14.0000	1.3080
12	12.4450	18.0330	12.9920	1.3630
13	12.1020	16.5540	14.0000	1.2730
14	12.9420	17.7650	13.7880	1.4230
15	26.0000	7.7680	10.0000	2.1660
16	26.0000	6.6690	10.0000	2.1150
17	26.0000	6.6650	10.0000	2.0740
18	26.0000	7.3420	10.0000	2.0950
19	26.0000	9.8900	10.0000	2.0960
20	26.0000	7.8590	10.0000	1.8910
21	26.0000	8.0100	10.0000	1.8960
22	26.0000	6.4850	10.0000	1.8020
23	20.5000	2.0810	10.0000	1.1880
24	11.1000	3.2280	10.0000	0.4540

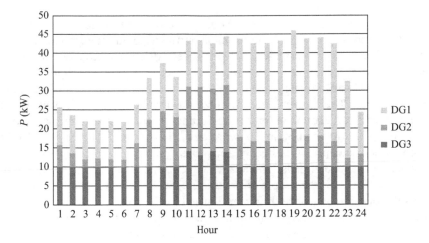

Figure 7.14　Solution of the OPF module

however, setting the generators at the economic optimum would produce the immediate operation of primary and secondary regulation to correct the frequency error produced by the unbalance, giving raise to an operational point which will not be, in general, at the predicted set point.

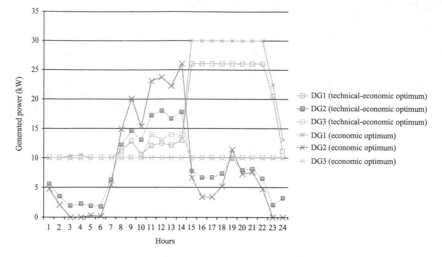

Figure 7.15   Comparison of economic optimum and technical optimum around the economic optimum

7.6   Conclusions

This chapter proposes an overview about the most recent topics in energy management in smart-grids. Then, the focus is restricted to an interesting application field concerning energy management in islanded MGs. For these systems, a two-level architecture for tertiary regulation in a particular type of smart-grid, the MG. These two levels are devoted to different time scales and solve different optimization problems. The higher level optimization, the energy management module, only takes care of economic issue, neglecting technical constraints as well as optimized operation minimizing power losses. It can be deployed in a 'rolling horizon' mode and provides hourly set points for energy storage systems, generators and manageable loads. The lower level optimization takes advantage from a precise modelling of the whole infrastructure between generation and loads. It models the dependency of both generation and loads from voltage and frequency, giving as a result a frequency-constrained operational asset, with minimized losses. The results show that economic optimum cannot account for technical constraints and by itself will give rise to the operation of lower level regulators that will produce, however, an operational asset with set points that are neither optimized economically nor technically. The inclusion of a heuristic-based OPF module easily solves the problem and produces reliable set points for optimized operation.

Appendix A

The results of the scheduled appliances washing machine (WM) and dishwasher (DW) in the flats (apt1 to apt30) for each load are presented in Tables A.1 and A.2.

Table A.1 Scheduled appliances in load 1

Flat	App	Hour																					
		1	2	3	4	5	6	7	8	9	10	11	12	13	14	15	18	19	20	21	22	23	24
apt1	WM						1	1															
	DW								1	1													
apt2	WM											1	1										
	DW																			1	1		
apt3	WM											1	1										
	DW										1	1											
apt4	WM											1	1										
	DW																	1	1				
apt5	WM											1	1										
	DW													1	1								
apt6	WM									1	1												
	DW										1	1											
apt7	WM												1	1									
	DW								1	1													
apt8	WM											1	1										
	DW													1	1								
apt9	WM																			1	1		
	DW																	1	1				
apt10	WM									1	1												
	DW								1	1													
apt11	WM								1	1													
	DW									1	1												
apt12	WM											1	1										
	DW								1	1													
apt13	WM								1	1													
	DW									1	1												
apt14	WM							1	1														
	DW								1	1													
apt15	WM				1	1																	
	DW									1	1												

(Continues)

Table A.1   (*Continued*)

Flat	App	Hour																					
		1	2	3	4	5	6	7	8	9	10	11	12	13	14	15	18	19	20	21	22	23	24
apt16	WM											1	1										
	DW								1	1													
apt17	WM													1	1								
	DW																			1	1		
apt18	WM																			1	1		
	DW								1	1													
apt19	WM																			1	1		
	DW								1	1													
apt20	WM														1	1							
	DW								1	1													
apt21	WM										1	1											
	DW											1	1										
apt22	WM											1	1										
	DW														1	1							
apt23	WM											1	1										
	DW														1	1							
apt24	WM								1	1													
	DW								1	1													
apt25	WM																	1	1				
	DW								1	1													
apt26	WM					1	1																
	DW																	1	1				
apt27	WM											1	1										
	DW											1	1										
apt28	WM																	1	1				
	DW							1	1														
apt29	WM											1	1										
	DW																	1	1				
apt30	WM																			1	1		
	DW																			1	1		

Table A.2   Scheduled appliances in load 2

		1	2	3	4	5	6	7	8	9	10	11	12	13	14	15	18	19	20	21	22	23	24
apt1	WM														1	1							
	DW												1	1									
apt2	WM																		1	1			
	DW																1	1					
apt3	WM										1	1											
	DW																1	1					
apt4	WM												1	1									
	DW																				1	1	
apt5	WM						1	1															
	DW													1	1								
apt6	WM																				1	1	
	DW																				1	1	
apt7	WM																1	1					
	DW												1	1									
apt8	WM								1	1													
	DW												1	1									
apt9	WM																				1	1	
	DW																				1	1	
apt10	WM														1	1							
	DW														1	1							
apt11	WM												1	1									
	DW										1	1											
apt12	WM																1	1					
	DW																		1	1			
apt13	WM												1	1									
	DW																1	1					
apt14	WM														1	1							
	DW														1	1							
apt15	WM																		1	1			
	DW																1	1					

(Continues)

Table A.2   (*Continued*)

		1	2	3	4	5	6	7	8	9	10	11	12	13	14	15	18	19	20	21	22	23	24
apt16	WM														1	1							
	DW																	1	1				
apt17	WM														1	1							
	DW								1	1													
apt18	WM				1	1																	
	DW																1	1					
apt19	WM								1	1													
	DW										1	1											
apt20	WM																			1	1		
	DW														1	1							
apt21	WM								1	1													
	DW																			1	1		
apt22	WM												1	1									
	DW														1	1							
apt23	WM					1	1																
	DW																1	1					
apt24	WM																			1	1		
	DW																	1	1				
apt25	WM														1	1							
	DW																				1	1	
apt26	WM												1	1									
	DW																			1	1		
apt27	WM																1	1					
	DW														1	1							
apt28	WM												1	1									
	DW														1	1							
apt29	WM												1	1									
	DW																			1	1		
apt30	WM	1			1																		1
	DW														1	1							

References

[1] Sachs, J.D.: "Jeffrey D. Sachs on how to forge a grand bargain on energy," *The Wall Street Journal*, December 31, 2015, available at http://jeffsachs. org/2015/12/jeffrey-d-sachs-on-how-to-forge-a-grand-bargain-on-energy/

[2] Kim, J.H., Powell, W.B.: "Optimal energy commitments with storage and intermittent supply", *Oper. Res.*, 2011, **59** (6), pp. 1347–1360.

[3] Harsha, P., Dahleh, M.A.: "Optimal sizing of energy storage for efficient integration of renewable energy", *Proceedings of the 50th IEEE Conference on Decision and Control and European Control Conference*, Orlando, FL, USA, December 2011, pp. 5813–5819.

[4] Zhou, Y., Scheller-Wolf, A., Secomandi, N.: "Managing wind-based electricity generation with storage and transmission capacity", 2011, Tepper Working Paper 2011-E36.

[5] Harsha, P., Dahleh, M.: "Optimal management and sizing of energy storage under dynamic pricing for the efficient integration of renewable energy", *IEEE Trans Power Syst.*, 2015, **30** (3), pp. 1164–1181.

[6] Fu, R., Wu, Y., Wang, H., Xie, J.: "A distributed control strategy for frequency regulation in smart grids based on the consensus protocol", *Energies*, 2015, **8**, pp. 7930–7944.

[7] Geidl, M., Koeppel, G., Favre-Perrod, P., *et al.*: "The energy hub – a powerful concept for future energy systems", *Proceedings of the Third Annual Carnegie Mellon Conference on the Electricity Industry*, Pittsburgh, PA, March 2007, pp. 1–10.

[8] Divya, K.C., Østergaard, J.: "Battery energy storage technology for power systems – an overview", *Electric Power Syst. Res.*, 2009, **79** (4), pp. 511–520.

[9] Krause, T., Andersson, G., Frohlich, K., Vaccaro, A.: "Multiple-energy carriers: modeling of production, delivery, and consumption," *Proc. IEEE*, 2011, **99** (1), pp. 15–27.

[10] Proietto, R., Arnone, D., Bertoncini, M., *et al.*: "A novel heuristics-based energy management system for a multi-carrier hub enriched with solid hydrogen storage", *Proceedings of the Fifth International Confernce on Future Energy Systems*, Cambridge, UK, June 2014, pp. 231–232.

[11] Proietto, R., Arnone, D., Bertoncini, M., *et al.*: "Mixed heuristic-non linear optimization of energy management for hydrogen storage-based multi carrier hubs", *Proceedings of the IEEE International Energy Conference (ENERGYCON)*, Cavtat, Croatia, May 2014, pp. 1019–1026.

[12] Toliyat, A., Kwasinski, A.: "Energy storage sizing for effective primary and secondary control of low-inertia microgrids", *Proceedings of the Sixth IEEE International Symposium on Power Electronics for Distributed Generation Systems (PEDG)*, Aachen, June 2015, pp. 1–7.

[13] Meng, L., Riva Sanseverino, E., Luna A.C., *et al.*: "Microgrid supervisory controllers and energy management systems: a literature review", *Renew. Sustain. Energy Rev.*, 2016, **60**, pp. 1263–1273.

[14] Katiraei, F., Iravani, R., Hatziargyriou, N., Dimeas, A.: "Microgrid management", *Proc. IEEE Power Energy Mag.*, 2008, **6** (3), pp. 54–65.

[15] De Brabandere, K., Vanthournout, K., Driesen, J., *et al.*: "Control of microgrids", *Proceedings of the IEEE Power Engineering Society General Meeting*, Tampa, FL, 2007, pp. 1–7.

[16] Piagi, P., Lasseter, R.H.: "Autonomous control of microgrids", *Proceedings of the IEEE Power Engineering Society General Meeting*, Montreal, Que, 2006, pp. 1–8.

[17] Bidram, A., Davoudi, A.: "Hierarchical structure of microgrids control system", *IEEE Trans. Smart Grid*, 2012, **3**, pp. 1963–1976.

[18] Guerrero, J.M., Vasquez, J.C., Matas, J., *et al.*: "Hierarchical control of droop-controlled ac and dc microgrids – a general approach toward standardization", *IEEE Trans. Ind. Electron.*, 2011, **58**, pp. 158–172.

[19] Olivares, D.E., Mehrizi-Sani, A., Etemadi, A.H., *et al.*: "Trends in microgrid control", *IEEE Trans. Smart Grid*, 2014, **5** (4), pp. 1905–1919.

[20] Kundur, P.: *Power System Stability and Control*. New York: McGraw Hill Education, 2006.

[21] Tsikalakis, A.G., Hatziargyriou, N.D.: "Centralized control for optimizing microgrids operation", *Proceedings of the IEEE Power Engineering Society General Meeting*, San Diego, CA, July 2011, pp. 1–8.

[22] Olivares, D.E., Canizares, C.A., Kazerani, M.A.: "A centralized optimal energy management system for microgrids", *Proceedings of the IEEE Power Engineering Society General Meeting*, San Diego, CA, July 2011, pp. 1–6.

[23] Vaccaro, A., Popov, M., Villacci, D., Terzija, V.: "An integrated framework for smart microgrids modeling, monitoring, control, communication, and verification", *Proc. IEEE*, 2011, **99**, pp. 119–132.

[24] Zamora, R., Srivastava, A.K.: "Controls for microgrids with storage: review, challenges, and research needs", *Renew. Sustain. Energy Rev.*, 2010, **14**, pp. 2009–2018.

[25] Tsikalakis, A.G., Hatziargyriou, N.D.: "Centralized control for optimizing microgrids operation", *IEEE Trans. Energy Convers.*, 2008, **23**, pp. 241–248.

[26] McArthur, S.D.J., Davidson, E.M., Catterson, V.M., *et al.*: "Multi-agent systems for power engineering applications – Part I: concepts, approaches, and technical challenges", *IEEE Trans. Power Syst.*, 2007, **22** (4), pp. 1743–1752.

[27] McArthur, S.D.J., Davidson, E.M., Catterson, V.M., *et al.*: "Multi-agent systems for power engineering applications – Part II: technologies, standards, and tools for building multi-agent systems," *IEEE Trans. Power Syst.*, 2007, **22** (4), pp. 1753–1759.

[28] Niyato, D., Xiao, L., Wang, P.: "Machine-to-machine communications for home energy management system in smart grid", *IEEE Commun. Mag.*, 2011, **49** (4), pp. 53–59.

[29] Qiu, R.C., Hu, Z., Chen, Z., *et al.*: "Cognitive radio network for the smart grid: experimental system architecture, control algorithms, security, and microgrid test bed", *IEEE Trans. Smart Grid*, 2011, **2**, pp. 724–740.

[30] Koval, D.O., Chowdhury, A.A.: "Base load generator unit operating characteristics", *Proceedings of the IEEE Industrial and Commercial Power Systems Technical Conference*, Irvine, CA, May 1994, pp. 225–230.

[31] Koval, D.O., Chowdhury, A.A.: "Generating peaking unit operating characteristics", *IEEE Trans. Ind. Appl.*, 1994, **30** (5), pp. 1309–1316.

[32] Sanseverino, E.R., Di Silvestre, M.L., Ippolito, M.G., *et al.*: "Execution, monitoring and replanning approach for optimal energy management in microgrids", *Energy*, 2011, **36**, pp. 3429–3436.

[33] "IEC. International Standards and Conformity Assessment for all electrical, electronic and related technologies", http://www.iec.ch/smartgrid/background/explained.htm, accessed 23 March 2016.

[34] Genova, K., Guliashki, V.: "Linear integer programming methods and approaches – a survey", *J. Cybern. Inf. Technol.*, 2011, **11** (1), pp. 1–25.

[35] Khan, A.A., Naeem, M., Iqbal, M., *et al.*: "A compendium of optimization objectives, constraints, tools and algorithms for energy management in microgrids", *Renew. Sustain. Energy Rev.*, 2016, **58**, pp. 1664–1683.

[36] "FrontlineSolvers. Solver Technology – Mixe_integer and Constraint Programming", http://www.solver.com/mixed-integer-constraint-technology, accessed 23 March 2016.

[37] Forner, D., Erseghe, T., Tomasin, S., Tenti, P.: "On efficient use of local sources in smart grids with power quality constraints", *Proceedings of the IEEE International Conference on Smart Grid Communication*, Gaithersburg, MD, USA, October 2010, pp. 555–560.

[38] Elrayyah, A., Sozer, Y., Elbuluk, M.: "A novel load flow analysis for particle-swarm optimized microgrid power sharing", *Proceedings of the IEEE APEC*, Long Beach, CA, USA, 2013, pp. 297–302.

[39] Lu, F.-C.; Hsu, Y.-Y.: "Fuzzy dynamic programming approach to reactive power/voltage control in a distribution substation", *IEEE Trans. Power Syst.*, 1997, **12** (2), pp. 681–688.

[40] Levron, Y., Guerrero, J.M., Beck, Y.: "Optimal power flow in microgrids with energy storage", *IEEE Trans. Power Syst.*, 2013, **28** (3), pp. 3226–3234.

[41] Lam, A.Y.S., Zhang, B., Dominguez-Garcia, A., Tse, D.: "Optimal distributed voltage regulation in power distribution networks", http://arxiv.org/abs/1204.5226v1, accessed 23 April 2012.

[42] Lavaei, J., Tse, D., Zhang, B.: "Geometry of power flows in tree networks", *Proceedings of IEEE Power Energy Society General Meeting*, San Diego, CA, USA, July 2012, pp. 1–8.

[43] Lu, D., François, B.: "Strategic framework of an energy management of a microgrid with a photovoltaic-based active generator", *Proceedings of IEEE Eighth International Symposium on Electromotion 2009*, EPE, Lille, France, July 2009, pp. 1–6.

[44] Kanchev, H., Lu, D., Colas, F., *et al.*: "Energy management and operational planning of a microgrid with a PV-based active generator for smart grid applications", *IEEE Trans. Ind. Electron.*, 2011, **58**, pp. 4583–4592.

[45] Di Silvestre, M.L., Graditi, G., Ippolito, M.G., *et al.*: "Robust multi-objective optimal dispatch of distributed energy resources in micro-grids", *Proceedings of IEEE Powertech 2011*, Trondheim, Norway, June 2011, pp. 1–6.

[46] Corso, G., Di Silvestre, M.L., Ippolito, M.G., *et al.*: "Multi-objective long term optimal dispatch of distributed energy resources in micro-grids", *Proceedings of Universities Power Engineering Conference UPEC 2010*, Cardiff, UK, September 2010, pp. 1–5.

[47] Bruno, S., Silvia, L., Giuseppe, R., *et al.*: "Unbalanced three-phase optimal power flow for smart grids", *IEEE Trans. Ind. Electron.*, 2011, **58** (10), pp. 4504–4513.

[48] Yan, S., Nai, S., Zhi, Y.: "Power flow calculation method for islanded power network", *Proceedings of Asia Pacific Power and Energy Engineering Conference, APPEEC 2009*, Wuhan, Asia-Pacific, March 2009, pp. 1–5.

[49] Vandoorn, T.L., De Kooning, J.D.M., Meersman, B., *et al.*: "Automatic power-sharing nodification of P/V droop controllers in low-voltage resistive microgrids", *IEEE Trans. Power Delivery*, 2012, **27** (4), pp. 2318–2325.

[50] Tabatabaee, S., Karshenas, H.R., Bakhshai, A., Jain, P.: "Investigation of droop characteristics and X/R ratio on small-signal stability of autonomous microgrid", *Proceedings of the Second Power Electronics, Drive Systems and Technologies Conference*, Tehran, February 2011, pp. 223–228.

[51] Schiffer J., Ortega R., Astolfi A., *et al.*: "Conditions for stability of droop–controlled inverter-based microgrids", *Automatica*, 2014, **50** (10), pp. 2457–2469.

[52] Dong J., Zhang C.J., Meng X.M., *et al.*: "Modeling and stability analysis of autonomous microgrid composed of inverters based on improved droop control", *Proceedings of the International Power Electronics and Application Conference and Exposition*, Shanghai, November 2014, pp. 1310–1315.

[53] Krishnanand, K.N., Ghose, D.: "Glowworm swarm based optimization algorithm for multimodal functions with collective robotics applications", *Multiagent Grid Syst.*, 2006, **2** (3), pp. 209–222.

[54] Riva Sanseverino, E., Quang, N.N., Di Silvestre, M.L., Guerrero, J.M., Chendan, L.: "Optimal power flow in three-phase islanded microgrids with inverter interfaced units", *Electric Power Syst. Res.*, 2015, **123**, pp. 48–56.

Chapter 8

Smart distribution system

*Jasrul Jamani Jamian[1], Muhammad Ariff Baharudin[1],
Mohd Wazir Mustafa[1] and Hazlie Mokhlis[2]*

The main idea in smart-grid concept is the integration of active communication in the power system. Traditionally, the communication in the power system is more toward the one-way approach. All the instructions of operations are given by the utility and will be operated by the controller at the load side, either by using supervisory control and data acquisition or by other simple means of communication. However, in a smart-grid topology, the load should be able to give the information to the utility or even be able to make decisions based on the feedback provided by the end user. In this chapter, the discussion on the concept of communication integration in balancing the system frequency is discussed. The load demand will adjust the power consumption by changing their operation mode, depending on their frequency condition. The technology in telecommunications has advanced tremendously within the recent decades; we do not have to reinvent the wheel. The same technology, such as the network protocols and standards, can be implemented to the existing power system to add the smart element with the aim to make the system more efficient and robust. Furthermore, the whole world is moving toward connectivity and ubiquity; the best way forward is to embrace this change and synergize the power system components with the telecommunications components to create a smarter and efficient power control system.

8.1 Introduction—key components of the smart distribution system network

Distribution system is the last part in the power system sectors that connects the power supply to the end user. In the traditional distribution network, no other power sources are injected into the network besides the substation (only one power source). This concept of single supply makes the protection system in the distribution network simpler compared to the protection scheme in the transmission line. However, due to technological advancements, small generator units based

[1]Faculty of Electrical Engineering, Universiti Teknologi Malaysia
[2]Faculty of Engineering, University of Malaya, Malaysia

on renewable energy such as photovoltaic (PV), mini-hydro, wind energy, as well as the non-renewable energy based diesel generator have been introduced. These generators have been used to supply some of the load in the distribution network directly and they are known as the distributed energy resources (DER).

There are several advantages by connecting DER in the distribution network. Since the location of DER is close to the user's location, it allows the DER to reduce the power loss in the system. For a passive network (without DER), the total power consumption in the distribution network is solely supplied by the grid, located far away from the load demand. As a result, the power losses will also increase when the current increases. This situation worsens if the R/X ratio in the distribution system is high, which indirectly causes the I^2R power loss to become very high. On the contrary, with the presence of DER, some of the required power source can be supplied by the DER units. This reduces the amount of transmitted power from the grid, which indirectly causes the total amount of current to lessen. This is the reason why many researchers have applied optimization methods to obtain the optimal DER output such as by using particle swarm optimization [1,2], evolutionary programming [3] and artificial bee colony [4] to reduce power losses in the network, either for a single DER unit or for multiple DER units. In general, most of the authors have used meta-heuristic techniques or other indicators, such as voltage sensitive analysis, for conducting the DER placement analysis and subsequent optimization of the DER output. However, the placement of the DER cannot be done anywhere in the system. This is due to the fact that the DER placement should be done considering some limitations such as the point of energy resources, especially for the renewable energy. Besides that, the objective function of the optimal DER is not only limited to the power loss reduction and voltage improvement, but it can also be set to improve the stability of the system [5], lowering the cost of total generation [6], minimizing the total harmonic distortion [7], controlling the stability margin for the network [8], and many more.

The term "smart grid" is becoming more popular when researchers start to integrate advanced communication technology into the power system. In the smart grid, two-way communication between the utility side and the load side can be established. This two-way communication approach will enhance the utilities' control over the power system compared to a single-way communication approach. However, to achieve this, the smart grid requires good data communication management and architecture. This chapter will be focusing on the use of the smart communication system to enhance the control among power utility, distributed generation (focus on renewable energy), as well as smart load devices. The proposed concept is more on the communication aspect that are required in order to control the load demand in a smart house.

There are several important components that should exist in the smart distribution system. One of the main components is the capability to communicate, from the source (utility) until the load (end user). However, traditionally, the load equipment does not have this ability (to communicate). Therefore, some modifications and enhancements on the load design need to be done. The smart-grid system will fully utilize all DER technologies such as PV, smart metering, as well as smart load. All these components will communicate to achieve the same goal, improving the power system performance.

8.2 Distributed energy resources

In the modern community, PV, electric vehicle (EV), and other energy devices in the home system are able to act as the DER in the distribution network. This external source is able not only to help in stabilizing the frequency of the system, but also to generate some money from the power supplied to the grid. For a deregulated power system, the electricity price is varied depending on the load condition. Thus, when the produced power selling price is higher than the power buying price, these energy resources can be sold to the utility. On the other hand, when their price is lower than the buying price, the DER can supply its power to its load (home) and get some support from the power system. If the house is totally disconnected from the utility due to some problems, these external resources still could supply the household with the smart appliances' (SA) collaboration. Hence, full knowledge on DER and SA categories is required, so that a good communication between DER and SA can be established. The following section discusses the different characteristics of DER in the distribution network and their operation method.

8.2.1 PV system

The characteristics of PV as a semi-conductor material have the ability to convert solar energy to electrical energy [9,10]. However, there are several difficulties in using PV systems. One of the complications is in terms of low-energy conversion efficiency of PV cells and the efficiency keeps reducing during the operational period due to the increase in the PV temperature by 4%–17% [11,12]. In general, the power output by the PV panel can be calculated using the following formula:

$$P_{PV} = G \times A_{PV} \times \eta_{PV} \qquad (8.1)$$

where G is the solar irradiance (kW/m^2), A_{PV} is the the PV array area (m^2), and η_{PV} is the PV module efficiency.

The performance of the PV system depends on the input of irradiance from sunlight and other factors such as temperature, weather, tilt angle, shading, battery efficiency, and many more. To increase the electrical efficiency of PV cells, the solar panel must absorb the maximum amount of solar radiation. However, not 100% of the solar radiation absorbed by the PV will be converted into electricity. Some of the solar radiation is transformed into thermal energy and it will reduce the electrical efficiency of the PV cell [12].

Figure 8.1 shows the comparison of voltage and current between two PV systems that are receiving different irradiation values. The increment in the irradiation value is done by installing a light reflection system near to the PV panel. The voltage curve of solar panel with the light reflection method is slightly higher than the voltage of conventional solar panel (without light reflection). The same goes to the power in Figure 8.2. The power of light reflection solar panel is higher than the conventional panel. The increased efficiency of light reflection solar panel is 25.4% more than that of the conventional solar panel.

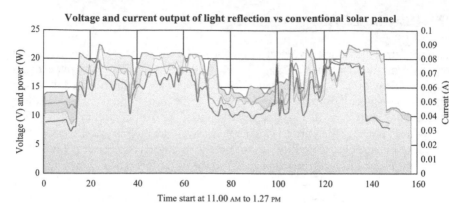

Figure 8.1 Voltage and current output of light reflection vs conventional solar panel

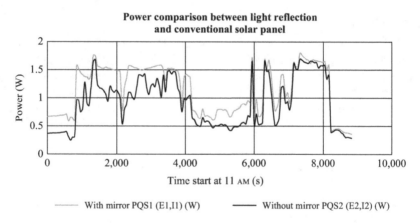

Figure 8.2 Power output of light reflection vs conventional solar panel

Higher PV power output can be obtained by reducing the temperature of the PV panel, which is due to the indirect increase in the efficiency. Figure 8.3 shows that the power generated using solar panels with a cooling system and light reflection is slightly higher than the system without a cooling system. The total energy of the cooling-light reflection method is 177.792 W, while for the light reflection without the cooling system it is 170.682 W.

From here, it can be seen that the PV power output is greatly influenced by many factors and it is very difficult to get consistent output. The use of energy storage system or battery is required to support the PV installation. However, with a good communication technology in the smart grid, the load might able to respond to the PV output and the power balancing can be done to ensure that the system frequency remains constant.

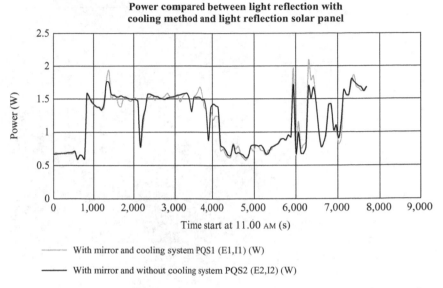

Figure 8.3 *Output of cooling-light reflection vs light reflection without cooling system solar panel*

8.2.2 Wind system

The selection of a renewable energy farming approach depends on the availability and the potential of the location. For example, Malaysia have higher solar irradiation reading throughout the year, on the other hand, the average wind speed in Malaysia is very low as compared to other countries. Figure 8.4 shows a one-year wind speed profile in Malaysia, where a monthly average value is in the range of 1.98–3.28 m s^{-1} [13]. Thus, similar to PV power, the power output from wind turbine (WT) generation also fluctuates depending on the average wind speed and can be calculated using the following equation:

$$P_{WTG} = \begin{cases} 0, & V < V_{ci} \\ a \times V^3 - b \times P_r, & V_{ci} \leq V \leq V_r \\ P_r, & V_r \leq V \leq V_{co} \\ 0, & V > V_{co} \end{cases} \qquad (8.2)$$

where P_r is the wind turbine's rated power, V_{ci} is the cut-in wind speed, V_r is the rated wind speed, and V_{co} is the cut-out wind speed.

8.2.3 Electrical vehicle

In this twenty-first century, environmental issues have become one of the most important concerns for environmental researchers. In addition, the fluctuation on the world fuel price has put some pressure on the gasoline vehicle (GV) users. For this reason, the EV can become an interesting option due to the non-dependency of the

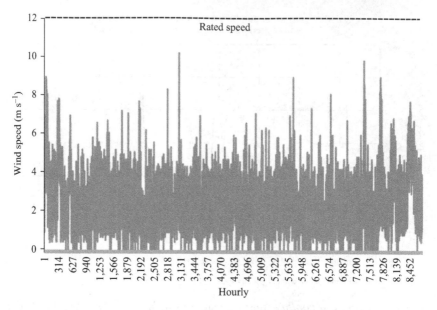

Figure 8.4   The yearly Malaysia wind speed data

Table 8.1   Comparison between EV and GV performance

	EV (e.g., Tesla—Model S)	GV (Sedan car)
Environmental impact	No emission	Greenhouse gases
Travelling distance	±480 km (85 kW battery)	±500 km
Maximum speed	210 km/h	240 km/h
Average travelling cost*	$0.107/km	$0.019/km
Refuel/recharge duration	9 hours 26 minute (level 2)	Minute to refuel

* Based on the 8,000-km distance calculation:
Cost of fuel: $3.80/gallon.
Cost of electricity: $0.11/kWh.

vehicle to the fuel price. Not only that, the usage of EV can also reduce vehicle emissions, improve the air quality [14–16], as well as help in controlling the fuel price [17–19]. Therefore, most of the giant vehicle companies such as Ford, BMW, Toyota, Honda, and Hyundai have started to do the research in improving the EV performance. As a result, unlike in the early stages of the EV (in 1912), the advancement on DC motors and battery technologies has made the EV performance nearly similar to the GV. Table 8.1 shows the performance of EV (Tesla—Model S) [20] compared to conventional GV.

However, from the distribution network's perspective, the EV charging process can be represented as a new load in the system. With high penetration of charging process, the performance of distribution network will be affected.

On the other hand, the connection of EV in the distribution system can also act as a DER. If the power system requires additional power, the connected EV is able to sell or inject their energy into the system. Thus, several factors need to be considered in implementing the strategy, as follows:

1. The country must use the competitive electricity market in their system. Without a variation in the electricity price, customer's charging behavior will not be influenced to help the utility in balancing the power.
2. Effective communication technology (in broadcasting the price and user selection) is required between utility and each EV customer.
3. The utility provider also needs to have intelligent computational capability to ensure that the customer and system gain advantages from the strategy.

8.3 Distribution grid architecture

The traditional power system requires at least three main components for the system to send power from the generator side to the consumer side. Starting from the "generation" component, the power will be sent through the "transmission" component, and dispensed via the "distribution" component. From this traditional scheme, the power is only flowing in one direction and the distribution system has the highest power loss, due to the lower X/R ratio, lower voltage level, radial configuration, and others. Studies have shown that more than 70% of power losses are due to the distribution system [21]. Many approaches have been introduced to improve the distribution network performance. Capacitor bank allocation [22–24] and reconfiguration [25–27] are the examples of proposed techniques that can be used to improve the voltage profile and power loss for distribution network. Other than these two techniques, the most significant improvement can be seen from the concept of DER. By placing the DER closer to the consumer side (end load), some of the demanded loads will be supplied by the DER, while other loads will receive the power from the transmission level.

Besides that, from the protection point of view, the use of high-speed protection and automatic restoration in the existing distribution system is able to improve the efficiency of the system. Thus, before the introduction of smart-grid concept, the intelligent schemes are already existing to protect the distribution system. Many analog measurements have been changed to digital types to get a more accurate reading. The use of advanced metering infrastructure has given further improvement on measurement instrument and control utility side. Currently, the infrastructure is more focusing on one-way communication—instead of two ways.

With the integration of advanced communication between the supply and the load side, many applications of the smart distribution grid can be done such as advanced demand response scheme, accurate fault location, and service restoration. These applications will require technologies, such as sensors, telecommunication infrastructure, analysis, and optimization, to facilitate real-time decisions and to meet growing customer expectations. Thus, several components are introduced in this discussion, such as smart server (SS), the smart regional server (SRS), the

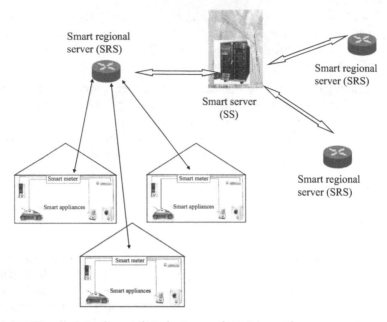

Figure 8.5   Smart distribution architecture with components

smart meter (SM), and the SA, to make communication and data management successful between the smart home system and the utility system. The network topology of the system distribution is illustrated in Figure 8.5. As the communication and data transfer in this smart system involves the SSs down to the SAs, these entities are required to have the capability to communicate and send data in both ways.

8.3.1   Smart server and smart regional server

The SS and SRS are introduced to transfer all the two-way communication between load and supply sides as well as to make some decision based on the optimization analysis. If any disturbance occurs in the distribution system, the SS will analyze and determine the appropriate action that must be taken—either from generation or from load sides. Next, the SS will communicate with SRS to determine the suitable load that can be involved in solving the problem before the final instruction is given to the SM. All the communication among SS, SRS, and SM can be done via physical network connections such as ADSL and fiber optics or Wireless technologies. Among these two technologies, the implementation based on wireless backhaul connection might able to give more advantages and reduce some installation cost as shown in Table 8.2. For instance, WiMAX can be used because its coverage is very wide (around 50 km radius) with very fast bit rates (up to 30 Mbps or more). Even at the cell edge, the speed is still around 1–4 Mbps, which is more than enough to send the data collected by the SM.

Table 8.2   Comparison between physical and wireless technologies performance

	Wired	**Wireless**
Speed	High	Lower than wired, but improving with LTE-A and 802.11ad
Cost	High due to cable installation	Lower than wired, less cable needed
Deployment	Slow/cumbersome	Fast
Mobility	Limited, only for areas covered by the wired network	Not limited, operates in the entire wireless network
Transmission medium	Copper wires, optical fiber, Ethernet	Electromagnetic waves/infrared/radio waves
Network coverage extensions	Limited expansion. Requires hubs, switches, and cables for network coverage extensions	More area is easily covered since it is wireless

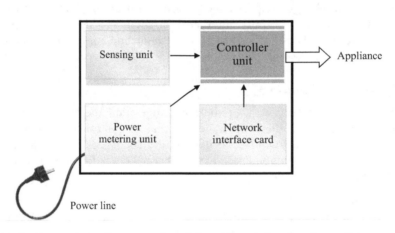

Figure 8.6   Conceptual module to be implemented into SAs

8.3.2   Smart appliances and smart meter

In general, the SA is categorized based on their working principles; either the appliances operate based on thermal, time, or operation mode. In order to allow the consumer to take part in the power management, the appliances must be able to communicate with the components at the power supply side. The SA is defined as appliances which are embedded with a module that would enable the home appliances (HA) itself to detect the state that it is in and schedules its task completion according to the decision by the SM.

The structure of the proposed module is as depicted in Figure 8.6, where it consists of the power metering unit, the sensing unit, the network interface card

(NIC), and the controller unit. The function of each component in this module can be summarized as follows:

Power metering unit: The entity that calculates the power consumption of the appliance and sends the data to the controlling unit.

Sensing unit: The entity that senses the current condition of the appliance—either "on" or "off".

NIC: The entity that will modulate and send the data in the uplink direction and demodulate and transfer the information received in the downlink direction to the SA's application.

Controller unit: Act as the brain for this module. It will get the data from the power metering unit and send it to the SM via the NIC. It will also sense via the sensing unit and react accordingly.

Furthermore, the SA module will receive commands from the SM via the NIC and executes the changes required by the SM. With the implementation of this module, the system will be fully automated. In addition, users can set their preferences in these modules so that when power reduction command is received, it will act according to the user preferences.

8.3.3   Types of load in SAs

All appliances in the house have their own operation method and it is very important to categorize all these appliances, so that the controller unit in SA will be able to give accurate instructions. In general, the appliances can be divided into three, which are time operation based (TOB), thermal operation based (ThOB), and mode operation based (MOB).

- **Time operation based**

    There are several house appliances working based on the TOB principle. Washing machine and dish washer are the examples of electrical devices that can delay or postpone their operation without giving a big impact to the consumer. Therefore, with the adoption of communication module in TOB appliance, it makes the SM able to control their operation, either by pausing or by totally stopping the operation when it is needed. With this action, the power system is able to implement indirect load shedding in the distribution load, especially when the power supply is not enough at the distribution side.

- **Thermal operation based**

    Some electrical appliances are operated based on thermal storage capacity, such as refrigerator and water heater. Thus, it can be turned on and off for certain seconds up to certain minutes without affecting their temperature. If the network losing generation sources, the controller unit in the SA module can turn off these units for a certain time until the system regenerates. However, the SM also needs to consider the behavior of equipment, such as marginal time before the load can

Table 8.3 Example of appliance with operation mode

Type of appliance	Category
Refrigerator	ThOB
Air conditioner	MOB
Washing machine	TOB
Dish washer	TOB
Water heater	ThOB
Computer with UPS	TOB
Electric vehicle	TOB
Intelligent fan	MOB
Intelligent lamp (dimmer application)	MOB

be turned on again. For instance, in order to lengthen the life span of the refrigerator, it cannot be turned on immediately after the turning off process.

• **Mode operation based**

For the load that can function in multi-mode operating conditions, the SM can request all these SA to change their mode when it is needed. However, the operation mode changes depending on the consumer settings. Air conditioner is one of the examples of the SA unit under this category. So, the user can set the range of the operation, for instance, from 19 up to 25 °C during day time, and the SM will run the analysis to decide the suitable temperature that will be allowed. The higher the temperature setting for the air conditioner, the lesser the power consumption from the grid will be required; therefore, the adjustment of the operating mode able to give extra available power in the system.

Table 8.3 shows the example of appliance and their categories that can be used in SA to determine the suitable action when it is needed.

8.4 Key challenges in smart distribution system

Data credibility of information communicated between the components of the smart grid is one of the most vital parts in ensuring that the proposed scheme can work successfully. Loss in data can cause disruptions and inaccuracies on the smart grid. The data collected from the power system components (generation, distribution, etc.) can be divided into two types: critical data such as amount of power available, stability index, and frequency and non-critical data such as voltage profile and current flow. This categorization is very important in order to choose a suitable protocol to ensure the reliability of the data passed to the SS.

8.4.1 Supervisory control and data acquisition

Supervisory control and data acquisition or in short SCADA is a system used for remotely monitoring and controlling a remote station with coded signals.

However, usually each remote station requires one dedicated channel; this is not very efficient since the number of channels needed increases with the number of remote stations within the system. Besides that, the system itself is isolated from the Internet; the system manager can only have access from the controller at the center. This is very inefficient and inconvenient.

Nevertheless, that is in the past. With the advent of new technologies such as wireless sensor network (WSN) and Internet of Things (IoT), it is possible to create a system that may be centralized or decentralized, a system that can be connected to the Internet and can be accessed securely anytime and anywhere. In WSN, the remote stations can be connected to the center wirelessly, while in IoT, each component in the network has the capability to connect to the Internet. In these cases, the wireless access technologies and network protocols are paramount in ensuring a secure, robust, and reliable network.

8.4.2 Protocol stacks of the proposed smart-grid system

In general, the internals of a network node (e.g., servers, SSs, routers, SMs, etc.) can be divided into several stacks according to their corresponding functions, which is determined using the OSI layer [28]. However, here, for simplicity, the transmission control protocol/Internet protocol (TCP/IP) stack [29] as shown in Figure 8.7 is adopted for ease of discussion. The IP stack is divided into five layers consisting of the application layer, transport layer, Internet layer, data link layer, and physical layer.

> **Application layer:** This refers to the protocols that sit behind the users and utility applications based on any OS and software installed in the device. Some examples are SMTP, POP, FTP, and HTTP.
>
> **Transport layer:** This layer determines the reliability of the data transmission by providing end-to-end data recovery. This layer also controls the data flow by integrating the data from the application layer into a single stream if there are more than one application running. Transport protocols such as TCP, user datagram protocol (UDP), and stream control transmission protocol (SCTP) resides in this layer.
>
> **Internet layer:** The network routing and path selection as well as the logical addressing of a device is controlled on this layer. Routers, switches, and IP operate on this layer.

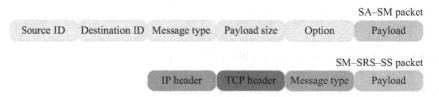

Figure 8.7   Example packet structure used for (a) in-home communication and (b) out-of-home communication

Data link layer: This layer formats the data sent from the upper layers into a format that can be sent through physical network connection as well as dealing with the error detection and correction. Some examples are the Ethernet and the Wi-Fi.

Physical layer: This is the physical part on a device that is used to connect itself to the network. This physical connection can be a copper wire, a fiber optic cable, or even an antenna to communicate wirelessly.

From these layers, the transport layer plays a major role in maintaining the reliability of the data communicated between network nodes. For example, smart HAs can create communication sessions among themselves via power line communication [30] using a modified version of the X10 protocol [31]. In this case, both critical and non-critical data can be transmitted using the same protocol as the network is small with a limited number of devices. However, this is not enough when dealing with disruption such as packet loss and high end-to-end delay in a larger network. Due to the huge amount of clients accessing and sending data to the designated server, congestion and collision which causes the packets to be dropped are bound to happen. Thus, suitable transport layer protocols which can cope with these issues are needed.

For non-critical data, the UDP is enough since it is fast and support real-time applications. Nevertheless, it does not support acknowledgments and retransmissions. Hence, for critical data, the TCP is more suitable, as it supports retransmissions with additional congestion avoidance features. Nonetheless, TCP does not support real-time applications. One alternative that has the combination of TCP and UDP features is the SCTP. Therefore, utilization of protocols such as SCTP is desirable in maintaining the reliability of the data transmitted as well as the timeliness of the data arrival.

8.4.3   Message exchange in the smart system

The transport protocols in the previous section are used as the tool to pave the road (connection) between one device with another device or the SS. However, the language used to communicate is a different thing. Usually, messages are used for machine-to-machine communication, based on the messaging protocol used. Different sets of messages were designed depending on the entities involved in the conversation. These sets can be divided into three categories:

1.   Messages between SA and SM
2.   Messages between SM and SRS
3.   Messages between SRS and SS

In previous works, the format of the communication packet was developed for the network layer. Two types of packet structure were introduced to cater for in-home communication (SA–SM communication) and out-of-home communication (SM–SRS–SS communication). For the in-home environment, the transport layer is not needed, as two-way communications is only done between the SA and SM. The source and destination are determined by the device id (e.g., media access

Table 8.4 Packet descriptions in messaging sequence

Packet type	Description
CONN_REQ	Request to make connection from all smart devices (SS, SRS, SM, or SA)
CONN_ACK	Acknowledgment of connection
ΔL_REQ	Request for load changes
ΔL_ACK	Acknowledgment for load changes
SM_ID_REQ	Request for transaction change from SM to SA
SA_STATE_REQ	SA update operation: total power consumption, current state appliance (on–off), state-level available, timing available, etc.
SM_AVAI_UPDATE	SM updates the total adjustable load
SRS_TOT_AVAI	SRS updates the total adjustable load

control address or other proprietary format). As for the out-of-home environment, the transport (TCP header) and network layer (IP header) is needed in order for the SM to communicate with SRS and SS via the Internet. The list of usable messages is shown in Table 8.4.

8.5 Case study: frequency balance in smart-grid architecture

The principle in the smart-grid approach to balance the system frequency is similar to load shedding method—however, with the two-way communication technologies, it is possible to have full control on the load condition. The main component required by a power system to maintain a stable frequency is by balancing active power supply and demand. Without this balancing mechanism, the frequency of the system might collapse and cause system to black-out. Commonly, load balancing is automatically conducted by the load frequency controller in a generator. However, the balancing mechanism could be limited by some constraints such as stability of the generators, transmission line capacity limit, and power losses [32–35]. Another technique of load frequency control is by controlling the total load consumption in a system. In this technique, the load is not connected or disconnected such as the load shedding method but instead, the load is controlled so that its power consumption can be adjusted whenever necessary in order to maintain the frequency stability. Many researchers have started to gain interest in this technique due to the introduction of controllable load and external energy sources such as EV and renewable energy (PV panel).

8.5.1 Frequency increase and drop operation

The SS is used to sense the frequency changes that occur in the system by using the method proposed in Reference 36. After analyzing and obtaining the amount of load to be increased or reduced, the SS will refer to its database and determine the available amount of load that can be adjusted in each SRS and send the data signal to select SRSs as shown in Figure 8.8. For the frequency drop operation, the SRS

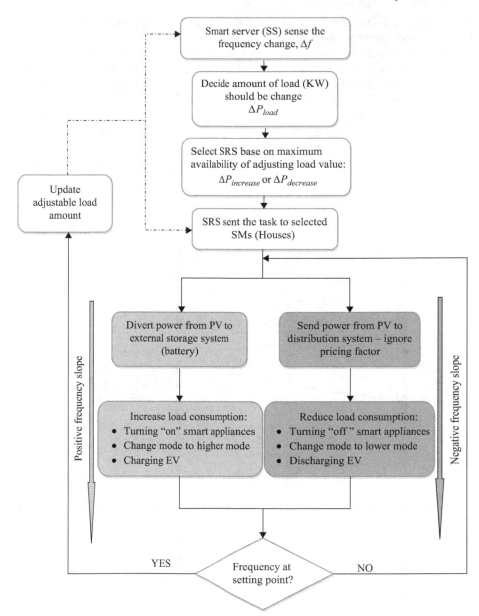

Figure 8.8   Proposed smart communication protocol in adjusting system frequency

receives the command to reduce the power consumption in its own region. Thus, the SRS will identify the suitable SM from its database and forward the message to the selected SMs (houses). The selection process of SM is based on their availability of load reduction and maximum reduction on power losses. After the SM gets the signal from SRS for adjusting their power consumption, it will ask the SA to reduce their power consumption in order to improve the frequency of the system. Not only that, external energy storage system, such as EV, will also support the system by supplying generated power to the distribution system. Once SA takes action, the SA will update the latest condition to SRS through SM. The SS will also know the changes in SRS and update its own database.

The same technique and communication management among SS, SRS, and SM will be used for load increment when the frequency in the system increases. The only difference is the condition of power consumption needed to be adjusted. In this condition, the loads will be commanded to increase their power consumption

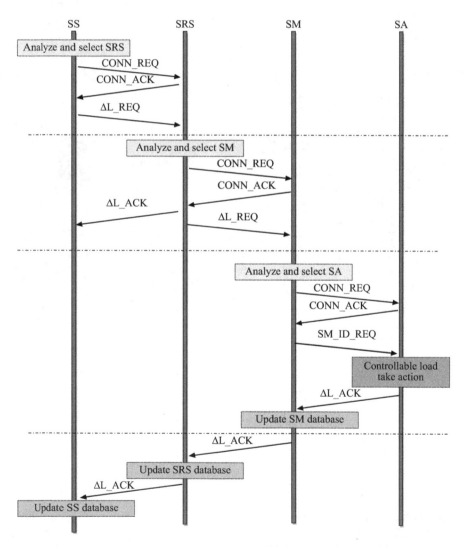

Figure 8.9 Messaging sequence in frequency drop operation

such as charging the EV and turning "on" the appliances which have delayed their task before this. The output from PV will not be supplied to the distribution system in this condition but it will be stored either by charging the EV or external batteries.

8.5.2 Message exchange in frequency operation

An example of message exchange between the entities within the smart system is shown in Figure 8.9. The SS will first communicate with SRS which, in turn, will analyze and select the most suitable SMs that can be adjusted. The SRS must give an acknowledgment to SS after the SRS–SM connection established. If not, the SS will communicate with other SRS. Next, the SRS will send a load adjustment request to the selected SMs, either to increase or to reduce the load depending on the frequency condition. If it is a frequency drop operation, the SRS will ask SM to send a "turn off" command to the related SAs and vice versa. Again, these SAs need to send acknowledgment messages to the SM before turning off and this message will be forwarded toward the SS, so that the SS can update its database for further analysis. With this approach, the status of the system can be updated in real time, allowing the system to maintain its stability and efficiently solve any problems or failures.

8.6 Conclusion

With the inclusion of two-way communication elements within the power system, it is possible to monitor the condition of the system in real time. This is especially useful in predicting the upcoming problems or failures; consequently, effectively avoid any system failures. The system discussed in this chapter is one of the early steps toward a more proficient power system management and maintenance. Currently and in the future, the monitoring approach can be further enhanced with the addition of WSNs, which is a platform toward the IoT, where all entities within the power system can be accessed via Internet and system engineers can effectively monitor the status of the system. Furthermore, automated approaches can also be implemented creating a self-organized, self-healing power system.

References

[1] El-Zonkoly, A. M. "Optimal placement of multi-distributed generation units including different load models using particle swarm optimization," *IET Generation, Transmission and Distribution* 2011; 5: 760–771. DOI: 10.1049/iet-gtd.2010.0676.

[2] Mohammadi, M. and Nasab, M. A. "PSO based multiobjective approach for optimal sizing and placement of distributed generation," *Research Journal of Applied Sciences, Engineering and Technology* 2011; 3: 832–837.

[3] Celli, G., Ghiani, E., Mocci, S., and Pilo, F. "A multiobjective evolutionary algorithm for the sizing and siting of distributed generation," *IEEE Transactions on Power Systems* 2005; 20: 750–757. DOI: 10.1109/TPWRS. 2005.846219.

[4] Abu-Mouti, F. S. and El-Hawary, M. E. "Optimal distributed generation allocation and sizing in distribution systems via artificial bee colony algorithm," *IEEE Transactions on Power Delivery* 2011; 26: 2090–2101. DOI: 10.1109/TPWRD.2011.2158246

[5] Alonso, M. and Amaris, H. "Voltage stability in distribution networks with DG," *IEEE Bucharest PowerTech*. June 28–July 2, 2009. Bucharest, Romania: IEEE. 2009. 1–6.

[6] Renani, Y. K., Vahidi, B, and Abyaneh, H. A. "Effects of photovoltaic and fuel cell hybrid system on distribution network considering the voltage limits," *Advances in Electrical and Computer Engineering* 2010; 10(4): 143–148.

[7] Alinejad-Beromi, Y., Sedighizadeh, M., and Sedighi, M. "A particle swarm optimization for sitting and sizing of distributed generation in distribution network to improve voltage profile and reduce THD and losses," *International Universities Power Engineering Conference*. September 1–4, 2008. Padova, Italy: IEEE. 2008. 1–5.

[8] Calderón-Guizar. J. G. and Tovar-González. E. A. "Impact on generator reactive power limits on a static voltage stability," *Advances in Electrical and Computer Engineering* 2011; 11(4): 105–110.

[9] Dobrzański, L. A., Wosińska, L., Dołżańska, B., and Drygała, A., "Comparison of electrical characteristics of silicon solar cells," *Journal of Achievements in Materials and Manufacturing Engineering* 2006; 18(1–2): 215–218.

[10] Masters, G. M. *Renewable and Efficient Electric Power Systems*. Hoboken, New Jersey: John Wiley & Sons, 2013.

[11] Hosseini, R., Hosseini, N., and Khorasanizadeh, H. "An experimental study of combining a photovoltaic system with a heating system," *World Renewable Energy Congress*, Linkoping, Sweden, Vol. 8, 2011.

[12] Elnozahy, A., Abdel Rahman, A. K., Ali, A. H. H., Abdel-Salam, M., and Ookawara, S. "Performance of a PV module integrated with standalone building in hot arid areas as enhanced by surface cooling and cleaning," *Energy and Buildings* 2015; 88: 100–109.

[13] Mukhtaruddin, R., Rahman, H., Hassan, M., and Jamian, J. "Optimal hybrid renewable energy design in autonomous system using iterative-pareto-fuzzy technique," *International Journal of Electrical Power & Energy Systems* 2015; 64: 242–249.

[14] Binggang, C. "Current progress of electric vehicle development in China," *Journal of Xi'an Jiaotong University* 2007; 41(1): 114–118.

[15] The Electric Vehicle Company Ltd. "Organisers announce changes to the official results of the 2011 RAC future car challenge," 2012. Available: http://evc.gg/blogs/news/5733482-organisers-announce-changes-to-the-official-results-of-the-2011-rac-future-car-challange [accessed on 11-04-2013].

[16] Sikai, H. and Infield, D. "The impact of domestic plug-in hybrid electric vehicles on power distribution system loads," *International Conference on Power System Technology*, October 24–28, 2010. Zhejiang, China: IEEE. 2010. 1–7.

[17] Green, R. C., Lingfeng, W., and Alam, M. "The impact of plug-in hybrid electric vehicles on distribution networks: A review and outlook," *IEEE Power and Energy Society General Meeting.* July 25–29, 2010. Michigan, USA: IEEE. 2010. 1–8.

[18] Marano, V., Tulpule, P., Gong, Q., Martinez, A., Midlam-Mohler, S., and Rizzoni, G. "Vehicle electrification: Implications on generation and distribution network," *International Conference on Electrical Machines and Systems.* August 20–23, 2011. Beijing, China: IEEE. 2011. 1–6.

[19] Pieltain, F. L., Roman, T. G. S., Cossent, R., Domingo, C. M., and Frias, P. "Assessment of the impact of plug-in electric vehicles on distribution networks," *IEEE Transactions on Power Systems* 2011; 26(1): 206–213.

[20] Tesla Motors Inc. "Model S – TESLA," 2013. Available: http://www.teslamotors.com/goelectric# [accessed on 17-04-2013].

[21] Lund, T. "Analysis of distribution systems with a high penetration of distributed generation," Ph.D. Thesis. Technical University of Denmark; 2007.

[22] Farahani, V., Vahidi, V., and Abyaneh, H. A. "Reconfiguration and capacitor placement simultaneously for energy loss reduction based on an improved reconfiguration method," *IEEE Transactions on Power Systems* 2012; 27(2): 587–595.

[23] Kasaei, M. J. and Gandomkar, M. "Loss reduction in distribution network using simultaneous capacitor placement and reconfiguration with ant colony algorithm," *Asia-Pacific Power and Energy Engineering Conference.* March 28–31, 2010. Chengdu, China: IEEE. 2010. 1–4.

[24] Neelima, S. and Subramanyam, P. S. "Optimal capacitor placement in distribution networks for loss reduction using differential evolution incorporating dimension reducing load flow for different load levels," *IEEE Energytech.* May 29–31, 2012. Case Western Reserve University: IEEE. 2012. 1–7.

[25] Jianming, Y., Zhang, F., and Dong, J. "Distribution network reconfiguration based on minimum cost of power supply," *International Conference on Sustainable Power Generation and Supply.* April 6–7, 2009. Nanjing, China: IEEE. 2009. 1–4.

[26] Tsai, M. S. and Chu, C. C. "Applications of hybrid EP-ACO for power distribution system loss minimization under load variations," *International Conference on Intelligent System Application to Power Systems.* September 25–28, 2011. Hersonisso, Greece: IEEE. 2011. 1–7.

[27] Jamian, J. J., Lim, Z. J., Dahalan, W. M., Mokhlis, H., Mustafa, W. M., and Abdullah, M. N. "Reconfiguration distribution network with multiple distributed generation operation types using simplified artificial bees colony," *International Review of Electrical Engineering* 2012; 7(4): 5108–5118.

[28] Forouzan, B. A. *TCP/IP Protocol Suite.* Second Edition. New York, USA: McGraw-Hill, 2003.

[29] Park, S., Kim, H., Moon, H., Heo, J., and Yoon, S. "Concurrent simulation platform for energy-aware smart metering systems," *IEEE Transactions on Consumer Electronics* 2010; 56(3): 1918–1926.

[30] Varma, M. K., Jaffery, Z. A., and Ibraheem. "Advances of broadband power line communication and its application," *2015 Annual IEEE India Conference (INDICON).* New Delhi, 2015. 1–6. DOI: 10.1109/INDICON. 2015.7443584

[31] Kim, J. E., Boulos, G., Yackovich, J., Barth, T., Beckel, C., and Mosse, D. "Seamless integration of heterogeneous devices and access control in smart homes," *Eighth International Conference on Intelligent Environments (IE), 2012.* Guanajuato, 2012. 206–213. DOI: 10.1109/IE.2012.57

[32] Verma, V., Singh, B., Chandra, A., Al-Haddad, K. "Power conditioner for variable-frequency drives in offshore oil fields," *IEEE Transactions on Industry Applications* 2010; 46(2): 731–739.

[33] Soder, L., "Explaining power system operation to nonengineers," *Power Engineering Review, IEEE* 2002; 22(4): 25–27.

[34] Fountas, N. A., Hatziargyriou, N. D., Orfanogiannis, C., and Tasoulis, A. "Interactive long-term simulation for power system restoration planning," *IEEE Transactions on Power Systems* 1997; 12(1): 61–68.

[35] Rajagopal, V., Singh, B., and Kasal, G. K. "Electronic load controller with power quality improvement of isolated induction generator for small hydro power generation," *Renewable Power Generation* 2011; 5(2): 202–213.

[36] Moslehi, K. and Kumar, R. "A reliability perspective of the smart grid," *IEEE Transactions on Smart Grid* 2010; 1(1): 57–64.

Chapter 9

Smart consumer system

Abdul R. Beig

Smart-grid technology will have a big impact on the way we generate, transport and consume electrical energy. Consumers will be the one who will be benefited most from the smart-grid technology. This chapter will discuss about the types of consumers, their role, their responsibilities and their awareness about smart-grid technology, relationship between the energy provider and consumer, and related issues. Also, this chapter presents a few case studies and their results. Consumers are grouped into three categories, namely industrial, commercial and residential. The demand-side management, the benefits of demand-side management in smart-grid technology and technological requirements to implement it are presented in detail. The consumer behaviour, their level of awareness about smart-grid technology, how to enhance the awareness and the man power training needs are discussed. The various government policies across the world and changes needed for the success of smart-grid technology are presented. Few case studies about the challenges faced and success of smart-grid technology will be useful for the further expansion of smart-grid technology.

9.1 Introduction

Smart grid is going to have a major impact on the way electric energy is generated, transported, distributed and utilized [1–3]. All involved in the energy industry agree that smart grid will improve the way we use energy in the future. This will help us to deal with rising demand, encourage to incorporation of renewable energy sources, which in turn help in reducing CO_2 emission. Improved life style, increased penetration of communication and electronic gadgets, increase in per capita income and growing economies in different parts of the world are demanding more and more energy especially electrical form of energy. In the next few decades, the electric vehicles (EVs) and plug-in hybrid electric vehicles (PHEVs) will demand additional electric energy from the grid. The smart-grid technology will directly address the way we use energy mainly focusing on efficiency and reliability. The smart grid will look in to how existing electricity infrastructure will be used to meet

The Petroleum Institute, Abu Dhabi, UAE

the growing demand without adding new generation, new transmission lines and distribution network. The smart grid will promote the use of renewable energy sources and thus address the environmental and pollution issues.

At the end of the line, the electricity is consumed or put to work by the consumers and it is those who will be benefited by the smart-grid technology. Unfortunately, consumers are most difficult to understand the smart-grid technology and its benefits. Smart-grid technology will have a major impact on the way consumers are engaged with the electric power generation, transportation and utilization. For the success of a smart grid, consumer's active participation or engagement is a must. It is important for the consumers to understand the need for a smart way of utilizing electric energy, reduce wastage, efficient management and their responsibility in reducing CO_2 emission.

The electric energy consumers can be grouped into industrial users, commercial users, including municipality and community users, and residential users. The policy on electricity generation, transportation and distribution, and the tariff structure varies from region to region, country to country and also among consumer types. Also, the nature of usage of electricity, demand on electricity and consumer behaviour varies from region to region and country to country. All these factors have a direct impact on the success of smart-grid technology. Compared to customers with telecommunication and television network, the electricity consumers are not provided with either the service options or price and time-based consumption option. The smart grid will provide the consumers this option.

Section 9.2 details about the different types of consumers that come into the smart grid. Section 9.3 deals with the methodologies of smart grid and Section 9.4 details the benefits of smart grid on different consumers. Section 9.5 deals with the consumer behaviour, power tariffs and policies and how consumers can be engaged in the smart grid. Section 9.6 gives a brief summary of the consumer-level infrastructure requirements. Finally, Section 9.7 lists out some case studies.

9.2　Types of smart-grid consumers

For the proper operation of the power system network, the generation should match the load in order to keep the voltage and frequency constant and also avoid damaging the expensive infrastructure. With the increased concern about efficiency, reliability and power quality, accurate prediction of the load is important. The load demand-side management (DSM) is the key for the success of the smart grid. For efficient DSM, it is necessary to have good load models. Unregulated use of load will result in large peak load which in turn lower the efficiency of the power system [2]. Electricity is consumed by a range of customers, including industrial, service/commercial and residential. This section will detail about the types of consumers, their share in electric energy consumption, their role in the smart grid and how much they will contribute in the smart grid. The type of consumers and their share on electric consumption varies from region to region and country to country. Figure 9.1 shows the worldwide average electric energy consumption of

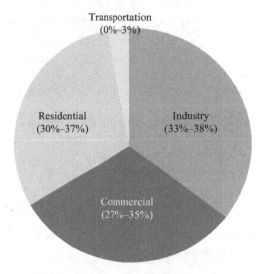

Figure 9.1 Worldwide average electric consumption of different types of consumers

different types of consumers [1,2]. The transportation sector electric energy usage varies from 1% to 3% from country to country. The usage is highest in this sector in the European region and Japan, mainly used for public transportation so can be classified under commercial type of consumers. The main purpose of showing this as a different category in Figure 9.1 is that it is predicted that with the emergency of EVs and PHEV, the electric consumption for transportation will be 8%–10% by 2030 [3].

9.2.1 Industrial consumers

Industrial consumers are mainly the bulk power consumers. The manufacturing industries such as oil and gas industry, cement industry, textile industry, paper industry, automobile manufacturing industry, foundries, cold stores/refrigeration companies, dryers, industrial furnaces, greenhouses, street lighting, air-conditioning/ventilation, electric heating and large IT systems fall in this category [4,5]. The electric power demand will be in the range of a few hundreds of kilowatts to few gigawatts with majority in the range of few megawatts. The process industry and manufacturing industry which run in shifts have almost constant electric load throughout the day. The industries that work only during the normal day shifts will have constant load during the daytime and around 1% of the full load during the night-time. The participation of industrial consumers in the smart grid is important due to their consumption footprint, heavy energy use and complexity in the implementation of smart-grid technologies. In a large industrial power system, network already has built in sophistication, intelligence and automation. The power system automation system already has data and processing capability. The main attractive features of industrial network which will make it a most easy candidate for the smart grid are

- High power density, large power consumed in a small area of power distribution network.
- Few interconnection points with the grid, a less number of points of common coupling.
- Single customer with large power consumption.
- High level of understanding of energy efficiency, reliability, power quality, standards and environmental issues. Regulation on pricing, power factor-based tariff and power quality penalties are already in place.
- Generation: The majority of these networks have their own generation plans. Many of these use their own generation to meet the peak demand. Some of them have provision to install large generation based on renewable energy sources.
- Bidirectional power flow capability, some of the existing industrial systems have already installed this feature.

With this environment, it is easier to understand and implement the smart grid. The availability of electric power in a reliable manner is the very important factor for industrial customers. The other most important factors are efficiency and power quality. As the smart grid is on improving these features, the industrial systems will easily adapt to the smart grid [4]. Additional advantages that can be available for the industrial consumers are

- Careful design of the network to achieve controlled short-time overloading of the system.
- Alternative supply paths and isolation of a possible faulty part of the network. In many cases, parallel common coupling points and main transformers can be used.
- If the sum of the fault currents exceeds permissible limits, current limiters may be used between parallel systems or sufficient operational values may be selected.
- Own power generation can be used for bidirectional power flow capability.

9.2.2 Commercial/service consumers

This is generally small- and medium-type enterprises such as non-resident lighting, commercial and general non-industry, agricultural industry, service providing facilities such as universities, shopping malls, sewage treatment plants, government sectors, and religious and social structures. Common uses of energy associated with this type of consumers include space heating, ventilation and air conditioning (HVAC), water heating, lighting, refrigeration and cooking [5–8]. Street lighting is one type of commercial loads which is managed by the municipality. The load pattern is somewhat constant with full load between 6 p.m. and 6 a.m. and almost zero load rest of the time. Typical load profile of shopping malls is constant load from 8 a.m. to 6 p.m. and higher load due to lighting from 6 p.m. to 12 a.m. and light load from 12 a.m. to 8 a.m. The typical office load will be 8 a.m. to 5 p.m.

and light load from 5 p.m. to 8 a.m. A typical university load with campus housing can be divided into university load and housing units as residential loads. Commercial loads will have some peak demands at regular intervals due to switching on and switching off of some of refrigerators, cold storage units in shopping malls, switching on and off of some of the appliances in restaurants, etc. Some of these loads can be programmed based on shift in time or shift in load demand, etc. Like industrial consumers, most of the commercial sectors, customer knowledge of energy management is high and technologies to enable demand response or energy efficiency are well known, mature and driven by cost saving. A smart energy consumption approach needs to be looked in two different angles in this sector. Some of the old establishments need to look for ways to reduce electric energy consumption and dependence by using smart-grid technology. In many countries, the new building codes have already made these buildings energy efficient and smart-grid technology will further improve the usage of electricity. These are detailed in the following section. Football and cricket stadium have potential to install photovoltaic (PV) panels in the stadium roof tops and can generate electricity in a few megawatts range. These sources can be tapped and connected to the grid. In the future, the vast parking lots in the urban area, near the shopping malls and university parking lots and community centre parking lots, can have solar chargers and the excess energy in these charger can be used to export energy to the grid.

9.2.3 Domestic consumers

Domestic or residential customers play a vital role in the smart energy consumption. The energy usage in residential sector varies from region to region as shown in Figure 9.2. For example, bulk of the energy is used for HVAC in developed countries which largely fall in the northern and southern hemisphere such as the USA, Canada, Europe, Australia, Japan and Russia. In these countries, the bulk of the energy consumption is used for heating and cooling purposes. The EVs and

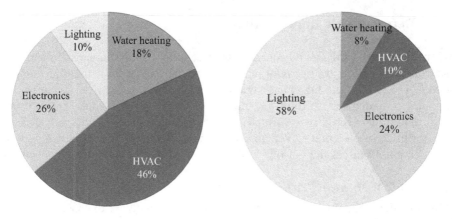

Figure 9.2   Energy consumption by domestic consumers in different geographical regions: (a) developed countries and (b) developing countries

PHEVs will form 10% of the overall electricity consumption by 2050 [1–3]. Smart-grid technology can enable intelligent management of vehicle charging [9].

The stored energy in the EV, when demand is low, can be used to pump back the electric energy to grid (V2G technology) [9] during peak power demand. Similarly, small renewable energy systems such as residential PV systems or wind energy systems can be used to generate and feed the energy when generation is high and or can be used to store energy and use when peak demand on the grid is high. With nearly one-third of the load is consumed by domestic consumers, there is large potential to expand the smart-grid technology to this type of consumers. The studies have revealed that the domestic consumers can be grouped into three groups based on their behaviour or attitude towards energy consumers [10].

Careless consumers: Some consumers do not bother about the month energy bills. The planned discounts on off-peak load consumption or penalties for peak time usage or any other rate-based policies have little influence with this type of consumers. The utilities and governments need to develop innovative ways to persuade this type of consumers to participate in smart energy consumption.

Cost conscious consumers: The majority of the consumers in domestic sector come under this category. Price incentive-based schemes are very useful and with little effort to educate these consumers they will be motivated to participate in the smart grid.

Environment conscious consumers: This type of consumers are highly motivated. They are well informed and educated about the need for reduced energy consumption, their responsibility towards environment, society and CO_2 emission. They will volunteer to participate in the smart grid, invest in a smart-grid apparatus and will actively engage in the energy-saving and DSM schemes.

In developing economies with large population still below the poverty line, the energy bills are highly subsidized by the government. Often this is misused and leads to careless utilization or excess consumption of energy. There are no proper methods to check if these schemes are really effective or reach the deserved people.

9.3 Demand-side management

Electric load DSM focuses on changing the electricity consumption patterns of end-use customers through improving energy efficiency and optimizing allocation of power [1–3,9–14]. The DSM has three different options, namely onsite generation of electricity, demand shifting and demand reduction (DR). The changes in the consumption of electricity are usually in response to the tax benefits, price incentives or price of electricity over time which is explained in the following section. This section discusses the various DSM methods for different types of consumers.

9.3.1 DSM in industrial consumers

DSM in industrial consumers will be different from other types of consumers. Most of the industrial plants already have the onsite generation which can be tapped as the first option for the DSM. Demand shifting and DRs are the other options when the onsite generation cannot meet the requirement. Figure 9.3 shows a functional

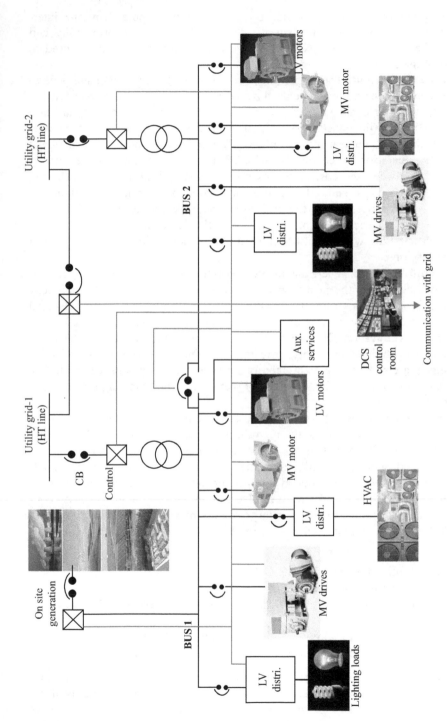

Figure 9.3   Industrial power distribution system with DSM

block diagram of a large industrial power system with more than one inter-connection point with the utility supply. Multiple feeder points form utility will increase the reliability. The existing distributed control system can be used to implement the DSM.

Onsite generation: The large industrial plants have their own onsite genera-tion. This onsite generation can be used by the industry to supply peak load demand and, thus, can offer constant load to utility [4,5]. The consumer can bargain for a better price by offering constant load. A large constant load is helpful for the power system with large nuclear power plants. The industrial consumer can adjust this onsite generation based on the price of electricity supplied by the grid. A comparison between the cost of the own generation and the cost of purchasing electricity from grid is required to decide on the amount of onsite generation. The consumer can export energy to the grid during the peak load demand based on the pricing of the electricity. In many countries, the large industrial consumer is already participating with the utility in bidirectional power flow and power import or export options.

Demand shifting: When the onsite generation is not available or not eco-nomically viable, then the next step is to shift the non-critical loads to non-peak hours [4]. Wherever possible the process can be scheduled to non-peak demand hours. This requires a careful planning of the production cycle or process.

Demand reduction: DR will be usually the last option for the industry [4]. Production constraints, inventory constraints, maintenance schedules and crew management are some of the many factors that have to be taken into account before one or more processes can be temporarily shut down. The following are the three schemes used in DR.

Incentive-based DR: A set of DR commands are issued by the utility or the DR service provider, and transmitted to the participating customers. These signal either voluntary DR requests or mandatory commands. Various types of resources can be utilized such as directly controllable loads (DCL) by the utility, loads that can be switched off or reduced upon command from the utility. Some of the residential loads may fit well within the DCL program. The command-based con-trollable loads are more suitable for large-scale commercial and industrial custo-mers [10]. The factors that need to be considered while implementing DSM are as follows

1. advance notice for receiving the DR signal,
2. maximum duration for the event,
3. the maximum number of times that the customer may receive a DR signal in a day or month or year,
4. incentive or fee payable upon compliance with a DR event and
5. penalty fee payable to the utility upon failure to comply with it.

Rate-based DR programs: The price of electricity is changed at pre-determined times or dynamically based on load demand, generation and available reserve margin [10]. Customer will pay high prices for peak hours and lowest prices for off-peak hours. The prices can be set in advance if the load profile is constant or known and remains more or less constant over a season or couple of months or can be dynamic on hourly basis, or in real time. The customer needs to respond voluntarily based on his/her necessity of loads and price he/she is going to pay.

DR bids: In this the customers participate in the bids and place bids to the utility [10]. This program encourages mainly large customers to provide load reductions at prices for which they are willing to be curtailed, or to identify how much load they would be willing to curtail at the posted price. Mainly suitable for bulk customers or customers with large reserved generation units will participate in this.

9.3.2 DSM commercial consumers

In commercial establishments, the need for smart energy management is gaining popularity to prevent high costs and bills associated with peak energy costs, and for the utilities to satisfy growing energy demand with the existing infrastructure [5]. For the existing commercial buildings, the building automation system can be imported with the building energy management and control system in order to implement DSM and include the loads. For new buildings, a separate approach where energy-saving and energy management and control aspect should be incorporated in the design stage itself. Figure 9.4 shows the DSM in a commercial power system.

Heating, ventilating and air conditioning: This forms a large portion of electricity consumption in commercial building. The thermal inertia or storage effect of inside environment of the buildings allows one to switch off the HVAC units for some time without altering the room temperature much. Pre-planned and

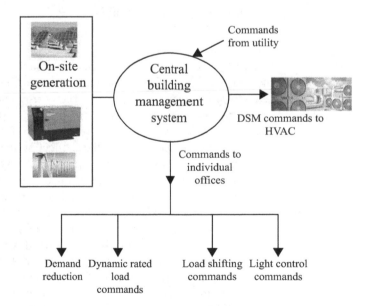

Figure 9.4   DSM in commercial buildings

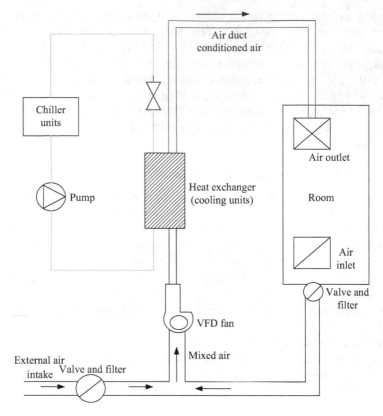

Figure 9.5   HVAC system

automated operational modes of HVAC control will help in energy saving. HVAC electric load is dynamic and sensitive to weather conditions, occupancy and other factors but has direct correlation with outside temperature. The energy consumption in HVAC can be reduced by automating control of HVAC based on occupancy, time and room temperature. Most of the office buildings are not functional in the night and the room temperature can be higher without affecting the equipment inside the rooms. The functional block diagram of a typical HVAC system is shown in Figure 9.5. In Figure 9.5 only one room air outlet and air inlet system is shown; however, in the actual HVAC system, several such units are connected in parallel to the cooling unit. The blinding HVAC consists of a central chiller plant that distributes chilled water to several heat exchangers. The heat exchanger cools the air and sends it to the air condition air duct. A variable frequency drive (VFD)-controlled fan controls the pressure and volume and circulates the air. The pump circulates the water in the chiller unit. Both the pump and VFDs consume electricity.

Based on the inputs received by the building automation unit about the pricing, peak rate energy price, etc., the HVAC should be adjusted to conserve energy. The command inputs can be given to (i) directly adjust the fan speed through control of the VFD, (ii) adjust the supply pressure/mass flow set point or (iii) by adjusting the thermostat set point in the office rooms. The first two approaches have to be implemented by the building automation unit and the third approach allows more flexibility where critical areas (e.g. rooms with servers or any temperature-sensitive equipment) can have lower temperature, whereas common office rooms, meeting rooms, etc. can have a slightly higher temperature. For example, the pilot project executed by the Sharjah municipality in the UAE with the help of Honeywell demonstrated that the energy consumption in Masjids can be reduced by 37% if the HVAC is used to set the required temperature of 22 °C just 10 minutes before the prayer time and after the prayer time when the Masjid is not occupied by the people the room temperature is set 5 °C above the normal room temperature. The controller has the capability to adjust the prayer time based on its geographical location.

Other approaches for reducing energy consumption by HVAC units are outlined below. The HVAC energy demand varies from region to region. In areas where there is high solar radiation, the heating effect due to solar irradiation will increase the load on air-conditioning units. In such regions, using large glass windows to get daylight or natural light will increase the energy consumption. The following are some of the approaches used to reduce the heating effect:

- Building orientation to reduce direct solar light penetration.
- Use proper design by incorporating harnessing of wind for energy generation, increased performance of thermal buoyancy in façade cavities, atria and voids and typically higher thermal mass from massive building structure and envelope. Examples of buildings that incorporate such strategies include the Pearl River Tower in Guangzhou and the Bahrain World Trade Centre, which incorporate large wind turbines into their façades [5].
- Glazed or screened corridor can be opened or closed to suit outside temperature conditions, thus allowing cross ventilation.
- Using thermal insulation of roof, wall and window elements can significantly reduce cooling loads of buildings in hot and arid climates. Insulating glass units can be used to reduce the heating effect. The electric energy consumption can be reduced by 40% in this approach.
- High thermal mass in external walls and roofs will be particularly effective in office buildings which are un-occupied at the nighttime.
- Shading of windows in hot and arid regions. Architectural shading can include correct orientation and opening sizing as well as colonnades, balconies, roof overhangs and planted mesh.

Lighting: Lighting is the next significant load in commercial buildings. Daylight can be used instead of electric lighting in the daytime [6,7].

- For the existing old buildings, some methods such as replacing the windows with glass windows and using reflectors to reflect daylight to rooms can be employed.
- Several countries have implemented building codes to allow maximum daylight or natural light and ventilation in the buildings [6]. Direct sunlight on building walls or windows will have a negative impact on air-conditioning units. A cost-effective method is using the window covers or shades, curtains

and tinted glasses. Building heat can be reduced by using thick glazed glasses for windows and increasing the area of windows, using glass or reflective tiles as cover to the outer walls. This approach is used in modern buildings in the Middle East and other developed countries.

- An expensive but very effective way of getting maximum daylight with minimum direct sunlight on the building wall is to use a fully automated protective moving shield around the building outer cover. An example is Al Bahar building at Abu Dhabi, UAE. The outer shield is moved to avoid the direct sunlight and the shield has umbrella-like elements that can be opened and closed to keep the sun off the glass building as it moves across the sky but also let in daylight. This approach has reduced the heating effect due to the direct sunlight by 50%.

- Reducing lighting also reduces room temperature, thus reducing load on air-conditioning energy. Lighting has safety implications, hence reducing lighting should be carried out selectively and carefully, considering the tasks in the space and ramifications of reduced lighting levels for the occupants. Another way of reducing lighting consumption is to use occupant-responsive lighting. The technology consisted of a workstation-specific lighting system, dimmable ballasts, occupancy sensors at each WS luminaire and a lighting management control system that coordinated these components. In a commercial building, this approach has resulted in 40%–65% saving in electric energy with a payback period of less than 7 years. Zone switching, fixture switching, lamp switching, stepped dimming and continuous dimming are some approaches implemented in modern building.

Other plug loads: In commercial office buildings and governmental buildings, the DSM of loads is bit restricted compared to residential consumers. The two main load components HVAC and lighting are the direct candidates for DSM. However, there are other loads which can also be considered under DSM [7]. This requires a different approach. One approach is to have a building-level energy management and control system (BLEMCS) and the BLEMCS will be communication with the office-level energy management and control system (OLEMCS). The office-level control will take a decision on which the plug loads in that particular office can participate in the DSM. The HVAC and lighting loads do not come under the OLEMCS. They will be controlled by the BLEMCS directly.

Own generation: Most of the large commercial buildings, shopping malls, have standby power generation units and storage systems (uninterruptible power supplies). This can be used to generate own electricity during the peak power demand. However, proper policy must be in place as this will add to urban pollution due to diesel run generators.

Shift in time: The majority of the office, government office, university loads exist between 8 a.m. and 5 p.m. Wherever possible the load can be shifted by changing the office hours. However, this has only a limited scope.

Intelligent loads: In large buildings and shopping malls, the elevators and escalators can be more cost efficient by using regenerative drives. The potential energy associated with the pumped water in the high-rise building can also be used to generate power.

Use PV generation units: Building tops need to be covered to reduce the heating effect due to direct sun irradiation. Solar panels can be used in conjunction with the concentrated solar units to cover the building tops. Proper orientation of the solar panels will increase the power generation. Building facades with solar panels can be used in addition to the building top panels. A further increase in the solar power generation can be achieved if the solar panel-based facades are movable and always orient towards the sun so that the PV panels are directly facing the sun and receiving the maximum irradiation. Another alternative option to enhance the power generation is to use solar concentrators. However, these ideas are in the infant stage only, more research is required to make these ideas economically viable.

9.3.3 DSM in residential consumers

In developed countries nearly 35%–40% of the electric consumption is by electrical appliances such as washing machines, dryers, dish washers, water heaters, ovens, mobile chargers and more recently EVs [9]. So, there is ample scope for demand management at the consumer end. Consumers in this category are not aware of energy saving, energy efficiency and smart-grid concept. The utilities have to invest on educating and creating awareness on DSM. The most effective method of DSM in residential consumers is pricing policy. The following pricing policy can be implemented.

- Time of use pricing: different types of pricing such as peak rate, off-peak rate and dynamic pricing, including critical day pricing, day ahead pricing, real-time pricing, probability spike rates, are suggested. In time of use, consumers are encouraged to shift the non-essential loads to off-peak time. Consumers can schedule or preprogram the use of some of the loads such as cloth washer, dryers, water pump and water heater. Recently, major manufacturers of home appliances such as LG, General Electric, Whirlpool, Panasonic, Electrolux, Bosch and Samsung have invested in the production of smart appliances which support automatic DR by networking customers with utilities. Wireless communication will be very handy in these applications. The residential energy management and control unit can be programmed to time the application of these appliances on a predetermined schedule time of each day or can start these appliances based on the load on the grid.
- Essential loads: some of the loads such as cookers, ovens and lighting load cannot be shifted or scheduled, proper provision and planning is required to supply load to this.
- Consumers can shift to energy-efficient lighting such as LED lights to reduce the energy usage.
- Water heater with high thermal inertia can be used and these water heaters can retain the heat for the whole day and can be switched on in the nighttime when the load demand is low and hot water will be available throughout the day.

- The HVAC system can be programmed to reduce electric consumption. Turning off the HVAC for a short interval will not alter the room temperature much.
- Alternative energy sources should be encouraged. Wherever possible, the solar water heater can be used instead of electric water heaters. PV-based electric energy can be used. Residential PV generator units should be interfaced with utility and consumer should be given the option to export the excess energy available in these units.
- In the near future, EV charging will alter the load demand in the residential consumers sector. Slow charging during off-peak time is the best option available for EV charging.

Figure 9.6 shows how the peak load can be reduced by shifting the non-essential loads. Figure 9.7 shows the functional block diagram of DSM in residences. As more and more electronic appliances and LED lighting use DC power

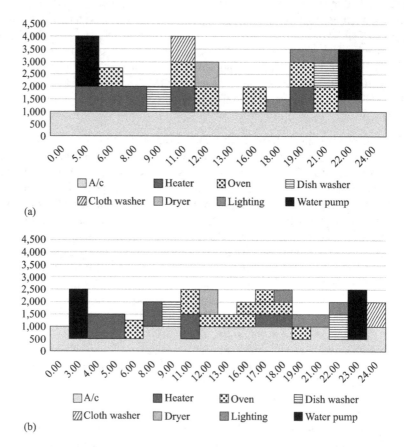

Figure 9.6   Load demand in a typical residential unit (a) without DSM and (b) with DSM

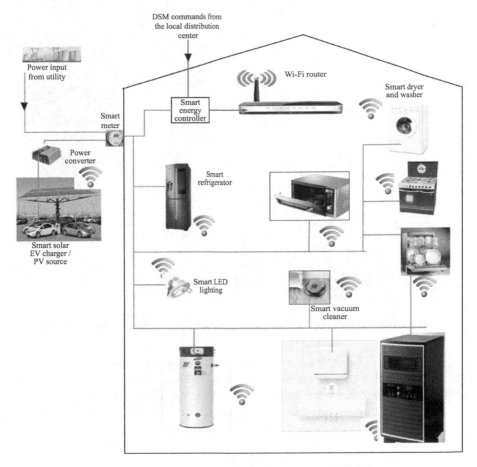

Figure 9.7 Smart residential system with DSM

supply, the residential energy efficiency can be increased if the number of power converter units is reduced. So, a hybrid distribution network as shown in Figure 9.8 is suggested for the residential distribution system. In future, the residential distribution system will have both AC and DC distribution systems.

9.4 Benefits from smart consumers

The following are the benefits of DSM by the smart consumers.

9.4.1 Reducing the peak load

By reducing peak load demand, the existing infrastructure can be used to supply more and more new consumers. The grid infrastructure can be operated with almost constant load with reduced power variations. For example, the study has

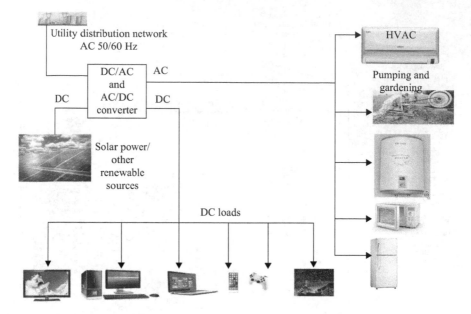

Figure 9.8   Smart residential system with hybrid (DC and AC) network

revealed that in cities like California, if 30% of the existing vehicles switch over to EV and if the charging is not regulated, then the peak load demand will double which, in turn, requires the utility infrastructure capacity to be doubled, which is not possible due to limitation in space for the new infrastructure, cost and environmental concerns.

9.4.2   Reducing power generation

With the increased use of communication gadgets, electrical and electronic appliances, introduction of EVs, the power demand may increase drastically. By properly managing the load through DSM, the requirement of new energy generation may be delayed. The need to install new power generation can be delayed. In summary, it can be said that the increase in power generation will be less compared to the increase in the new load on the utility.

9.4.3   Postponing the new grid infrastructure

With proper DSM, the smart grid will help in reducing the transmission and distribution losses, will ensure generation of grid with high efficiency, high power factor and with much more or less constant load. So, the existing infrastructure such as generators, transmission lines, towers, transformers, substations and circuit breakers can be utilized efficiently. This will help in delaying the installation of additional facility in the grid to meet the growing energy needs. Adding renewable energy sources will also help with finding a solution.

9.4.4   *Efficient and reliable operation of grid*

Smart-grid technology will enable the consumers to manage their usage of electricity properly, thus reducing the peak demand on utility. This will ensure that the load on the utility will be almost constant, hence reducing variations. This will result in operation of the grid with constant voltage and constant frequency. The system disturbance will be lesser in magnitude so that the system will be more stable, resulting in efficient and reliable operation of the utility grid.

9.4.5   *Positive environment impact – reducing CO_2 emission*

In the present way of energy usage, the CO_2 emission is estimated to double by 2050 [3]. So, the current trends of energy supply and usage are economically and environmentally not sustainable. So, work needs to be done in two fronts, namely (i) reduce energy usage and (ii) use renewable energy sources. The bulk consumers like large industries can set up nuclear power stations or large solar generation units. The commercial consumers and residential consumers can take measures to reduce electric energy consumption through DSM and can also set up alternate energy sources such as small-scale PV sources, biogas plants and solar water heaters. This approach will reduce the consumption of fossil fuel and thus help the environment through reduced CO_2 emission. The roadmap by international energy agency (IEA) is to achieve 47% of the electric energy generation by renewable sources [12]. Department of Energy (DOE) USA is predicting that with the present trend and technology, the electric energy generation from the renewable sources will be 80% by 2050. Denmark has 30% of its electric energy from the wind energy and it is expected to reach 50% by 2030, this was possible with the development of the smart grid. A large number of off-shore wind generators are connected through the HVDC link and smart-grid technology enabled the operational grid under fluctuating wind speeds. European Commission is aiming 50% electricity from renewable sources by 2030.

9.4.6   *Cost saving*

DSM by smart consumers will result in reduced energy bill. This will enable the consumer to save the cost on energy. The properly managed load will reduce the peak load on the utility grid. The electric generation from alternate sources at the consumer end will reduce the load demand on utility. So, the utility will save cost. A study requested by Federal Energy Regulatory Commission (FERC) reported that a moderate amount of demand response could save about $7.5 billion annually by 2010 [5]. In the current pricing policy where the consumer is charged with constant charge irrespective of their consumption, the smaller consumers are charged more than the large or careless users. The DSM with smart metering based on dynamic pricing or consumption-based pricing will benefit the small consumers. For example, the deployment of smart meters and home displays in Italy encouraged 57% of the involved customers to change their behaviours [3]. Time-based rates are expected to reduce energy consumption by 5%–10% and shift 1% of the energy demand to low peak load times. Also, smart meters reduced the errors in billing and also reduced the response time significantly in addressing the issue of wrong bills [3].

9.4.7 Distributed generation and distributed storage

Smart-grid technology will encourage distributed generation. Smart inverters are required with fast response time, bidirectional power exchange to integrate the distribution generation to the grid. With smart consumer technology even a small amount of surplus power available can be exported to the grid. Example is vehicle-to-grid (V2G) technology. Grid may not be in a position to absorb the available surplus energy at the consumer end all the time. So, the surplus energy can be stored by low-energy consumers in lithium-ion batteries, lead acid batteries, some types of flow batteries, and by bulk energy consumers in thermal storage, flywheels, supercapacitors and hydrogen storage. This will result in distributed storage in the grid.

9.4.8 Microgrid and virtual power plants

Smart-grid technology with distributed generation will also result in micrograms. With the integration of renewable generations, the voltage and frequency may fluctuate [12]. The smart grid can take a decision on islanding and operating the unstable area of the grid as microgrid. A microgrid with a weak link with the grid will operate like a virtual power plant.

9.4.9 Opportunity for research, training and manufacturing

Smart consumers require smart technology in their premises. So, the future home appliances, the HVAC system, need to be smart with the built-in intelligence system and wireless communication capability. The dynamic pricing requires smart metering, communication between smart meter and loads and also communication between smart meters, load and central energy management and control unit in the premises of the consumers. The onsite energy management unit should communicate with the utility energy management system in real time. So, the existing meters and loads need to be upgraded with information and communication technology (ICT) capability. The load isolation or turning off requires remotely operated controllers. This will give an opportunity for new products and development of industry. Also, the DSM will motivate consumers to install alternate energy sources, even a small amount of excess energy produces need to be exported to the grid. This requires efficient, reliable, compact and smart power converter circuits with bidirectional power flow capability. Smart consumers and smart grid ensure reliable and clear electrical energy. So, the loads will be of high quality with low harmonic distortion. Smart loads and smart loads manufacturing, commissioning and maintenance industry require trained man power in the area of smart-grid technology. This will create opportunity in new training courses on emerging technology. All these changes can be achieved only through research and development. So, there is ample opportunity for research and development in this area for next couple of decades. For example, several research projects already awarded by European Commission and the USA to universities and the trend is growing in other countries also. Examples are Low-Carbon Networks Fund and smart meter roll out project in Great Britain, etc. Large off-shore wind farms in northern Europe have enabled HVDC technology for bulk power transfer. Also, the volume of flexible AC transmission systems is increasing. Research opportunities exist in the area of distributed generation, electricity storage and HVDC technology [11]. Small consumer like residential consumers will make the best use of their onsite renewable energy sources if cost-effective storage solutions are available. From the reliability, cost, size and weight point of view, tremendous research is required in battery technology.

9.5 Challenges in smart-grid operations

Currently, the smart grid is not implemented in a large scale. Several pilot projects and experimental trials are conducted [1,2,11–13]. However, to reap the benefits of smart grid, it should be implemented in a large scale for a long period of time. In order to implement the smart grid, several obstacles or challenges need to be addressed. These are explained below.

9.5.1 Challenges faced by utilities

- Utility infrastructure upgradation: the power generation, transmission and distribution industry should be prepared to upgrade the existing infrastructure to meet the smart-grid requirements. The utility should have facility and capacity to collect the vast data from the consumers, ensuring reliable and guaranteed uninterruptible power to the consumers. So, the existing infrastructure should be combined with the ICT. This requires proper planning and capital investment. In countries like the USA where utility is owned by the private industry, the cost of system upgradation will be passed on to consumers, this may increase the energy bill in the beginning phase. In many other countries, the utility operate with subsidized pricing due to social and economic constraints. So, the cast starved utility requires funding for upgradation.
- Smart consumers will expect reliable and uninterruptible power supply. The supply outage for industrial units may cause production disruption, process shut down and equipment damage resulting in huge financial loss. The smart consumer will expect a higher level of reliability compared to the current power systems [10,14]. In many countries, there is a low level of trust on utility. So, there will be hesitation by consumers to accept the changes.
- Any noncompliance with the promised service may result in litigations between consumers and utility. So, proper laws and legal procedure need to put in place.

9.5.2   Government policies

In many countries, the power generation and distribution is highly regulated and under government managed and supported by state. Proper policy should be developed and put in place. Proper legal policies, producers, consumers and utility responsibility duties should be developed and put in action. Proper policy should be in place for economically weak consumers, government social responsibility in providing electricity to the mass population economically and socially weak sector. Studies have shown that some consumers will be lazy to adapt for the smart grid. The dynamic pricing and penalty for excessive usage of electricity will not deter them from using electricity in an uncontrolled way. Proper policy should be in place for the utility to penalize such consumers or disconnect the power supply to such customers.

9.5.3   Consumer engagement, education and concerns

- Consumer participation and active engagement is key for the success of the smart grid. For proper consumer engagement and participation, it is important that consumers are educated about the smart grid, its benefits and importance. Industrial consumers are already aware of the smart grid and it will be easy to implement. Commercial and residential consumers are still unaware of smart-grid technology. So, educating this group is very important for the success of smart grid. The utilities need to plan about educating the residential consumers about smart-grid technology, their responsibilities, the benefit of smart grid and importance of their participation in the smart grid. At the same time, proper survey and study on the consumer behaviour, consumer load use pattern, consumer attitude is required. The utilities are required to conduct educational awareness on DSM. According to a questionnaire distributed to about 2,000 customers within the GAD project [3], 62% of consumers would modify their behaviour if they were notified when the current energy production came from renewable energies. More than 55% of the consumers would also modify their habits if the energy price varied during different hours of the day. Another survey conducted at the end of the GAD project among 300 advanced users revealed that 65% would use the system in the short term if the cost did not exceed €500. The smart grid will result in two classes of small consumers – active and non-active. The active consumers will benefit from the subsidies and the supplier will be compensated through higher rates for non-active consumers [3].
- Use of smart meters, concerns about data privacy, safety: with the use of network systems for data exchange and wireless connectivity, the consumers will be concerned about the data privacy, data storage and safety of this data. This will be a major concern especially for industrial consumers and some of the commercial consumers. Proper policy should be developed to address this concern. A detailed information about electricity use could be used by insurers, market analysts, or even criminals to track the daily routine of consumers. A survey has shown that 35% customers would not allow the utility to control thermostats in their homes at any price [12].

- Smart home appliances, concerns about efficiency and reliability of these appliances: There is a general concern with the residential consumer about the reliability of the new smart home appliances and their remote control or operation. DSM will ensure reliable electric supply compared to the current supply situation. The consumers must be explained on this factor and proper training and education should be given in this direction.
- Health concerns: Wireless control of smart home appliance will result in more electromagnetic waveform in the residence. Some of the communication gadgets such as cell phones, Wi-Fi router and repeater will be self-charged through electromagnetic waves. Proper methods need to be developed such that the electromagnetic radiation should be within the safe limits set by standards. Consumers must be made aware of the standards and safety limits.
- Motivating the consumers to change their behaviour towards shifting of noncritical loads, such as using dish washer, water heaters and dryers. There will be initial hesitation to adapt to the change, adjusting the life style for the controlled load usage. There will be concerns about the impact on comfort especially in the control of HVAC, freezer units, etc. Proper training and education need to be given to the consumers. This can be done by publishing the success stories, through TV advertisement, community engagement, regular training courses, etc. Age: aged people are more interested in interacting through meter display or TV. New generation prefers online portal through mobile applications. Using price as a tool to control consumption may not have much impact on less technically savvy people.
- Wherever possible use natural methods, for example use sunlight for cloth drying instead of dryer, use small solar power supplies to charge electronic gadgets such as mobiles, use solar water heaters to heat water. Use biogas in rural areas for cooking and lighting. Some of these usages are not as convenient as electricity but proper training is required to motivate the people to switch to alternate energy sources.
- Changes in life style, for example encouraging to avoid repeated reheating of food in small quantities by encouraging all family members to consume food together on prescribed time schedule, avowing frequent use of dish washer or cloth washer by using these with full load with reduced frequency, use water heater optimally, reduce the consumption of water thus reduce the requirement on water pumping.
- Educate on the importance of the smart grid on energy saving, the responsibility on environmental impact, role in reducing CO_2 emission, etc. Educate on the need for renewable energy sources. Smart-Grid consumer Collaborative (SGCC) is an initiative taken by energy utilities in the USA for educating and engaging their customers about how to better control the energy they use, the resulting costs they incur and the benefits of shifting their consumption [10].
- Reducing subsidy: In most of the countries, electricity is subsidized by the governments. It may be difficult to completely remove the subsidy as the governments need to look into the social impact on the people. However,

the governments should put an action plan to make the subsidy more effective, avoid proliferation or misuse of the facility, gradually reduce the subsidy, discourage waste of electricity. Many of the countries have started implementing the same. For example, the countries in the Middle East have an action plan to reduce the government subsidy. The developing economies such as India, Pakistan, Bangladesh, Indonesia and Brazil have limited the subsidy to low-income families.

- Loans: The governments should extend soft loans for the research and manufacturing of smart-grid apparatus such as smart meters, control units, smart grid friendly appliances, energy-efficient gadgets and renewable energy sources.
- Tax incentives: The governments should put a plan to extend tax benefits to the smart-grid apparatus manufacturing units, tax incentives for consumers participating in smart grid, renewable energy sources.
- Price discounts: Encourage consumers to participate in the smart grid by giving electricity price discounts. For example, countries such as India and China give price discounts on electricity consumption bill if the consumers use alternate sources such as solar water heaters and PV sources. Developed countries such as Europe and the USA are encouraging consumer participation by give price discounts on the use of electricity based on time.
- Penalties: Discourage the wastage of electricity or reduce the energy consumption by charging more for consumers. For example, many countries give subsidized or reduced price for basic minimum usage or electricity and charge very high price if the consumption is above the prescribed limit.
- Force the consumer to pay: Labour laws need to be embedded to remove the energy bills from the list of benefits to employees. Employer should not be allowed to pay the energy bill, the consumer should be encouraged to pay the energy bill from his monthly income. This will make the consumer to be more cost aware.
- New skill requirements – training. New job profiles: high level of flexibility, adaptability, customer-focused approach, sales skills, regulatory expertise. Need for investment in the development of relevant undergraduate, postgraduate and vocational training to ensure the building of a sufficient pipeline of the next generation, smart-grid savvy electrical engineers [14]. Adequate training, re-skilling, up-skilling of the workforce is essential. Homeowners also get email, text or phone alerts when events are called. The customers can access historical and real-time consumption data on their electricity consumption via meter display, TV, smartphone/tab devices or an online portal access information on the energy consumption of their appliances in an engaging user interface as a web application running on smartphone. There is the need for smart-grid investors to look beyond national borders. Especially, China and India need to have infrastructure to train the young generation with the new skill requirements. Europe, Japan and the USA have low workforce and they depend on these countries for the skilled manpower.

- Collect the best practice on consumer feedback and use it to improve pilot projects: monthly electricity bills can be used as feedback. More rigorous and methodical research and evaluation is needed to obtain consumer feedback and pricing or incentives and the effect of enabling technologies on results. History of consumer feedback policies, variety in customer types and preferences and the specifics of the service options are also important. Scientific research on consumer feedback by collecting and comparing the results of advanced metering, real-time pricing and consumer feedback demonstration. Establish a community of practice internationally to develop standard methods and analytic tools for estimating the consumer behaviour change benefits of smart grids. It is necessary to carefully choose and involve consumers early on in trials and demonstrations, before moving to full-scale deployment, in the initial stages give consumers the freedom to choose their level of involvement. Special attention needs to be devoted to the needs of vulnerable consumers. Proper measures to be taken to counter negative consumer perceptions and to build trust and understanding among the consumer, the utility/consortia and smart technologies.

9.5.4   Technological challenges in smart-grid operations

- Lack of predictive real-time management tools, predictive optimal power flow control software and hardware, and DMS technology.
- Lack of funding of the development of predictive grid management and control technology for the deployment of smart control software and hardware with tax incentives are options.
- Lack of predictive control signals to operate devices and lack of energy storage devices affect the deployment of smart devices. Funding is required to spur innovations of new technologies in this area for the smart grid.
- Need for smart home appliances. Recently, major manufacturers of home appliances such as LG, General Electric, Whirlpool, Panasonic, Electrolux, Bosch and Samsung have invested in the production of smart appliances which support automatic DR by networking customers with utilities. However, it will take few more years for the wide spread use of these appliances.
- Necessary infrastructure for centralized control system in industrial systems, building managements systems, home networking, data managements, data security, wireless communication, etc. The industrial systems already have the central DCS system and necessary data network. It will be easy to further enhance this network to implement smart-grid functions. However, the infrastructure in commercial buildings and residential systems need to be developed. Many countries have already implemented new building codes to incorporate intelligent energy managements system by using occupant-responsive lights, low-energy consumption lights such as LED lighting, regenerative escalators and elevators. The residential smart distribution system is still in the infant stage. More work has to be done on developing this system.

- Own generation: Commercial buildings and residential units should be encouraged to have own generation especially from the solar PV panels and wind energy systems. Bidirectional power flow should be incorporated. Even small generating units should be in position to send the energy back to the grid. Vast football fields, cricket stadiums have large capacity solar panels at the stadium roof and these can be integrated into the grid. University complexes and shopping malls should be encouraged to install PV units. Concentrated solar power generation units should be used instead of conventional PV panels.
- Electric vehicles: Proper charging regulation should be in place for the future EVs and PHEVs. It is expected that 10% of the residential power consumption will be by EV by the year 2030. Slow charging and off-peak charging should be allowed at residential units. Solar chargers should be encouraged. Urban car parking areas, university car parking area, shopping mall car parking area and other places where large yards are used for car parking must be fitted with hybrid power sources, namely a combination of solar chargers and utility-connected chargers. These solar chargers should be connected to the grid and excess electric energy can be exported to the grid. Large parking lots can also be used to store energy and export energy when needed from the EVs using V2G technology.
- Wireless communication towers: With the 4G LTE and future 5G, the power consumption in wireless communication base stations will increase. Wherever possible these base stations should be powered by renewable energy sources and excess energy can be exported to the grid. These base stations can be supplied from the grid during the off-peak period, for example during mid-night to dawn.
- More research need to be required on solar panels, line-interactive inverters and storage batteries. Still these are very expensive compared to the price electricity from the fossil fuel.
- Oil prices: More and more energy-efficient usage and dependence on renewable energy may bring down the cost of the fossil fuel such as oil and gas. Low oil process may move the customers away from smart-grid practices. Proper plans must be in place to motivate the consumers to continue to use renewable sources. They should be educated about their responsibility on environment and CO_2 emission.

9.6 Consumer data protection, privacy and system reliability

Smart grids are to improve reliability, stability and performance but the increased levels of user access and access points may become more vulnerable to attack. At its worst, cyber-attacks may lead to complete collapse of a grid. Smart meter can be tampered with and false data may be fed to the system. This will give wrong information to the control centre. Direct impact may be the financial loss but indirect impact may be the impact on system operation and stability. Cyber-attacks on the smart grid for corporate espionage, creating false blackouts to hamper production, virus attack on system, electronic data theft, etc., will create huge financial loss and worst-case grid collapse. International-level cyber security measures and standards should be developed. The requirements of a smart grid are

- The grid should continue to work even during cyber intrusion. Smart consumers should not be affected.
- Consumer data must be protected. Consumer protection issues associated with remote disconnection functions made possible by smart grids. Smart grid and smart meter deployments create large amounts of detailed customer-specific information, while energy providers gain a new medium for customer interaction.
- Confidentiality, privacy, ownership and security issues associated with the availability of detailed customer energy consumption data must be maintained.
- Customer acceptance and social safety net issues associated with dynamic pricing must be maintained.
- Proper policy on sale, disclosure or transfer of customer data, restrictions, permissions, terms and conditions related to this need to be put in place.
- Vulnerable consumers such as those who cannot adjust their usage patterns as a result of pricing, those who cannot change their usage behaviour or those who are socially and economically weak may create problem with smart-grid operation. Smart grids could allow quicker disconnection of service and negatively impact vulnerable consumers.

Further research is needed to identify the full range of consumer protection policies and make recommendations to governments on smart-grid-related consumer protection issues.

9.6.1   Reliability

The smart grid in the long run will improve the system reliability. The factors that will help in improving system reliability are

- DSM will ensure that the load is managed in an effective way and sudden changes in load or peak loads are avoided. This will reduce the sudden changes and associated transient in the system.
- Too many and large variations in the load will be reduced which will help the grid infrastructure.
- The grid will operate with constant voltage and frequency compared to the present scenario. This will improve the system reliability and stability.
- Small consumers and also the bulk users will have alternate power supply sources which will improve the reliability of electric energy availability.
- Smart grid will have distributed storage [12], which will help in enhancing the system reliability.
- Proper grid code must be developed and put in place for the safe, and reliable operation of the grid.

9.7   Case study

Smart-grid technology is still in early stages of implementation. Several pilot projects and actual projects covering a small region or area are implanted. Some of these are listed below:

- A GAD Project [3] based on hourly energy price has encouraged the consumers to react by modifying their loads in order to reduce their bills by 15%.
- In the Storstad Smart Metering Project in Sweden [3], the deployment of about 370,000 smart meters contributed to a significant change in customer interest in their electricity consumption.
- Increased consumer involvement and consumer engagements will benefit the smart grid. It is imperative to ensure tangible benefits, privacy and easy access for consumers; and to grant open access and fair competition among energy players.
- Perth Solar City, an Australian Government solar initiative launched in the year 2009, has made the households to collectively save about $1 billion in their energy bill [13]. The government and utility worked together to communicate with the consumers before the roll out of program. Eco-education was given through telephone and through feedback letters. Nearly, 9,000 houses were given eco-coaching and consultation. In all, 700 houses are fitted with solar PV systems and 1,100 houses are fitted with solar water heater.
- Oklahoma Gas & Electric (OGE) has launched a project to reduce energy consumption with an aim to postpone the need to set up new generation till 2020 [13]. The programme has used various marketing messages for different groups of participants and it has put people on a variable peak pricing plan, as well as giving homeowners a smart thermostat and a web portal.
- Several utility companies such as Newmarekt-Tay Power Distribution and Powerstream in Ontario, Kankas City Power and Light, Hydro One, BC Hydro Toronto Hydro in Canada achieved 100% smart metering over a span of 5–6 years [14]. Prior to roll out, the camping and advertisement was done mainly through web. The utility directly communicated with the consumer through web and other social media and particular attention was paid to anticipating concerns and addressing them [14]. Many more examples of smart meter deployment are available in Reference 14.
- The Power shift Atlantic project successfully implemented load shifting devices to balance customer demand with variable wind generation. The project controlled remotely the water heaters, boilers and refrigeration units without disrupting participating residential and commercial customers [13].
- Singapore already has a smart electric grid and its 'Intelligent Energy System' project is expanding its grid communications network using a combination of radio communications, fibre optics and broadband-over-powerline with a view to supporting intelligent buildings, vehicles and sensors. The DSM technology has in home displays, web portals and dynamic pricing on 30-minute intervals. The DSM pilot has reduced peak residential loads by 3.9% and total energy consumption by 2.4% [12].

- The success stories of the case studies in smart metering and smart-grid initiative in the USA and also some of the international projects are available in the report published by US Energy Information Admiration [1].
- The smart grid customer engagement success stories by US utilities is available in Reference 10.

9.8 Conclusions

The smart-grid technology will change the way electricity is generated, transmitted, distributed and consumed. Both utility and consumers need to understand the concept of smart grid, their responsibilities and duties towards energy efficiency and CO_2 reduction. The smart grid will allow the existing infrastructure to power the new loads in an effective, efficient and reliable way. In order to see the full benefits of smart-grid technology, the smart grid should be implemented in a large scale and constant manner. In order to do this, the supplier and consumers should understand the smart-grid technology and should be ready to accept it. DSM is key for the success of smart grid. Consumer active engagement is very important for the success of DSM. Implementation of DSM will be easy in industrial consumers; however, lot more need to be done in commercial consumers and credential consumer sector. Residential consumers need to be educated about the smart-grid technology. Governments should develop regulatory policies to implement the smart grid. The communication and intelligence built into the loads, a proper protocol for interaction with the utility control centre needs to be put in place. The smart-grid technology will encourage the consumers to use more and more renewable energy sources, thus will have a positive impact on the environment. It can be predicted that the smart grid will be an enabling technology in the near future. This will give a new boost to the manufacturing sector for the manufacturing of smart meters, smart communicating instruments and smart consumer loads. Also, there exist need to educate the consumers, need for training courses and developing man power with smart-grid technology know how. More research is required to make the smart grid consumer friendly and reliable.

References

[1] "EIA – Smart Grid Legislative and Regulatory Policies and Case Studies," available at https://www.eia.gov/analysis/studies/electricity/ (accessed 3–23/6/2016).

[2] "Technology Roadmap – Smart Grids, Report by IEA," available at https://www.iea.org/publications/freepublications/publication/smartgrids_roadmap.pdf (accessed 18/6/2016).

[3] Vincenzo Giordano, Flavia Gangale, Gianluca Fulli, Manuel Sánchez Jiménez, "Smart Grid Projects in Europe: Lessons Learned and Current Developments", available at ses.jrc.ec.europa.eu/sites/ses/files/documents/smart_grid_projects_in_europe.pdf (accessed 14/6/2016).

[4]   "Industrial Smart Grid, ABB," available at https://library.e.abb.com/public/ 18f20edbba8069e5c1257b4a004a943c/Industrial%20Smart%20Grid_EN.pdf (accessed 14/6/2016).

[5]   N. Motegi, M. A. Piette, D. S.Watson, S. Kiliccote, and P. Xu, "Introduction to Commercial Building Control Strategies and Techniques for Demand Response," Lawrence Berkeley National Laboratory, Berkeley, CA, May 2007 [Online], available at http://gaia.lbl.gov/btech/papers/59975.pdf (accessed 12-14/6/2016)

[6]   Peter St Clair, "Low Energy Designing in the United Arab Emirates – Building Design Principles," available at http://www.seedengr.com/Low% 20energy%20Design%20Guide.pdf (accessed 14/6/2016).

[7]   "Occupant Responsive Lighting," available at http://www.gsa.gov/graphics/ pbs/OccupantResponsiveLighting_508c.pdf (accessed 14/6/2016).

[8]   Daniel Arnold, Michael Sankur, and David M. Auslander, "An Architecture for Enabling Distributed Plug Load Control for Commercial Building Demand Response," *IEEE PES Innovative Smart Grid Technologies (ISGT)*, 2013, pp. 1–6, 2013.

[9]   Shengnan Shao, Manisa Pipattanasomporn, and Saifur Rahman, "Demand Response as a Load Shaping Tool in an Intelligent Grid with Electric Vehicles, *IEEE Transactions on Smart Grid*, vol. 2, no. 4, 624–631, 2011.

[10] "SGCC Smart Grid Customer Engagement Case Studies," by SGCC, available at http://smartgridcc.org/sgcc-smart-grid-customer-engagement-case-studies/ (accessed 15–23/6/2016).

[11] Salman Mohagheghi, James Stoupis, Zhenyuan Wang, Zhao Li, and Hormoz Kazemzadeh, "Demand Response Architecture: Integration into the Distribution Management System," *First IEEE International Conference on Smart Grid Communications (SmartGridComm)*, 501–506, 2010.

[12] "Smart Grids and Renewables: A Guide for Effective Deployment," IRENA, available at www.irena.org/documentdownloads/publications/ smart_grids.pdf (accessed 13/6/2010).

[13] "Smart Grid: A Great Consumer Opportunity – A Report by SmartGrid GB," available at www.engage-consulting.co.uk/.../pdf/Smart-Grid-A-Great-Consumer-Opportunity.pdf (accessed 10–18/6/2016).

[14] "Smart Grid Consumer Engagement – A Report by Indeco," available at www.indeco.com/www.nsf/.../Smart+grid+consumer+engagement+final+ report.pdf (accessed 10–13/6/2010).

Chapter 10

Importance of energy storage system in the smart grid

Faisal Mumtaz[1], Islam Safak Bayram[1,2] and Ali Elrayyah[2]

Recent advances in energy storage and power electronics technologies are offering promising solutions to improve the grid resilience and allow higher renewable energy penetration. Energy storage systems (ESSs) act as energy buffers to aid the operations and lifetime of the grid assets and bridge the gap between supply and demand for renewable energy generation. Currently, there are more than 650 active ESS projects around the globe with a total capacity of 3.83 GW, representing a significant market potential for companies. To that end, this chapter aims to provide a comprehensive overview and classification of ESSs, underlying technologies and working principles, current and future applications, and economic analysis.

10.1 Introduction

Over the last decade, the stress on the power grids has intensified due to growing peak demand and the grid assets have become less secure as the share of renewable energy production increases. This trajectory is expected to continue and pose more pressure on the grid operations because policy-makers have set serious targets (e.g., 20% in EU, 33% in California both by 2020, and similar targets around the world) on renewable energy penetration. The trend toward renewables also boosts the need for grid regulation, extra spinning reserves, and load ramping capability to handle the unexpected fluctuations in the renewable output. Energy storage systems (ESSs), on the other hand, can help network operators to tackle the aforementioned problems. ESS can provide a buffer zone and decouple the time of generation and demand by capturing the energy off-peak hours and delivering it during the times when it is needed.

In line with the motivation given above, the need for energy storage technology has emerged as a key enabler to restructure the power grid operations. To that end, there are more 650 active ESS projects around the world with a total capacity

[1]College of Science and Engineering, Hamad Bin Khalifa University, Doha, Qatar
[2]Qatar Environment and Energy Research Institute, Hamad Bin Khalifa University, Doha, Qatar

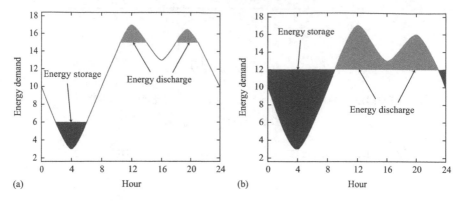

(a)   Hour

(b)   Hour

*Figure 10.1   Daily load curve. (a) Peak shaving of daily load curve and
(b) load leveling of daily load curve*

of 3.83 GW [1]. ESSs provide a multitude of benefits to power system operations. ESS can support the grid at various levels, most importantly in the bulk generation, transmission and distribution, and end-user applications and the major benefits include peak shaving, load leveling, grid support, voltage stability, reduced ramping impacts, and frequency stability. Besides, ESS can offer several financial benefits such as cost reduction and deferring the need for construction of new power plants. Two of the most popular applications, peak shaving and load leveling, are illustrated in Figure 10.1(a) and (b), respectively.

A confluence of drivers, including the shift towards the smart grids, microgrids, and distributed generation, have accelerated the interest in ESS. Renewable energy resources (RESs) are now substantial parts of the modern power systems. However, there are significant challenges that need to be tackled before RESs can be exploited to the maximum. One such challenge is the intermittent output of the renewables. For example, the fluctuations in wind speed result in a variable output of a wind turbine. Similarly, the variable sunlight intensity (global horizontal irradiance) and weather conditions, such as dust, fog, and extreme heat, could lead to reductions in the solar output, even endanger the security of the supply. Therefore, ESS is an essential tool to smoothen the fluctuating power output and play a vital role in the successful implementation of RESs. Moreover, the RESs can gradually replace conventional fossil fuels over the next decades, however, a phenomenon known as the "duck curve" should be carefully addressed. This problem occurs when a substantial amount of customer demand is met by the renewables during the daytime and around the sun set time the solar output will quickly go to zero. Hence, there will be a mismatch between the supply and the demand, requiring need for fast response resources. The energy storage units, along with demand side management and network investments, play a key role in the solution process.

In this chapter, we provide a detailed analysis on the importance of ESSs in the smart grids. We performed a holistic classification methodology and grouped the ESS according to the stored form of energy, the applications they are employed, and their storage capabilities (see Section 10.2). This methodology summarizes the main features, allowing readers to establish a final evaluation for their specific applications. Furthermore, we list the benefits of storage units for different applications of the grid and provide basic economic analysis. In the rest of the chapter, a detailed analysis of the working principles for storage units is provided (see Sections 10.3–10.11). Moreover, Section 10.15 presents an overview of analytical research problems and we conclude the chapter with a case (Section 10.17) study which shows the application of an ESS in a wind farm.

10.2 Classification of ESS

10.2.1 Classification by form of energy

Our first classification methodology is based on the form of energy that the storage unit uses. As we will discuss in the next section, the form of energy determines specifications of the storage unit. For instance, applications requiring long discharge duration typically store energy in the mechanical storage, while electric field storage can provide high volumes of power in a very short duration. Moreover, the classification of storage units based on the form of energy is presented in Figure 10.2, and the details are given below.

Electrochemical storage: ESS that uses electrochemical reactions to store energy is categorized into electrochemical storage devices. These batteries are further categorized into secondary and flow batteries which are discussed in detail in Sections 10.3 and 10.4, respectively. The literature on batteries also include the category of chemical storage, where chemical storage is then further divided into three types: electrochemical (secondary and flow batteries), chemical, and thermochemical storage [2].

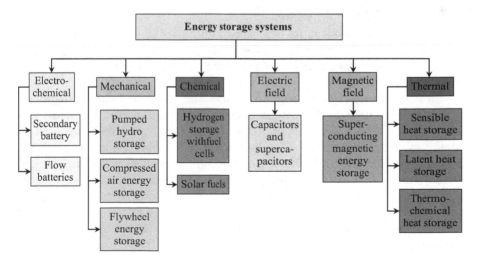

Figure 10.2 Classification of ESS based on form of energy

Mechanical storage: Storage technologies that store energy in mechanical forms such as a rotating axis or compressed air are classified as mechanical ESSs. Pumped hydro storage (Section 10.5), compressed air energy storage (Section 10.6), and flywheel energy storage (FES, Section 10.7) are the examples of mechanical ESSs.

Chemical storage: Hydrogen storage with fuel cells and solar fuels are the examples of chemical ESSs [3]. Solar fuels store energy in thermochemical form. A detailed discussion of hydrogen storage and solar fuels is presented in Sections 10.8 and 10.9, respectively.

Electric field: Capacitors store energy in the electric field. They directly store the electrical energy without any conversion from chemical, mechanical, or thermal. A parallel plate capacitor with dielectric between its plates is an example of such electrical storage devices. To store a large amount of energy, supercapacitors are used. They have much higher capacitance and high energy density. Section 10.10 explains the working principle, advantages, and limitations of supercapacitors.

Magnetic field: Just like a capacitor store electrical energy in the electric field, a magnetic coil can store electrical energy in the magnetic field. It is another example of direct storage of electrical energy. The working of a superconducting magnetic energy storage (SMES) is presented in Section 10.11.

Thermal energy storage (TES): TES stores thermal energy at very high or low temperatures so that the stored energy can be used later on. Thermal energy storage can be further classified into different categories. The classification can be based on temperature levels or material used. Based on temperature levels, TES can be categorized as high-temperature TES and low-temperature TES. In this chapter, the TES technologies are classified as sensible heat storage, latent heat storage, and thermochemical heat storage (discussed in detail in Section 10.12).

10.2.2   Classification by application

ESSs have different applications depending on the place of integration in the power grid. Some ESSs are suitable for end-users, i.e., behind the meter, while other ESSs have applications on the generation side. Next, we classify the ESS based on the applications as below.

Bulk energy storage to support system and large renewable integration: The ESS applications in this category are the wholesale market, renewable integration, and ancillary services. In wholesale markets, large ESSs are used for bidding into energy markets. ESS deployment will enable higher penetration of renewables to reduce the hydrocarbon emissions associated with electricity generation. Moreover, it will help utility operators to satisfy regulatory requirements regarding renewable portfolio targets. ESS can also provide protection against the system contingencies such as failures and outages in the power plant. Moreover, ESS can provide ancillary services such as frequency regulation and load following.

Energy Storage for ISO fast frequency regulation: The applications in this group include utility frequency control and deferral of capital cost improvements. A gap between the supply and demand will lead grid frequency to deviate from its nominal value (50 Hz in Europe and 60 Hz in the US). In order to fill the missing energy, ESS can be used and the system frequency will remain in the safe zone. Therefore, ESS defers the need to purchasing extra generators.

Energy storage for utility T&D grid support applications: The ESS applications in this group include utility transmission and distribution substation support, peak shaving, and regional transmission operator market participation. Currently, dozens of successful implementations show that ESS can provide significant savings by aiding the substation operations and extending the useful life of the existing grid.

Energy storage for commercial and industrial applications: ESS can offer energy management for commercial industrial customers. Typically, such customers pay a significant amount of money for the demand charges which is proportional to their peak demand. ESS can be used at peak durations to reduce the electricity bill. Moreover, ESS can improve the power quality which is measured by the percentage and duration of outages, deviation from the nominal voltage and frequency levels, and if it maintains unity power factor.

Distributed energy storage near pad-mounted transformer: Storage unit at the distribution network will improve the reliability and defer the investments for transformer upgrades. Also called as neighborhood storage systems, ESS at this level could be employed for peak-shaving and frequency regulation.

Energy storage for residential energy management applications: The deployment of ESS has a multitude of benefits including power quality, power reliability, reduced time of use charges, and deployment of distributed renewables. Moreover, ESS could be used in energy trading among the neighbors or selling electricity back to the grid.

10.2.3   Classification by storage capability

Our final classification methodology is based on the storage capacities in terms of rated power and energy. For example, pumped hydro storage (PHS) and FES systems could provide enormous storage capacities (in the order of several MWh). On the other hand, secondary batteries and fuel cells have relatively small storage capacity. In addition, some energy storage devices have very high charge and discharge rates which are used in applications where instant energy is required. Other types of storage devices have low charge and discharge rates. Such ESDs are useful where energy is required for longer periods. Based on storage capability ESSs can be classified as (i) large capacity with high discharge rate (i.e., FES), (ii) large capacity with low discharge rate (i.e., PHS), (iii) small capacity with high discharge rate (i.e., capacitors), and (iv) small capacity with low discharge rate (i.e., batteries). The positioning of several ESSs depending on the capacity and discharge ratio is given in Figure 10.3 [4]. In the next sections, we provide a detailed analysis on the working principles and the dynamics of different storage units.

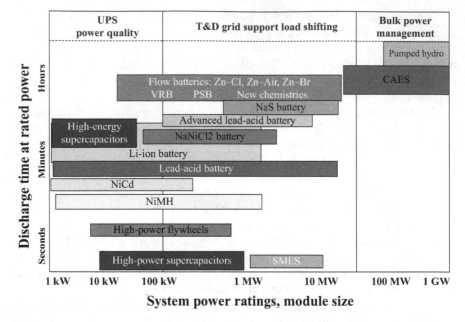

Figure 10.3   Placement of different ESSs depending on the capacity and the discharge rate

10.3   Battery storage technologies

Rechargeable (secondary) batteries are the oldest form of electricity storage technology. They convert the chemical energy into electrical energy via electro-chemical reactions. A battery is made up of a cell (or cells) which consist of electrodes together with an electrolyte. An electrolyte can be a liquid, solid, or paste. Each cell has two electrodes, the anode and the cathode. Reactions occur at both anode and cathode. During discharge, the electrons move from the anode and are collected at the cathode. During charging, reverse reactions occur and the battery is charged by connecting it to a power source. The output voltage of a cell is very low, therefore, they are connected in series and/or parallel combinations to achieve the desired capacity and voltage. Batteries have enormous applications, for example, power quality, fast response, power stability, low standby losses, energy management, and high efficiency [2]. However, they have low energy density, high maintenance costs, limited discharge capability, and short life. Furthermore, most of the secondary batteries have toxic materials, therefore, safety precaution is necessary for the disposal process [2]. The types of the secondary batteries that are discussed in detail in the next subsections include lead acid batteries, sodium-sulfur batteries, sodium nickel chloride batteries, nickel cadmium batteries, nickel-metal hydride, and lithium-ion batteries.

10.3.1   Lead acid batteries

Lead acid batteries are the oldest and most widely used battery types. The battery (in charged state) consists of an anode which is made of lead (Pb), a cathode which is made of the lead oxide (PbO_2) and an electrolyte which is sulfuric acid (37% by weight). When the battery is discharged, both electrodes become lead sulfate and electrolyte turns into water. The reactions at the anode and the cathode are $Pb + SO_4^{2-} \leftrightarrow PbSO_4 + 2e^-$ and $PbO_2 + SO_4^{2-} + 4H^+ + 2e^- \leftrightarrow PbSO_4 + 2H_2O$, respectively [2]. Flooded battery, sealed maintenance free battery, and the valve regulated battery are the examples of lead acid batteries.

The efficiency of Lead acid batteries ranges between 70% and 90%. They are very reliable and low-cost batteries ($50–600/kWh) [5–7]. Other advantages include very small self-discharge rate (<0.1%) and fast response time. They are widely used in uninterruptible power supply (UPS) for power quality application and also in spinning reserve applications [8,9]. However, they have a very low energy density (50–90 Wh/L) and low specific energy (25–50 Wh/kg) [3,10]. Their performance is also affected by the temperature (performance degrades at low temperatures). They have a very short life of 500–2,000 cycles because the reversible redox reactions deteriorate the electrodes of batteries [9]. Deep discharges further reduce the life. Another major disadvantage of flooded type batteries is their periodic water requirements. Nevertheless, lead acid batteries are still very commonly used in commercial and energy management sector. Current research on lead acid batteries focuses on performance improvements, extending life cycles and improving deep discharge capabilities.

10.3.2   Sodium-sulfur (NaS) batteries

Sodium-sulfur (NaS) battery consists of a molten sulfur and a molten sodium as positive and negative electrodes, respectively. The electrodes are separated by a solid electrolyte (beta alumina electrolyte). Main parts of an NaS battery and its operation is shown in Figure 10.4 [11,12]. The battery cell is contained in a metallic cell case of cylindrical structure (Figure 10.4(a)). Figure 10.4(b) shows the operation of an NaS battery. During discharging, sodium is oxidized at the anode and the positive sodium ions (Na^+) are formed. At the cathode, sulfur is reduced and (S^{-2}) ions are released. Na^+ pass through the electrolyte and combines with sulfur ions to form sodium polysulfide. The reaction is given by; $2Na + 4S \leftrightarrow Na_2S_4$. During discharge, electrons flow through the external circuit and the sodium ions move through the electrolyte creating 2 volts at the battery terminals. During charging, the reversible reaction occurs and sodium sulfide releases Na^+ which move back through the electrolyte to recombine as sodium element [13].

NaS is a high-temperature battery with operating ranges between 300 and 350 °C. In this operating range, the resistivity of the battery is low (4Ω cm) [14,15]. The amount of heat required to carry out the reactions is provided by the reactions themselves. Hence, an external heat source is not needed. The efficiency is around 70%–90% and cycle life is typically 2,500 cycles. Main advantages are low maintenance, no self-discharge and recyclability (99%). However, they must be

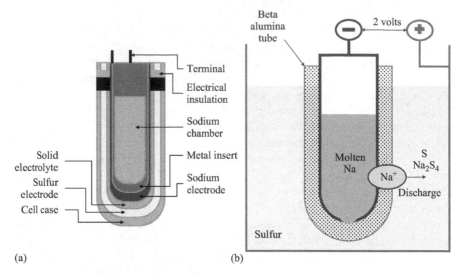

Figure 10.4   NaS battery. (a) Main parts of NaS and (b) NaS battery operation

kept between 300 and 350 °C and have a high cost ($2,000/kW) [2,9]. NaS batteries have applications in utility, i.e., peak shaving, load leveling, and power quality. Other applications include electric cars and hybrid vehicles.

10.3.3   Nickel cadmium (NiCd) batteries

Nickel cadmium (NiCd) batteries were initially developed for domestic-level use because of its several advantages including long life (>3,500 cycles), immune to deep discharges, low maintenance, high energy density (50–75 Wh/kg), robust, and temperature tolerance. However, it does have some major drawbacks which are the toxicity, high capital cost (ten times of lead acid batteries), and the periodic requirement to carry out a complete cycle (the capacity decreases dramatically if repeated charged after partial discharges called the memory effect). Nevertheless, life can be greatly increased, even more than 50,000 cycles at 10% depth of discharge can be achieved [3,9,16].

NiCd batteries use cadmium (Cd) and nickel hydroxide as the two electrodes. The alkaline solution (KOH) acts as the electrolyte. During discharging, $Cd(OH)_2$ is the active material, at the positive electrode, whereas $Ni(OH)_2$ is the active material, at the positive electrode. During charging, Cadmium (Cd) is the active material, at the negative electrode, whereas NiOOH is active material, at the positive electrode. The chemical reaction is given by: $2NiO(OH) + Cd + 2H_2O \leftrightarrow 2Ni(OH)_2 + Cd(OH)_2$ [2,17]. It is available in two forms: sealed form (for portable equipment) and flooded form (for industrial applications). However, in 2003, European Commission drew up a proposal for new directives including recycling targets of 75% of this type of batteries and their domestic use has been banned by Directive 2006/66/EC [9,16]. Hence, their future is uncertain because of high toxicity and unable to be integrated with renewable energy resources because of the memory effect.

10.3.4   Nickel-metal hydride batteries

Nickel-metal hydride (NimH) is another nickel based battery and a good alternative to replace NiCd batteries. It is similar to NiCd battery except for the fact that instead of cadmium, it uses hydrogen-absorbing alloy as the electrode. The specific energy of NimH battery is moderate (around 70–100 Wh/kg). It has several advantages over NiCd batteries including reduced memory effect, being more environmentally friendly, higher energy density (170–420 Wh/L) and improved performance. Its energy density is 25%–30% greater than NiCd battery. Since there are no toxic materials such as cadmium, mercury, or lead, NimH is benign to the environment. Although NimH batteries have several advantages over NiCd and lead acid batteries, their self-discharge rate is very high. It loses 5%–20% of its capacity within first 24 hours and hence cannot be used for long-term energy storage. It is also sensitive to deep discharge. The performance of these batteries starts to decrease after few hundreds of full cycles [3,18,19].

10.3.5   Lithium-ion (Li-ion) batteries

Lithium-ion (Li-ion) batteries were proposed in the 1960s and became commercially available in the early 1990s. Now, Li-ion batteries are most widely used batteries for small applications (cell phones, laptops, and other portable electronic devices) with their annual production more than 2 billion cells [20]. Li-ion batteries consist of an anode made of graphitic carbon (C_6), a cathode made of lithium metal oxide ($LiCoO_2$, $LiNi_2$, $LiMO_2$, etc.), and an electrolyte which is made of lithium salt ($LiClO_4$ or $LiPF_6$) dissolved in organic carbonates [2,9]. During charging, lithium atoms become ions and move towards anode through the electrolyte where they combine with the electrons and deposited as lithium atom. During discharge, lithium ions move toward cathode whereas the electrons move through the external circuit towards the anode. The charge and discharge cycle of a lithium-ion battery is shown in Figure 10.5.

The advantages of Li-ion batteries are numerous such as high energy density (80–1,500 Wh/kg), high specific energy (75–200 Wh/kg), high power density (500–2,000 Wh/kg) long cycle life (>1,500 cycles), very high efficiency (>97%–100%), reliability, very low self-discharge (maximum discharge rate is 5% per month), and good response time (order of milliseconds) [2,3,6]. Although Li-ion batteries have taken over 50% of the small portable electronic devices market, there are still some challenges for large-scale batteries. The major challenges are the high cost (>\$600/kWh); decreased cycle life due to deep discharges, decreased life and performance due to overcharging, complex protective circuitry, and overheating (which also decreases performance and cycle life) [15]. Current research on Li-ion batteries focuses on enhancing power capability and increasing specific energy with the help of nano-material and advanced electrode materials [3]. A comparison of all the batteries is given in Table 10.1.

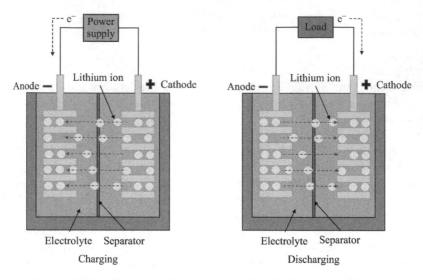

Figure 10.5   Charge and discharge cycle of a lithium-ion battery

Table 10.1   Comparison of different battery technologies

Properties	Energy price ($/kWh)		Power interface price ($/kW)		Gravimetric energy density (Wh/kg)		Volumetric energy density (Wh/L)	
Technologies	Low	High	Low	High	Low	High	Low	High
Li-ion	400	1,000	30	50	170	500	250	360
Lead acid	150	150	30	50	15	40	108	108
NaS/NaSCl	250	500	30	50	150	150	180	180
ZnBr	300	400	30	50	70	190	15.7	39
NimH	300	1,000	30	50	30	110	140	300
NiCd	400	700	30	50	35	60	50	150

Properties	Nominal cycles at 80% DOD		Round trip efficiency		Interface efficiency	Self-discharge	Maintenance	System
Technologies	Low	High	Low	High	Low	High	Low	High
Li-on	4,000	10,000	85.00%	95.00%	97.00%	0.01%	1.00%	87.29%
Lead acid	800	1,200	70.00%	85.00%	97.00%	0.33%	3.00%	74.92%
NaS/NaSCl	12,500	18,750	80.00%	95.00%	97.00%	0.00%	3.00%	84.88%
ZnBr	2,500	2,500	60.00%	70.00%	97.00%	0.07%	1.00%	63.01%
NimH	2,000	2,000	60.00%	70.00%	97.00%	1.00%	1.00%	62.42%
NiCd	5,000	10,000	60.00%	70.00%	97.00%	0.33%	1.00%	62.84%

10.4   Flow battery energy storage

Flow batteries, also known as redox flow batteries, are electrochemical batteries that store energy in solutions (liquid electrolyte). The electrochemical materials provide the necessary force (electromotive force) required for the

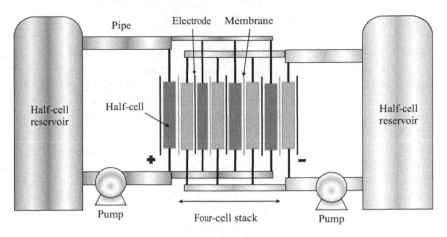

Figure 10.6   Schematic of a flow battery

oxidation–reduction reactions to charge and discharge the battery. In flow batteries, the liquid electrolytes are stored in a separate tank and are pumped into the reactor during the operation. During charging, one electrolyte is oxidized (at the anode), while the other electrolyte is reduced (at the cathode). During this process, electrical energy is converted into chemical energy. The process is reversed during the discharge phase. The reversible reaction allows the battery to be charged and discharged. Schematic of a flow battery is depicted in Figure 10.6 [2]. Flow batteries have several significant advantages, for example, the power of flow batteries is independent of their storage capacity. The power of flow batteries is determined by the size of the electrodes, whereas the storage capacity depends on the amount and concentration of the electrolyte. They are inherently more stable because the electrodes are separate from the electrolyte and do not participate in the reactions. The capacity is flexible because the reactants are stored in the separate tank. Flow batteries can continuously supply energy at high discharge rate for up to 10 h. Another advantage is that the bipolar electrode technology can be adopted [2,3,21,22]. On the other hand, the major challenges include shunt currents, low energy density, low performance, high cost, and complicated design [22,23]. The three types of flow batteries are described in as below:

Vanadium redox flow battery (VRB): The vanadium redox battery (VRB) is one of the most mature technologies in this group. The VRB stores energy in two liquid vanadium-based electrolytes and the chemical reaction is given by $V^{4+} \leftrightarrow V^{5+} + e^-$ and $V^{3+} + e^- \leftrightarrow V^{2+}$ [21]. The cell voltage is around 1.4–1.6 V requiring multiple series and parallel connections to achieve desired voltage and current levels. VRBs have long life cycles (10,000–16,000 cycles), fast response time (order of milliseconds), and the efficiency is around 85% [2,21]. Another advantage is their longer discharge duration. VRBs have applications in power quality improvement, UPS devices, supporting the variable output of renewables, load leveling, and power security. However, the high operation cost is still a big challenge.

Zinc bromine (ZnBr) flow battery: In ZnBr battery, two aqueous electrolyte solutions are pumped from two electrolyte tanks which are based on zinc and bromine. The anolyte (electrolyte in the anode) and catholyte (electrolyte in the cathode) are in contact through a separator. During charging or discharging, the anolyte and catholyte flow through the cell and the electrochemical reactions occur. During discharge, zinc and bromine combine to form ZnBr. During charge, zinc is deposited on one electrode and bromine comes out as a dilute solution on the other side of polyolefin membrane. The reactions are given by $2Br^- \leftrightarrow Br_2 + 2e^-$ and $Zn^{2+} + 2e^- \leftrightarrow Zn$ [2,3]. ZnBr flow batteries have high energy density (30–65 Wh/L) with a cell voltage of 1.8 V. Another advantage is the deep discharge capability. Life cycle is also long (10–20 year) with discharge duration up to 10 hours. However, it has several disadvantages including low efficiency (65%–75%), dendrite formation and corrosion. Their operating temperature range is also small [23].

Polysulfide bromine flow battery: Polysulfide bromine (PSB) uses two salt solution electrolytes (sodium bromide and sodium polysulfide). The electrolytes are separated by a polymer membrane. Only positive sodium ion moves across the membrane. The electrochemical reactions are given by $3NaBr + Na_2S_4 \leftrightarrow 2Na_2S_2 + NaBr_3$, $3Br^- \leftrightarrow Br_3^- + 2e^-$, and $2S_2^{2-} \leftrightarrow S_4^- + 2e^-$ [2,3]. 1.5 V is generated across the polymer membrane. Therefore, multiple cells are series and parallel to achieve the desired output. The main advantages of PSB flow batteries include fast response time (order of milliseconds) and low cost. PSB flow batteries have applications in frequency and voltage control. On the other hand, they may have environmental impacts and also the efficiency is low (75%) [3].

10.5 Pumped hydro storage

PHS is a method of storing energy in the form of potential energy of water. It is a large-scale ESS in which water from the downstream is pumped up into a reservoir during the time of low energy demand and low energy cost. This water is then used at the time of peak energy demand.

The reversible pump turbine acts as a generation during normal operation. When energy storage is required, the reversible pump turbine starts to act as a motor and pump the water from lower reservoir to upper reservoir (Figure 10.7). In some cases, the generating unit is separated from pumping unit (motor). The generating unit operates during normal operation when the energy is generated. The pumping unit is turned off during normal operation. When the energy storage is required, pumping unit is activated, and the generating unit is shut down.

PHS systems have high efficiency, lower storage cost ($kWh), and they can store large volume for very long periods. Considering the losses due to evaporation, approximately 70%–80% of electrical energy used to pump the water can be regained [6]. This technique is most effective in the case of existing hydropower plants. It is economically viable to store a large amount of energy for existing

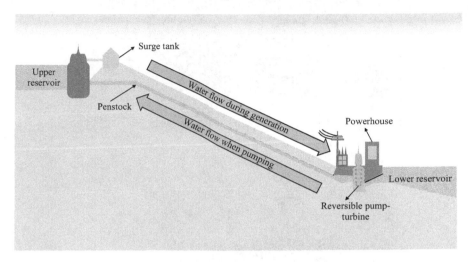

Figure 10.7   Pumped hydro storage

structures, and useful for high power applications. However, pumped hydro requires two reservoirs at different heights to create the necessary potential required to produce electrical energy. The reservoirs can be natural or man-made. A hydropower plant that cannot operate as pumped storage is conventional hydropower plants. The capital cost of a hydropower plant is very high. However, if the pumped storage is incorporated into existing conventional hydropower plants, it can be very economical because the required structure already exists. Open sea can also act as the lower basin (reservoir) [2]. Pumped hydro storage offers several advantages. It can help smooth the load curve. Another big advantage of pumped hydro is that it can respond to sudden load change. If the power demand increases suddenly, hydropower can respond quickly compared with thermal and nuclear power plants.

PHS rating varies from hundreds of megawatts to thousands of megawatts. It is the highest rating among all available ESSs, which makes it very useful for energy management, frequency control, and voltage stabilization. Despite many advantages, deployment of PHS is challenging. The major constraints restricting the exploitation of PHS is the very high capital cost (thousands of millions of US dollars). Furthermore, many environmental concerns are associated with hydropower which includes land loss, habitat destroyed and marine life is affected.

10.6   Compressed air energy storage

Compressed air energy storage (CAES) is the only technology that is commercially available, after PHS, that can deliver a large amount of energy [2]. During off-peak periods when the demand is low, energy is taken from the grid and used to compress air and pump it into sealed underground cavern under at a very high pressure.

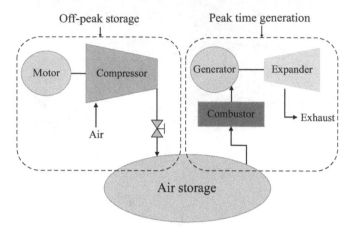

Figure 10.8   Components of CAES

Whereas, during peak hours when the energy demand is high, the compressed air can be used to drive turbines and energy can be generated. The schematic of CAES is shown in Figure 10.8. The major components of CAES include motor–generator, air compressor, turbines, a container to store compressed air, equipment controls, and other auxiliaries. CAES works on the principle of a gas turbine. However, in the case of CAES, the compression and expansion cycles are decoupled. Compression is done during energy storage whereas the expansion is done during energy generation. When the demand is low, energy is stored by compressing the air at high pressures, typically 40–80 bars. Large underground caverns made of rock or salt, or depleted oil or gas fields are used as containers to store energy. The compressed air is stored in caverns. At the time of peak demand, the compressed air is released and heated, and then expanded in a high-pressure turbine which captures the energy stored in compressed air. The air is then mixed with fuel and flamed in a low-pressure turbine which captures further energy. Both turbines are connected to an alternator to convert the mechanical energy into electrical energy. Currently, there exist only two operating CAES plants. World's first CAES plant started operating in 1978 in Huntorf, Germany, and the second CAES plant has been operating since 1991 in McIntosh, USA. The CAES in Germany can provide 290 MW for 2 h while 110 MW over 26 h.

The efficiency of a CAES system is around 70%–80% [24]. Self-discharge results in efficiency loss. CAES can only be deployed at large scales. The rating of a typical CAES system is 50–300 MW. A properly designed CAES system can store energy for more than a year [6]. Another advantage of CAES is the small up and down time (small start-up time of 12 min under normal conditions and 9 min in case of emergency start). Quick start-up and shutdown capabilities of CAES enable one to use it during peak times and store energy during off-peak times. It is also considered as environment-friendly. Furthermore, the initial cost of a CAES system is not very high if the natural geological formation is used. However, the existence of limited geological structures restricts the exploitation of CAES.

10.7 Flywheel energy storage

The flywheel is one of the oldest energy storage technologies. It is composed of a mass rotating about an axis and the rotating mass is coupled to an electric machine (motor–generator). Energy is stored in the form of rotating kinetic energy (angular momentum of the rotating mass). During the charging cycle, the electric machine acts as a motor and convert the electrical energy into mechanical energy. Whereas, during the discharge cycle, the machine acts as a generator and converts mechanical energy of the flywheel into electrical energy. The storage capacity depends on the size and the rotation speed of the flywheel. Therefore, larger size and higher speed flywheel can store more energy. Low-speed flywheels have operating speeds up to 6,000 rpm, whereas the operating speeds of high-speed flywheels can reach up to 50,000 rpm [6]. The flywheel assembly is kept inside a housing at very low pressure (typically $10^{-6}-10^{-8}$ atm) to minimize the self-discharge losses due to air friction [2]. It also helps reduce rotor stresses. The specific energy of a typical FES is 5 Wh/kg. However, lightweight and high-speed FES can achieve up to 100 Wh/kg of specific energy. Figure 10.9 shows the main components of a flywheel ESS. The components include magnetic bearings, motor–generator, vacuum pump, housing,

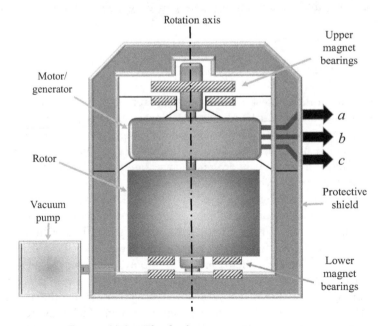

Figure 10.9 Flywheel energy storage system

and spinning flywheel. The magnetic bearings are at the top and the bottom of the assembly. These bearings also help reduce the rotor stress and minimize losses. The FES system is connected to a bus through a bi-directional converter which can be single stage or double stage. The converter is coupled with a power controller to control the system variables.

The main advantage of a flywheel system is their long operating life. Unlike batteries, flywheels are capable of providing several thousand to millions of full charge–discharge cycles, which is 10–1,000 times more than batteries [25]. Friction losses for a 200-ton flywheel are approximately 200 kW [26]. The efficiency of a flywheel can be as high as 95%. However, it can drop below 50% in one day, which is the main disadvantage of flywheels. The self-discharge rate of an FES is very high (typical discharge rate is higher than 20% per hour) [6]. Therefore, unlike batteries, long-term storage is not possible using FES. The application of FES is, therefore, limited to high power and short duration applications.

The disadvantage of high discharge rates can be put to advantage by integrating with renewable energy resources which have intermittent nature. For example, photovoltaic and wind energy resources have fluctuations on an hourly and daily basis. If the generation increases suddenly because of a sudden increase in wind speed, the extra energy can be stored in FES. Conversely, if the generation decreases suddenly because of the sudden decrease in wind speed or clouds causing shading of PV panels, the stored energy can be released. This can be very effective where there are higher fluctuations in power generation. Another application of FES is to provide ride through during interruptions. FES have also applications in power conditioning and providing power during fault or grid failure.

10.8   Hydrogen energy storage with fuel cell

Hydrogen is one of the lightest, cleanest, and most efficient fuels and can be used as a storage media. Figure 10.10 shows the process of energy storage using the hydrogen storage system [3]. This type of storage uses two separate processes for storing energy and producing electricity. The major components include electrolyzer (to produce hydrogen), hydrogen storage tanks, and hydrogen fuel cell to generate electricity from hydrogen. When the energy demand is low, hydrogen is produced and stored. Hydrogen produced by water electrolysis, which is one of the most common methods to produce hydrogen, can be stored in high-pressure containers. During the time of high energy demand, the stored hydrogen is used in a fuel cell to generated electricity. Fuel cells are the key technology in the hydrogen storage system. They convert chemical energy into hydrogen and oxygen (from the air) to electricity ($2H_2 + O_2 \rightarrow 2H_2O$ + energy). Heat and electrical energy are released during the reaction. There are several groups of hydrogen fuel cells including alkaline fuel cells, proton exchange membrane fuel cells, regenerative fuel cells, and phosphoric acid fuel cells. Other fuel cells include direct methanol fuel cells, molten carbonate fuel cells, and solid oxide fuel cells.

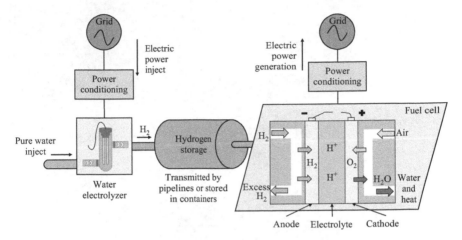

Figure 10.10   Topology of hydrogen storage and fuel cell

Hydrogen fuel cells have several advantages including high efficiency, less noise and pollution, independent system charge and discharge rate, and high energy density. In addition to this, these fuel cells have the ability to be implemented over a wide range of scales (from a few kilowatts to hundreds of megawatts). The disadvantages, on the other hand, include lower round trip efficiency (20%–50%) and high cost (estimated cost is between $500 and $10,000/kW) [2,12]. Hydrogen storage is approximately 4.5 times more costly than natural gas. Hence, the fuel cell systems are not cost-effective for lower levels of wind penetration [5,27,28]. However, hydrogen energy storage with fuel cell technology is in the development stages. Stationary power applications, i.e., primary electrical power, heating/ cooling or backup power) are relatively mature. The world's first utility-scale renewable energy system with hydrogen storage was installed in Norway [29]. In 2012, 2.8 MW biogas fuel cell power plant (which is one of the world's largest) was launched. However, cost reduction is essential for successful implementation of this technology in large-scale applications.

10.9   Solar fuels

Solar fuel technology uses the solar energy to generate heat. Figure 10.11 shows the working principle of a solar fuel [2]. The sunlight is concentrated in a small area with the help of reflectors. The heat captured from this sunlight is used in the endothermic chemical transformation to produce solar fuel. This fuel can then be stored for electricity generation at a later time. Off note, solar fuel technology is still in its development phase and practical deployments are very limited.

Instead of concentrated solar, electrical energy can also be used as an input for the solar fuel generation but it results in an overall lower efficiency. For this reason, concentrated solar energy is used for solar fuel generation. The major advantage of

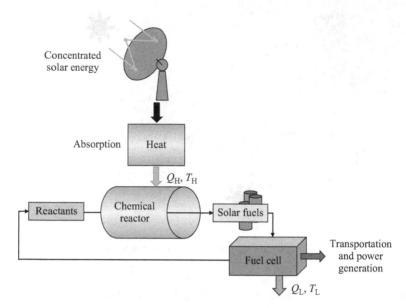

Figure 10.11   Schematic of a solar fuel

energy storage in the form of solar fuel is that the system efficiency remains the same with or without the use of solar fuel storage. Hence, very high storage efficiency (virtual storage efficiency of 100%) can be achieved [2]. Several solar fuels including solar hydrogen, solar metal, and solar chemical heat pipe can be produced using solar energy. Solar hydrogen fuel can be produced from several resources including water for the solar thermolysis, fossil fuels for the solar cracking, and solar gasification. Solar metal uses solar energy as a source of high temperature to produce metal (solar fuel) from their oxides. In the case of the solar chemical heat pipe, the products of high-temperature endothermic reversible reaction are stored and are used later when needed [2].

There are three approaches to produce and store solar fuels, namely, natural photosynthesis, artificial photosynthesis, and thermochemical approaches [3,30]. The first two approaches use the process of photosynthesis to capture solar energy. Sunlight converts carbon dioxide and water into oxygen and other elements. Artificial photosynthesis requires catalysts such as ruthenium, rhenium or palladium to split water into its constituents [30,31]. The third approach (thermochemical), for water splitting, requires very high temperatures which are achieved by thermal processes. Therefore, it required stronger sunlight. Concentrated heat from solar is used to carry out the endothermic reaction process. Hydrogen, carbon monoxide, and other products are the result of the chemical transformation [32].

Solar fuels capacity can reach up to 20 MW with specific energy ranging from 800 to 100,000 Wh/kg. One of the major promising advantages of this technology is the long storage capacity (up to several months). The major disadvantage, on the other hand, is the need for catalysts for artificial photosynthesis which are scarce and expensive. Another disadvantage is the large area requirements for the light collectors [3]. However, this area requires further exploration and a significant amount of research is still needed to realize the full potential of solar fuels.

10.10   Capacitors and supercapacitors

A capacitor is a device that can store energy in an electric field. A typical capacitor is consists of two plates (conductors) separated by a dielectric (thin film of insulation) material. When a voltage is applied across its plates, charges are accumulated on the plates, positive charges on one plate, while negative charges on the other. During the charging process, energy is stored between the plates in the electrostatic field [2,33]. The capacity (capacitance) of a capacitor is directly proportional to the area of the plates and inversely proportional to the distance between the plates that is given by $C = \varepsilon A/d$, where C is the capacitance, A and d are area and distance between the plates, respectively, and ε is the permittivity of the material. The stored energy (W) is given by $W = 1/2 \times CV^2$ (V is the voltage across the capacitor plates). They charge instantly and last longer. However, their capacities are very limited and are only suitable for storing a small amount of energy. The self-discharge losses are very high and the energy density is also low [2,33]. Capacitors have applications in voltage correction, energy recovery in mass transit systems, and smoothing the output of power supplies.

Supercapacitors, also known as ultracapacitors, are the high surface area activated capacitors. They consist of two electrodes, an electrolyte, and a porous membrane separator. As a result of this structure, supercapacitors have the characteristics of both capacitors and electrochemical batteries. Supercapacitors offer very high capacitance compared with a traditional capacitor. The separation of charge at an electric interface in a supercapacitor is in the measurement of fractions of nanometers, whereas in a traditional capacitor this separation is in the measurement of micrometers [6]. Depending on the material technology, supercapacitors can be classified into electrochemical double-layer (ECDL) supercapacitors, pseudo-capacitors, and hybrid capacitors. Figure 10.12 shows the classification of supercapacitors.

ECDL supercapacitors are the least expensive and the most common type of supercapacitors. Figure 10.13 shows the schematic of an ECDL supercapacitor [3]. They have a double-layer construction with a liquid electrolyte and porous carbon based electrodes. In addition, there is a separator immersed in the electrolyte. An individual cell can provide a voltage of up to 3 V. Hence, series and parallel combinations of supercapacitors are used to achieve the desired voltage. During the charging, the ions move from electrolyte towards the electrodes of opposite polarity. As a result, two charge layers are formed. Energy is stored between the electrolyte and the two electrodes. The energy densities of supercapacitors are less than electrochemical batteries but more than traditional capacitors. The main advantages of supercapacitors are their long cycling times ($>1 \times 10^5$) and high cycle efficiency (around 84%–97%) [2,34]. Other advantages include immunity to

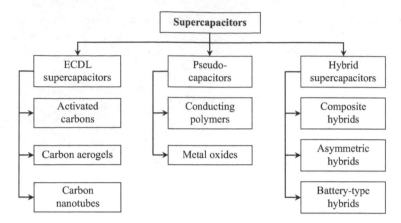

Figure 10.12   Types of supercapacitors

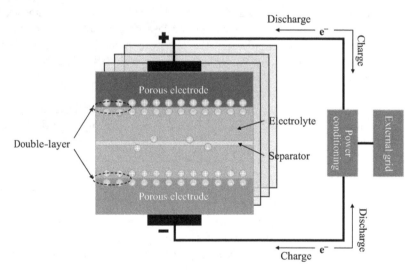

Figure 10.13   Schematic of an ECDL supercapacitor

deep discharge, extreme durability, less maintenance requirements, and the fast response time (tens to hundreds of milliseconds) [15,16]. On the other hand, the main disadvantages include high cost (6,000–9,500 $/kWh) and high self-discharge losses (around 5%–40%) [2,9,26,35]. Therefore, supercapacitors are not suitable for long-term storage, but they have many useful applications in short-term storage including pulse power, UPS devices, and bridging power to equipment.

Developments in supercapacitor technologies have shown that using vertically aligned, single-walled carbon nanotubes, instead of porous carbon electrodes, can result in significantly higher power density (100,000 Wh/kg), higher energy density (60 Wh/kg), and increased capacity. This is because of the very small width of carbon nanotube (few atomic diameters) and the increased surface area of electrodes [6]. Pseudo-capacitors and hybrid capacitors are relatively new technologies and are in development phases. Pseudo-capacitors consist of electrodes which are made of conducting polymers or metal oxides such as ruthenium oxide (RuO_2), nickel oxide, or iridium oxide. The high cost limits their applications to military operations. Hybrid supercapacitors can achieve even higher power and energy densities compared to ECDL supercapacitors or pseudo-capacitors.

10.11 Superconducting magnetic energy storage

SMES is relatively a new technology. In 1970, the first SMES system was built [36] in which the energy is stored in the magnetic field. The field is created by the flow of DC current in a superconducting coil with nearly zero resistance at very low temperature. The schematic diagram of an SMES system is shown in Figure 10.14 [3]. It consists of three main components: a superconducting coil (generally made of niobium–titanium (NbTi) filaments), a power conditioning system, and a refrigeration and vacuum system [9,37]. In general, electrical energy is dissipated in the form of heat when the DC current passes through an ordinary magnetic coil. However, in the case of a superconducting material in its superconducting state, i.e., at very low temperature, the resistance becomes zero. In this state, electrical energy can be stored without any losses. The amount of energy stored depends on the self-inductance of the coil and the current flowing through that coil. During discharge, AC current can be achieved by connecting the SMES system to a power converter.

Superconducting coils can be categorized as low-temperature superconducting (LTS) coils and high-temperature superconducting (HTS) coils. LTS coils operate around 5 K, whereas HTS coils operate around 70 K. SMES with LTS is relatively mature compared with SMES based on HTS. SMES with 0.1–10 MW of storage capacity are available commercially [3]. More than 30 SEMS systems are installed in different parts of United States. They provide UPS, improved power quality, and

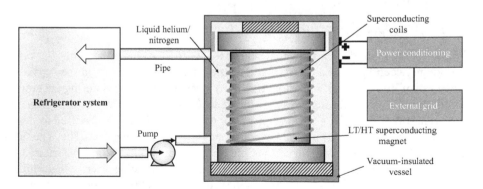

Figure 10.14   Schematic diagram of an SMES system

reactive power support [38]. The advantages of SEMS include high power density (around 4,000 W/L), high cycle efficiency (95%–98%), long life (up to 30 years), fast response time (order of milliseconds), durability, and reliability. Furthermore, unlike batteries, SEMS are immune to deep discharge [2,34]. On the other hand, the major disadvantage of this technology is the very high cost (up to 7,200 $/kW). Another disadvantage is the negative environmental impact because of the strong magnetic field [2,3,5]. However, the fast charge and discharge cycles make SEMS a potential candidate for integration with the intermittent renewable energy resources.

10.12   Thermal energy storage

TES can be defined as a temporary storage of energy at very high or low temperatures [12]. The fundamental principle of TES is the same for all its applications, i.e., store energy in TES (charging) when the surplus supply of energy is available, and obtain energy from TES (discharging) when needed. The classification of TES can be based on the temperature levels: high-temperature TES (if the storage temperature is higher than room temperature) and low-temperature TES (if the storage temperature is lower than room temperature). More precisely, TES can be categorized into industrial cooling (<18 °C), building cooling (0–12 °C), building heating (25–50 °C) and industrial heat storage (>175 °C) [2]. The classification methodology can also be based on the storage material as sensible heat storage, latent heat storage, and thermochemical heat storage (THS) [39]. TES uses heat engine cycles to generate electricity. The overall efficiency of TES systems is very low (typically 30%–60%) [2,12]. However, they have the potential in demand side management programs. They have the capability to shift load from peak to off-peak hours. TES systems are usually deployed for minimizing the mismatch between the energy demand and supply of energy. Other advantages include environment-friendly and energy security. Other media storage includes water, metals, stones, and ceramics [40]. The following subsections briefly describe the working principles of sensible heat system, latent heat system, and thermochemical heat system.

10.12.1   Sensible heat system

Sensible heat system uses bulk materials such as molten salt or sodium. The difference between latent heat system and sensible heat system is that in the sensible heat system, the material does not change its state. Heat is stored during accumulation and recovered during retrieval to produce steam, which drives the turbine. Hot water obtained from a thermal plant during off-peak hours can be used to heat the water supply during peak hours to generate extra electricity. The Themis station in France is an example of sensible heat system with the storage capacity of 40,000 kWh [26]. Another type of storage combines sensible heat with turbine (Figure 10.15 [26]). Bulk material is heated up to 14,000 °C during storage.

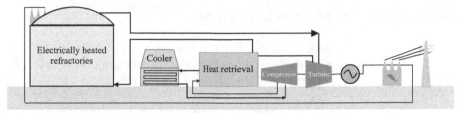

Figure 10.15   Sensible heat storage with turbine

This stored energy is then used to heat the air, which is then used in the combined cycle plant. Its efficiency is around 60%. Less self-discharge losses, less area requirement, and very small capital cost makes this type of thermal storage a good ESS to be invested in.

10.12.2   Latent heat system

Latent heat system uses liquid–solid transition of material at a constant temperature. These materials are also called phase change material (PCMs). PCMs shift from the solid state to liquid during accumulation and transfer back to solid during retrieval. Heat is transferred from the accumulator and to the environment through a special fluid called heat transfer fluid. The amount of the energy stored depends on the temperature at which it is being stored. More energy can be stored at higher temperatures. Furthermore, the PCMs offer higher energy storage densities compared with non-PCM. Sodium hydroxide is an example of PCM with a storage capacity of 1,332 MJ/m^3 [26]. Other examples of PCMs include sodium sulfate, sodium nitrate, and sodium nitrite. Latent heat system can find its applications in industrial processes especially in electric boilers for the steam generation where the need of electrical power varies significantly. However, PCM technology is still in development phases.

10.12.3   Thermochemical heat storage

Thermochemical heat storage (THS) is another technique for thermal energy storage. THS uses enthalpy of reaction (ΔH). Heat is stored during endothermic reactions (when the enthalpy is positive), and it is released during backward reaction when the enthalpy is negative. Thermochemical storage can be separated into thermochemical reactions and sorption processes. Sorption is the phenomena of capturing sorbate (gas of vapor) by using sorbent (solid or liquid) [41]. The term sorption includes both adsorption and absorption. The advantage of THS is its capability of energy storage over longer periods. However, some major challenges in the application of THS include cost, toxicity, energy density, corrosiveness, safety, reaction rate, reaction temperature, reactors configuration, and the heat exchanger design [42].

10.13   Overview of current projects

Previous sections summarized the detailed working principles of different ESSs. To that end, there has been a growing amount of interest in deploying storage systems. According to US Department of Energy, most of the projects target the following use cases: (1) Renewables capacity firming, (2) electric energy time shift, (3) electric bill management, (4) renewables energy time shift, (5) frequency regulation, (6) voltage support, (7) electric supply capacity, (8) onsite renewable generation shifting, (9) electric supply reserve capacity, and (10) on-site power applications. The top countries in terms of the number of projects are the United States with 351 projects, China with 54 projects, Spain with 40 projects, Germany with 35 projects, Japan with 35 projects, and the United Kingdom with 24 projects. Moreover, the global landscape for the energy storage projects over the years is depicted in Figure 10.16.

We proceed to provide an overview of energy storage projects and the required specifications based on the applications. In Section 10.2.2, we classify the storage systems into six based on the applications. In Table 10.2, we provide candidate storage units and their characteristics for each application type. The table shows that different storage types can be deployed for the same applications. The total cost calculations are conducted based on the assumptions given in Reference 4.

We proceed to provide an overview of the benefits for different players. Table 10.3 identifies 20 different benefits for 5 value chains and summarizes the monetary benefits in terms of both \$/kWh and \$/kW. As given in Reference 4, each benefit is computed in isolation with a battery configuration of 1 MW of discharge capacity and 2 MWh of energy storage capability. The economic analysis is conducted based on the assumption of 15 year life and a 10% discount rate. The net present value can be calculated by computing the present worth factor by $PW = \sum_{i=1}^{l} (1 + e)^{i-0.5} / (1 + d)^{i-0.5}$, where i is the year, l is the lifetime (in years), e is the annual price escalation rate, and d is the discount rate.

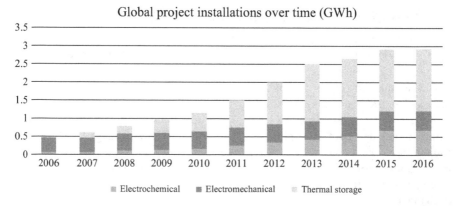

Figure 10.16   Overview of storage projects around the world

Table 10.2   Energy storage characteristics by application [4]

Technology Option	Capacity (MWh)	Power (MW)	% Efficiency/ (total cycles)	Total Cost ($/kW)/ Cost ($/kWh)
Bulk energy storage to support system and renewables integration				
Pumped hydro	1,680–14,000	280–1,400	80–82/ (>13,000)	1,500–4,300/ 250–430
CAES	2,700	135	70–80/ (>13,000)	60/ 1,250
Sodium-sulfur	300	50	75/ (4,500)	3,100–3,300/ 520–550
Advanced lead-acid	400	100	85–90/ (4,500)	2,700/ 675
Vanadium redox	250	50	65–75/ (>10,000)	3,100–3,700/ 620–740
ZnBr flow battery	250	50	60/ (>10,000)	1,450–1,750/ 290–350
Energy storage for ISO fast frequency regulation and renewables integration				
Flywheel	5	20	85–87/ (>100,000)	1,950–2,200/ 7,800–8,800
Li-ion	0.25–25	1–100	87–92/ (>100,000)	1,085–1,550/ 4,340–6,200
Advanced lead-acid	0.25–50	1–100	75–90/ (>100,000)	950–1,590/ 2,770–3,800
Energy storage for utility T&D grid support applications				
CAES	250	50	70–80/ (>10,000)	1,950–2,150/ 390–430
Advanced lead-acid	3/2/48	1–12	75–90/ (4,500)	2,000–4,600/ 625–1,150
Sodium-sulfur	7.2	1	75/ (4,500)	3,200–4,000/ 445–555
Zn/Br flow battery	5–50	1–10	60–65/ (>10,000)	1,670–2,015/ 340–1,350
Vanadium redox	4–40	1–10	65–70/ (>10,000)	3,000–3,310/ 750–830
Li-ion	4–24	1–10	90–94/ (4,500)	1,800–4,100/ 900–1,700
Energy storage for commercial and industrial applications				
Advanced lead-acid	0.1–10	0.2–1	75–90/ (4,500)	2,800–4,600/ 700–460
Sodium-sulfur	7.2	1	75/ (4,500)	3,200–4,000/ 445–555
Zn/Br flow battery	2.5	0.5	60–63/ (>10,000)	2,200/ 485–440
Vanadium flow battery	0.6–4	0.2–1.2	65–70/ (>10,000)	4,380–3,020/ 1,250–910
Li-ion	0.1–0.8	0.05–0.2	80–93/ (4,500)	3,000–4,400/ 950–1,900

(Continues)

Table 10.2   (Continued)

Technology Option	Capacity (MWh)	Power (MW)	% Efficiency/ (total cycles)	Total Cost ($/kW)/ Cost ($/kWh)
Distributed energy storage system (DESS) near pad-mounted transformer				
Advanced lead-acid	100–250	25–50	85–90/ (4,500)	1,600–3,725/ 400–950
Zn/Br flow battery	100	50	60/ (>10,000)	1,450–3,900/ 725–1,950
Li-ion	25–50	25–50	80–93/ (5,000)	2,800–5,600/ 950–3,600
Energy storage for residential energy management applications				
Lead-acid battery	10	5	85–90/ (1,500–5,000)	4,520–5,600/ 2,260
Zn/Br flow battery	9–30	3–15	60–64/ (>5,000)	2,000–6,300/ 785–1,575
Li-ion	7–40	1–10	75–92/ (5,000)	1,250–11,000/ 800–2,250

Table 10.3   Representative benefit present values (PVs) of selected energy storage benefits [4]

Value chain	Benefit		PV $/kWh		PV $/kW	
			Target	High	Target	High
End-user	1	Power quality	19	96	571	2,854
	2	Power reliability	47	234	537	2,686
	3	Retail TOU energy charges	377	1,887	543	2,714
	4	Retail demand charges	142	708	459	2,297
Distribution	5	Voltage support	9	45	24	119
	6	Defer distribution investment	157	783	298	1,491
	7	Distribution losses	3	15	5	23
Transmission	8	VAR support	4	22	17	83
	9	Transmission congestion	38	191	368	1,838
	10	Transmission access charges	134	670	229	1,145
	11	Defer transmission investment	414	2,068	1,074	5,372
System	12	Local capacity	350	1,750	670	3,350
	13	System capacity	44	220	121	605
	14	Renewable energy integration	104	520	311	1,555
ISO Markets	15	Fast regulation (1 h)	1,152	1,705	1,152	1,705
	16	Regulation (15 min)	4,084	6,845	1,021	1,711
	17	Spinning reserves	80	400	110	550
	18	Non-spinning reserves	6	30	16	80
	19	Black start	28	140	54	270
	20	Price arbitrage	67	335	100	500

For instance, the benefits for end-users are power quality, power reliability, retail TOU energy, and retail demand chargers. The power quality benefits are based on the statistics collected by utilities. For instance, the average power quality cost is $0.10/kW for residential customers, $0.42/kW for small commercial and industrial customers, and $1.40/kW for large commercial and industrial customers. Similarly, retail time of use savings can be calculated by the avoided peak consumption, e.g., assume that the on-peak and off-peak prices are $0.25 and $0.06, respectively. Then for two hours of peak hour, the customer can save ($0.25−$0.06) × 2 = $0.36 each day and net present value can be applied to calculate the benefits for the entire lifetime. Demand charges, on the other, depend on the user profile and more specifically on the peak user consumption. According to EPRI, average demand charge for small commercial and industrial customers is $15/kWh-month, while it is $12/kWh-month for large commercial and industrial customers. The details of the calculations are presented in Reference 4. The details of the energy storage benefits are described as below.

End-user benefits:

1. *Improved power quality*: It is very often that the customers experience poor voltage and blackouts. Small voltage fluctuations happen all the time, while, large-voltage sags can damage the electrical and electronic equipments. The impact of temporary partial blackouts and voltage fluctuations for the majority of residential customers is very small. However, poor power quality can cause significant damage to the commercial and industrial customers. Frequency voltage sags and power outages can cost millions to the commercial and industrial customers. ESSs can provide help in voltage regulation which results in improved power quality.

2. *Improved power reliability*: Energy storage devices can help provide a continuous supply of electrical energy, hence increasing the power reliability. They can continue to supply the energy in case of a temporary fault. Small ESSs can be used to ride through an outage by supplying power for a few minutes until the conventional backup generation can take over. On the other hand, large ESSs can supply energy for several hours.

3. *Reduce retail TOU energy charge*: One of the benefits of an ESS is the reduced cost of electricity by shifting the energy purchase from on-peak times to off-peak times. This way, customers store cheap electricity during off-peak hours and use it during peak-hours where the tariffs are higher.

4. *Reduce retail demand charges*: Commercial and industrial users usually pay a monthly demand charge which depends on the peak demand of that industry. Industrial customers can benefit from ESS by reducing their peak demand.

Distribution system benefits:

5. *Voltage support*: Voltage is dropped when there is a mismatch between the supply and the demand. This phenomenon is very common during peak times and ESSs can provide voltage support by providing the extra energy to meet the demand.

6. *Defer distribution investment*: A reduction in peak demand with the help of ESSs helps the distribution utilities to defer the investments in the distribution system that would be needed otherwise.
7. *Distribution losses*: Power losses in the distribution sector typically range from 2% to 6%. Deployment of ESSs in the network can reduce the losses by 30% [4].

Transmission system benefits:

8. *VAR support*: Reactive power support is required to maintain the grid stability. ESSs integrated with the transmission system can absorb and supply reactive power to maintain the voltage levels in the grid and hence they can help maintain the system stability.
9. *Reduce transmission congestion*: In transmission congestion, the physical limits of transmission lines are reached. In such cases, it is necessary to change the power flow to prevent overloading of transmission lines. As a result, customers are forced to buy power from a local higher priced power plant rather than a distant cheaper power plant. Properly sited ESSs can relieve the transmission congestion by shifting the peak demand.
10. *Reduce transmission access charges*: Transmission access charges are paid by transmission customers in return for transmission reservation. Transmission access is charged on a per-kilowatt basis. Energy storage can reduce the peak load demand of transmission customers and thereby reduce their transmission access charges [4].
11. *Defer transmission investment*: Due to increasing energy demands, transmission systems require periodic and costly upgrades. Nevertheless, deployment of ESSs can reduce the peak energy demand help system operators to defer the investments needed for transmission system upgrades.

System benefits:

12. *Provide local capacity*: Sometimes some areas of a grid cannot be served due to limited generation or transmission capacity. Local investors who own ESSs can enter the energy market and can sell power to the local areas when needed.
13. *System capacity*: System capacity, or sometimes referred as resource adequacy, is a commitment of system operators to provide enough generation capacity to meet the needs of consumers (usually determined as the peak demand plus 15% extra reserves. ESS can reduce the system capacity requirements through bulk energy store systems. This way, "Loss of Load Probability" will be reduced.
14. *Renewable energy integration*: Integration of renewable energy is not possible without an ESS. The intermittent solar and wind energy resources can only be integrated with the power system if they are incorporated with an ESS. A case study is presented in the last section.

ISO market benefits:

15. *Regulation*: Regulation is needed to match the short-term fluctuations in the generation, load, and frequency. The regulation is usually provided by the generators that can respond quickly to sudden changes in the demand. Some ESS can be a useful tool because of their fast response times and can be used for regulation purposes.

16. *Fast regulation*: Sometimes a very fast response time is a need for sudden huge change in the load. Some ESS have the capability to act very fast such load changes with the response time of the order of milliseconds. These ESS can be used to for the fast regulation.

17. *Spinning reserves*: In power system operations, spinning reserves are required to respond quickly to generation or transmission outages within minutes. The generation capacity is on-line but kept unloaded. Since ESSs do not require long start-up time, they can be used as spinning reserve i.e., ready to be dispatched whenever needed.

18. *Non-spinning reserves*: Non-spinning reserves is the backup generation that is not up and running but can be dispatched within 10–30 minutes, if needed. ESSs also have the capability to act as a non-spinning reserve.

19. *Black start*: Many power plants themselves require electrical energy to start their generation. Usually, this power is imported from the grid. However, in the case of a blackout, ESSs can be used to start the generators.

20. *Energy arbitrage*: Energy prices vary according to the daily load curve. These prices are high during the peak time and low during the off-peak periods. ESS can take advantage of this pricing and can be used to store energy during the off-peak periods and deliver that stored energy during the peak periods.

10.14 Power electronics for ESSs

Power electronics converters (PECs) are usually used to interface storage systems with various kinds of power systems [43] (as shown in Figure 10.17). These circuits contain semiconductor-based switches which are turned ON/OFF to convert the voltage and current on the input side of the circuits into a form that matches the power systems on the output side.

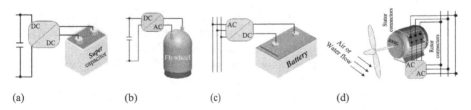

(a)   (b)   (c)   (d)

Figure 10.17   A schematic overview of power electronics converters. (a) DC/DC PEC, (b) DC/AC PEC, (c) AC/DC PEC, and (d) AC/AC PEC

Storage elements such as pumped hydro and compressed air storage systems could be interfaced with power systems through various kinds of electromechanical generators [44,45]. However, PECs provide more flexibility in controlling the power transfer between the various parts within the system, have a faster response to stabilize the system operation and have the capabilities to provide ancillary services to power systems to enhance the power quality [46]. The input and output currents to the PECs could be in either DC or AC form, accordingly PECs could be classified into four main types based on forms of the input and output currents which are DC/DC, AC/DC, DC/AC, and AC/AC [47]. In this section, these types of PECs are discussed along with their applications to interface storage systems with power systems.

DC/DC converters: DC/DC-type PECs are used to convert a certain DC voltage at their input into a higher or lower DC voltage. In certain cases, the DC/DC PECs are used to feed the power directly into a DC power system. On the other hand, these circuits could be utilized as an intermediate unit to produce a DC voltage that represents the input stage of another PEC.

Batteries and supercapacitors produce DC voltage at their outputs and in several applications their outputs are processed by DC/DC PECs. Examples of these applications are electric and hybrid vehicles and DC microgrids [48]. The battery voltage varies as with its state of charge (SoC) and range of this variation depends on the used technology. The ratio between the lowest and the highest operating voltages of the batteries could vary in the range 0.82−0.92 depending on the considered battery technology. The voltage supercapacitors vary in a much larger range as the ratio between the lowest and highest voltage could be as low as 0.4 [49]. In all these cases, a DC/DC converter is needed to maintain a regulated output voltage as the voltage of the storage element varies.

AC/DC converters: AC/DC PECs are usually referred to as rectifiers. Rectifiers are needed to convert the AC power produced by flywheel and possibly pumped hydro and compressed air energy storage into DC power [50]. The frequency of the AC signal of these sources could be in many cases different from the grid frequency. Therefore, rectifiers are used to convert the AC power from these elements into DC and then, a DC/AC PEC is used to produce an AC voltage at a frequency that matches the grid requirement. On the other hand, rectifiers are needed to charge DC-type storage elements, such as batteries and supercapacitors, using the AC power of the power grid.

DC/AC converters: DC/AC PECs are usually called inverters. The DC power supplied at the input of the inverters could be (i) the direct output of a storage element such as battery, (ii) the output power of DC/DC PECS that are fed from batteries of supercapacitors, (iii) the output of AC/DC PECs that are supplied from flywheels, pump hydro, or compressed air energy storage.

In all these cases, the inverters convert the DC power into an AC form with the frequency that matches the requirement of the items connected to their outputs. In some application, the inverters supply power to power grid where the AC voltage and frequency are already set and the inverters supply an amount of current depending on the power required to be supplied. In other applications, the inverters operate in OFF-grid mode where they are used to supply power to loads isolated from the power grid [51]. In this case, the inverters are responsible for setting the amplitude and the frequency of the voltage at their output and depending on the load value, the appropriate current will be supplied out from the inverters.

AC/AC converters: The AC/AC PECs are less common than the types discussed before. However, they still have application for ESSs. In pumped hydro and compressed air ESS, doubly fed induction generators (DFIG) are usually used especially for high power applications whose capacities could reach up to several hundred MW. In DFIG, the stored energy is extracted as mechanical energy to rotate the rotor. Based on the rotor speed, the AC/AC converter AC/AC PECs convert an AC voltage to fixed voltage and frequency into and AC voltage with variable voltage and frequency [52]. Usually, for storage application, the AC/AC converter output frequency is a few Hertz and when that is added to the rotating frequency of the motor the results becomes the same as the grid frequency. In DFIG, most of the power is transferred directly to the grid through the stator windings of the generators while a small amount of the power (around 25% of the total power) is processed by the PECs which improves the system efficiency and decreases the cost. In recent year, there is a trend to replace the AC/AC PECs in DFIG application by an AC/DC followed by a DC/DC PECs as they can be controlled more effectively.

10.15 Analytical problems

As given in the previous sections, ESS offers flexibility to the power grid both in planning and operations. Moreover, ESS operators are likely to join energy markets as virtual generators, or "virtual power plants", and promote renewable generation. However, due to their complex operating principles and high acquisition and maintenance costs, the development of novel mathematical modeling, control, and optimization methodologies is required to assess the technical and economic potential of energy storage devices.

One of the critical analytical problems is the optimal energy storage sizing which naturally arise due to the high capital cost of storage technology. More specifically, over-provisioning storage units will lead to costly and under-utilized assets, while under-provisioning may reduce the operating lifetime due to frequent exceeding of allowable depth-of-charge (DoD), heating, and fast cycling. The sizing requirements, on the other hand, depends on the application. In the case of energy large industrial customers who seek to use ESS as for peak shaving, load leveling, and spinning reserves may be in need of storage with 250–300 kWh. The work in Reference 53 solves the energy sizing problem for a single industrial customer with a given consumption pattern and the goal is to maximize the monetary benefits which are the reductions in the demand charges minus the acquisition and operating cost. Moreover, a dynamic programming problem is formulated and solved to find the optimal operating schedule for the ESS.

ESS also provides a buffer zone to alleviate the intermittencies introduced by the integration of renewables. ESS can be employed to shape the generated power and provide a constant output, in which ESS can also compensate the forecast errors. For this case, the analysis requires probabilistic methods and uncertainty quantification. The work presented in Reference 54 provides a probabilistic ESS sizing approach for wind farms and evaluates the model with real data collected from wind farms.

Even though the cost of energy storage units decline, the monetary benefits for residential customers or end-users for peak-shaving applications is still not viable due to high acquisition cost [55,56]. Instead, sharing-based energy storage units are proposed to cover the cost of the storage and maximize the benefits. To that end, the works presented in References 55 and 56 models the usage of the appliance with Poisson process and solve the optimal sizing problem by calculating the probability distribution function for "underflow events" which represents the case in which supply cannot meet the demand. The sizing problem is also necessary for microgrids with renewable integration operating under both islanded and grid-connected modes [57]. Also, storage devices can play a vital role in active distribution networks by improving the system reliability and efficiency and improve the utilization of assets utilizing optimal load management. To that end, optimal location and operation of storage units pose new analytical challenges for distribution system operators [58].

10.16   Case study: renewable energy integration

Integrating renewable power generation (e.g., wind and solar) facilitates to decreasing the stress on the grid. However, such generation options are intermittent in nature and the uncertainty associated with predicting the future generation poses more challenges in smart grid applications. For instance the production can fall less during peak hours, or it can be high during low demand regime. However, as the ESS become a more viable option, it can be used to improve the integration of renewable energy resources. For instance energy storage can be used to shape the generated power and provide a constant output charging infrastructure. For instance, stochastic renewable generation can be fed into the ESS, and depending on the charge level ESS can provide a regulated constant output. By regulating the system output, our model can still be applied under stochastic power generation regime. A conceptual overview is illustrated in Figure 10.18(a).

There exist case studies that substantiate the observation above. In References 59 and 60, authors present a case study for dispatching a wind farm using battery ESSs. An 8 MW ESS (max 4 hour discharge) is employed to dispatch a 60 MW wind farm. The average power output for the next hour (P_{set}) is assumed to be forecasted with 10% mean absolute error, and the difference is compensated with the energy storage. Then the power used from the storage P_{ESS} can be found by $P_{ESS} = P_{set} - P_{wind}$ where P_{wind} is the output of the renewable generation. Hence, the total power sent to charging station can be computed by $P_{output} = P_{wind} + P_{ESS}$.

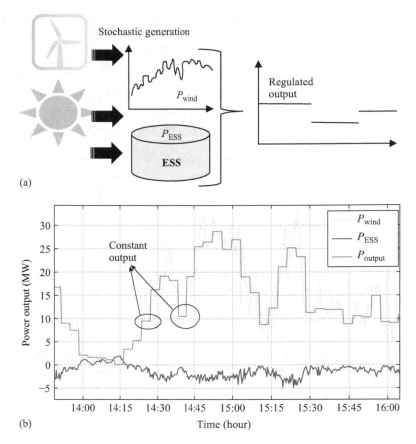

Figure 10.18   Case study on renewable integration. (a) A conceptual overview and (b) ESS dispatch performance

The assumptions on the power electronics of the systems include 3% AC/DC converter loses for the ESS, and the State of the Charge (SoC) of the ESS varies between 30% and 100%. The results depicted in Figure 10.18(b) shows that ESS technologies can be used to smoothen the stochastic generation.

References

[1]   DOE Global Energy Storage Database. Available from http://www.energystorageexchange.org/. [Accessed March 2016].
[2]   Haisheng Chen, Thang Ngoc Cong, Wei Yang, Chunqing Tan, Yongliang Li, and Yulong Ding. "Progress in electrical energy storage system: A critical review". *Progress in Natural Science*, 19(3):291–312, 2009.
[3]   Xing Luo, Jihong Wang, Mark Dooner, and Jonathan Clarke. "Overview of current development in electrical energy storage technologies and the

application potential in power system operation". *Applied Energy*, 137: 511–536, 2015.

[4] D. M. Rastler. *Electricity Energy Storage Technology Options: A White Paper Primer on Applications, Costs and Benefits*. Electric Power Research Institute, 2010.

[5] Marc Beaudin, Hamidreza Zareipour, Anthony Schellenberglabe, and William Rosehart. "Energy storage for mitigating the variability of renewable electricity sources: An updated review". *Energy for Sustainable Development*, 14(4):302–314, 2010.

[6] Ioannis Hadjipaschalis, Andreas Poullikkas, and Venizelos Efthimiou. "Overview of current and future energy storage technologies for electric power applications". *Renewable and Sustainable Energy Reviews*, 13 (67):1513–1522, 2009.

[7] J. Kondoh, I. Ishii, H. Yamaguchi, *et al.* "Electrical energy storage systems for energy networks". *Energy Conversion and Management*, 41(17): 1863–1874, 2000.

[8] Jeffery B. Greenblatt, Samir Succar, David C. Denkenberger, Robert H. Williams, and Robert H. Socolow. "Baseload wind energy: Modeling the competition between gas turbines and compressed air energy storage for supplemental generation". *Energy Policy*, 35(3):1474–1492, 2007.

[9] Francisco Díaz-González, Andreas Sumper, Oriol Gomis-Bellmunt, and Roberto Villafáfila-Robles. "A review of energy storage technologies for wind power applications". *Renewable and Sustainable Energy Reviews*, 16(4):2154–2171, 2012.

[10] John Baker. "New technology and possible advances in energy storage". *Energy Policy*, 36(12):4368–4373, 2008.

[11] SMART GRID. Energy storage – a key enabler of the smart grid. 2009. The U.S. Department of Energy (Technical report). Available at https://www. netl.doe.gov/File%20Library/research/energy%20efficiency/smart%20grid/ whitepapers/Energy-Storage_2009_10_02.pdf.

[12] Tarik Kousksou, Pascal Bruel, Abdelmajid Jamil, T. El Rhafiki, and Youssef Zeraouli. "Energy storage: Applications and challenges". *Solar Energy Materials and Solar Cells*, 120:59–80, 2014.

[13] Faizur Rahman, Shafiqur Rehman, and Mohammed Arif Abdul-Majeed. "Overview of energy storage systems for storing electricity from renewable energy sources in Saudi Arabia". *Renewable and Sustainable Energy Reviews*, 16(1):274–283, 2012.

[14] Robert Alan Huggins. *Energy Storage*, vol. 391. Berlin: Springer, 2010.

[15] Sam Koohi-Kamali, V. V. Tyagi, N. A. Rahim, N. L. Panwar, and H. Mokhlis. "Emergence of energy storage technologies as the solution for reliable operation of smart power systems: A review". *Renewable and Sustainable Energy Reviews*, 25:135–165, 2013.

[16] Helder Lopes Ferreira, Raquel Garde, Gianluca Fulli, Wil Kling, and Joao Pecas Lopes. "Characterisation of electrical energy storage technologies". *Energy*, 53:288–298, 2013.

[17] Michel Broussely and Gianfranco Pistoia. *Industrial Applications of Batteries: From Cars to Aerospace and Energy Storage*. Amsterdam: Elsevier, 2007.

[18] Wenhua H. Zhu, Ying Zhu, Zenda Davis, and Bruce J. Tatarchuk. "Energy efficiency and capacity retention of Ni–Mh batteries for storage applications". *Applied Energy*, 106:307–313, 2013.

[19] M. A. Fetcenko, S. R. Ovshinsky, B. Reichman, *et al.* "Recent advances in nimh battery technology". *Journal of Power Sources*, 165(2):544–551, 2007.

[20] A. G. Ritchie. "Recent developments and future prospects for lithium rechargeable batteries". *Journal of Power Sources*, 96(1):1–4, 2001.

[21] Zhenguo Yang, Jianlu Zhang, Michael C. W. Kintner-Meyer, *et al.* "Electrochemical energy storage for green grid". *Chemical Reviews*, 111(5): 3577–3613, 2011.

[22] M. Stanley Whittingham. "History, evolution, and future status of energy storage". *Proceedings of the IEEE*, 100(Special Centennial Issue):1518–1534, 2012.

[23] Trung Nguyen and Robert F. Savinell. "Flow batteries". *Electrochemical Society Interface*, 19(3):54–56, 2010.

[24] Brian Elmegaard and Wiebke Brix. "Efficiency of compressed air energy storage". In *24th International Conference on Efficiency, Cost, Optimization, Simulation and Environmental Impact of Energy Systems*, 2011.

[25] Matt Lazarewicz and Jim Arseneaux. "Flywheel-based frequency regulation demonstration projects status". In *Proceedings of EESAT Conference*, San Francisco, CA, pp. 1–22, 2005.

[26] Hussein Ibrahim, Adrian Ilinca, and Jean Perron. "Energy storage systemsâ€" characteristics and comparisons". *Renewable and Sustainable Energy Reviews*, 12(5):1221–1250, 2008.

[27] Niels J. Schenk, Henri C. Moll, José Potting, and René M. J. Benders. "Wind energy, electricity, and hydrogen in the Netherlands". *Energy*, 32(10):1960–1971, 2007.

[28] Dennis Anderson and Matthew Leach. "Harvesting and redistributing renewable energy: On the role of gas and electricity grids to overcome intermittency through the generation and storage of hydrogen". *Energy policy*, 32(14):1603–1614, 2004.

[29] T. Nakken, L. R. Strand, E. Frantzen, R. Rohden, and P. O. Eide. "The utsira wind-hydrogen system–operational experience". In *European Wind Energy Conference*, pp. 1–9, 2006.

[30] Solar Fuels and Artificial Photosynthesis. "Science and innovation to change our future energy options". *RSC*, January 2012.

[31] Stenbjörn Styring. "Artificial photosynthesis for solar fuels". *Faraday Discussions*, 155:357–376, 2012.

[32] A. Steinfeld. "Solar hydrogen production via a two-step water-splitting thermochemical cycle based on Zn/ZnO redox reactions". *International Journal of Hydrogen Energy*, 27(6):611–619, 2002.

[33] F. C. Lin, Xin Dai, Z. A. Xu, J. Li, and Z. G. Zhao. "High density capacitors". *High Power Laser Part Beams*, 1:94–6, 2003.

[34] Steven C. Smith, P. K. Sen Sr., and Benjamin Kroposki Sr. "Advancement of energy storage devices and applications in electrical power system". In *Power and Energy Society General Meeting-Conversion and Delivery of Electrical Energy in the 21st Century, 2008 IEEE*, pp. 1–8. IEEE, 2008.

[35] Alexander Kusko and John Dedad. "Short-term, long-term, energy storage methods for standby electric power systems". In *Industry Applications Conference, 2005. Fourtieth IAS Annual Meeting. Conference Record of the 2005*, vol. 4, pp. 2672–2678. IEEE, 2005.

[36] Paulo F. Ribeiro, Brian K. Johnson, Mariesa L. Crow, Aysen Arsoy, and Yilu Liu. "Energy storage systems for advanced power applications". *Proceedings of the IEEE*, 89(12):1744–1756, 2001.

[37] M. H. Ali, B. Wu, and R. A. Dougal. "An overview of SMES applications in power and energy systems". *IEEE Transactions on Sustainable Energy*, 1(1):38–47, April 2010.

[38] Peter J. Hall and Euan J. Bain. "Energy-storage technologies and electricity generation". *Energy Policy*, 36(12):4352–4355, 2008.

[39] Harald Mehling and Luisa F. Cabeza. *Heat and Cold Storage with PCM*. Berlin: Springer, 2008.

[40] Nicole Pfleger, Thomas Bauer, Claudia Martin, Markus Eck, and Antje Wörner. "Thermal energy storage – Overview and specific insight into nitrate salts for sensible and latent heat storage". *Beilstein Journal of Nanotechnology*, 6(1):1487–1497, 2015.

[41] Andreas Hauer. "Sorption theory for thermal energy storage". In *Thermal Energy Storage for Sustainable Energy Consumption: Fundamentals, Case Studies and Design*, pp. 393–408. Dordrecht: Springer, 2007.

[42] Jean-Christophe Hadorn and Groupe Berney-BASE Consultants SA. "IEA solar heating and cooling programme task 32: Advanced storage concepts for solar and low energy buildings". In *Ecostock 2006 – The 10th International Conference on Thermal Energy Storage*, 2006.

[43] S. Vazquez, S. M. Lukic, E. Galvan, L. G. Franquelo, and J. M. Carrasco. "Energy storage systems for transport and grid applications". *IEEE Transactions on Industrial Electronics*, 57(12):3881–3895, December 2010.

[44] P. D. Cavazzini. "Technological developments for pumped-hydro energy storage". *EERA Report*, 2014.

[45] Brian Elmegaard and Wiebke Brix. "Efficiency of compressed air energy storage". In *24th International Conference on Efficiency, Cost, Optimization, Simulation and Environmental Impact of Energy Systems*, 2011.

[46] Mahdi Johar, Ahmad Radan, Mohammad Reza Miveh, and Sohrab Mirsaeidi. "Comparison of DFIG and synchronous machine for storage hydropower generation". *International Journal of Pure and Applied Sciences and Technology*, pp. 48–58, 2011.

[47] Muhammad H. Rashid. *Power Electronics Handbook: Devices, Circuits and Applications*. Cambridge, MA: Academic Press, 2010.

[48] Zahra Amjadi and Sheldon S. Williamson. "Power-electronics-based solutions for plug-in hybrid electric vehicle energy storage and management systems". *IEEE Transactions on Industrial Electronics*, 57(2):608–616, 2010.

[49] P. J. Grbovic, P. Delarue, P. Le Moigne, and P. Bartholomeus. "The ultra-capacitor-based regenerative controlled electric drives with power-smoothing capability". *IEEE Transactions on Industrial Electronics*, 59(12):4511–4522, December 2012.

[50] G. O. Suvire and P. E. Mercado. "Active power control of a flywheel energy storage system for wind energy applications". *IET Renewable Power Generation*, 6(1):9–16, January 2012.

[51] David Velasco De La Fuente, César L. Trujillo Rodrguez, Gabriel Garcerá, Emilio Figueres, and Rubén Ortega Gonzalez. "Photovoltaic power system with battery backup with grid-connection and islanded operation capabilities". *IEEE Transactions on Industrial Electronics*, 60(4):1571–1581, 2013.

[52] Roberto Cárdenas, Rubén Peña, Salvador Alepuz, and Greg Asher. "Overview of control systems for the operation of DFIGs in wind energy applications". *IEEE Transactions on Industrial Electronics*, 7(60):2776–2798, 2013.

[53] A. Oudalov, R. Cherkaoui, and A. Beguin. "Sizing and optimal operation of battery energy storage system for peak shaving application". In *Power Tech, 2007 IEEE Lausanne*, pp. 621–625, July 2007.

[54] H. Bludszuweit and J. A. Dominguez-Navarro. "A probabilistic method for energy storage sizing based on wind power forecast uncertainty". *IEEE Transactions on Power Systems*, 26(3):1651–1658, August 2011.

[55] I. S. Bayram, M. Abdallah, A. Tajer, and K. A. Qaraqe. "A stochastic sizing approach for sharing-based energy storage applications". *IEEE Transactions on Smart Grid* (99), 2015. Available at http://ieeexplore.ieee.org/document/7217850/?arnumber=7217850&tag=1.

[56] I. S. Bayram, M. Abdallah, A. Tajer, and K. Qaraqe. "Energy storage sizing for peak hour utility applications". In *IEEE International Conference on Communications*, pp. 770–775, June 2015.

[57] S. X. Chen, Hoay Beng Gooi, and MingQiang Wang. "Sizing of energy storage for microgrids". *IEEE Transactions on Smart Grid*, 3(1):142–151, 2012.

[58] Ahmed S. A. Awad, Tarek H. M. El-Fouly, and Magdy Salama. "Optimal ESS allocation for load management application". *IEEE Transactions on Power Systems*, 30(1):327–336, 2015.

[59] J. Castaneda, J. Enslin, D. Elizondo, N. Abed, and S. Teleke. "Application of STATCOM with energy storage for wind farm integration". In *Proceedings of the IEEE PES Transmission and Distribution Conference and Exposition*, pp. 1–6, New Orleans, LA, April 2010.

[60] Sercan Teleke. "Energy storage overview: Applications, technologies and economical evaluation". Technical Report, Quanta Technology, January 2011.

Chapter 11

Control and optimisation for integration of plug-in vehicles in smart grid

Zhile Yang and Kang Li

The plug-in vehicles are one of the new participants in power systems. Mass roll-out of plug-in vehicles, on one hand, will significantly challenge the existing power system scheduling strategies and infrastructures. On the other hand, it will also create many opportunities for power-system operators or users. The aggregated energy capacity of a large number of plug-in vehicles can provide ancillary services to improve power system reliability and power quality, in addition to the environmental benefits in reducing emissions from the transportation sector. Advanced control and optimisation methods have been utilised to integrate plug-in vehicles with smart grid by providing optimal charging and discharging profiles. In this chapter, the key features of plug-in vehicles and their impacts on the grid are firstly detailed, followed by the review of state-of-the-art optimisation and control techniques which have been used in the plug-in vehicles scheduling. Then, a case study of unit commitment integrating plug-in vehicles is presented and numerical results are illustrated.

11.1 Introduction

Global warming due to extensive consumption of fossil fuels has been one of the key challenges for human beings in the past decades and the situation is still deteriorating with fast industrialisation of developing countries. The global COP21 agreement forged in the Paris Climate Conference 2015 set a target to limit the maximum temperature rise within 2 °C, and ultimately, realise zero net carbon emissions by the end of this century [1]. To achieve this goal, reducing the green-house gas (GHG) emissions from both the power generation and transportation is the key and emergent issue [2]. Among various energy consumers, power generation and transportation are two major fossil fuel users and the largest contributors of GHG and pollutant emissions [3]. According to the report of International Energy Agency, coal, oil and

School of Electronics, Electrical Engineering and Computer Science, Queen's University Belfast, Belfast, BT9 5AH, United Kingdom

natural gas consumption accounted for 80.6% of total energy supply in 2014; transportation accounted for 45% of oil consumption in 1973 and rises to 63.8% in 2013 [4]. In terms of GHG emissions, power generation contributes 37% of global emissions [5], whereas transportation produces over 20% of carbon dioxide emissions [6]. To tackle these problems, various policies, strategies and technical approaches have been proposed, among which to intelligently integrate electric vehicles (EVs) in the smart grid environment is a crucial yet challenging issue.

Transportation electrification is an alternative approach to reduce fossil fuel consumption from the transportation sector. In particular, internal combustion engine (ICE) based vehicles are one of the major carbon emission producers, and EVs have long been viewed as a promising option to replace traditional ICE-based vehicles, though the popularity of EVs was hindered by the battery technologies, including shortage capacity, weight and price of batteries suitable for EV applications. Fortunately, the situation is changing along with the continuous technical breakthroughs by leading automotive and battery manufacturers as well as strong incentives from governments and authorities. The successful commercialisation boosts the number of both EVs and EVs supply equipment, providing active interaction opportunities between the plug-in EVs (PEVs) and smart grids.

The large number of PEVs is a two-edged sword for the grid. On one hand, the stochastic charging behaviours of EV users significantly challenge the existing power system infrastructure. They may bring unexpected peaks of charging power during peak-load periods, which calls for extra spinning reserve for maintaining system-load balance. On the other hand, coordinated charging and vehicle to grid (V2G) scheme of EVs enable them to become distributed energy storages, providing energy buffers for system operators to intelligently schedule EVs power load and for individual users to realise the arbitrage and reduce charging cost. Moreover, the increasing penetration of wind power, solar power and other non-carbon emission energy sources introduce remarkably undispatchable generations for the power system, which needs energy storages such as EVs for seamless integration and load shifting.

In this chapter, we focus on the discussion of optimisation and control methods for integrating PEVs with the smart grid. The categories of EVs and their impact on the smart grid will be discussed first, followed by the review of control and optimisation techniques which have been used in the specific PEVs integrating utilisation. Then, a case study is presented evaluating an economic-unit-commitment problem integrated with charging and discharging of PEVs. The final section concludes the chapter.

11.2   Modelling of plug-in electric vehicle for smart grid

Days back to the first model developed in 1827; EVs have long been viewed as an alternative option in transportation field. In the past few decades, typical popular utilisation of EVs lie on some short mileage, light loaded and environmental sensitive applications such as golf carts, campus buses or airport commuters. These applications are normally of small scale and charging temporal insensitivity, which may not have noticeable impact on the power grid. However, the commercial success of electric cars is a complete new story. Considering the 1.5 billion global car stocks, only a small percentage of cars being replaced by the EVs would introduce over a hundred thousand MW power demand into the power system. Moreover, batteries in these vehicles are portable energy storages and bidirectional power flow controllable, which has the potential to be important participants in power grid.

11.2.1 Categories of electric vehicles

Literally, a vehicle which utilises an electric motor to drive could be categorised as an EV. The power of an EV could be provided by batteries such as lead-acid and Li-ion batteries, flywheel, super capacitor as well as hydrogen etc.

The battery-based EVs could be categorised as pure battery EVs (BEVs), hybrid EVs (HEVs) (normally non-plug-in), and plug-in HEVs (PHEVs) [7], with both BEVs and PHEVs being referred to as PEVs. The word 'plug-in' here indicates that the vehicle has electric chargeability from the external sources, rather than that a physical charging cable is required to be plugged in. Therefore, the wireless-charging-based EVs which employ the fast developing wireless charging technique should also be categorised as a specific type of 'plug-in'. Figure 11.1 illustrates the structure of both types of PEVs.

As shown in Figure 11.1, the generic structure for a BEV includes a battery charger for power injection (bidirectional for future V2G application), a battery pack for storing driving energy with the association of a battery management system, a DC/DC converter and a DC/AC inverter for supporting auxiliary devices and electrical motor respectively, as well as a mechanical transmission system. In addition, the components of ICE-based vehicles, such as the ICE, an engine control unit and a fuel tank need to be parallel installed in a PHEV. Note that the electric motor and ICE are coupled and coordinated either in an electrical pathway by a generator or a mechanical coupling approach, as shown ① and ② in Figure 11.1 respectively.

11.2.2 Modelling of key components of plug-in electric vehicles

Among various components of PEVs, the electric motor, the power inverter and the battery system play key roles and build a bridge of power and energy flow. The mathematical formulations of these key components are briefly reviewed in this section.

(1) Modelling of electric motor

Electric motor is the core component of PEV, aiming to replace the ICE and provide torque output for mechanical transmission system. A number of types of electric motor have been adopted in PEV models such as DC motor, switched reluctance motor, permanent magnet synchronous motor and induction motor (IM) [8]. Though DC motor is more reliable and efficient, the AC based IM is the

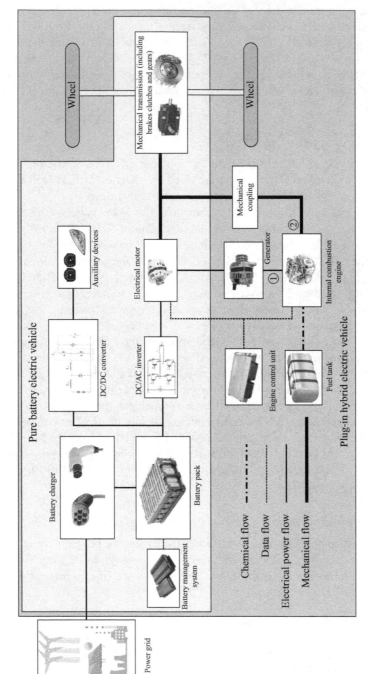

Figure 11.1 The structure of BEV and PHEV

Pure battery electric vehicle

Plug-in hybrid electric vehicle

Wheel

Wheel

Mechanical transmission (including brakes clutches and gears)

Mechanical coupling

Auxiliary devices

Electrical motor

Generator ①

Internal combustion engine ②

DC/DC converter

DC/AC inverter

Engine control unit

Fuel tank

Battery charger

Battery pack

Battery management system

Power grid

Chemical flow

Data flow

Electrical power flow

Mechanical flow

current first choice for automotive manufactures due to the low cost and maturity of frequency control techniques. An IM model adopting complex space-vector description [9] is shown as below:

$$u_{sK} = r_s i_{sK} + T_N \frac{d\psi_{sK}}{dt} + j\omega_K \psi_{sK} \tag{11.1}$$

$$0 = r_s i_{sK} + T_N \frac{d\psi_{rK}}{dt} + j(\omega_K - \omega_m)\psi_{rK} \tag{11.2}$$

$$\psi_{sK} = l_s i_{sK} + l_M i_{rK} \tag{11.3}$$

$$\psi_{rK} = l_r i_{rK} + l_M i_{sK} \tag{11.4}$$

$$\frac{d\omega_m}{dt} = \frac{1}{T_M}\left[Im\left(\psi_{sK}^* i_{sK}\right) - m_L\right] \tag{11.5}$$

where a frame index K denotes the rotating coordinate system and u_s, i_s, i_r, ψ_s and ψ_r are the stator voltage, stator current, rotor current, stator flux linkage and rotor flux linkage vectors respectively; ω_K and ω_m denote the rotating angular speed and mechanical angular speed; m_L represents the load torque; l_s, l_r and l_M are the inductances of stator, rotor and magnetism; T_N and T_M are the nominal frequency time constant and the mechanical time constant, respectively. The vector-control method of IM could be categorised as scalar-based and vector-based controllers which have been well developed in various automotive models.

(2) Modelling of inverter

In a PEV system, an inverter connects the DC battery and AC electric IM by inverting DC to AC in driving propulsion mode and converting AC to DC in braking mode. The voltage and current output of a single-phase DC–AC inverter v_{out} and i_{out} could be modelled as below [10],

$$v_{out}(t) = \sum_{k=1}^{\infty} \left[\frac{4V_{in}}{k\pi}\sin(k\omega t)\right] \quad k = 1, 3, 5, \ldots \tag{11.6}$$

$$i_{out}(t) = \begin{cases} \sum_{k=1}^{\infty} \left[\frac{4V_{in}}{k\pi R}\sin(k\omega t)\right] & k = 1, 3, 5, \ldots \quad \text{for pure resistive load} \\ \sum_{k=1}^{\infty} \left[\frac{4V_{in}}{k^2\pi\omega L}\sin\left(k\omega t - \frac{\pi}{2}\right)\right] & k = 1, 3, 5, \ldots \quad \text{for pure inductive load} \end{cases} \tag{11.7}$$

where V_{in} is the battery output voltage and the input voltage of the inverter; $\omega = 2\pi f$ is the angular frequency of the output voltage; R and L are the resistor and inductance, respectively. The direction of the current through inverter is determined by the motor status which acts either resistive or inductive load.

(3) Modelling of battery system

Electric battery system is the major energy source for PEV and the key technique to replace fossil fuel consumption in transportation sector. Accurate battery models provide indispensable information for battery management system to intelligently control the charging and discharging behaviours. Though many different types of batteries have been used in the PEV, Li-ion battery is the most popular option due to its large energy and power density. Categorised by modelling structure, white-box based electrochemical model, grey-box based equivalent circuit model and empirical model, as well as black-box based neural network model have been intensively developed [11]. Grey-box based equivalent circuit model utilises an equivalent electric circuit to represent the complex chemical reaction in the battery and only a few electric components need to be identified. The basic equivalent circuit model [12] is shown in Figure 11.2 and the status equations are shown below:

$$SOC(k) = SOC(k-1) - \frac{\eta[I_O(k) + I_O(k-1)][T(k) + T(k-1)]}{2C} \quad (11.8)$$

$$U_o = U_{oc} - R_i I_o - R_p I_p \quad (11.9)$$

where SOC is the state-of-charge of the battery; η is the coulombic efficiency; C is the nominal capacity of the battery; T is the sampling instance and k denotes the time index. In equivalent circuit model shown in Figure 11.2, the I_o represents load current, and I_p is the battery polarisation current; U_o and U_{oc} are the battery output voltage and open circuit voltage of battery, respectively; R_i and R_p are the battery internal resistor and polarisation resistor. The initial SOC is calculated by open circuit voltage method [13]. Apart from the SOC model, state of health (SOH) and thermal models are also important to illustrate the battery statues [14,15].

11.2.3 Foundation of plug-in vehicle facilities for smart grid integration

Due to the continual technical development, the capacity of single PEV battery pack is quickly increasing. Ideally, a single-charge capacity of a PEV's battery pack

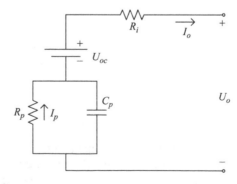

Figure 11.2   Equivalent circuit model

should be able to support the equivalent driving mileage or even longer than a full tank of ICE based vehicles. For the majority of normal or compact PEV models, 1 kW h of battery capacity is able to support about 4 miles travelling distance. Leading automobile manufacturers have made a huge amount of effort to advance the EV technologies [16] and Table 11.1 lists the statistics of different PEV categories based on the data published in 2009 National Household Travel Survey (NHTS) [17].

Table 11.1 shows that approximately half of the vehicles are automobiles or cars, followed by the pickup trucks and sports utility vehicles (SUVs). Note that the NHTS was taken in the United States where the low fuel price encourages the adoptions of large displacement vehicles, such as pickup trucks and SUVs. Therefore, the proportion of economic cars could be much higher in countries of Europe or Asia due to the higher fuel cost. Moreover, typical PEV popular models from different manufactures are selected in column 4 associated with the official claimed battery capacity in column 5 of Table 11.1. Assuming that, all the ICE based vehicles are replaced by PEVs, the accumulated battery capacities of the PEVs are then calculated in column 6. Totally 11,271 MW h battery capacity is available for supporting the daily transportation millage as well as potential to play the roles of energy storages for providing various support to the power grid.

On the other hand, a power grid is a large complex network system which integrates a number of generation units, transmission utilities as well as end users.

Table 11.1   Typical vehicle types and their PEV models

Type number	Vehicle type	Vehicle number	Typical PEV model	Battery capacity (kW h)	Total battery capacity (MW h)
1	Automobile/car/ station wagon	151,238	Nissan Leaf	24	3,629
2	Van (mini, cargo, passenger)	24,145	Renault Kangoo ZE	44	1,062
3	Sports utility vehicle	58,718	Tesla Model X	85	4,991
4	Pickup truck	60,308	VIA Extended-Range Electric Truck	23	1,387
5	Other truck	1,152	Condor Electric Truck	50	57
6	RV (recreational vehicle)	2,193	Not available	Not available	Not available
7	Motorcycle	10,202	Zero Motorcycle	12.2	124
8	Golf cart	292	Yamaha G30EK	1.86	0.5
9	Other	915	Not available	20	18.3
Total/ average		309,163		36	11,271

New participants such as intermittent and undispatchable renewable generations, large capacity grid-tied energy storages, as well as PEVs make the power system further complicated. On one hand, large penetration of uncoordinated PEVs brings considerable power demand increase and high uncertainty to the grid, which significantly challenges the current scheduling strategy and system infrastructure. On the other hand, the large distributed battery storage and bidirectional charging and discharging options provide many opportunities for PEVs to contribute to the smart grid operations. Researchers have analysed the utilisation of energy storage capacity of PEVs for shifting the peak load to valley period [18–21], providing V2G service [22] during peak load, acting as ancillary service providers for frequency regulation [23], power loss minimisation [24], power reserve [25] and so on. Some studies also investigate the coordination of PEVs charging and discharging with renewable generations [26–28].

To help analyse the impact of PEVs, the statistics data of vehicle type, vehicle number and daily average miles from NHTS 2009 which are listed in Table 11.2 can be used, where the energy cost of corresponding type is calculated by using the claimed standard millage of the popular models which have been shown in Tables 11.1 and 11.2 further links the model types with the energy cost, daily energy cost and load contribution, assuming that all the surveyed vehicles are PEVs.

It could be seen from Table 11.2 that though the cars take the largest utilisation percentage among all vehicle types, the pickup trucks produce the largest daily-load contribution. This is due to the much higher energy cost per mile. However, full electrification of the pickup trucks might not be realistic in the foreseen future due to the longer travelling distance and heavy duty tasks. It is also shown that the

Table 11.2   Typical EV models and daily load

Type number	Vehicle type	Vehicle number	Per cent (%)	Daily average miles	Energy cost (kW h/mile)	Daily energy cost (kW h/d)	Daily load contribution (MW h)
1	Automobile/car/ station wagon	151,238	48.92	23.93	0.22	5.27	796
2	Van (mini, cargo, passenger)	24,145	7.81	27.46	0.35	9.61	232
3	Sports utility vehicle	58,718	18.99	29.26	0.32	9.36	550
4	Pickup truck	60,308	19.51	23.89	0.57	13.62	821
5	Other truck	1,152	0.37	37.26	0.43	16.02	18
6	RV (recreational vehicle)	2,193	0.71	7.04	Not available	Not available	Not available
7	Motorcycle	10,202	3.30	6.90	0.10	0.69	7
8	Golf cart	292	0.09	2.65	0.15	0.40	0.1
9	Other	915	0.30	11.70	0.32	3.75	3
Total/ average		309,163	100	24.52	0.32	7.86	2,428

daily load of PEVs is accumulated to 2,428 MW h, which presents a significant load to the distribution level grids. Given such a huge load, coordinated charging strategies would undoubtedly be necessary and important for seamlessly integration of the PEVs into the power system.

Due to the fact that the scheduling strategy is only applicable to PEVs when they are plugged in, the number of 'on-line' power system connected PEVs needs to be estimated for the power system scheduling. According to the travelling data published in NHTS 2009, the number of on road vehicles among the total 30,357 activated (used during the survey day) vehicles within the 24 h-time horizon is shown in Figure 11.3, assuming that all the vehicles are PEVs and all off-road vehicles are plugged into the gird. The on-road number of different types (shown as bars) and the total number of off-road activated vehicles (shown as plugged-in curve) in each hour are illustrated accordingly.

According to Figure 11.3, the majority of the activated PEVs are online along the 24 h horizon. The online percentage ranges from over 99% during off-peak time to 65% within peak time. Moreover, the statistics data in Figure 11.3 only considers activated vehicles which have been used on the single day. If we consider the total number of 309,163 registered vehicles in the survey, the maximum percentage of vehicles on road would be less than 4%. In another word, over 96% of vehicles are plug-in and ideally available for online scheduling for most time of the day. This statistical data further reveal that it is possible for the grid operators to intelligently schedule the charging and discharging of PEVs to benefit power system operation and control as well as EV owners.

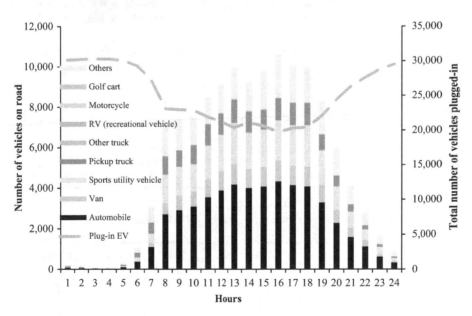

Figure 11.3   Number of on road unplug-in and plug-in PEVs in one day horizon of part of vehicles in NHTS 2009

The penetration of PEVs is expected to grow continuously. However, considering the large charging power, even a small number of PEVs would exert significant impact on the operation and control of micro-grids or distributed level power systems. At present, the charging standards for PEVs are quite diverse for different countries and regions. The charging power of a single PEV ranges from 1.4 to 20 kW for a home charger, but could be as much as 120 kW in a Tesla supercharging station. The International Electromechanical Commission (IEC) EV standards IEC61851 for plugs and socket-outlets and IEC15118 for power electronics devices have been widely adopted in PEV charging. Some other popular standards such as SAE J1772 in North America and JEVS G105-1993 in Japan have also been used for PEV charging. Tables 11.3 and 11.4 list the charging modes adopted in many European countries and Tesla charging modes in the United States.

Comparing the PEV charging power with other household appliances such as 3 kW air conditioner and 2 kW refrigerator, the PEV is an occasional but huge load contributor especially in fast charging modes. The charging behaviour will significantly impact the overall household load curve. Till now, only the charging facilities have been widely adopted. The discharging facilities are, however, still at the research stage due to a number of factors, including the lack of willingness of the PEVs users to participate. In the future, the discharging facilities are potential to be merged with the charging facility to allow the bidirectional power flow between the grid and individual PEVs.

In addition to the various charging standards, geographical distribution of PEVs means that their charging is remarkably different among different countries, or even among cities in the same country. In mega metropolitan cities such as London and New York, the charging load of PEVs tends to boost in daytime during

Table 11.3   PEVs charging mode and power in many European countries [29]

Level	Type	Electrical	Resulting charging	Time to charge	Power (kW)
Level (Mode) 1	Standard (Domestic)	230 V 16 A1 or 3 Phase	100%	6–8 h	3–10
Level (Mode) 2	Opportunity	400 V 32 A	50%	30 min	22
Level (Mode) 3	Emergency	400 V 32 A	20 km	10 min	22
Level (Mode) 4	Range extension	400 V 63 A	80%	30 min	44

Table 11.4   PEVs charging mode and power of Tesla motor [30]

Outlet details	Electrical	Miles of range per hour of charge	Power (kW)
NEMA 5–15 Standard Outlet	110 V 12 A	3	1.4
NEMA 5–20 Newer Standard Outlet	110 V 15 A	4	1.8
NEMA 14–50 RVs and Campsites	240 V 40 A	29	10
Wall connector	240 V 80 A	58	20
Super charging station	Parallel chargers	340	120

working hours but will drop in the evening. In the rural areas nearby the cities, however, the charging load of PEVs will likely increase when the commuting PEVs arrives home in the evening, and on the contrary, the charging load will dramatically decrease when the PEVs are outbound during the day time.

As a result, the massive load and huge charging power of PEVs will inevitably exert noticeable impacts on the power grid. The long plug-in time and large aggregated capacity of battery packs, on the other hand, provide many opportunities for PEVs and their coordinators to flexibly schedule the charging behaviours, provide ancillary services and play other available roles in future power system. This calls for advanced control and optimisation techniques to intelligently integrate the PEVs into the smart grid.

11.3 Control and optimisation techniques for plug-in vehicle integration with grid

Numerous control and optimisation methods have been developed and employed for power system operation and scheduling, playing crucial roles in endowing the grid with 'smartness'. The integration of PEVs requires more intelligent methods to be employed in this area. The scheduling strategies of PEVs are developed and analysed from the perspectives of power generator, operator, PEVs aggregator as well and individual user levels, associated with various objectives and constraints [31,32]. Given the problem formulations at different participant levels of the grid, various conventional mathematical approaches, intelligent meta-heuristic algorithms (MAs) as well as intelligent control methods have been utilised in achieving the proposed goals.

11.3.1 Objectives and constraints

In order to seamless integrate, the PEVs with the grid, power system generators, operators, PEVs aggregators as well as individual users propose their own objectives respectively from economic or environmental perspectives to schedule PEVs charging and discharging in various time scales and make decisions. Figure 11.4 illustrates the objectives which consider the participation of PEVs.

(1) Generation objectives

From the generation side, the high penetration of PEVs is likely to increase the overall load of a certain district. The original infrastructure capacity is required to be expanded to accommodate the load increase under different levels of penetration and power size of PEVs. It is necessary to consider the capacity expansion cost to meet the PEVs load in the long-term system planning [33]. An optimisation formulation minimising the expansion cost is given below:

$$\text{Minimise } C_E = \sum_{i \in \psi = \chi \cup N} c_i^{\text{BLD}} y_i^{\text{BLD}} \qquad (11.10)$$

$$\text{Subject to} \qquad \sum_{i \in \chi} k_i + \sum_{i \in N} k_i y_i^{\text{BLD}} \geq \left(1 + R^{\text{RM}}\right) L^{\text{PEAK}} \qquad (11.11)$$

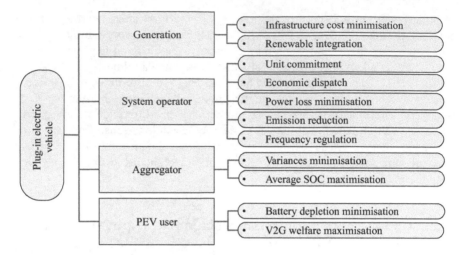

Figure 11.4   Objectives for PEVs integration

In the objective function, C_E is the total cost of expansion capacity, and k_i is the capacity of the ith unit. Y and N denote the original units and potential expansion units, respectively. The cost constant and binary decision variables of expanded units are represented as c_i^{BLD} and y_i^{BLD}, respectively. If y_i^{BLD} is '1', the corresponding candidate unit will be expanded, and vice versa. The constraint stipulates that the accumulated capacity of units should be able to provide the predicted peak load L^{PEAK} as well as the spinning reserve scaled by reserve margin R^{RM}.

In addition to the PEVs, the capacity expansion of the generation should also consider the penetration of renewable generations. The high intermittence of wind power significantly challenges the ramping rate of power generators fuelled by cheap resources such as coal and peat. The gas fired generators ramp much faster, therefore are more capable of meeting challenges of variable generations from intermittent renewable energy resources, however the cost is higher. It is therefore important to simultaneously consider the unit commitment (UC) task when deciding the generator expansion, thus the system operator has a key role to play.

(2) Operator objectives

Power system operator coordinates power supply of the generation units, maintains the fulfilment of power demand and keeps the load balance. Moreover, the ancillary services provided by the grid participants are also scheduled by the operator. Therefore, the key scheduling strategies of PEVs are implemented by the system operator in solving problems such as economic dispatch, UC, the power loss minimisation, environmental cost minimisation as well as ancillary services dispatch including frequency regulation and power reserve.

One of the most important tasks for the operator is to schedule the UC. It is a mixed-integer optimisation problem, aiming to minimise the generation fossil fuel cost and unit start-up/shut-down cost [34]. The constraints of the problem include generation limit, power demand limit, power spinning reserve limit and minimum up/down time limit. Charging and discharging scheduling of PEVs in combination with the UC problem was first considered in Reference 35 where 50,000 PEVs were integrated with a 10 unit system. PEVs are assumed to be available to either be charged or discharged in each hour and the numbers of charged and discharged PEVs are dispatched together with the UC. An additional constraint for UC considering PEVs is that the accumulated charging energy of PEVs should be affordable for supporting the daily PEVs usage. To further demonstrate the problem formulation and solving techniques, a numerical case study of UC problem considering flexible PEVs charging and discharging is presented in Section 11.4.

In the UC considering PEVs integration, the objectives could be both economic and environmental cost minimisation. Besides, the PEVs scheduling strategy is developed in association with the handling of the stochastic wind and solar power generation to increase the penetration of renewable energy or with dispatchable hydro and gas power to realise the load levelling. The intermittent renewable energy is commonly modelled as probability distribution and sampled as scenarios or solved using stochastic optimisation [36]. The conventional UC task is a day-ahead scheduling that determines the generation profile within 24 h. An optimal profile for 24-h PEVs power delivery derived from solving the UC problem provides instruction for the aggregator to intelligently control the real time charging and discharging of PEVs.

UC is not the only task which needs to consider PEVs scheduling in problem formulation. System operator is also required to consider dynamic Economic/environmental dispatch under different PEVs scheduling [37,38], utilisation of PEVs for frequency regulation [39] and power loss minimisation [40], etc. Due to the page limit, the specific issues are not discussed in details.

(3) Aggregator objectives

The PEVs aggregator accumulates PEVs power delivery and directly interacts with the power grid on behalf of belonging PEVs. It also plays a key role in scheduling and balances each individual PEV charger to minimise the economic cost and to produce the best trade off of demands for different users.

An important objective is to maximise the overall state of charge while maintaining the fairness among users. The objective function is formulated as follows [41]:

$$\text{Maximise } J(k) = \sum_i w_i(k) \cdot \text{SOC}_i(k+1) \tag{11.12}$$

$$w_i(k) = f\left(\text{Cap}_{r,i}(k), T_{r,i}(k), D_i(k)\right) \tag{11.13}$$

$$\text{Cap}_{r,i}(k) = (1 - \text{SOC}_i(k)) \cdot \text{Cap}_i \tag{11.14}$$

where $\text{SOC}_i(k+1)$ and $\text{SOC}_i(k)$ denote the state of charge of the ith PEV at time step $k+1$ and k, respectively. $w_i(k)$ is the weighting term for the corresponding PEV at time step k. $\text{Cap}_{r,i}(k)$ and $\text{Cap}_i(k)$ is the remaining battery capacity required to be filled and the rated battery capacity (Ah) for the ith PEV at time step k. $T_{r,i}(k)$

is the remaining time for charging, whereas $D_i(k)$ is the price difference between the real-time energy price and the price that a specific customer at the ith PEV charger is willing to pay at time step k. The system constraints of this optimisation problem include the charging rate of all and individual chargers, battery capacities as well as state of charge limits.

Another important problem for the aggregator is to dispatch the charging and discharging of PEVs by following an optimal load curve. Due to the direct inter-action with the grid and individual PEVs, aggregator strategies appear to be more powerful on fulfil the whole system dispatching for shaving peaks and filling val-leys of the load curve as well as financial profit of system operators, however, sacrifice the charging freedom and the further financial benefit of PEV owners.

(4) PEV user objectives

The individual PEV users decide the PEVs charging and discharging behaviour. They may wish to have a complete freedom of charging and be conditionally willing to join in the V2G service if the benefit outweighs the loss. A home-based PEVs charging controller is expected to be capable of scheduling the charging profile intelligently according to the real-time power price and user necessity. Moreover, demand side management methods have been studied in some studies [42]. The major load demand of home electric appliances such as air conditioner, boiler and refrigerator are coordinated with PEV charging. They will be analytically scheduled by being switched off during PEVs charging time to avoid dramatic household load increase. However, such straightforward strategy is sometimes difficult to imple-ment and sacrifice the comfort of users. The PEVs user charging strategy should, therefore, be coordinated in the development of smart home/building optimal strategy, offering the best trading off for the user comfort and economic perspective.

As a result, a number of optimisation objectives from multiple levels of system participants are formulated for seamless integration of PEVs with the grid. The new problems considering PEVs are high dimensional, non-linear, non-convex and multi-objective, which call for advanced control and optimisation tools.

11.3.2   Control and optimisation tools

Up to date, a number of methods have been utilised to develop the optimal sche-duling strategy of PEVs including conventional methods such as linear program-ming (LP) [34], quadratic programming (QP) [43], mixed-integer programming (MIP) [44] and dynamic programming (DP) [33], game theory (GT) [45], queuing theory (QT) [46] as well as meta-heuristic methods. These methods aim at specific priority applications and are implemented in embedded mathematical environments and commercial solvers, shown in Table 11.5.

(1) Linear programming

LP formulates the problem using a first order polynomial and some equality/inequality linear constraints, offers a simple and effective optimisation approach to model and solve problems with a low-computation cost. It is easy to implement and convenient to solve through multiple platforms or solvers. The basic formulation

Table 11.5   PEVs charging mode and power in many European countries

Method	Advantage vs. disadvantage		Priority applications	Preferred solvers or platforms
Linear programming	A	Easy implementation Multiple platform support Low computational cost	Operational cost minimisation with determined power price	CPLEX, GAMS
	D	Limited framework Low flexibility		
Quadratic programming	A	Higher flexibility Multiple platform support	Variances minimisation Power loss minimisation ED problem without valve-effect Welfare maximisation for PHEV users	CPLEX, IPOPT, CONOPT
	D	Limited framework Specific solver requirement		
Mix-integer programming	A	High flexibility Convenient for PHEV charging status	Unit commitment Generation cost minimisation Operational cost minimisation for aggregators	Xpress-MP, CPLEX, GAMS
	D	High computational cost for high dimensional problems Specific solvers and methods Requirement		
Dynamic programming	A	Time varying parameters Fast for low dimension problems	Operational cost minimisation with dynamic price for system operators Welfare maximisation for PHEV users	MATLAB®, C++
	D	'Dimension curse' for high dimensional problems		
Game theory	A	Economic interaction consideration	Price negotiation for PHEV users, aggregators and operators	MATLAB, C++
	D	Limited application		
Queuing theory	A	Scientific stochastic statistics for modelling PHEV charging	Modelling the stochastic PHEV charging load	MATLAB, C++
	D	Limited application		
Meta-heuristic methods	A	High flexibility Multiple platform support simple implementation	Large scale unit commitment Optimal power flow Operational cost minimisation with significant non-linear terms Other problems with highly non-linear terms Multi-objective problems	MATLAB, C++
	D	Higher computational cost Comparatively low performance Parameter configuration Difficulty		

and some popular variants such as robust and stochastic LP have been used in scheduling the PEVs integration.

One of the priority applications of LP is for the aggregator. It schedules the charging and discharging behaviours given the piece-wise power price and user-necessity configuration, minimising the charging cost and maximising the revenue [47]. The power delivery plan of a day-head time is achieved by solving the LP problem. The weakness of the LP is that the limitation of linear formulation sometimes fails to model practical non-linear problems.

(2) Quadratic programming

QP is a technique that allows formulations of second-order terms in describing the PEVs scheduling problem. It has been used in modelling fossil fuel cost for system operators, the variances between the PEVs dispatched power curve and a target curve for aggregator, as well as many other power-system problems considering PEVs. In the variance minimisation, the optimal target curve may refer to the average load of one-day time, a desired curve received from the upper level operators [37] or load levelling curves [48]. QP improves the accuracy of some problems solved by the linear approximation, but is not alternative to solve more complex optimisation problem with strong non-linear or integral parameters.

(3) Mixed-integer programming

Numerous optimisation problems for PEV charge scheduling involve integer, discrete variables or piecewise constraints such as the charging/discharging number of PEVs [49] and the on/off status of a generation unit [50] in UC, as well as the charging/discharging status of PEV battery charging [51] for PEVs aggregator. The MIP problems could be solved by branch and bound (BB) or branch and cut (BC) method in multiple mathematical coding environments or BB and BC embedded commercial solvers. It should be noted that the BB and BC methods work efficiently in low dimensional problems but suffer high computational cost in dealing with high dimensional problem. Meta-heuristic methods are widely used in solving high dimensional MIP problems, especially, considering PEVs integration.

(4) Dynamic programming

DP splits the whole optimisation process into a series of timeslots and seeks the solutions in each time step, thus being practical and useful to model time-varying scenarios. It has also been used to optimise PEV charge scheduling in order to solve time-varying dynamic optimisation with multiple systems such as power system and devices such as PEV chargers [52]. The DP methods can also be utilised for solving the UC and other linear, non-linear or MIP problems, associated with Lagrangian relaxation for dealing with constraints. The weakness of DP also lies on the so called 'curse of dimension' that the computational burden boosts in high dimensional cases.

(5) Meta-heuristic algorithm approaches

MAs are pure and popular optimisation tools for solving complex and non-smooth power system scheduling problems especially for problems of non-smooth, high dimensional and other intractable features. The MAs maintain stochastic generated solutions and move them towards optimum through heuristic trials. It is not guaranteed for MAs to achieve the global optimum, but they are available to provide alternative good solutions to the recently introduced problems of PEVs with uncertainty and controllable charging loads.

Meta-heuristic methods can be categorised as the binary-based approaches and real-value based approaches depending on the variable formulation, or as trajectory-based methods and population-based methods based on the evolutionary logic [27]. Many MA methods have been utilised in solving PEVs scheduling problems such as genetic algorithm [53], particle swarm optimisation [54], estimation of distribution algorithm [55], simulated annealing [56], teaching–learning-based optimisation [16], etc. The heuristic-based process requires considerable computational cost, which proportionally increases with the dimensions. This prevents the MA methods from the real-time applications.

(6) Intelligent-control methods

In order to follow intermittent renewable energy and provide fast response ancillary service such as load-frequency regulation and voltage regulation, intelligent control methods have also been utilised in coordinating PEVs charging for aggregators. In a particular case, a proportional-integral controller is designed in PEV aggregator to control the PEVs regulation up or down following the real-time wind generation [57]. The online power deviations feed back to the controller and generate control signal to the actuator (i.e. PEVs chargers). Given different control-time intervals, some other optimisation-based control methods such as model prediction control are also used to follow the online adaptive changing results, which are achieved by the real-time optimisation. Generally speaking, the control-based methods are suitable for distributed operator or PEVs aggregators for directly controlling the behaviours of PEVs chargers in accordance with fast and intermittent renewable generations.

(7) Game theory

Electricity market is a large scale complex system, where many participants, such as generators, system operators and PEVs aggregators, are keen to gain the best revenues, respectively. However, the optimal decisions provided by optimisation are zero-sum games which call for methods to provide marginal strategies for each participant. GT can analyse this conflict and help to make interactive decisions. During the system operating process, a third party operator is adopted to coordinate the optimisations of two participants such as system operator and PEVs aggregator, stop the process of both once a Nash equilibrium between the two roles is achieved [58]. The application of GT guarantees the fairness of the electricity market and helps all the participants to make the scientific decisions.

(8) Queuing theory

QT is an analytic approach for studying queues. It is a very useful method to model the charging sequence in scheduling the PEVs charging for aggregators and distributed operators. The queue lengths and waiting time can be predicted according to the customer's behaviours, providing the optimal charging profiles for system schedulers rather than simple-random charging based on a single-probability distribution [59].

(9) Platforms and simulation tools

Numerous optimisation and control methods have been used to model and solve the PEV scheduling problems of various objectives from multiple perspectives. Due to the strong compatibility and employment of well know methods, some commercial software platforms and solvers such as CPLEX, GAMS, gPROMS, IPOPT, CONOPT and MATLAB® optimisation toolbox are available to solve the majority of problems. In terms of some specific complex problem such as mixed integer non-LP, problems with absolute values, trigonometric function and exponential terms, specific algorithms or MAs have been used for individual cases. One advantage of MAs is that they do not rely on a specific platform. There is no difficulty to model the problems and implement MAs to solve them in MATLAB, Microsoft Visual Studio with C, C++, C# and Java etc. Intelligent-control algorithms prefer MATLAB Simulink®, DIgSCILENT and some other commercial software with mature model blocks and environment for strategy simulation. GT and QT are also independent of simulation tools.

11.4   Unit commitment with plug-in vehicle integration: a case study

As an important step in power-system operation, the UC is a large scale non-convex mix-integer optimisation problem. The on/off status of each unit and the delivered power of online units need to be simultaneously determined to minimise the generation cost. Numerous methods have been proposed to tackle the UC problem including conventional methods such as priority list and DP, meta-heuristic methods such as simulated annealing, genetic algorithm and binary particle swarm optimisation (BPSO), as well as some hybrid method including Lagrangian relaxation combined with evolutionary programming and evolutionary algorithm etc. The UC problem becomes even more challenging when a large number of PEVs are integrated into the system [60].

In this chapter, a novel model, namely PEV-UC problem, is formulated which integrates the conventional UC problem with flexible scheduling of PEV charging/discharging as shown in Figure 11.5. The PEVs are utilised as 'energy sponge' and get charged and discharged in a flexible way, whereas the daily energy necessity is maintained by the renewable generation. PEV aggregator is designed to have the capability of providing power reserve to the grid with the support of PEVs battery storage.

To solve the mixed-integer problem, a modified binary particle swarm optimisation (PSO) method is employed to determine binary on/off status of unit. A self-adaptive differential evolution (SaDE) algorithm is utilised to optimise the real-value

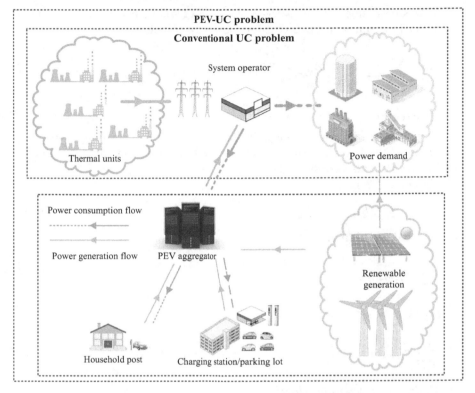

Figure 11.5   Power flow of PEV-UC problem

charging/discharging power of PEVs, followed by a Lagrangian relaxation method to cooperatively solve the PEV-UC problem. Then, a bidirectional energy flow scenarios combined both grid to vehicle (G2V) mode and V2G mode is implemented to evaluate the economic impact of 50,000 PEVs on the 10-unit power system.

11.4.1   Problem formulation

(1) Objective function

The objective function is the economic cost of power generation, composed of two parts as fossil fuel cost and start-up cost, respectively.

Fuel cost

$$F_{j,t}(P_{j,t}) = a_j + b_j P_{j,t} + c_j P_{j,t}^2 \tag{11.15}$$

Start-up cost

$$SU_{j,t} = \begin{cases} SU_{H,j}, & \text{if } MDT_j \leq TOFF_{j,t} \leq MDT_j + T_{\text{cold},j} \\ SU_{C,j}, & \text{if } TOFF_{j,t} > MDT_j + T_{\text{cold},j} \end{cases} \tag{11.16}$$

In this case study, the battery-depletion cost for both G2G/V2G service is ignored and the final objective cost function is given below:

$$\min \sum_{t=1}^{T} \sum_{j=1}^{n} \left[F_j(P_{j,t}) u_{j,t} + \mathrm{SU}_{j,t}(1 - u_{j,t-1}) u_{j,t} \right] \quad (11.17)$$

(2) Constraints

The new PEV-UC problem integrates the PEVs into the power system. Some constraints of the inherent power system, as well as limitations of G2V/V2G service of PEVs are also considered.

Generation limit

$$u_{j,t} P_{j,\min} \leq P_{j,t} \leq u_{j,t} P_{j,\max} \quad (11.18)$$

Power-demand limit

$$\sum_{j=1}^{n} P_{j,t} u_{j,t} + P_{\mathrm{EVD},t} = P_{D,t} + P_{\mathrm{EVC},t} \quad (11.19)$$

Power-reserve limit

$$\sum_{j=1}^{n} P_{j,\max} u_{j,t} + P_{\mathrm{EVD},t} \geq P_{D,t} + R_{\mathrm{The},t} + R_{N\mathrm{The},t} + P_{\mathrm{EVC},t} \quad (11.20)$$

Minimum up/down-time limit

$$u_{j,t} = \begin{cases} 1, & \text{if} \quad 1 \leq \mathrm{TON}_{j,t-1} < \mathrm{MUT}_j \\ 0, & \text{if} \quad 1 \leq \mathrm{TOFF}_{j,t-1} < \mathrm{MDT}_j \\ 0 \quad \text{or} \quad 1, & \text{otherwise} \end{cases} \quad (11.21)$$

Charging/discharging power limit

$$\sum_{t=1}^{T} P_{\mathrm{EVC},t} - \sum_{t=1}^{T} P_{\mathrm{EVD},t} = E_{\mathrm{EV,total}} \quad (11.22)$$

$$P_{\mathrm{EVC},t,\min} \leq P_{\mathrm{EVC},t} \leq P_{\mathrm{EVC},t,\max} \quad (11.23)$$

$$P_{\mathrm{EVC},t,\min} \leq P_{\mathrm{EVC},t} \leq P_{\mathrm{EVC},t,\max} \quad (11.24)$$

where all the above novel formulation of PEV-UC problem retains key formulations of the conventional UC problems, whereas the power demand and power-reserve limits are updated and PEV charging/discharging limits are also incorporated.

11.4.2 Methodology

The complicated PEV-UC problem calls for powerful computational techniques. Basic binary PSO has been employed in some early researches [31] in association with integer PSO. However, basic BPSO endures low-convergence speed and is easy to be trapped within local optimum. In this chapter, a modified BPSO is employed in which the sigmoid-probability function is redesigned to a symmetric shape to speed up the convergence speed.

(1) Modified binary particle swarm optimisation

The BPSO is an important variant of PSO and has been widely used for solving the UC problem [61,62]. The original BPSO uses updated velocities to achieve binary status from the sigmoid-probability function [55], which determines the binary variable comparing with a uniformly distributed random number. The velocities are updated as follows:

$$v_i(t+1) = w(t) \cdot v_i(t) + c_1(t) \cdot \text{rand}_1 \cdot \left(p_{\text{lbest},i} - x_{\text{bi}}(t)\right)$$
$$+ c_2(t) \cdot \text{rand}_2 \cdot \left(p_{\text{gbest}} - x_{\text{bi}}(t)\right) \tag{11.25}$$

where $v_i(t+1)$, $v_i(t)$ and $x_{bi}(t)$ are the updated velocity, current velocity and the binary variable of the ith binary particle at tth iteration. The $w(t)$, $c_1(t)$ and $c_2(t)$ represent the weighting, social and cognitive coefficients, respectively. $p_{\text{lbest},i}$ and p_{gbest} are the binary local and global best solutions. Two random numbers ranging from 0 to 1 are denoted as rand_1 and rand_2, respectively. The original sigmoid-probability function converges slowly and is easy to become pre-mature. This is partly due to the fact that when the value of updated velocity is small, the probability of the binary variable in the corresponding position should not be changed. While in the original function, the probability is 0.5 when v_i is 0, leading to an unsteady status for the optimal solution. To remedy this drawback, the probability is redesigned as (11.26), where an absolute value operator is utilised to convert the probability distribution P to be symmetric [63] as follows:

$$P(v_i(t+1)) = 2 \times \left| \frac{1}{1 + e^{-v_i(t+1)}} - 0.5 \right| \tag{11.26}$$

According to this probability, the new iteration of binary variable x_i is generated as:

$$x_{bi}(t+1) = \begin{cases} 1, & \text{if } \text{rand}_3 < P(v_i(t+1)) \\ 0 & \text{otherwise} \end{cases} \tag{11.27}$$

In terms of the parameter selection for (11.25), the original configuration is remained for implementation. This algorithm variant is named as modified BPSO (MBPSO).

(2) Self-adaptive differential evolution

Differential evolution (DE) is another popular heuristic optimisation method and has also been widely used in various applications and engineering fields [64]. Two-key phases are employed in DE namely mutation and crossover. The original DE has the advantage in exploitation, but is weak in terms of convergence speed.

To overcome this drawback, the SaDE [65] is adopted to optimise the PEV charging/discharging power where two different DE variants, namely rand/1/bin and current to best/2/bin are introduced. The selection of the variants is determined by the probability p_s.

The mutation process of SaDE is denoted as follows:

$$V_{i,G} = \begin{cases} X_{r1,G} + F \cdot (X_{r2,G} - X_{r3,G}), & \text{if } p_s < p_1 \\ X_{i,G} + F \cdot (X_{\text{best},G} - X_{i,G}) + F \cdot (X_{r1,G} - X_{r2,G}), & \text{otherwise} \end{cases}$$

(11.28)

where $V_{i,G}$ represents the mutation vector of the ith particle in the population at Gth iteration and F is the mutation factor. $X_{r1,G}$, $X_{r2,G}$, $X_{r3,G}$, $X_{\text{best},G}$ and $X_{i,G}$ denote three random particles $r1$–$r3$, best particle so far and the ith particle in the whole population at Gth iteration. When the determined probability p_s is less than p_1, the rand/1/bin is selected, and the current to best/1/bin variant is selected vice versa. The p_1 is calculated by the equation as follows:

$$p_1 = \frac{\text{ns}_1 \cdot (\text{ns}_2 + \text{nf}_2)}{\text{ns}_2 \cdot (\text{ns}_1 + \text{nf}_1) + \text{ns}_1 \cdot (\text{ns}_2 + \text{nf}_2)}$$

(11.29)

where ns_1 and ns_2 denote the number of trail vectors that have been adopted by the next generation of both variants, whereas the nf_1 and nf_2 represent the failure times. The trail vectors are generated by a crossover operation and as shown below,

$$u_{j,i,G} = \begin{cases} v_{j,i,G}, & \text{if rand}_4 < \text{CR} \quad \text{or} \quad j = j_{\text{rand}} \\ x_{j,i,G}, & \text{otherwise} \end{cases}, \quad j = 1, 2, \ldots, n \quad (11.30)$$

with the $u_{j,i,G}$, $v_{j,i,G}$ and $x_{j,i,G}$ representing the trail vector, mutation vector and original vector of jth position in the ith particle of Gth iteration, respectively. The mutation vector is adopted as part of trail vector when a random number rand_4 is less than crossover rate CR. The selection operation aims to determine whether to adopt the trail vector for the algorithm and it is denoted as follows,

$$X_{i,G+1} = \begin{cases} U_{i,G}, & \text{if } f(U_{i,G}) < f(X_{i,G}) \\ X_{i,G} & \text{otherwise} \end{cases}$$

(11.31)

The corresponding number in equation (11.29) will be modified according to the adoption of this operation. It should be noted that the variables in the conventional DE method are continuous real-valued.

(3) Proposed algorithm for PEV-UC

The MBPSO for binary optimisation and the SaDE for real-valued optimisation are combined to solve the PEV-UC problem. The whole optimisation process of the proposed algorithm is illustrated in Figure 11.6. There are five steps in the proposed algorithm for solving the PEV-UC problem.

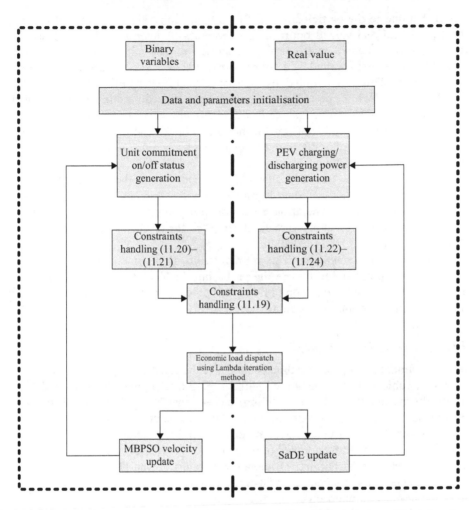

Figure 11.6 Optimisation process of the proposed meta-heuristic method for the PEV-UC problem

Step 1: initialisation. At this step, all the power system data such as the maximum and minimum generation power, cost coefficients, minimum up/down time, start-up cost as well as PEV data including maximum charging power, total energy necessity should be imported. Algorithm parameters are initialised, respectively.

Step 2: generate new solutions. A new iteration of solutions is generated at this step for both binary and real-value optimisation. The binary variables denoting the on/off statuses of units are generated based on (11.26)–(11.27) whereas the PEV variables are updated based on the initialisation values or updated values from the previous iteration according to (11.28)–(11.31).

Step 3: constraints handling. Power unit constraints and PEV constraints are handled in this step to adjust the initial solutions obtained in step 2. It should be noted that due to the participation of PEVs, some constraints are coupled in both the binary and real-valued optimisation processes. A hybrid handling process is, therefore, implemented in the proposed algorithm. The minimum up/down time limit (11.21) is first handled by a heuristic-based method proposed in Reference 66 where the violation is avoided by heuristic check. Then, PEV limits (11.22)–(11.24) are handled in the real-value optimisation procedure. Finally, the coupled constraint lies on the power reserve limit (11.20) where the commit and de-commit decisions are made depending on the heuristic check. These decisions take both binary on/off status of units and the charging/discharging power of PEVs into account at the same time.

Step 4: economic load dispatch (ELD). According to the decision variables adjusted by Step 3, a conventional Lambda iteration method [67] is applied to solve the ELD sub-problem. Generation power of on-line units are determined with the limits (11.18) and (11.19) being relaxed. Based on the results of ELD, the values of fitness function (11.17) are calculated.

Step 5: algorithm updates. The optimal results of the fitness function achieved from Step 4 are utilised to update the probability in MBPSO according to (11.25) and solutions by SaDE based on (11.31). The process goes back to Step 2 until the maximum iteration number is achieved.

11.4.3   Numerical study

In this section, the proposed method is applied to the standard 10-unit UC problem as shown in Table 11.6 [68], integrated with charging/discharging intelligent-switching mode. In the practical PEV scheduling, bidirectional charging and discharging is promising to reduce the economic cost while maintaining the charging and discharging necessity of users at the same time. A total of 50,000 PEVs with 15 kW h battery in each vehicle are adopted to integrate into the 10-unit system and 50% SOC and 85% charging/discharging efficiency are assumed. The limit of charging and discharging power is assumed as 20% of the total battery number [43]. Therefore, the maximum and minimum charging/discharging power is calculated as 15 kW h $\times$ 50,000 $\times$ 50% $\times$ 85% $\times$ 20% = 63.75 MW h. The total power necessity is 0 MW h when PEVs are charged by renewable generations and only act as the energy storage to shift the power loads. The priority list of charging and discharging is to deliver the charging power in an ascending sequence of power demand (off-peak first) and allocating the discharging power in a descending order of power demand (peak first). One original scenario C1 of 10% power reserve without PEVs and three scenarios S1–S3 of different power reserve levels are studied with 10 independent runs, respectively. The statistics results are shown in Table 11.6. The power reserves of S1–S3 are assumed to be provided by thermal plants at different rates of zero, 5 and 10% and the rest of the reserves are supported by V2G service from PEVs and renewable generation. The proposed algorithm was implemented in the MATLAB 2014a on an Intel i5-3470 CPU at 3.20 GHz and 4 GB RAM personal computer.

It could be observed from Table 11.7 that the proposed MBPSO-SaDE method achieved a best cost as 556,796 $/day. On the other hand, the S1 scenario spends

Table 11.6 Ten-unit system data

Parameters	U1	U2	U3	U4	U5	U6	U7	U8	U9	U10
P_{max} (MW)	455	455	130	130	162	80	85	55	55	55
P_{max} (MW)	150	150	20	20	25	20	25	10	10	10
a ($/h)	1,000	970	700	680	450	370	480	660	665	670
b ($/MW h)	16.19	17.26	16.6	16.5	19.7	22.26	27.74	25.92	27.27	27.79
c ($/MW h^2)	0.00048	0.00031	0.002	0.00211	0.00398	0.00712	0.0079	0.00413	0.00222	0.00173
MUT (h)	8	8	5	5	6	3	3	1	1	1
MDT (h)	8	8	5	5	6	3	3	1	1	1
SU_H ($)	4,500	5,000	550	560	900	260	260	30	30	30
SU_C ($)	9,000	10,000	1,100	1,120	1,800	520	520	60	60	60
T_{cold} (h)	5	5	4	4	4	2	2	0	0	0
Initial states (h)	8	8	−5	−5	−6	−3	−3	−1	−1	−1

Table 11.7   Numerical results for UC considering G2V/V2G service

Scenarios of spinning reserve	MBPSO-SaDE cost ($/day)		
	Best	**Worst**	**Mean**
C1: 10% reserve, No PEVs	563,937	563,977	563,964
S1: 0% reserve from thermal generation	548,456	549,488	548,925
S2: 5% reserve from thermal generation	550,316	551,114	550,785
S3: 10% reserve from thermal generation	556,796	557,211	556,964

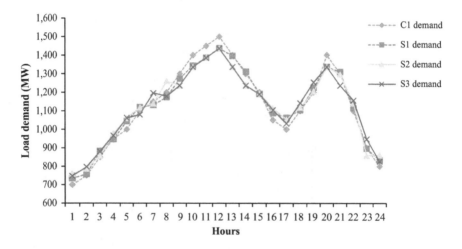

Figure. 11.7   Load demand and its combinations with three PEVs dispatching results

more in the order of 8,340 and 1,860 $/day than S3 and S2, respectively. Compared with the original no PEVs scenarios denoted as C1, the scenario S3 reduces 7,141 $/day with the aid of PEVs. The accumulation curves of the best results of PEVs dispatched power and the predicted power demand of 10 units are illustrated in Figure 11.7.

All the three scenarios have utilised the flexible PEVs charging and discharging ability to relieve the peak load and fill the load valley. Among the three sets of dispatching results, S3 shows the strongest ability to reduce peak load during 9:00–15:00 and 20:00–22:00 and move these load to off-peak time. The scheduling results of two scenarios S2 and S3 are listed in Tables 11.8 and 11.9, respectively for comparison, where G2V power is referring to positive value and V2G power is denoted as negative value in the both tables.

As it can be seen from Table 11.8, V2G services are fully committed during peak-load period from 9:00 to 14:00 and 20:00 with 63.75 MW power in each hour feeding back to the grid. The expensive units 6, 7 and 9 are less committed and these loads are shifted to the cheaper generation units like units 2, 3, and 5 in the

Table 11.8 Best unit scheduling result of **S2**: 5% reserve from thermal generation

Hour	U1 (MW)	U2 (MW)	U3 (MW)	U4 (MW)	U5 (MW)	U6 (MW)	U7 (MW)	U8 (MW)	U9 (MW)	U10 (MW)	G2V/V2G (MW)	Demand (MW)	Thermal generation reserve (MW)
1	455	305.33	0	0	0	0	0	0	0	0	60.33	760.33	149.67
2	455	330.94	0	0	0	0	0	0	0	0	35.94	785.94	124.06
3	455	401.88	0	0	0	0	0	0	0	0	6.88	856.88	53.12
4	455	455	0	0	66.89	0	0	0	0	0	26.89	976.89	95.11
5	455	450.57	130	0	25	0	0	0	0	0	60.57	1,060.57	141.43
6	455	455	130	0	78.38	0	0	0	0	0	18.38	1,118.38	83.62
7	455	455	130	0	99.44	0	0	0	0	0	-10.56	1,139.44	62.56
8	455	455	130	130	89.61	0	0	0	0	0	59.61	1,259.61	72.39
9	455	455	130	130	66.25	0	0	0	0	0	-63.75	1,236.25	95.75
10	455	455	130	130	141.25	0	25	0	0	0	-63.75	1,336.25	80.75
11	455	455	130	130	162	29.25	25	0	0	0	-63.75	1,386.25	110.75
12	455	455	130	130	162	69.25	25	0	10	0	-63.75	1,436.25	115.75
13	455	455	130	130	146.25	20	0	0	0	0	-63.75	1,336.25	75.75
14	455	455	130	130	66.25	0	0	0	0	0	-63.75	1,236.25	95.75
15	455	452.84	130	130	25	0	0	0	0	0	-7.16	1,192.84	139.16
16	455	373.34	130	130	25	0	0	0	0	0	63.34	1,113.34	218.66
17	455	312.67	130	130	25	0	0	0	0	0	52.67	1,052.67	279.33
18	455	378.23	130	130	25	0	0	0	0	0	18.23	1,118.23	213.77
19	455	444.38	130	130	25	20	0	0	0	0	4.38	1,204.38	207.62
20	455	455	130	130	146.25	20	0	0	0	0	-63.75	1,336.25	75.75
21	455	455	130	130	106.9	20	0	0	0	0	-3.1	1,296.9	115.1
22	455	455	130	0	100.49	0	0	0	0	0	40.49	1,140.49	61.51
23	455	403.97	0	0	0	0	0	0	0	0	-41.03	858.97	51.03
24	455	405.36	0	0	0	0	0	0	0	0	60.36	860.36	49.64

Total economic cost (550,316 $/day)

Table 11.9 Best unit scheduling result of **S3**: 10% reserve from thermal generation

Hour	U1 (MW)	U2 (MW)	U3 (MW)	U4 (MW)	U5 (MW)	U6 (MW)	U7 (MW)	U8 (MW)	U9 (MW)	U10 (MW)	G2V/V2G (MW)	Demand (MW)	Thermal generation reserve (MW)
1	455	293.39	0	0	0	0	0	0	0	0	48.39	748.39	161.61
2	455	341.86	0	0	0	0	0	0	0	0	46.86	796.86	113.14
3	455	400.25	0	0	25	0	0	0	0	0	30.25	880.25	191.75
4	455	455	0	0	56.34	0	0	0	0	0	16.34	966.34	105.66
5	455	452.21	130	0	25	0	0	0	0	0	62.21	1,062.21	139.79
6	455	455	130	0	39.9	0	0	0	0	0	−20.1	1,079.9	122.1
7	455	455	130	130	26.18	0	0	0	0	0	46.18	1,196.18	135.82
8	455	438.64	130	130	25	0	0	0	0	0	−21.36	1,178.64	153.36
9	455	455	130	130	46.25	20	0	0	0	0	−63.75	1,236.25	175.75
10	455	455	130	130	121.25	20	25	0	0	0	−63.75	1,336.25	160.75
11	455	455	130	130	161.25	20	25	10	0	0	−63.75	1,386.25	165.75
12	455	455	130	130	162	59.25	25	10	10	0	−63.75	1,436.25	170.75
13	455	455	130	130	121.25	20	25	0	0	0	−63.75	1,336.25	160.75
14	455	455	130	130	46.25	20	0	0	0	0	−63.75	1,236.25	175.75
15	455	450.08	130	130	25	0	0	0	0	0	−9.92	1,190.08	141.92
16	455	364.08	130	130	25	0	0	0	0	0	54.08	1,104.08	227.92
17	455	291.26	130	130	25	0	0	0	0	0	31.26	1,031.26	300.74
18	455	402.47	130	130	25	0	0	0	0	0	42.47	1,142.47	189.53
19	455	455	130	130	58.8	0	25	0	0	0	53.8	1,253.8	163.2
20	455	455	130	130	121.25	20	25	0	0	0	−63.75	1,336.25	160.75
21	455	455	130	0	151.25	20	25	0	0	0	−63.75	1,236.25	130.75
22	455	455	130	0	95.73	20	0	0	0	0	55.73	1,155.73	126.27
23	455	455	0	0	37.34	0	0	0	0	0	47.34	947.34	124.66
24	455	371.5	0	0	0	0	0	0	0	0	26.5	826.5	83.5

Total economic cost (556,796 $/day)

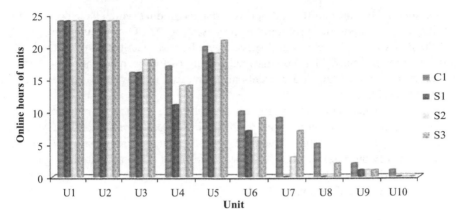

Figure 11.8   Online hours of units for the scenarios

hours of 1:00, 5:00, 8:00, 16:00 and 24:00. In some lower power reserve periods such as 3:00 and 23:00, G2V power is less committed or even V2G is launched to reduce the start-up cost of unit 5 which is used to provide spinning reserve. Similar scheduling result for S3 is illustrated in Table 11.9 where the spinning reserve provided by thermal units is 10%.

It could be seen from Table 11.9, peak-load periods including 9:00–14:00 and 20:00–21:00 are committed as full power discharging to reduce the commitment of expensive thermal units, whereas heavy charging tasks are scheduled in some large power reserve scenarios such as 1:00–3:00, 5:00, 7:00 16:00–19:00 and 22:00–23:00. In some special time period such as 6:00, V2G is utilised to reduce the power demand to avoid an extra start-up cost of unit 4. In addition to the specific dispatching results, Figure 11.8 illustrates the online hours of each units in the system.

Figure 11.8 shows that the base-load units U1 and U2 are online all 24 h. The occasional units U3–U7 are variously committed, and the S3 has successfully reduced the online hours of more expensive units U6–U7 compared with C1 under the same spinning reserve levels, shifting these hours to U3 and U5. In terms of the commitment agenda of peak units U8–U10, S3 also sees noticeably online hour reductions. Comparing scenarios S1–S3, more units are offline due to the less burden of spinning reserves. In a result, the PEVs play important roles in saving economic cost and relief the pressure of the generations by providing extra spinning reserves.

11.5   Conclusion

The fast development of PEVs significantly challenges the security, economy and stability of the power-grid operation. Advanced control and optimisation algorithms enable the PEVs to provide opportunities for improving flexibility and smartness of grid. This chapter first introduces the categories and features of EVs, and the impact of the plug-in vehicles on the grid is briefly analysed. Then, a survey on the state-of-the-art control and optimisation techniques which have been used in solving PEVs scheduling problems is proposed. A 10-UC case study integrating 50,000 PEVs is implemented and solved by a hybrid meta-heuristic method. The numerical results show that the smart scheduling of PEVs could noticeably reduce the economic cost of UC and presents an alternative to provide power reserve for the power system.

The electrification of transportation is an ultimate solution for reducing carbon emissions and fossil fuel consumption from the transportation sector. The wired or wireless charging based plug-in vehicles would be one of the most important and portable energy storages or hubs in future smart grids. Such an ambitious goal still calls for the development and implementation of intelligent scheduling strategies, which would be the future work.

Symbols and abbreviations

a_j, b_j and c_j	fuel cost coefficients of unit j
$E_{EV,total}$	Total energy necessity from power grid
$F_{j,t}$	fuel cost of unit j at time t
MDT_j	minimum down time of unit j
MUT_j	minimum up time of unit j
n	number of units
$P_{j,min}$	minimum power limits of unit j
$P_{j,max}$	maximum power limits of unit j
$P_{j,t}$	determined power of unit j at time t
$P_{D,t}$	predicted power demand
$P_{EVC,t}$	charge power for plug-in electric vehicle at time t
$P_{EVD,t}$	discharge power from plug-in electric vehicle at time t
$P_{EVC,t,max}$	maximum charge power for plug-in electric vehicle at time t
$P_{EVC,t,min}$	minimum charge power for plug-in electric vehicle at time t
$P_{EVD,t,max}$	maximum discharge power from plug-in electric vehicle at time t
$P_{EVD,t,min}$	minimum discharge power from plug-in electric vehicle at time t
$R_{NThe,t}$	reserved power amount provided by non-thermal plant at time t
$R_{The,t}$	reserved power amount provided by thermal plant at time t
$SU_{C,j}$	cold-start cost of unit j at time t
$SU_{H,j}$	hot-start cost of unit j at time t
$SU_{j,t}$	start-up cost of unit j at time t
T	scheduling hours
$T_{cold,j}$	cold-start hour of unit j
$TOFF_{j,t}$	off-line duration time of unit j
$TON_{j,t}$	off-line duration time of unit j

Acknowledgements

This work was financially supported by UK EPSRC under grant EP/L001063/1 and China NSFC under grants 51361130153, 61673256, 61533010, 51607177 and 61273040. Zhile Yang would like to thank EPSRC Studentship for financially supporting his research.

References

[1]   United Nations Conference on Climate Change, 2016 [Online]. <http://www.cop21.gouv.fr/en/learn/> [Accessed Jan 2016].

[2]   International Energy Agency, Key Trends in CO2 Emissions, 2015 [Online]. <http://www.iea.org/publications/freepublications/publication/CO2Emissions Trends.pdf>; 2015 [Accessed Jan 2016].

[3]   A. Ipakchi, F. Albuyeh. "Grid of the future". *IEEE Power and Energy Magazine*, 2009, 7(2): 52–62.

[4]   International Energy Agency, Key World Energy Statistics 2015 [Online]. <http://www.iea.org/publications/freepublications/publication/KeyWorld_Statistics_2015.pdf> [Accessed Mar 2016].

[5]   World Nuclear Association, Comparison of Lifecycle Greenhouse Gas Emissions of Various Electricity Generation Sources, 2011 [Online] <http://www.world-nuclear.org/uploadedFiles/org/WNA/Publications/Working_Group_Reports/comparison_of_lifecycle.pdf> [Accessed Mar 2016].

[6]   International Energy Agency, Global EV Outlook 2013, International Energy Agency, 2014 [Online]. <http://www.iea.org/publications/globale-voutlook_2013.pdf> [Accessed Dec 2014].

[7]   C. Chan. "The state of the art of electric, hybrid, and fuel cell vehicles". *Proceedings of the IEEE*, 2007, 95(4): 704–718.

[8]   M. Zeraoulia, M. Benbouzid, D. Diallo. "Electric motor drive selection issues for HEV propulsion systems: a comparative study". *IEEE Transactions on Vehicular Technology*, 2006, 55(6): 1756–1764.

[9]   G. Buja, M. Kazmierkowski. "Direct torque control of PWM inverter-fed AC motors-a survey". *IEEE Transactions on Industrial Electronics*, 2004, 51(4): 744–757.

[10] W. Liu. *Introduction to Hybrid Vehicle System Modeling and Control*. Hoboken, New Jersey: John Wiley & Sons, 2013.

[11] C. Zhang, K. Li, S. Mcloone, Z. Yang. "Battery modelling methods for electric vehicles – a review". *Control Conference (ECC)*, 2014 European. IEEE, 2014: 2673–2678.

[12] Z. Xu, S. Gao, S. Yang. "LiFePO4 battery state of charge estimation based on the improved Thevenin equivalent circuit model and Kalman filtering". *Journal of Renewable and Sustainable Energy*, 2016, 8(2): 024103.

[13] K. Ng, C. Moo, Y. Chen, Y. Hsieh. "Enhanced Coulomb counting method for estimating state-of-charge and state-of-health of lithium-ion batteries". *Applied Energy*, 2009, 86(9): 1506–1511.

[14] C. Zhang, K. Li, J. Deng. "Real-time estimation of battery internal temperature based on a simplified thermoelectric model". *Journal of Power Sources*, 2016, 302: 146–154.

[15] C. Zhang, K. Li, L. Pei, C. Zhu. "An integrated approach for real-time model-based state-of-charge estimation of lithium-ion batteries". *Journal of Power Sources*, 2015, 283: 24–36.

[16] Tesla Motors, Battery, Performance and Drive Options, 2014 [Online]. <http://www.teslamotors.com/models> [Accessed Jan 2016].

[17] US Department of Transportation, National Household Travel Survey 2009, 2011 [Online] <http://nhts.ornl.gov/> [Accessed Jun 2015].

[18] K. Clement-Nyns, E. Haesen, J. Driesen. "The impact of charging plug-in hybrid electric vehicles on a residential distribution grid". *IEEE Transactions on Power Systems*, 2010, 25(1): 371–380.

[19] L. Zhang, F. Jabbari, T. Brown, S. Samuelsen. "Coordinating plug-in electric vehicle charging with electric grid: valley filling and target load following". *Journal of Power Sources*, 2014, 267: 584–597.

[20] A. Foley, B. Tyther, P. Calnan, B. Gallachóir. "Impacts of electric vehicle charging under electricity market operations". *Applied Energy*, 2013, 101: 93–102.

[21] Z. Yang, K. Li, Q. Niu, Y. Xue, A. Foley. "A self-learning TLBO based dynamic economic/environmental dispatch considering multiple plug-in electric vehicle loads". *Journal of Modern Power Systems and Clean Energy*, 2014, 2(4): 298–307.

[22] W. Kempton, J. Tomić. "Vehicle-to-grid power fundamentals: calculating capacity and net revenue". *Journal of Power Sources*, 2005, 144(1): 268–279.

[23] S. Deilami, A. Masoum, P. Moses, M. Masoum. "Real-time coordination of plug-in electric vehicle charging in smart grids to minimize power losses and improve voltage profile". *IEEE Transactions on Smart Grid*, 2011, 2(3): 456–467.

[24] C. White, K. Zhang. "Using vehicle-to-grid technology for frequency regulation and peak-load reduction". *Journal of Power Sources*, 2011, 196(8): 3972–3980.

[25] P. Sanchez-Martin, S. Lumbreras, A. Alberdi-Alen. "Stochastic programming applied to EV charging points for energy and reserve service markets". *IEEE Transactions on Power Systems*, 2016, 31(1): 198–205.

[26] J. Wang, C. Liu, D. Ton, Y. Zhou, J. Kim, A. Vyas. "Impact of plug-in hybrid electric vehicles on power systems with demand response and wind power". *Energy Policy*, 2011, 39(7): 4016–4021.

[27] P. Denholm, M. Kuss, R. Margolis. "Co-benefits of large scale plug-in hybrid electric vehicle and solar PV deployment". *Journal of Power Sources*, 2013, 236: 350–356.

[28] D. Dallinger, M. Wietschel. "Grid integration of intermittent renewable energy sources using price-responsive plug-in electric vehicles". *Renewable and Sustainable Energy Reviews*, 2012, 16(5): 3370–3382.

[29] A. Foley, I. Winning, B. Gallachóir. "State-of-the-art in electric vehicle charging infrastructure". *Vehicle Power and Propulsion Conference (VPPC)*, 2010 IEEE. 1–6.

[30] Tesla Motors, Charging Basics, 2015 [Online]. <https://www.teslamotors.com/models-charging#/basics> [Accessed Jan 2016].

[31] Z. Yang, K. Li, A. Foley, C. Zhang. "Optimal scheduling methods to integrate plug-in electric vehicles with the power system: a review". *Proceedings of the 19th World Congress of the International Federation of Automatic Control (IFAC'14)*, Cape Town, South Africa. 2014: 24–29.

[32] Z. Yang, K. Li, A. Foley. "Computational scheduling methods for integrating plug-in electric vehicles with power systems: a review". *Renewable and Sustainable Energy Reviews*, 2015, 51: 396–416.

[33] A. Weis, P. Jaramillo, J. Michalek. "Estimating the potential of controlled plug-in hybrid electric vehicle charging to reduce operational and capacity expansion costs for electric power systems with high wind penetration". *Applied Energy*, 2014, 115: 190–204.

[34] N. Padhy. "Unit commitment – a bibliographical survey". *IEEE Transactions on Power Systems*, 2004, 19(2): 1196–1205.

[35] A.Y. Saber, G.K. Venayagamoorthy. "Plug-in vehicles and renewable energy sources for cost and emission reductions". *IEEE Transactions on Industrial Electronics*, 2011, 58(4): 1229–1238.

[36] J. Zhao, F. Wen, Z.Y. Dong, Y. Xue, K.-P. Wong. "Optimal dispatch of electric vehicles and wind power using enhanced particle swarm optimization". *IEEE Transactions on Industrial Informatics*, 2012, 8(4): 889–899.

[37] Z. Yang, K. Li, Q. Niu, Q.C. Zhang, A. Foley. "Non-convex dynamic economic/environmental dispatch with plug-in electric vehicle loads". *Computational Intelligence Applications in Smart Grid (CIASG), IEEE Symposium on. IEEE*, 2014: 1–7.

[38] Z. Yang, K. Li, Q. Niu, Y. Xue, A. Foley. "A self-learning teaching-learning based optimization for dynamic economic/environmental dispatch considering multiple plug-in electric vehicle loads". *Journal of Modern Power System and Clean Energy*, 2014, 2(4): 298–307.

[39] S. Han, S. Han, S.K. Sezaki. "Development of an optimal vehicle-to-grid aggregator for frequency regulation". *IEEE Transactions on Smart Grid*, 2010, 1(1): 65–72.

[40] K. Clement-Nyns, E. Haesen, J. Driesen. "The impact of charging plug-in hybrid electric vehicles on a residential distribution grid". *IEEE Transactions on Power Systems*, 2010, 25(1): 371–380.

[41] W. Su, M. Chow. "Performance evaluation of an EDA-based large-scale plug-in hybrid electric vehicle charging algorithm". *IEEE Transactions on Smart Grid*, 2012, 3(1): 308–315.

[42] S. Shao, M. Pipattanasomporn, S. Rahman. "Challenges of PHEV penetration to the residential distribution network". *IEEE Power & Energy Society General Meeting*, 2009: 1–8.

[43] L. Jian, H. Xue, G. Xu, X. Zhu, D. Zhao, Z. Shao. "Regulated charging of plug-in hybrid electric vehicles for minimising load variance in household smart microgrid". *IEEE Transactions on Industrial Electronics*, 2013, 60(8): 3218–3226.

[44] R. Sioshansi, J. Miller. "Plug-in hybrid electric vehicles can be clean and economical in dirty power systems". *Energy Policy*, 2011, 39(10): 6151–6161.

[45] Z. Ma, D. Callaway, I. Hiskens. "Decentralized charging control for large populations of plug-in electric vehicles". In: *2010 49th IEEE Conference on Decision and Control (CDC)*, 2010: 206–212.

[46] K. Turitsyn, N. Sinitsyn, S. Backhaus, M. Chertkov. "Robust broadcast-communication control of electric vehicle charging". In: *2010 First IEEE International Conference on Smart Grid Communications (Smart-GridComm)*, 2010: 203–207.

[47] E. Sortomme, M. El-Sharkawi. "Optimal scheduling of vehicle-to-grid energy and ancillary services". *IEEE Transactions on Smart Grid*, 2012, 3 (1): 351–359.

[48] W. Su, M. Chow. "Computational intelligence-based energy management for a largescale PHEV/PEV enabled municipal parking deck". *Applied Energy*, 2012, 96: 171–182.

[49] E. Talebizadeh, M. Rashidinejad, A. Abdollahi. "Evaluation of plug-in electric vehicles impact on cost-based unit commitment". *Journal of Power Sources*, 2014, 248: 545–552.

[50] P. Zhang, K. Qian, C. Zhou, B. Stewart, D. Hepburn. "A methodology for optimization of power systems demand due to electric vehicle charging load". *IEEE Transactions on Power System*, 2012, 27(3): 1628–1636.

[51] M. Honarmand, A. Zakariazadeh, S. Jadid, M. Honarmand, A. Zakariazadeh, S. Jadid. "Self-scheduling of electric vehicles in an intelligent parking lot using stochastic optimisation". *Journal of the Franklin Institute*, 2015, 352 (2): 449–467.

[52] J. Xu, V. Wong. "An approximate dynamic programming approach for coordinated charging control at vehicle-to-grid aggregator". Smart Grid Communications (SmartGridComm), 2011 *IEEE International Conference on. IEEE*, 2011: 279–284.

[53] M. Shaaban, Y. Atwa, E. El-Saadany. "PEVs modeling and impacts mitigation in distribution networks". *IEEE Transactions on Power Systems*, 2013, 28(2): 1122–1131.

[54] G. Venayagamoorthy, P. Mitra, K. Corzine, C. Huston. "Real-time modeling of distributed plug-in vehicles for V2G transactions". *Energy Conversion Congress and Exposition, ECCE 2009*. IEEE, 2009: 3937–3941.

[55] W. Su, M. Chow. "Performance evaluation of an EDA-based large-scale plug-in hybrid electric vehicle charging algorithm". *IEEE Transactions on Smart Grid*, 2012, 3(1): 308–315.

[56] J. Soares, T. Sousa, H. Morais, Z. Vale, P. Faria. "An optimal scheduling problem in distribution networks considering V2G". *Computational Intelligence Applications in Smart Grid (CIASG), 2011 IEEE Symposium on. IEEE*, 2011: 1–8.

[57] J. Pillai, B. Bak-Jensen. "Integration of vehicle-to-grid in the western Danish power system". *IEEE Transactions on Sustainable Energy*, 2011, 2(1): 12–19.

[58] X. Xi, R. Sioshansi. "Using price-based signals to control plug-in electric vehicle fleet charging". *IEEE Transactions on Smart Grid*, 2014, 5(3): 1451–1464.

[59] G. Li, X. Zhang. "Modeling of plug-in hybrid electric vehicle charging demand in probabilistic power flow calculations". *IEEE Transactions on Smart Grid*, 2012, 3(1): 492–499.

[60] Z. Yang, K. Li, Q. Niu, A. Foley. "Unit commitment considering multiple charging and discharging scenarios of plug-in electric vehicles". Neural Networks (IJCNN), 2015 International Joint Conference on. IEEE, 2015: 1–8.

[61] Z. Gaing. "Discrete particle swarm optimization algorithm for unit commitment". *Power Engineering Society General Meeting* 2003. IEEE, vol. 1, 424.

[62] P. Chen. "Two-level hierarchical approach to unit commitment using expert system and elite PSO". *IEEE Transactions on Power Systems*, 2012, 27(2): 780–789.

[63] H. Nezamabadi-Pour, M. Rostami Shahrbabaki, M. Maghfoori-Farsangi. "Binary particle swarm optimization: challenges and new solutions". *CSI Journal on Computer Science and Engineering*, 2008, 6(1): 21–32.

[64] S. Das, P.N. Suganthan. "Differential evolution: a survey of the state-of-the-art". *IEEE Transactions on Evolutionary Computation*, 2011, 15(1): 4–31.

[65] A. Qin, P. Suganthan. "Self-adaptive differential evolution algorithm for numerical optimization". *Evolutionary Computation*, 2005, *IEEE Congress on. IEEE*, 2005, 2: 1785–1791.

[66] Y. Jeong, J. Park, J. Shin, K. Lee. "A thermal unit commitment approach using an improved quantum evolutionary algorithm". *Electric Power Components and Systems*, 2009, 37(7): 770–786.

[67] Q. Jiang, B. Zhou, M. Zhang. "Parallel augment Lagrangian relaxation method for transient stability constrained unit commitment". *IEEE Transactions on Power Systems*, 2013, 28(2): 1140–1148.

[68] T. Ting, M. Rao, C. Loo. "A novel approach for unit commitment problem via an effective hybrid particle swarm optimization". *IEEE Transactions on Power Systems*, 2006, 21(1), 411–418.

Chapter 12

Multi-agent based control of smart grid

K.T.M.U. Hemapala[1], S.L. Jayasinghe[1] and A.L. Kulasekera[2]

Controlling the power system has become challenging task with increasing power demands with limited resources. Therefore, energy utilities are looking advanced techniques for controlling their grid. In this context, multi-agent systems (MASs) are popular due to their inherent benefits. In the first section of this chapter, the functionality of MAS is presented with inherent advantages. Then, the applications of MAS are discussed in four aspects of power systems including power-system restoration, electricity trading, optimization, and smart-grid control. At the end two case studies, MAS in a microgrid and application of MAS in smart-grid transmission/generation have presented.

12.1 Introduction

Controlling the power system has become very complicated with increasing power demands with limited resources. This growth in demand for electrical power is becoming a major challenge for electrical supply authorities worldwide with the present energy crises. Therefore, energy utilities are looking toward new concepts in electricity generation, transmission, distribution, and utilization [1]. The present network which was built over the last decade is not capable of handling the problems faced by the industry. In this context, the power industry requires a smarter grid with advanced capabilities to meet present and future demands [1–3]. Converting the existing network into a smarter network cannot be achieved within few days, months or years. This is a process which is currently happening throughout the world utilizing advanced control and communications technologies.

In traditional power systems, centralized control played a vital role in controlling available resources. However, with highly intermittent renewable energy sources, the dynamic behavior of the system has increased considerably. A single fault can result in total blackout of the system, affecting the reliability of the system [4]. Power quality and reliability are the most important factors which are

[1]Department of Electrical Engineering, University of Moratuwa, Moratuwa, Sri Lanka
[2]Department of Mechanical Engineering, University of Moratuwa, Moratuwa, Sri Lanka

considered for measuring the efficiencies of distribution licensees. Therefore, control and automation are very important for the utilities for automating fault detection, isolation and restoration processes [5]. Various approaches have so far been proposed to obtain target configurations, such as expert systems, heuristics, soft computing, and mathematical programing (MP) [6]. Therefore, power systems have to become smarter with advanced control features. As an innovative method in providing this flexible distributed control requirement, multi-agent systems (MASs) are stepping forward providing more flexibility to the power network [7].

12.2   What is MAS and its functionality

An agent system is a hardware or software entity that possesses social coordination and communication capabilities and capable of achieving a larger overall goal by tackling smaller individual tasks. Agents have the ability to operate within its environment to achieve its goal by sharing knowledge with other agents or taking initiative. Even a single agent can operate as a system, by reacting and interacting with its environment. Such an agent will require specific programming to provide itself with individual knowledge, actions, and goals within a larger picture.

A collection of social, collaborative agents constitute a MAS. All the agents will inherit a set of common goals, and react to change in the environment, make decisions to achieve those goals, or help other agents in achieving their goals [8]. To understand, how agents can be used in a power system the following basic example is illustrated.

Figure 12.1 shows a decentralized meter reading system approach with the collaboration of the following agents who perform their defined work.

- Meter monitor agent.
- Meter access agent.

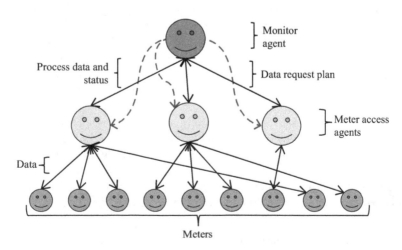

Figure 12.1   Decentralized meter reading system approach

- Data base agent.
- Meter communication driver.

Meter monitor agent is the main control agent in the system. It is located at the sites where meter reading monitoring is required. Meter-access agents are distributed everywhere in the utility power network. Data base agents are also distributed everywhere in the utility network.

When meter access agents are running, each broadcasts its availability. Then, meter-monitoring agents create meter-data downloading plans and send to the available meter-access agents. According to that plan, meter-access agents get meter details, as meter-communication IP addresses, meter-serial numbers etc. Then, meter-access agents connect to the relevant meters and download data through meter-communication drivers. Then, downloaded data will be sent back to meter-monitoring agents.

According to the response time of meter-access agents, meter-accessing plans will be optimized. This system facilitates the sharing of the meter-access agents' duties in the case of unresponsive agents. Therefore, the system becomes resilient to the failure of several agents.

12.3 Advantages of MAS

Most conventional power systems rely on Supervisory Control and Data Acquisition (SCADA) systems for control and communications. The control and communication architecture provided by such SCADA systems is built upon proprietary communication and signaling protocols. In terms of a smart grid, the aim is to allow for flexible integration of multitude of different distributed energy resources and/or nonconventional renewable energy sources. However, the vendor-specific-control protocols—used in commercially available SCADA systems for smart grid—restrict flexible interconnectivity and interoperability of diverse systems. Such restrictions can result in increased deployment costs and prevent smart-grid rollout. Use of MASs to provide distributed control capabilities to the smart-grid applications offers various advantages over conventional SCADA systems.

MAS development platforms are available as Free and Open Source Software applications and most of them are based on Java, making them platform independent. They can also be combined with external programming to interface with different external hardware and software systems. These advantages make MASs more suited for smart-grid control. In addition, MASs also have the following advantages in context of power system control:

- MAS can take rapid autonomous decisions during faults by creating microgrids and protecting critical loads by load shedding.
- MAS allows for a complex task to be broken down into several smaller tasks assigned to a team of agents. This allows for easier handling of a larger problem.
- As agents are taking decisions, the amount of the data coming to the central server will be reduced. This will help to overcome the server-overloading issues in larger systems providing greater scalability.

12.4   Multi-agent systems and applications

MASs are becoming more and more ubiquitous control systems in the world. Recently, MASs are being implemented in many fields such as mathematics, physics, engineering sciences, and social sciences [9]. They have been used in a wide array of applications ranging from computing and processing [10–13], robotics and control systems [14–18], transportation [19–22], aerospace, and nuclear engineering in recent literature. Most importantly, several implementations can be found on several areas of power systems [9] such as power system restoration, distributed control, electricity trading, optimization, microgrid and smart grid. Some of these are detailed in the following section.

12.4.1   Power system restoration

Faults in the distribution system reduce the availability of electricity. Distribution automation systems are introduced to automate the fault detection, isolation, and restoration process to improve the reliability. Various approaches have so far been proposed such as expert systems, heuristics, soft computing, and MP. Heuristics and expert systems have problems in achieving the optimal solution, whereas MP is able to obtain optimum solutions. In the case of soft computing, it is easy to implement, but is difficult to obtain the optimal solutions unless with high computational time.

Each of the above systems relies on SCADA systems. For large scale power systems, when all decisions are being taken by the control center, it will be very difficult to handle all these calculations and communications. And also the vendor-specific-control protocols—used in commercial SCADA systems for power systems—restrict the interconnectivity and interoperability and this results in increasing deployment costs and prevents power system rollout. Intelligent distribution automation will be required by means of decentralized power management as well as smarter information and communication technologies to overcome the issues discussed above and ultimately to actualize grid modernization. Therefore, MAS-based-power-system restoration methods have become more popular.

Several MAS approaches are available in literature [23–26]. The basic MAS structure shown in Figure 12.2 can be used for power-system restoration. The given architecture consists of a number of circuit breakers (CBAs), bus bar agents (BAs), and a single facilitator agent (FA). A main function of the CBA is the status update of the distribution network and giving the command to the motorized CBAs to operate according to the identified configurations by FA for restoration. BA is developed to update the FA after a fault occurs by interacting with other BAs. The operation is illustrated using a two-feeder X and Y distribution system, which is shown in Figure 12.3.

Figure 12.3(a) illustrates a simple fault occurred between Isolator B and Isolator C on feeder X. During the restoring process, B and C isolators (CB agents) communicate with the BA and update their status. Then according to the faulty section, FA takes the decision and sends commands to the B and C isolators to isolate the BC section. Figure 12.3(b) illustrates another fault occurring in the feeder Y while the fault between Isolator B and Isolator C remain. Figure 12.3(c) illustrates power

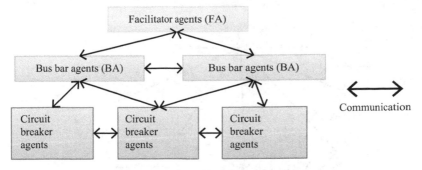

Figure 12.2   MAS architecture for meter reading

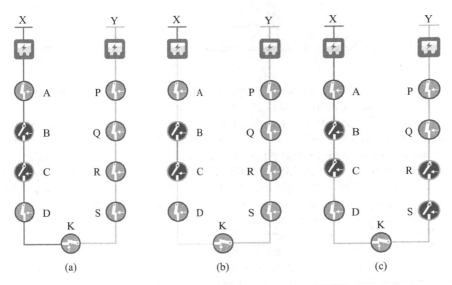

Figure 12.3   The fourth configuration of faulty sections and power restoration

restoration for the faultless sections by opening and closing appropriate isolators after communicating with relevant agents and taking the correct decisions.

12.4.2   Electricity trading

During the last 20 years, a process has been undergoing worldwide to restructure electrical power facilities and liberalize the markets for services based on these facilities. This process moves the electricity industry from vertically integrated monopolies to multiple independent companies and replaces the centralized cost-based market with a supply and demand-based competitive market. The major goal of this reform is to promote energy conservation and alternative energy technologies and to reduce oil and gas consumption through technology improvement and regulations. In the restructuring process, a power system is divided into multiple

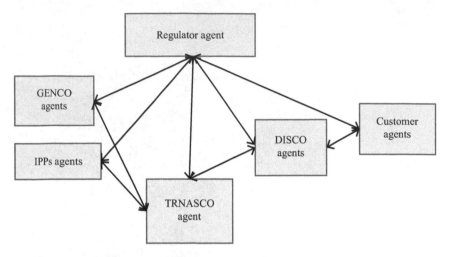

Figure 12.4   Agent-based energy market in a structured power network

components including generation, transmission, distribution and system operation, which are the major participants of an energy market.

Agent-based simulation has been a widespread method in modeling and analyzing electricity markets [27,28]. Several agent-based techniques are being presented with several general purpose agent-based simulation tools. MAS architectures may not be able to fully operate in the electricity market with the presence of advanced communication protocols. These advanced communication structures are very important in electricity trading. Based on the electricity structure in the country, the agent network will be different. An example agent-based energy market in a structured power network is shown in Figure 12.4.

The shown example consists six different agents for representing generation companies (GENCO), individual power producers (IPPs), transmission companies (TRANSCO), distribution companies (DISCO), a regulator and consumers. GENCO agents and IPP agents are bidding into a pool market, that the price is depended on the production, maintenance, and other costs. Electricity markets are very dynamic and highly uncertain with load changes. Therefore, the decision-making process for GENCO and IPP agents can be complex. As generation agents do not have the actual and perfect knowledge about other agents in the market with them, the final pricing will be depended on the TRANSCO agent. The individual GENCO and IPP agents decide their pricing based on the available data as shown in Figure 12.5.

The whole transmission is controlled by a single agent called TRANSCO agent. It makes the decision process simple and straightforward.

In this model (Figure 12.6), TRANSCO agent is directly controlled by the regulator agent. Buying price is defined by the regulator based on demand prices, maintenance cost, profit, and the efficiency achieved by TRANSCO. Consumer agents determine the buying price for specific time duration while DISCO agents try to maximize their profit by setting a maximum selling price. Even within this

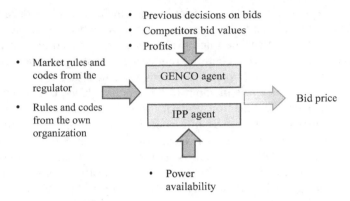

Figure 12.5   GENCO agent decision process

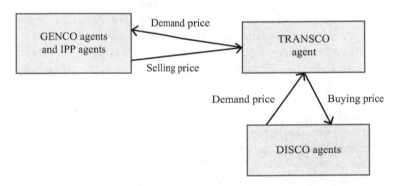

Figure 12.6   TRNSCO agent communication network

bilateral contract between consumer agents and DISCO agents, the regulator agent is controlling the process by setting a price window.

12.4.3   Optimization

MAS-based controllers have been used in for optimizing power-network operations. As loads and sources within a microgrid can be diverse and distributed, real-time reaction and distributed generation (DG)-source management are critical in preventing local power outages, MAS-based optimization will play a critical role in successful microgrid operations [29]. Various architectures and algorithms can be used to optimize the operation against different characteristics.

● Cost optimization: These systems consider the characteristics of the source or load types and self-regulates with other agents in order to globally optimize in terms of cost and efficiency.
● Load optimization: MAS is able to determine the optimal operation of a solar powered microgrid considering the load demand, environmental requirements, and source capacities.

In both optimizations, a competitive pricing mechanism is also introduced in order to serve the consumers at a reduced price and to provide better revenues.

Figure 12.7 shows a simple microgrid architecture for load optimization. The model presents a distributed control architecture operating in two layers. The primary (strategic) layer comprises the control agent who makes the run-time decisions. The control agent is in control of a secondary layer, overseeing the load control, and microgrid connection control capabilities. The control agent is the center of the primary layer holding influence over the DG agent, controlling load agents, and interfacing with the user agent. The DG agent collects several information related to the DG: (i) availability, (ii) connection status, (iii) power rating, (iv) energy source availability, and (v) cost of energy. During an upstream fault, the low voltage (LV) agent islands the microgrid, by tripping the CBA at the point of common coupling (PCC), if the control agent allows it. During islanded operation, the load shedding is carried out by load agents to ensure that power is delivered to the most critical loads on priority basis set by the control agent. This is implemented by load agents opening CBAs at each of the controlled local loads.

The user agent behaves as the gateway for the user to interact with the system, to obtain real-time information and to set system goals. A new agent such as an auction agent can be introduced to coordinate the power purchasing for cost optimization [30].

The MAS is developed for the load management during the islanded phase. As local generation capacities will most definitely be limited with respect to local loads, load shedding becomes unavoidable during islanding. Proper load management during load shedding provides better protection to the most critical loads. By assigning priority values to the local loads, the most critical loads can be identified for protection during islanded mode. By allowing priority level updates during operation, the control over load shedding/management can be improved.

Power production is maximized with respect to production cost and DG unit constraints. Demand side management (DSM) is also utilized via the load-control

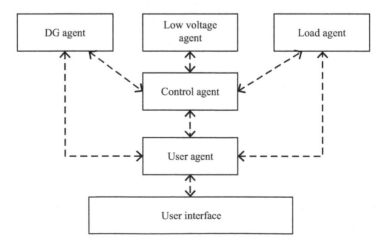

Figure 12.7   Microgrid architecture for load optimization

agent. The MAS is able to monitor energy resources and schedule generation (dispatch control) for optimized microgrid operation.

12.4.4 Smart-grid control

According to the US Department of Energy's Modern Grid Initiative, a smart-grid integrates advanced sensing technologies, control methods, and integrated communications into current electricity grid both at transmission and distribution levels. MAS can be used to gain more benefits and improve the characteristics of smart grids. The following sections describe the MAS applications in smart grid.

12.4.4.1 PHEVs and PEVs

Plug-in hybrid electric vehicles (PHEVs) are more popular among the people because of their low-operating cost and lower emissions compared to their traditional internal combustion-engine based counterparts. Most big automotive companies have already started the making of PHEVs for mass market. PHEVs give more advantages for customers than using electric vehicles (EVs). Plug-in electric vehicles (PEVs) cannot run for long distances and also it cannot guarantee the low emission as they are using electricity from the fossil fuel based plants. The operation of PHEVs and PEVs in a distribution system will be a challenging task for the utilities in balancing the power demand. This becomes further complicated as most of the utilities face difficulty in DSM during peak hours. As EV battery charging current is significant and it is difficult to predict its consumption time, it becomes much more difficult to predict their effect on the demand. This can make severe damages to the network due to rapid voltage and frequency fluctuations as well as ending in total blackouts. Therefore, there is a need for real-time monitoring and control of the network. This can be achieved by the use of MAS with a smart-grid communication network as shown in Figure 12.8.

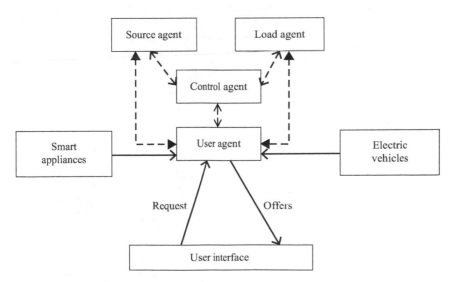

Figure 12.8   Smart appliances and EVs in MAS architecture

Smart appliances and EVs are connected to the user agent. User agent collects the necessary current data such as voltage, frequency and current of the user. User can send the advanced request from the user agent for the use of EVs and smart appliances. Then, the control agent is responsible for sharing the energy among the connected user agents by looking at the current capabilities [31].

12.4.4.2   Demand side management and smart meters

DSM and smart meters can be used for better power distribution among the consumers in a communication enabled distribution network. DSM promotes the efficient usage of power, while focusing on the network stability and reliability. It monitors the real-time power consumption of users and automatically distributes the excess power of the system, while controlling the power usage of the users to keep the network stability. In this context, the power systems have to become smarter, more reliable, and more robust in taking on renewable energy sources without losing stability and efficiency. Matching the demand during an islanded operation is very critical in terms of system stability. Therefore, monitoring and automation of the microgrid is essential and demand-side energy management is becoming a trending topic in the field. Figure 12.9 shows how the DSM and smart meters can be used in an MAS-controlled environment efficiently.

Individual smart meters act as separate agents, and they will send the real-time power consumption details of each consumer to the main server agent through the communication channel. The controlling process runs according to the gathered information, and the same time updates the web site through the web-server agent. The main disadvantage of this system is that it relies on the central main server agent. If the central server crashes, the whole system will crash. But this scenario can be avoided by using a distributed server agent system. One of the foreseen fact is the devices can be developed to gain the direct DSM. That is direct control of household appliances through smart-meter agents. Another fact to be stated is, as the data transmission plays a vital role, enhancing the data security becomes crucial. That can be achieved by using advanced data encryption, digital certificates, and intrusion detection systems.

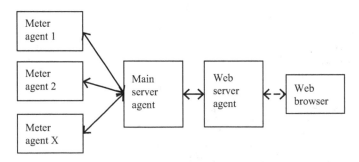

Figure 12.9   Smart meters in MAS-controlled environment

12.5 Case studies

12.5.1 MAS in a microgrid

The application of MAS in microgrid has been checked and validated through a simulation test-bed, and further tested through physical implementations. A dual-layered novel architecture has been introduced, and it is presented in Figure 12.10. The proposed dual-layered MAS comprises six agents.

1. Control agent: Control agent monitors the main grid for faults and/or outages and controls the secondary layer in the MAS. It oversees the operation of the secondary layer which is tasked with handling loads and the LV connection. It also informs the agents in the MAS whether the system is islanding capable or not. Ultimately, the control agent will be the center of the primary layer holding influence over the DG agent.
2. LV agent: It controls the islanding operation by opening the CBA at the PCC.
3. Load agent: It stores information regarding each load and controls the connect/disconnect operation.
4. DG agent: DG agent is responsible for storing information regarding the DG unit, metering and controlling the output and connecting/disconnecting the unit to/from the microgrid and controlling the LV agent and load agents.
5. User agent: User agent behaves as the gateway for the user to interact with the system, to obtain real-time information and set system goals.
6. DB agent: Database agent stores system information and data and messages shared between the agents.

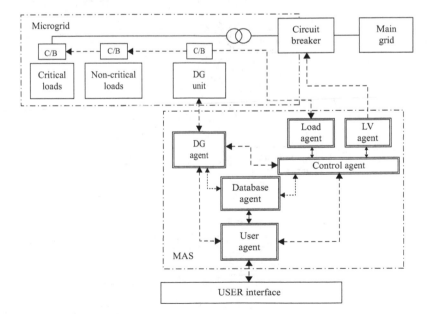

Figure 12.10 Dual-layered architecture

The dual-layered MAS is developed using JADE [32], and it can be broken into the following main steps:

1. Agent specification: Here, the specifications, requirements, and tasks for the agents in MAS are defined.
2. Application analysis: Here, the target is to develop a MAS that achieves the objectives. In order for the agents to follow the objectives, how they collaborate with one another must be clearly defined.
3. Application design: The detailed structure of the ontology for the dual-layered MAS is given in Figure 12.11.
4. Implementation: The MAS is implemented by developing the control algorithm depicted in Figure 12.12.

After implementing the simulation test-bed with the required MAS, several case studies has been done. It is assumed that unlimited power is available from the grid and local embedded generator has unlimited fuel to withstand an outage of infinite duration. The embedded generator is assumed to be always online-supplying part of the local demand, and capable of instantly changing the output depending on changes in local critical demand. Operation of each of the objectives is simulated and the success of the MAS is verified.

The MAS attempts to control physical entities or a simulation of the same. This puts the MAS and the simulation into two different domains. In order for the MAS

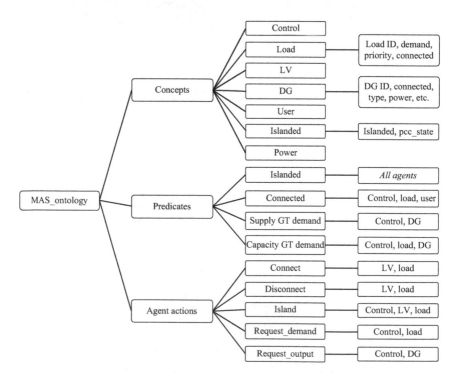

Figure 12.11   Ontology for the dual-layered MAS

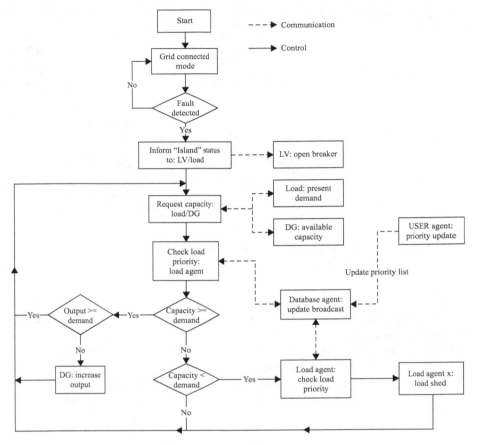

Figure 12.12   Control algorithm for multi-agent system operation

to work on the simulated environment establishing communication between the two domains is required. The MAS runs on a JAVA platform, whereas the simulation is run on a simulating environment (MATLAB®/Simulink®). Therefore, in order to establish communication between the two domains for sensing and control, transmission control protocol and the Internet protocol (TCP/IP) is used. A third party TCP/IP server [33] is utilized as an interface to MATLAB and it connects to another middle server which allows multiple simultaneous TCP/IP connections. Each agent requires a separate connection to the test-bed. The interface allows only a single connection at a time; therefore a separate middle server is required. This process is depicted in Figure 12.13.

The middle server allows various agents in the MAS to establish multiple simultaneous TCP connections and multiplexes all of them through the single TCP connection to the TCP server. Both servers can be implemented on the same PC running the simulation or on a separate PC on a connected network.

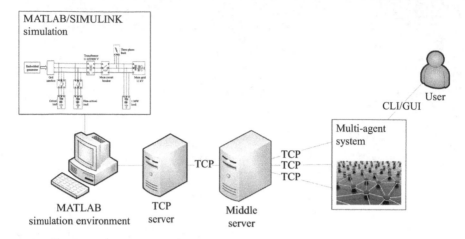

Figure 12.13   Establishing communication between simulation environment and MAS

Three case studies are developed to verify the capability of the proposed MAS to provide control and protection for the microgrid. Within conventional power systems, frequency control is handled by rotating inertial masses of large generators. This causes a problem within smaller microgrids. Therefore, an inverter based system is selected, where the frequency control is handled by the inverter interface.

The simulations are carried out on an inverter-based microgrid test-bed developed using MATLAB/Simulink. The test-bed is built mainly using the SimPower Systems block set available in Simulink. The test-bed based on the simplified block diagram, shown in Figure 12.14, is used for the simulation case studies.

Simulated microgrid test-bed is shown in Figure 12.15. The test-bed comprises an embedded generator, acting as a DG unit, supplying part of the local demand, and some critical and noncritical loads. It is assumed that the DG unit can operate at full capacity without fuel limitations during any outage. It is also assumed that the DG unit can instantly increase output using battery storage. As an inverter-based microgrid is considered, a grid interface unit comprising an inverter, low-pass filter, pulse width modulation and CBA is used in the case of a DC source. The microgrid is connected to a main utility grid (11 kV) across a transformer (11 kV/400 V) and a main CBA.

12.5.1.1   Case study 1: intentional islanding during an upstream fault

In this case study, an upstream fault is introduced on the main grid system at 0.05 s into the simulation. Upon the detection of the fault, the MAS switches from grid connected mode to islanded mode of operation. The transition is carried out within 0.02 s of detecting the fault. This verifies the capability of the MAS in rapidly initiating a seamless transition. The messages exchanged between the agents during

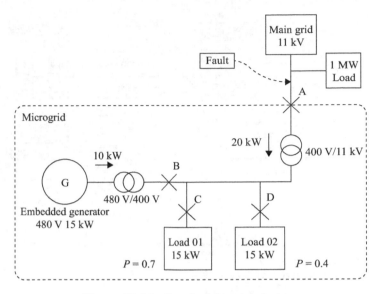

Figure 12.14   Test-bed for simulations

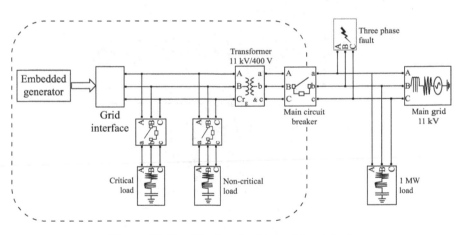

Figure 12.15   Simulated microgrid test-bed

islanding are shown in Figure 12.16. The simulation results presented in Figure 12.17 depict the successful islanding process.

All agent operations are carried out rapidly, from detecting fault, opening the main breaker, connecting the local source, and shedding loads, to stabilize the microgrid within 0.02 s. Therefore, the system is able to disconnect from the main utility grid and maintain the supply to the critical loads without suffering a brownout and/or blackout.

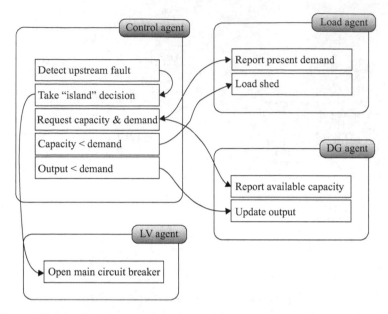

Figure 12.16   Key messages exchanged between agents during islanding

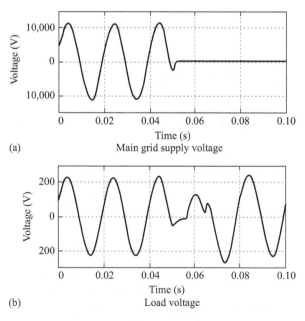

Figure 12.17   Line-to-line voltages during intentional islanding: (a) line-to-line voltage across the noncritical load at "A" and (b) line-to-line voltage at breaker "B"

12.5.1.2 Case study 2: protecting critical loads during intentional islanding

During an outage or fault on the main grid, the microgrid has to cater the local demand based solely on the local sources. In order to meet the local demand and maintain the stability of the system at the same time, some loads have to be shed. This is done by, shedding lower priority noncritical loads and protecting the higher priority critical loads. This is completed within 0.02 s of detecting the fault and initiating islanding transition. This verifies the capability of the MAS in rapidly protecting critical loads during intentional islanding.

The second case study addresses the capability of the MAS to protect/secure power to critical loads during intentional islanding. This should happen immediately following the transition to islanded mode. If power supply from available local DG unit is limited and insufficient to supply the total local demand, load shedding has to be carried out rapidly in order to secure supply to the most critical loads. The test-bed is same as the one used in case study 1, with only the following changes. Two local loads are 15 kW, each with assigned priorities of 0.7 and 0.4, respectively. The total local demand is 30 kW; initially the DG unit supplies 10 kW whereas the other 20 kW is provided from the main grid. Similar to case study 1, an upstream fault is introduced to the system at $t = 0.05$ s in to the simulation. At the beginning of the simulation breakers A, B, C, and D are all closed. The results are shown in Figure 12.18.

All agent operations are carried out rapidly; from detecting fault, opening the main breaker, connecting the local source, and shedding loads, to stabilize the microgrid within 0.02 s. Therefore, the system is able to disconnect from the main utility grid and maintain the supply to the critical loads without suffering a brownout and/or blackout.

12.5.1.3 Case study 3: managing critical loads during islanded operation

The third case study addresses the capability of the MAS to provide load management capabilities during islanded operation. As the available local capacity is limited and insufficient to cater total local demand during islanded mode, non-critical (lower priority) loads have to be dynamically shed in order to protect the stability of the microgrid. Simulation results (Figure 12.19) show the ability of the MAS to allow for dynamic update of load priority levels.

Therefore, the control agent commands the load agents to shed the load 01 of 25 kW to match the DG capacity. Thus, the supply to the new most critical load, load 02 is provided by reclosing "D" (see Figure 12.19(c)). The load agent at the new lower priority load, load 01, sheds it from the system by opening breaker "C" (see Figure 12.19(b)). This process is shown in Table 12.1.

When the load priorities are revised by a user via the user agent, the MAS is able to reconfigure the system, in order to provide power to the new most critical load. The reconfiguration is also done within 0.02 s of the revision and successfully reconnects the new critical load. The results show the capability of the MAS to safely island and maintain the supply to its critical loads, while allowing for the critical loads to be dynamically revised.

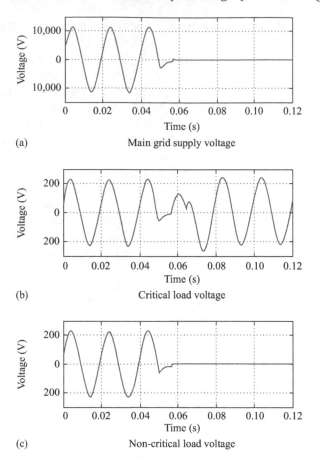

Figure 12.18   Line-to-line voltages during securing critical loads: (a) line-to-line voltage across the non-critical load at "A," (b) line-to-line voltage at breaker "C," and (c) line-to-line voltage at breaker "C"

12.5.2   Application of MAS in smart-grid transmission/generation

The Sri Lankan Electric Power System is evolving into a complicated network. The meter reading technology of the distribution network is being updated to an automated meter reading (AMR) system. Development of AMR systems has begun a new era for power distribution networks. In Sri Lanka at the moment, a very traditional distribution network is available. These distribution networks need to be upgraded to a smart grid. Upgrade cannot be done overnight and needs multiple innovations for quick and inexpensive rollout. One such innovation currently developed is remote meter reading through direct dialing via general packet radio service (GPRS) and AMR system. In terms of the distribution network, distribution utilities started testing fully agent-based operation

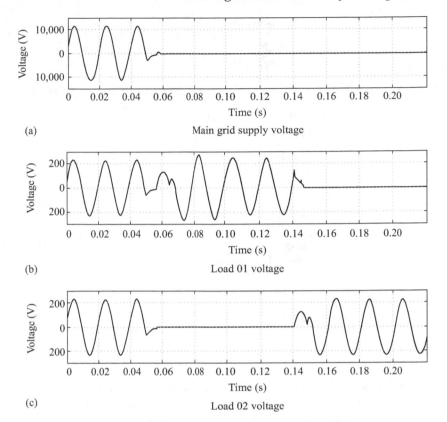

*Figure 12.19   Line to line voltages during islanded operation. (a) Main grid
voltage at breaker "A," (b) load 01 measured at "C": switches
from high priority to low priority, (c) load 02 measured at "D":
switches from low priority to high priority*

Table 12.1   Priority revisions during islanded mode

	Grid-connected mode	Islanded mode	
		Initial priority	**Revised priority**
Load 01	Priority = 0.7	Priority = 0.7	Priority = 0.7
	Connected	**Connected**	*Disconnected*
Load 02	Priority = 0.4	Priority = 0.4	Priority = 0.9
	Connected	*Disconnected*	**Connected**

of the distribution network as an experiment. Accordingly, four major agent-based projects were started as illustrated in Figure 12.20.

- Project 1: physical meter agent communication
- Project 2: network resource planning and optimization

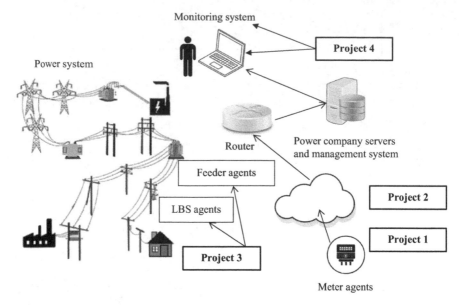

Figure 12.20   Agent based distribution network

- Project 3: agent based network reconfiguration and restoration
- Project 4: agent based monitoring system. Monitoring system of the network needs to be done as agent base monitoring system to be compatible with all other systems to run full agent base distribution network

Upon the study of currently available AMR systems, it was identified that in order to implement a proper monitoring system, analysis of available meter data was essential. Based on this requirement, agents were defined to represent the actual distribution network components virtually. Agent architecture was designed to implement the monitoring system. Agents were represented by separate graphical user interface (GUIs), and an individual online voltage monitoring chart was used for each bulk consumer. A separate database was developed to maintain outage data and calculate power quality measurement indexes more accurately. A demonstration model to represent area distribution network was designed and implemented the data acquisition system to database. Based on this model data, agent system was synchronized and GUIs were run to show the real outages. For the selected case study, 34 nos of meters were stored in to a database through GPRS direct calling method. These voltages were processed with the monitoring system to find out changes of the area network graphically.

AMR system is used to read energy meters and record the meter data to central data base (Figure 12.21). Real-time voltage of the relevant energy meter was sorted out from main data base and stored into a separate data base.

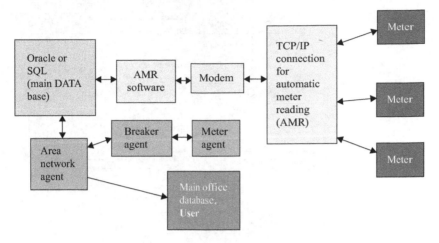

Figure 12.21   Schematic diagram of the whole data processing system

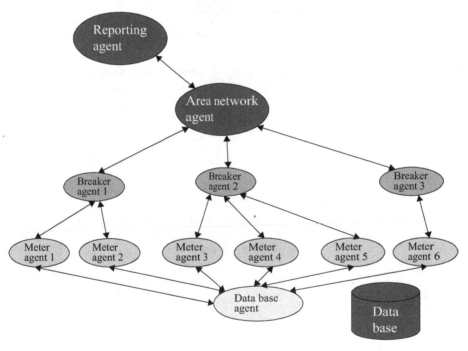

Figure 12.22   Proposed agent architecture

To simplify the database operations, a partial area was selected from the Sri Lankan network.

Vertically, layered two-pass architecture was selected for this project considering the data-flow pattern between the agents (Figure 12.22). The agent architecture

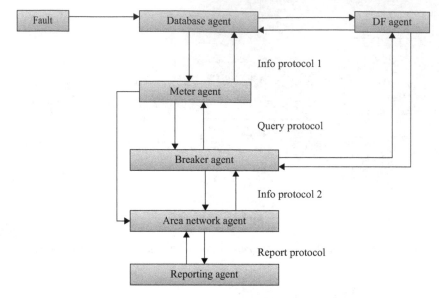

Figure 12.23   Fault processing diagram

is important when creating flexible and extensible NASs. Agents of that can be explained as below:

- Data base agent
- Meter agent
- Breaker agent
- Area network agent
- Reporting agent

The power distribution network is considered to have a tree structure. The power is supplied from the primary station and is completely drawn by the downstream loads. An outage is considered when a line tripping leads to the loss of the corresponding downstream loads. An outage can be caused by any type of fault event that triggers protection devices to isolate the fault.

Once the fault occurs in the system, the respective data will be sent to the data base through the AMR system. Then, the database agent will start its process to through info protocol 1 to inform to all meter agents. The detailed fault processing diagram is shown in Figure 12.23.

12.6   Conclusions

Due to increasing complexities in the power system, MAS has become a dominant research area in electricity generation, transmission, distribution and utilization. MAS will be a solution for emerging complexities with bidirectional power flows in the power system. Multiple DG units and storage elements (i.e., battery storage, water pumping, hydrogen generation etc.) can be integrated into the microgrid to increase reliability, reduce emissions and to improve sustainability. As most non-conventional renewable energies tend to be of a highly fluctuating output, energy storage would smooth out the variation. With the integration of multiple/various DG units, the user (utility or consumer) can select the source which offers the optimal energy solution. These improvements will require the development of responsibilities of an agent in charge of the DG.

The following improvements can also be incorporated in future.

1. Agent behaviors can be improved by utilizing fuzzy/neural control logic.
2. The effect of increasing number of agents on controller stability and service latency can be studied to evaluate long-term extensibility of MAS.
3. The requirements of communication network redundancy and aspects of how to improve data network security and preventing cyber-attacks and intrusions should also be addressed in future work.

Practical application of MAS for smart grid is difficult with existing incompatibility hardware, proprietary protocols used by the manufacturers and security risk. Anyway, practical tests of MAS applications for smart-grid control by the industry is needed to show the applicability.

References

[1] Li F., Qiao W., Sun H., *et al.* 'Smart transmission grid: vision and framework'. *IEEE Transactions on Smart Grid*. 2010, vol. 1(2), pp. 168–177.

[2] Lasseter R.H., Piagi P. 'Microgrid: a conceptual solution'. *Proceedings of the Power Engineering Specialists Conference (PESC'04)*; Aachen, 2004, pp. 4285–4290.

[3] Moslehi K., Kumar R. 'A reliability perspective of the smart grid'. *IEEE Transactions on Smart Grid*. 2010, vol. 1(1), pp. 57–64.

[4] Piagi R.H., Lasseter P. 'Microgrid: a conceptual solution'. *Proceedings of the Power Electronics Specialists Conference*; Aachen, 2004, pp. 20–25.

[5] Andreou A.S., Labridis G.T., Bakirtzis D.P., Bouhouras A.G. 'Selective automation upgrade in distribution networks towards a smarter grid'. *IEEE Transactions on Smart Grid*. 2010, vol. 1(3), pp. 278–285.

[6] Sasaki H., Nagata T. 'A multi-agent approach to power system restoration'. *IEEE Transactions on Power Systems*. 2002, vol. 17(2), pp. 457–642.

[7] Dou C., Hao D., Jin B., Wang W., An N. 'Multi-agent-system-based decentralized coordinated control for large power systems'. *International Journal of Electrical Power & Energy Systems*. 2014, vol. 58, pp. 130–139.

[8] Bellifemine F.L., Caire G., Greenwood D. 'Developing Multi-Agent Systems with JADE'. John Wiley & Sons, 2007.

[9] Lu J., Chen G., Yu X. 'Modelling, analysis and control of multi-agent systems: a brief overview'. *Proceedings of the IEEE International Symposium Circuits and Systems, ISCAS*; Rio de Janeiro, 2011, pp. 2103–2106.

[10]   Li Z., Duan Z., Chen G., Huang L. 'Consensus of multiagent systems and synchronization of complex networks: a unified viewpoint'. *IEEE Transactions on Circuits and Systems I: Regular Papers*. 2010, vol. 57(1), pp. 213–224.

[11]   Acampora G., Loia V., Gaeta M. 'Exploring e-learning knowledge through ontological memetic agents'. *IEEE Computational Intelligence Magazine*. 2010, vol. 5(2), pp. 66–77.

[12]   Meng Z., Ren W., Cao Y., You Z. 'Leaderless and leader-following consensus with communication and input delays under a directed network topology'. *IEEE Transactions on Systems, Man, and Cybernetics, Part B (Cybernetics)*. 2011, vol. 41(1), pp. 75–88.

[13]   Semsar-Kazerooni E., Khorasani K. 'Optimal consensus seeking in a network of multiagent systems: an LMI approach'. *IEEE Transactions on Systems, Man, and Cybernetics, Part B (Cybernetics)*. 2010, vol. 40(2), pp. 540–547.

[14]   Mousavi M.S.R., Khaghani M., Vossoughi G. 'Collision avoidance with obstacles in flocking for multi agent systems'. *Proceedings of International Conference on Industrial Electronics, Control & Robotics (IECR)*; Orissa, 2010, pp. 1–5.

[15]   Lepuschitz W., Koppensteiner G., Barta M., Nguyen T.T., Reinprecht C. 'Implementation of automation agents for batch process automation'. *Proceedings of IEEE International Conference on Industrial Technology (ICIT)*; Via del Mar, 2010, pp. 524–529.

[16]   Barbosa J., Leitao P. 'Modelling and simulating self-organizing agent-based manufacturing systems'. *Proceedings of 36th Annual Conference on IEEE Industrial Electronics Society, IECON*; Glendale, AZ, 2010, pp. 2702–2707.

[17]   Dong Q., Bradshaw K., Chaves S., Bai L., Biswas S. 'Multi-agent based federated control of large-scale systems with application to ship roll control'. *Proceedings of Fourth International Symposium on Resilient Control Systems (ISRCS)*; Boise, ID, 2011, pp. 142–147.

[18]   Tan S., Jiang Y., Zhang H., Liu J. 'Research on continuous rolling process control system based on multi-agent'. *Proceedings of Ninth World Congress Intelligent Control and Automation*; Taipei, 2011, pp. 75–78.

[19]   Claes R., Holvoet T., Weyns D. 'A decentralized approach for anticipatory vehicle routing using delegate multiagent systems'. *IEEE Transaction on Intelligent Transportation System*. 2011, vol. 12(2), pp. 364–373.

[20]   Chen B., Cheng H.H. 'A review of the applications of agent technology in traffic and transportation systems'. *IEEE Transaction on Intelligent Transportation System*. 2010, vol. 11(2), pp. 485–497.

[21]   Gokulan B.P., Srinivasan D. 'Distributed geometric fuzzy multiagent urban traffic signal control'. *IEEE Transaction on Intelligent Transportation System*. 2010, vol. 11(3), pp. 714–727.

[22]   Seow K.T., Lee D. 'Performance of multiagent taxi dispatch on extended-runtime taxi availability: a simulation study'. *IEEE Transaction on Intelligent Transportation System*. 2010, vol. 11(1), pp. 231–236.

[23]   Chen X., Kong B., Liu F., Gong X., Shen X. 'System service restoration of distribution network based on multi-agent technology'. *Proceedings of the*

Fourth Conference on Digital Manufacturing and Automation (ICDMA); Qingdao; 2013, pp. 1371–1374.

[24]   Lim I.H., Sidhu T.S., Choi M.S., *et al.* 'Design and implementation of multiagent-based distributed restoration system in das'. *IEEE Transactions on Power Delivery*. 2013, vol. 28(2), pp. 585–593.

[25]   Solanki J.M., Khushalani S., Schulz N.N. 'A multi-agent solution to distribution systems restoration'. *IEEE Transactions on Power Systems*. 2007, vol. 22(3), pp. 1026–1034.

[26]   Lo Y.L., Wang C.H., Lu C.N. 'A multi-agent based service restoration in distribution network with distributed generations'. *Proceedings of the 15th International Conference on Intelligent System Applications to Power Systems*; Curitiba, 2009, pp. 1–5.

[27]   Kantamneni A., Brown L.E., Parker G., Weaver W.W. 'Survey of multi-agent systems for microgrid control'. *Engineering Applications of Artificial Intelligence*. 2015, vol. 45, pp. 192–203.

[28]   North M., Conzelmann G., Koritarov V., Macal C., Thimmapuram P., Veselka T. 'E-laboratories: agent-based modeling of electricity markets'. *Proceedings of the American Power Conference*; Chicago, 2002.

[29]   Maity I., Rao S. 'Simulation and pricing mechanism analysis of a solar powered electrical microgrid'. *IEEE Systems Journal*. 2010, vol. 4(3), pp. 275–284.

[30]   Bhuvaneswari R., Srivastava S.K., Edrington C.S., Cartes D.A., Subramanian S. 'Intelligent agent based auction by economic generation scheduling for microgrid operation'. *Proceedings of the Conference on Innovative Smart Grid Technologies, ISGT*; Gaithersburg, MD, USA, 2010, pp. 1–6.

[31]   Masoum A.S., Deilami S., Moses P.S., Masoum M.A.S., Abu-Siada A. 'Smart load management of plug-in electric vehicles in distribution and residential networks with charging stations for peak shaving and loss minimisation considering voltage regulation'. *IET Generation, Transmission & Distribution*. 2011, vol. 5(8), pp. 877–888.

[32]   Jade – Java Agent DEvelopment Framework, 2014 [Online]. Available: http://jade.cselt.it/

[33]   Zimmermann W. IOlib – Hardware input/output library module for MATLAB and Simulink, 2011. [Online]. Available: http://www.it.hs-esslingen.de/~zimmerma/software/index_uk.html

Chapter 13

Compressive sensing for smart-grid security and reliability

Mohammad Babakmehr[1], Marcelo Godoy Simões[1] and Ahmed Al-Durra[2]

This chapter aims to introduce the applications of a newly born theorem in signal processing and system identification, widely known as *compressive sensing–sparse recovery* (CS–SR), in smart power grid monitoring, security, and reliability. We will discuss how the sparse nature of the electrical power networks can be exploited to mathematical model and reformulate some of the most famous monitoring and security problems in power engineering as compressive system identification (CSI) problems. First, a short background on CS–SR theorems and techniques is presented. Next, the state-of-the-art in CS–SR applications in smart grid technology will be discussed, and finally, three distinctive monitoring problems are specifically addressed in detail, within comprehensive mathematical descriptions, and their specific features are explored through variety of case studies.

13.1 Introduction

Due to the expansion of structural and functional complexity of power systems, the monitoring and control of large-scale dynamic power networks (PNs) represent a critical and challenging issue for the future of the power industry. Smart-grid (SG) technology, regarded as the future generation of PNs, merges the traditional layer of electricity flow to a second layer of information flow in order to create an efficient automated energy delivery system as stated in References 1–3. Along the information layer, new measurements and sensing technologies bring new possibilities to address many of the conventional monitoring issues. However, as we get into the era of large-scale SGs, the amount of data we demand from sensing infrastructures continue to explode. Not only collecting, storing and transferring this huge amount of data brings new challenges to SG technology, but also, the conventional control and monitoring frameworks act dramatically complicated and slow for modeling and tracking

[1]Department of Electrical Engineering and Computer Science, Colorado School of Mines, Golden, CO, USA
[2]Department of Electrical Engineering, The Petroleum Institute, Abu Dhabi, UAE

the dynamic behavior of the new generation of power systems. In this chapter, we aim to introduce the potential applications of a new field in signal processing, system control, and identification, that it is named as *compressive sensing–sparse recovery* (CS–SR) in the literature [4–6], for SGs control and monitoring.

CS is a new technique in signal processing that has revolutionized the data analysis science. Due to its time efficiency, the CS–SR framework has found a critical role in large-scale data/system modeling/analyzing and significantly affected the era of Big Data. From the signal-processing viewpoint, CS is a novel sensing/sampling technique that abjures the prevalent wisdom in signal and data acquisition. Based on the Shannon sampling theorem and regarding the Nyquist sampling rule, the minimum sampling frequency (number of samples per cycle) needed for capturing the complete information from a phenomenon should exceed two times of the phenomenon bandwidth (please also refer to Reference 7). Roughly speaking, the CS theorem tries to answer the following question:

Can a sensor be designed that records fewer samples of a very high-dimensional signal ($f \in R^N$), whereas missing information (or samples) can be reconstructed from the collected ones?

The answer is "yes" as long as the phenomenon under study has an inherent information level that could be well modeled utilizing only K parameters or degrees of freedom.[1] A comprehensive study in CS literature has revealed the fact that practically most of the natural and industrial N-sample signals have far smaller than N degrees of freedom. This fact suggests that, under appropriate conditions, one may exploit the signal's information level K instead of its ambient dimension N, to design an efficient measuring, and modeling framework for high-dimensional signal analysis. In other words, the fundamental idea that underlies CS–SR theorem is that

> if a signal has an intrinsic information level K it can be possible to record a small number of linear measurements of a signal, M, proportional to K; "compressive sensing," and from those measurements reconstruct the complete set of all, N, samples that a conventional sensor would have recorded, "sparse recovery"

as stated in Reference 8.

Not only low-dimensional behavior can be captured in many real world signals, but also many system models rely on a low-dimensional (sparse) structure. As a result, another branch in the CS–SR field has been developed so far, widely known as CSI, as termed in Reference 9, in which an inherent sparsity can be observed in system structure or model. In compressive systems, the SR techniques can be utilized in order to solve large-scale system modeling problems in a fast and time efficient manner. A famous class of CSI problems is related to graph structures analysis. In the rest of this chapter, we will first present an introduction to the CS–SR theorem. Next, by exploiting the inherent sparse structure of the corresponding PN graphs, a couple of possible new applications of this technique are described for SG monitoring and control.

---

[1] In general, the number of K can be much fewer than N.

13.2   Mathematical modeling of a compressive sensing or sparse recovery problem

In the CS–SR area, we consider sensing mechanisms such that, the underlying information about a signal $f \in R^N$ is obtained by the following standard linear functional recording framework:

$$y_i = \langle f, \phi_i \rangle, \quad i = 1, \ldots, M. \tag{13.1}$$

Simply, a set of waveforms, ϕ_i, is correlated with the desirable object to acquire the corresponding signal information. For example, considering the sensing waveforms to be a collection of Dirac delta functions (spikes train), the resulting signal y is composed of an $M \times 1$ vector of sampled values of f in the time domain. Commonly, y is called the measurement vector. Thanks to the linearity of the measurements, one can represent (13.1) by the following matrix-vector format:

$$y = \Phi f; \quad \Phi = [\phi_1, \ldots, \phi_M]^T; \quad \phi_i \in R^N \quad \text{for } i = 1, \ldots, M \tag{13.2}$$

where the $M \times N$ matrix Φ is referred to as the sensing matrix, whereas the $N \times 1$ vector f represents the hypothetical signal samples (Figure 13.1 illustrates a visualization of (13.2)).

Based on what we have mentioned so far, we are interested in recovering the original object f in an underdetermined mathematical situation in which the number M of the accessible measurements (number of equations) is much fewer than the original dimension N of the signal f^2 (number of unknowns). Such problems commonly occur in a variety of engineering fields for different reasons. For instance, in many practical cases the number of sensors installed in natural locations may be limited (such as underwater places), the measurement technology can be expensive as in neural extracellular micro-electrodes, the experimental sensing

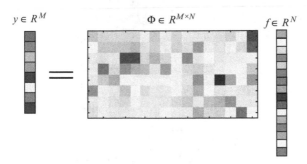

$y \in R^M$ $\Phi \in R^{M \times N}$ $f \in R^N$

Figure 13.1   A visualization of (13.2)

---

[2]A compressed set of samples has been collected: compressed sampling, compressive sensing.

time of an object can be slow so that we may only record the phenomenon a few times as in Gene regulatory network or we may be limited in the number of measurements that can be transferred into a communicational channel at each time slot, and so on.

From the algebraic viewpoint, due to the overcomplete nature of the problem infinitely, many possible solutions exist for this problem. However, it has been shown in CS–SR literature that if signal f can be represented with a low-dimensional model in some space (domain), it is possible to recover its original N-dimentional model using SR techniques. Figure 13.2 represents a visualized interpretation of low dimensionality of a signal. The corresponding low-dimensional model of f determines the appropriate technique that should be used for recovering this signal from a set of compressed (or compressive) measurements y and a sensing protocol Φ. Roughly speaking, all CS recovery techniques could be considered as a searching algorithm that tries to find the one target signal that coincides the low-dimensional pattern the best among all possible solutions of the linear system of equations represented by $y = \Phi f$.

If a signal f happens to be sparse in nature (in its original recorded domain), we can directly apply the SR methods to recover it from y and Φ. However, for most of the natural and industrial signals, the inherent low-dimensional model can be obtained by implementing a mathematical transform (Figure 13.2). It has been shown that for many signals, the Fourier and wavelet decomposition basis are well-chosen options to reveal their low-dimensional structure. If the basis is taken appropriately, it is often seen that a considerable fraction of the resulting expansion (transform) coefficients, x, are small and there will be only a few large ones (say, K).

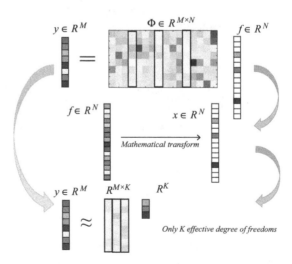

Figure 13.2   A signal f can have a low-dimensional (sparse) representation in its original domain, or we may reveal its low dimensionality by transforming it into another domain, x

13.2.1 Sparse modeling

Sparsity can be interpreted as the fact that most of the high-dimensional signals can actually be projected into a space with much lower degrees of freedom. Once a signal of interest is decomposed in some basis or dictionary, the nonzero coefficients can be expressed as aforementioned degrees of freedom. For example, consider a typical sinusoidal voltage signal f_v with some unwanted harmonic terms that may occur because of the nonlinearity in load (Figure 13.3(a)). Expanding such a signal under the Fourier basis will give another representation of the signal in the frequency domain as has been illustrated in Figure 13.3(b). Almost all of the voltage magnitude samples in the time domain representation of this signal are nonzero (except in crossing points), despite the fact that almost all of the Fourier coefficients of the same signal are zero except the fundamental frequency and those related to harmonic frequencies.

Definition 13.1 (sparse signal): *A signal $x \in R^N$ of length N is called a K-sparse signal if all its entries are equal to zero except a small portion K, where $K < N$ (in many cases $(K \ll N)$).*

Definition 13.2 (compressible signal): *Keeping the K largest entries of x and setting the rest of them to 0, one may acquire the nearest K-sparse signal to x, named x_K. Whenever x is exactly K-sparse, then $x = x_K$. However, if the distance of x to x_K happens to be small (but not necessarily zero), x is termed as a compressible signal.*

Using Definition 13.1 for a Fourier expansion of a common sinusoidal voltage signal f_v, it can be considered as a sparse (or at least compressible) signal, if

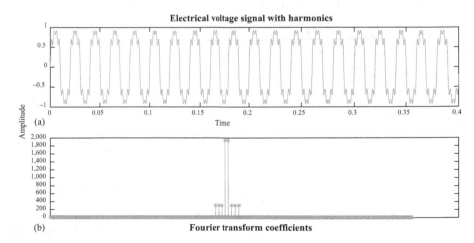

Figure 13.3 (a) A sinusoidal voltage signal with harmonics and (b) its corresponding Fourier coefficients

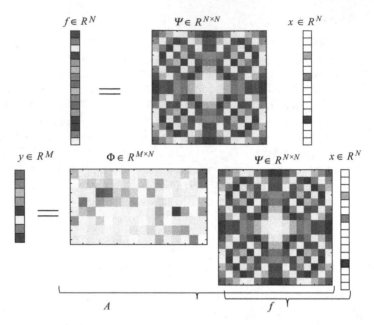

Figure 13.4   A visualization of (13.3) and (13.4)

the voltage signal f_v can be represented using a low-dimensional ($K = 5$ in this example) model in the Fourier basis. Now suppose a signal f can sparsely be expanded in some basis:

$$f = \Psi_x \tag{13.3}$$

where Ψ stands for the transformation operation matrix or dictionary (such as the Fourier transform basis), and x is the corresponding sparse vector of transform coefficients (such as the Fourier coefficients). Inserting (13.3) into (13.2), we have

$$y = \Phi f = \Phi\Psi_x = Ax \tag{13.4}$$

where A is an $M \times N$ matrix termed a measurement protocol or matrix. Figure 13.4 illustrates a visualization of (13.3) and (13.4).

Now, one can change the direct recovery problem of signal f to recovering the *sparse* signal x from a set of underdetermined measurements y and reconstructing the original f from $f = \Psi_x$. It has been shown in the literature that despite the overcomplete[3] nature of this recovery problem, under certain conditions (guarantee conditions) on the measurement protocol A, the *sparse-based recovery methods* can be guaranteed to efficiently find the candidate solution x that is well sparse. Beyond the necessary guarantee conditions on A, the most famous ones are the restricted

---

[3]The number of unknowns exceeds the number of equations or measurements equivalently.

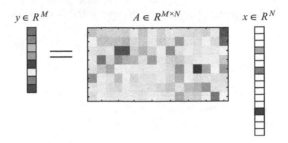

$y \in R^M$ $A \in R^{M \times N}$ $x \in R^N$

Figure 13.5   A visualization of a general sparse recovery problem

isometry property (RIP), the exact recovery condition, and mutual coherence. Roughly, what these conditions are trying to say is that within the measurement matrix A, each pair of subsets of columns must be almost orthogonal to each other, so one is able to guarantee that the SR algorithms will obtain the correct sparsest solution. Figure 13.5 illustrates a visualization of a general SR problem (SRP). In the next section, we introduce the SRP, definitions and theorems. For more discussion and technical proofs and details, please refer to References 4–12 and references therein.

13.2.2   Sparse-recovery problem

The SRP is an optimization problem that tries to recover a K-sparse signal $x \in R^N$ from a set of observations $y = Ax \in R^M$, where $A \in R^{M \times N}$ is the corresponding measurement matrix with $M < N$[4] as stated in References 4–6. According to the underdetermined nature of this recovery problem (due to $M < N$), we have infinite candidate solutions for this problem. However, if the measurement matrix A satisfies the corresponding SR guarantee conditions, and the number of available measurements M is proportional to $K \log(N/K)$, the K-sparse signal x can be recovered. It has been shown in the CS literature that, in practice, the number of measurements, M, can be much smaller than N. Actually, this value is only greater than the fundamental limit K by a logarithmic factor.[5] Remarkably, in the case of noise-free measurements and considering the signal f to be exactly sparse, the recovered sparse solution is an exact match. In the noise attendance or considering the signal f to be nearly sparse (compressible), the recovery result is incontrovertibly robust.

Definition 13.3 (pairwise coherence): *Consider a pair of orthobases (Φ, Ψ) of R^N including (1) Φ that is the sensing protocol that is used to measure the object f as in (1), and (2) Ψ that is used as an alternative transform to represent the original*

---

[4]In many cases, $M \ll N$.
[5]Without loss of generality, one may consider this logarithmic factor as the price we should pay because of not knowing the exact positions of the nonzero elements in advance.

signal f within another space. The pairwise coherence between these two bases is defined as follows:

$$\mu(\Phi, \Psi) = \sqrt{N} \times \max_{1 \leq i, j \leq N} |\phi_i, \psi_j|. \tag{13.5}$$

In plain language, the coherence measures the largest pairwise correlation among all of the two elements of Φ and Ψ; see also [4,8]. The coherence will have a small value except, Φ and Ψ contain highly correlated elements.

Definition 13.4 (coherence): *The coherence of an $M \times N$ matrix A with normalized columns $a_1, a_2, \ldots, a_N$ is defined as follows:*

$$\mu_A = \max_{1 \leq i, j \leq N, \, i \neq j} \frac{|\langle a_i, a_j \rangle|}{||a_i||_2 ||a_j||_2}. \tag{13.6}$$

The coherence of an M-by-K matrix A is a value in the interval $1/\sqrt{N}, 1$.

Definition 13.5 (RIP): *An M-by-N sensing matrix A satisfies the RIP of order K if*

$$(1 - \delta_K)||x||_2^2 \leq ||Ax||_2^2 \leq (1 + \delta_K)||x||_2^2 \tag{13.7}$$

is satisfied over all K-sparse vectors x with cardinality $||x||_0 \leq K$, where $\delta_K \in (0, 1)$ is termed as the isometry constant of order K.

Theorem 13.1 [5]: *If an M-by-N sensing matrix A satisfies the RIP condition of the order 2K with any isometry constant, the following optimization problem can recover the original K-sparse signal (vector) x from the set of underdetermined measurements $y = Ax$.*

$$P_0 : \hat{x} = \text{argmin}_{x'}||x'||_0 \quad \text{subject to} \quad y = Ax'. \tag{13.8}$$

In general, this l_0-minimization[6] problem is known to be NP-hard. Significantly, there is a relaxed version of this l_0-minimization problem, which can still

---

[6]l_p-norms:
For a vector $x \in \mathbb{C}^N$, the l_p norm of x is defined as follows:

$$||x||_{l_p(\mathbb{C}^N)} = \begin{cases} \left(\Sigma_i |x_i|^p \right)^{1/p} & 0 < p < \infty \\ \max_i |x_i| & p = \infty \\ \Sigma_i 1_{x_i \neq 0} & p = 0 \end{cases}$$

here x_i is the ith entry of the vector x and, **1** denotes the indicator function. Without loss of generality we simplify the norm notation from $||x||_{l_p(\mathbb{C}^N)}$ to $||x||_p$. Actually, for $0 < p < 1$, the above measures are not defined as norm operators. Moreover, roughly speaking, l_0 norm counts the number of nonzero elements for a vector $x \in \mathbb{C}^N$. This l_0 norm will play an important role in compressive sensing and sparse recovery theorems and concepts.

guarantee the recovery of sparse signal, named "l_1-minimization."

$$P_1 : \hat{x} = \text{argmin}_{x'} ||x'||_1 \quad \text{subject to} \quad y = Ax', \tag{13.9}$$

while the l_1-norm is a convex operator, this resulted into an obedient convex optimization problem, which is widely known as basis pursuit (BP).

Theorem 13.2 [6]: *In the case of noisy measurements, one can solve the following equivalent problem instead:*

$$NP_1 : \hat{x} = \text{argmin}_{x'} ||x'||_1 \quad \text{subject to} \quad ||y - Ax'||_2 < \eta. \tag{13.10}$$

This problem is widely known as basis pursuit de-noising (BPDN). Now, assuming the measurement matrix A satisfies the RIP condition of the order 2K with an isometry constant $\delta_{2K} < 0 : 4651$. Let $y = Ax + n$ be a noisy set of measurements from any signal (vector) x. If $\eta \geq ||n||_2$, then the solution $\hat{x}$ to (13.10) obey

$$||x - \hat{x}||_2 \leq C_3 \frac{||x - x_K||_1}{\sqrt{K}} + C_4 \eta, \tag{13.11}$$

where C_3 and C_4 depend only on δ_{2K}. Moreover, if A satisfies the RIP conditions, the BPDN can provide a stable recovery for sparse signals. In general, it is hard to check if a sensing matrix A satisfies the RIP condition (14). Thus, an alternative approach has been introduced to check the suitability of a sensing matrix to be used in an SRP.

Theorem 13.3 [10–13]: *Consider a K-sparse vector x with a vector of available measurements $y = Ax$. P_1 can recover the original K-sparse vector x from the set of measurements y, if the coherence of the matrix A remains bounded to $\mu_A < (1/2K - 1)$.*

It has been shown in the literature that the minimum number of measurements needed for perfect recovery is related to both the original dimension of the signal x, N, and the sparsity level K; specifically, M must be at least proportional to $K \log(N/K)$. Roughly speaking, the smaller the coherence of A, the larger the authorized value of K, and as a result, the broader the type of the vectors that can be perfectly recovered within SR technique.

Remark 13.1. *In order to implement the CS technique in a sample data acquisition system, the sensing protocol Φ should be selected on the basis of the appropriate low-dimensional model (transform) Ψ of the signal such that the overall coherence of the measurement matrix A (4) is suitably low so the SR of the sparse representation, x, is possible.*

Remark 13.2. *In general, the standard techniques for solving SRPs can be categorized into two major classes. The first class contains the greedy-based algorithms such as famous matching pursuit, orthogonal matching pursuit (OMP) stated in Reference 12, and compressive sampling matching pursuit stated in Reference 11.*

The second class, widely known as convex optimization based algorithms, includes methods such as famous least absolute shrinkage and selection operator (LASSO) [13]. Once a problem is formulated as an SRP, SRP solvers are needed to recover the sparse vector x. As we proceed with mathematical sparse formulation of a couple of famous SG monitoring problems within the rest of this chapter, we briefly introduce different types of SRP solvers that are used to evaluate the performance of the proposed methods. Since OMP and LASSO are the most popular and fundamental SRP solvers, a description about their operational framework is given. In a nutshell, the OMP method tries to correct the least-square (LS) sequence of errors for the estimated sparse solution of the vector x emerged from minor possible correlations beyond the columns of the measurement matrix A. To identify the correct support (location of nonzero elements) of the K-sparse vector x, OMP runs an iterative procedure called support selection step.[7] The OMP is detailed in Algorithm 13.1.

The LASSO estimator is one of the most well-known optimization-based SRP solvers, that is defined as the following unconstrained version of the famous BPDN:

$$\hat{x} = \text{argmin}_{x'} ||y - Ax'||_2^2 + \lambda ||x'||_1 \tag{13.12}$$

Algorithm 13.1 Orthogonal Matching Pursuit (OMP)

---

require: measurement matrix A, vector of measurements y, stopping criterion
initialize: $r^0 = y, x^0 = 0, l = 0, SUP^0 = \phi$
  repeat
    1. match: $h^l = A^T r^l$
    2. identify support indicator:

    $sup^l = \{argmax_j |h^l(j)|\}$

    3. update the support:

    $SUP^{l+1} = SUP^l \cup sup^l$

    4. update signal estimate:

    $x^{l+1} = argmin_{z:\ SUP(z) \subseteq SUP^{l+1}} ||y - Az||_2$
    $r^{l+1} = y - Ax^{l+1}$
    $l = l + 1$

  till stopping criterion met
output: $\hat{x} = x^l$

---

[7]In a variety of sparse recovery algorithm design literature, this support selection step has been modified to adapt the sparse recovery technique to many practical situations. We will introduce and use couple of these modified techniques in the future sections.

This optimization problem can be considered as the Lagrangian form of the BPDN (3.10) and is a convex quadratic program. The l_2 term in the objective function tries to minimize the fitting error, whereas the l_1 term declares the sparsity constraint on the solution, and λ plays the tuning parameter role in this problem (13.12). Least angle regression (which has been initiated in Reference 13) is the most popular approach widely used in the literature to solve this convex optimization problem.

13.3   Applications of CS–SR in smart grid

CS–SR techniques have been applied in a diverse variety of signal-processing and data analysis fields, from communication and radar systems [14,15] to biomedical engineering [16], geophysics [17], social media [18], and target detection [19]. By the same token, CS–SR-based signal-processing frameworks have been already developed to design a variety of protocols in order to address different data collection and transmission issues in SGs [20,21]. Although CS–SR has been initially developed for signal processing and data compression applications, their use is not limited to signal sensing, modeling, and analysis. Specifically, the SRP setup has found lots of other useful applications as an individual mathematical technique. This chapter outlines how to use CS–SR in SG applications.

Roughly speaking whenever sparsity exists in the nature of a problem, one may explore a way in mathematically reformulating the problem as an SRP and efficiently solve the problem within a short amount of time using SRP solvers. As has been mentioned before, another branch in the CS–SR field has been developed so far, widely known as CSI. A famous class of CSI problems is related to graph structure analysis. In the real world, there are many networks that can be modeled as a graph structure: communication networks, sensor networks, social networks, gene networks, neural networks, and especially *PNs*.

Although some of these PNs are dense in structure, most of them follow a sparse composition (the nodal connectivity level[8] of the graph is a small fraction of total system size). Sparsity in structure makes them great candidates to be considered as *sparse or compressive systems*. The inherent sparsity in compressive systems can be used in order to model many of their corresponding problems under SR formulation. Famous graphical problems such as shortest path identification, topology identification, anomaly recognition, and so on can be taken as examples. Within SG monitoring and control issues, SR has been used in attack recognition [22], communicational protocol design [20,21], power signals pattern recognition and disturbance classification [23–25]. Moreover, CS–SR techniques have been exploited to propose novel methods for single and simultaneous fault location in smart distribution and transmission networks [26,27], distribution system state

---

[8]In graph theory, the nodal connectivity level is defined as the maximum number of connected edges to an individual node of a connected graph.

estimation [28], and line outage identification [29,30]. Also, in Reference 31 a pioneering sparse-based framework has been developed for power-flow modeling.

In the rest of this chapter, we exploit the sparsity in PNs structure in order to introduce new mathematical formulations for a couple of famous power system control and monitoring problems, including power line outage identification (POI-SRP) [32,33], PN topology identification (PNTI-SRP) [34,35], and line parameters dynamic modeling (LPDM-structured SRP (SSRP)) [36]. These three problems are presented in Sections 13.4, 13.5, and 13.6, respectively. Before discussing the aforementioned problems and their corresponding sparse formulations, we need to give a brief description about some mandatory graph theory concepts and a famous load flow system of equations known as the DC power flow model that are used as fundamental mathematical notations in our SG-related SRP formulations.

13.3.1   Power network model: a sparse interconnected graph

In the rest of this chapter, the PN is considered a typical graph $G(\mathcal{N}, E)$, consisting of a set of N nodes: $\mathcal{N} = \{1, \ldots, N\}$, in which each node i represents an arbitrary electrical bus of the PN, and a set of L edges or transmission lines: $E = \{l_{(i,j)} | i, j \in \mathcal{N}\}$, which connect the buses through the network and form the general structure or topology of the grid. For instance, Figure 13.6 illustrates the underlying structure of the corresponding graph of the IEEE Standard-118 Bus. This graph representation will be useful in our mathematical formulations, and the corresponding matrices of graphs will play an important role, specially, in POI-SRP, PNTI-SRP, and LPDM-SRP formulations.

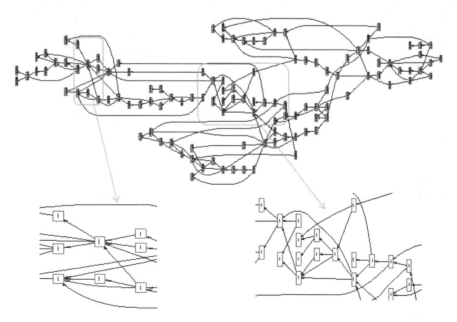

Figure 13.6   Corresponding graph of IEEE-118 Bus

For a graph, $G(\mathcal{N}, E)$, with a set of N nodes, $\mathcal{N}$, and a set of L edges, E, there are four important matrices that can roughly describe the whole structure of the graph:

✓ Adjacency matrix $\text{Adj}(G(\mathcal{N}, E))$:
An $N \times N$ matrix that represents which nodes of a graph are adjacent to which other nodes.
✓ Incidence matrix $\text{Inc}(G(\mathcal{N}, E))$
An $N \times L$ matrix (where N and L are the number of nodes and lines) that shows which nodes of a graph are connected to which edges.
✓ Degree matrix $\text{Deg}(G(\mathcal{N}, E))$
An $N \times N$ diagonal matrix which contains information about the degree of each node, that is, the number of edges attached to each nodes.
✓ Laplacian matrix $\text{Lap}(G(\mathcal{N}, E))$
The Laplacian matrix, nodal admittance matrix, Kirchhoff matrix or discrete Laplacian matrix is an $N \times N$ matrix representation of a graph that is defined as follows:

$$\text{Lap}(G(\mathcal{N}, E)) = \text{Deg}(G(\mathcal{N}, E)) - \text{Adj}(G(\mathcal{N}, E)) \tag{13.13}$$

One of the useful relationships between Laplacian and incidence matrices is as follows:

$$\text{Lap}(G(\mathcal{N}, E)) = \text{Inc}(G(\mathcal{N}, E)) \times \text{Inc}(G(\mathcal{N}, E))^T \tag{13.14}$$

From now on, it can be seen that the corresponding Laplacian matrix of the graph can be replaced by B, and the incidence matrix as M, so we have: $B = MM^T$. Further details are given in Reference 37, with more information on graph theory and corresponding matrices of a graph. A visualization of the incidence, Laplacian, and degree matrices for the corresponding graph of the IEEE Standard-118 Bus are presented in Figure 13.7. As can be seen from these structural matrices, a PN can be considered a sparse-interconnected system. This assumption is based on a survey of articles and standard PN models found in software and toolboxes such as MAT-POWER. We observe that the regular maximum connectivity level of a bus in a network is typically less than 5–10%, especially in very large power transmission systems. For example, the highest connectivity level[9] of a bus in IEEE 30 Bus is 6 (10%), in IEEE Standard-118 Bus is 9 (7%), in IEEE 300 Bus is 12 (4%), and in IEEE 2383 Bus is 9 (0.3%). We will see how this sparsity helps us to formulate POI, PNTI, and LPDM problems as an SRP, which as supported by the theory of CS, can be solved with a small set of measurements in a fast and accurate way using SRP algorithms. If the sparsity assumption is violated, however, the performance of the proposed algorithms and methods will suffer.

---

[9]Bus connectivity level: number of transmission lines directly connected to an individual bus. We also refer to this number as *bus in-degree*.

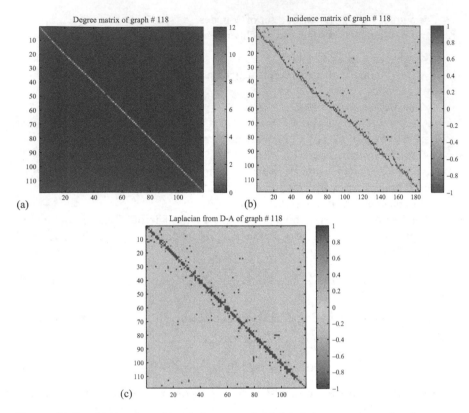

Figure 13.7   (a) Degree, (b) incidence, and (c) Laplacian matrices for the corresponding graph of IEEE Standard 118 Bus PN

13.3.2   DC power-flow model and weighted Laplacian matrix of PN

The AC power-flow model or the AC load-flow model is a numerical technique in power engineering applied to a power system in order to analyze the behavior of different forms of AC power (i.e., voltage amplitudes, voltage angles, active [real] power, and reactive [imaginary] power) under normal steady-state operation. In such a model, two distinct sets of nonlinear equations (called power balance equations) indicate the relationship between active and reactive power (injected at bus i from bus j) and the voltage magnitude and phase angle as follows [38]:

$$P_{ij} = g_{i,j}V_i^2 - g_{i,j}V_iV_j\cos(\theta_i - \theta_j) + b_{i,j}V_iV_j\sin(\theta_i - \theta_j) \qquad (13.15)$$

$$Q_{ij} = b_{i,j}V_i^2 - b_{i,j}V_iV_j\cos(\theta_i - \theta_j) + g_{i,j}V_iV_j\sin(\theta_i - \theta_j). \qquad (13.16)$$

Here, P_{ij} and Q_{ij} are the active and reactive powers injected from node j to node i, respectively; V_i and θ_i represent the ith node voltage amplitude and phase, respectively; $g_{i,j}$ is the real part of the admittance of the line $l_{i,j}$ (or the conductance); and $b_{i,j}$ is the imaginary part of the admittance of line $l_{i,j}$ (or the susceptance).

In Reference 32, it has been shown that when the system is stable under a normal operating condition, the phase angle differences are small, meaning the term $\sin(\theta_i - \theta_j) \approx \theta_i - \theta_j$, that is, for a small phase-shift the power flow is proportional to the phase shift, instead of the sine of this phase difference. Using this simplification and the DC power-flow method discussed in Reference 38, the power injected to a distinct bus i follows the superposition law. This means that for each bus i in the PN, we have the two following approximated summations over the active and reactive powers (however, in practical cases (13.18) and (13.20) cannot be validated exactly):

$$P_i = \sum^j P_{ij} = \sum_{j \in N_i} b_{i,j}(\theta_i - \theta_j) \tag{13.17}$$

$$Q_i = \sum^j Q_{ij} = \sum_{j \in N_i} b_{i,j}(V_i - V_j). \tag{13.18}$$

Here, N_i is defined as the set of neighbor buses directly connected to the bus i. It is useful to rewrite these summations in a matrix-vector format, where we have

$$p = B\theta, \qquad p \in R^N \tag{13.19}$$

$$q = Bv, \qquad q \in R^N. \tag{13.20}$$

In this notation, the voltage amplitude and voltage phase angle values of all of the PN buses are collected in two vectors $v, \theta \in R^N$, respectively. Also, the active and reactive power values of the all of the buses are stored within the two following vectors, p and q, respectively. The matrix $B \in R^{N \times N}$ is called the nodal-admittance matrix, describing a PN of N buses, and its elements can be represented in the following format:

$$B_{ij} = \begin{cases} -b_{i,j} & \text{if} \quad l_{i,j} \in S_E \\ \sum_{j \in N_i} b_{i,j} & \text{if} \quad i = j \\ 0 & \text{otherwise} \end{cases} \tag{13.21}$$

The nodal-admittance matrix of the IEEE Standard-118 Bus is depicted in Figure 13.7(c). We see that this matrix has a sparse structure.

Remark 13.3: *The literature [32–35,37,38] shows within a PN the nodal-admittance matrix can be decomposed into a weighted version of the Laplacian matrix of the corresponding graph $G(\mathcal{N}, E)$ as follows:*

$$B = \sum_l^L b_l m_l m_l^T = M W_b M^T, \tag{13.22}$$

where $m_l \in R^K$ for $l = 1, \ldots, L$ represents the *lth* column within the incidence matrix *M*, and W_b is formatted as the following $L \times L$ diagonal matrix:

$$W_b = \begin{bmatrix} b_1 & \cdots & 0 \\ 0 & \ddots & 0 \\ 0 & \cdots & b_L \end{bmatrix}, \tag{13.23}$$

where b_l indicates the susceptance along the *lth* line.

13.4 Sparse recovery-based power line outage identification in smart grid

Beyond the critical monitoring and control tasks in the future SGs, the real-time detection and localization of the power line outages arises as one of the most challenging issues [39–43]. Huge blackouts can be prevented by having online situational knowledge from the structure of the PN. Due to mathematical complexity and operational time cost, most of the conventional real-time approaches have been adapted in the cases of single, double, or small number of multiple outages [44–46]. Recently, sparse-based formulations for POI have been introduced in References 32 and 33, which can address the aforementioned issues by conducting SR techniques. In this section, three POI-SRP formulations are represented: Distributed-POI-SRP [32], Binary POI-SRP [47], and Structured-POI-SRP [33]. Next, OMP and LASSO algorithms are applied in order to solve the POI-SRP in the case of single, or multiple outages. Also, the modified clustered OMP (MCOMP) recovery algorithm is utilized, and it is discussed how the possible presence of a structured sparsity in the case of multiple power line outages helps the MCOMP to improve the final outage identification results.

An important practical issue in the case of POI-SRP formulation is the inherent high coherence in the corresponding measurement matrices. For the sake of finding the root of high coherence issue, the structure of the corresponding measurement matrices of IEEE Standard 118 Bus is studied. Moreover, the SR guarantee conditions that the measurement matrix should satisfy to be applicable within an SRP are discussed. Based on our observations, two categories of coherence modification approaches are suggested. The first category aims to decrease the coherence by making some meaningful changes in the structure of the measurement matrix, whereas the second category tries to improve the support selection step in the greedy OMP algorithm. On the other hand, an important advantage of the sparse POI formulation is its flexibility to be adjusted in the case of limited number of sensors installed in the PN (specifically phasor measurement units "PMU").

13.4.1 POI-SRP formulations

Consider the following matrix-vector notation for the DC power-flow model: $p = B\theta \in R^N$. Suppose that, following by a systematic fault, several line outages

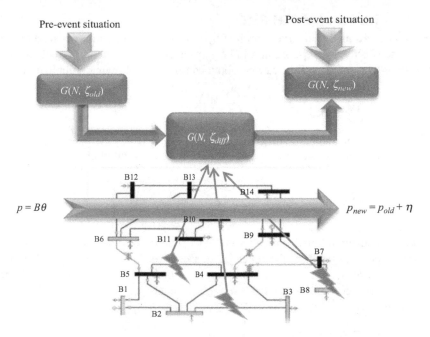

Figure 13.8   *Modeling an outage event using a set of three subgrids*

happen in the PN. Accordingly, a partial change occurs in the structure of the electrical grid and a new nodal-admittance matrix B_{new} accordingly. For the sake of clarity, we lay out a typical grid-wide fault using a set of three subgrids (Figure 13.8): (1) the pre-event or the old grid structure, contains N buses and L_{old} lines; (2) the post-event or new structure, contains N buses and L_{new} lines; and (3) finally, the difference or outage grid that again contains N buses but only $L_{diff} = L_{old} - L_{new}$ lines. The following notation is used in order to identify the corresponding specifications for each of these three networks: p_{old}, p_{new}, θ_{old}, θ_{new}, $\theta_{diff} \in R^N$ and B_{old}, B_{new}, $B_{diff} \in R^{N \times N}$. It has been shown in the literature that, following an outage incident, the new power vector is modeled as follows [48]: $p_{new} = p_{old} + \eta$, where η is defined as the $N \times 1$ perturbation or residual vector. Assuming $\theta_{diff} = \theta_{new} - \theta_{old}$, we can reach to (13.24). Moreover, filling in the format (13.22) for B_{diff} in (13.24), we can conclude (13.25) as well.

$$p_{new} = p_{old} + \eta \Rightarrow B_{old}\theta_{diff} = B_{diff}\theta_{new} + \eta \qquad (13.24)$$

$$B_{old}\theta_{diff} = \sum_{l \in \zeta_{diff}} x_l m_l + \eta, \qquad (13.25)$$

where ζ_{diff} stands for the set of outage line indices (so ζ_{diff} contains L_{diff} number of lines) and $x_l := m_l^T \theta_{new} b_l \; \forall \, l \in \zeta_{diff}$. Solving (13.25) with respect to θ_{diff}, we have

$$\theta_{diff} = B_{old}^{-1} \sum_{l \in \zeta_{diff}} x_l m_l + B_{old}^{-1} \eta. \qquad (13.26)$$

13.4.1.1   Distributed POI-SRP

In Reference 48 an exhaustive search (ES) approach has been implemented to iden-tify the index set ζ_{diff}. Suppose in a PN L_{diff} number of line outages has occurred. All possible post-event structures can be obtained from the following combination:

$$\zeta_{\text{diff}_{\text{all}}} = \{\zeta_{\text{diff}}^i\}_{i=1}^I \quad \& \quad I = \binom{L_{\text{old}}}{L_{\text{diff}}} \tag{13.27}$$

Within the ES approach the following l_2-norm of the least-squares error is defined as the cost function to be minimized over the line index number:

$$\hat{\zeta}_{\text{diff}} = \text{argmin}_{i=1:I} \left\{ \min_{\{x_l\}} \left\| \theta_{\text{diff}} - \left(\sum_{l \in \zeta_{\text{diff}}^i} B_{\text{old}}^{-1} x_l m_l \right) \right\|_2^2 \right\} \tag{13.28}$$

Due to high computational cost, this approach would not be feasible in the case of large-scale power grids. On the other hand, such a method is only applicable in the case of single or at most double line outages. In order to address this issue the fol-lowing sparse interpretation has been suggested and adapted in References 32 and 33.

Roll up over all $l \in \zeta_{\text{diff}}$ in (13.26), we have

$$\theta_{\text{diff}} = B_{\text{old}}^{-1} M x + B_{\text{old}}^{-1} \eta \rightarrow y = Ax + n. \tag{13.29}$$

Assuming that in general, the total number of simultaneous line outages, ζ_{diff}, may occur in a PN is a small subset of all lines ($L_{\text{diff}} \ll L$ or L_{old}) the vector $x \in R^L$ can be considered as a sparse vector (Definition 13.1—Figure 13.9(a)) that is defined as follows:

$$x = \begin{cases} x_l & \text{if} \quad l \in \zeta_{\text{diff}} \\ 0 & \text{otherwise} \end{cases}. \tag{13.30}$$

Given that in (13.29), the fat measurement matrix $A \in R^{N \times L}$ has the following format: $A = B_{\text{old}}^{-1} M, x$ is a sparse vector to be recognized, and n is the perturbation vector, solving (13.29) for sparse outage vector x can be construed as a SRP termed

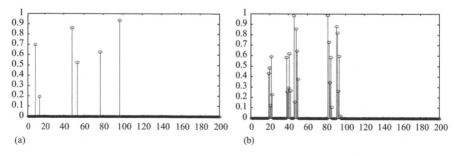

Figure 13.9   (a) Structure of a 6-sparse signal and (b) a (25,5)-clustered-sparse signal

POI-SRP or Distributed-POI-SRP. This sparse formulation can easily be extended in the case of limited number of PMUs. The change is happened in the number of measurements that we have and in the dimension of the corresponding sensing matrix A, accordingly. Consider a situation in which only N_I buses out of N total buses of the network have been guarded by a PMU. By removing the corresponding immeasurable bus phase angles from measurement vector y in addition to corresponding rows in the inverse of the nodal-admittance matrix B_{old}^{-1}, as stated in Reference 32, (13.26) can be reformed as follows:

$$\theta_{diff}^I = [B_{old}^{-1}]^I \sum_{l\in\zeta_{diff}} x_l m_l + [B_{old}^{-1}]^I \eta \tag{13.26. I}$$

where $[B_{old}^{-1}]^I$ is formed by the rows corresponding to N_I measurable electrical buses within the inverse matrix. In such a situation, the measurement matrix A will change its dimension to $R^{N_I \times L}$. From SR viewpoint, this means that we should solve an SRP with even a smaller number of measurements; however, from SR theorem, the minimum number of measurements needed for perfect recovery in this POI-SRP must be at least proportional to $(L_{diff} \log(L/L_{diff}))$. Thus, in the case of limited number of PMUs in the PN, we may be restricted with the simultaneous number of outages that can be recovered correctly within POI-SRP formulation.

13.4.1.2 Binary POI-SRP

Utilizing the previous formulations and a similar notation as POI-SRP, we introduce a formulation named Binary POI-SRP. Considering $\theta_{new} = \theta_{diff} + \theta_{old}$, $B_{new} = B_{old} - B_{diff}$ and using notation (13.22) for B_{diff}, we have

$$B_{diff} = \sum_{l\in\zeta_{diff}} b_l m_l m_l^T = MW_b \text{diag}(x_b) M^T \tag{13.31}$$

where $x_b \in \{0,1\}^L$ such that

$$x_b = \begin{cases} x_b[l] = 1 & \text{if } l\in\zeta_{diff} \\ 0 & \text{otherwise} \end{cases}, \tag{13.32}$$

fill in (13.29) for B_{diff} results in

$$B_{old}\theta_{diff} = B_{diff}\theta_{new} + \eta = MW_b \text{diag}(x_b) M^T \theta_{new} + \eta$$

$$= MW_b \text{diag}(M^T \theta_{new}) x_b + \eta. \tag{13.33}$$

Solving this equation with respect to θ_{diff}, we have

$$\theta_{diff} = B_{old}^{-1} MW_b \text{diag}(M^T \theta_{new}) x_b + B_{old}^{-1}\eta \tag{13.34}$$

Considering $A_b = B_{old}^{-1} MW_b \text{diag}(M^T \theta_{new})$, we can reach the following new sparse formulation for POI, in which the signal x_b to be recovered is a binary sparse signal.

$$y_b = A_b x_b + n. \tag{13.35}$$

Index b stands for binary representation of POI so we call this problem as Binary POI-SRP. This new formulation can be assumed as an alternative SRP to POI-SRP. We discuss how this new formulation can SRP solvers to deal with the dynamic range issue.[10]

13.4.1.3   Structured POI-SRP

By expanding the outages, the neighbor transmission lines become highly likely to get involved, rather than those once which are far. In other words, in the case of cascade faults, line outages are developed within a structural pattern. From SR perspective, this means that the corresponding outage vector x will have a clustered format.

Definition 13.6 (clustered-sparse signal [9]): *A signal $x \in R^N$ is called (K;C)-clustered sparse if it contains K nonzero coefficients such that, all of the nonzero coefficients are distributed inside C splintered clusters with arbitrary locations and sizes (Figure 13.9(b)).*

We may call such a situation as structured multiple outages and rename the POI-SRP into structured-POI-SRP, accordingly. In order to address such an SRP, one can utilize the clustered sparsity idea and the MCOMP SR algorithm [9,33]. MCOMP adapts the support selection step of the OMP for structured sparse signals as follows:

3.1. finding the boundaries of the support:

$$Upper\ Bound = min((sup^l + m - 1), 1))$$
$$Lower\ Bound = max((sup^l - m + 1), N))$$

3.2. modified extend support:

$$sup^l = \{Lower\ bound, \ldots, sup^l, \ldots, Upper\ Bound\}$$

---

where m is the maximum cluster size (please refer to References 9 and 33 for details on clustered sparse signals, clustered OMP (COMP), and MCOMP algorithms).

13.4.2   *Measurement matrix structure and its properties in POI-SRP*

Comprehensive study on a variety of IEEE standard test-beds has revealed that the high coherence of the corresponding measurement matrix A is an important practical issue within the POI-SRP formulation. Taking a look at the corresponding

---

[10]The dynamic range for a signal $x \in \mathbb{R}^L$ is defined as $|\max(x_i)|/|\min(x_i)|$. It has been discussed in the CS literature that, if this factor is happened to be large, as a result, it can significantly affect the SRP solver algorithms performance.

measurement matrix A of IEEE 118-Bus and its Grammian[11] matrix (Figure 13.10(a) and (c), respectively), one can clearly observe the high pairwise average correlation among the matrix columns. In this situation, the coherence (13.6) is equal to its highest possible value, means "1." In the POI-SRP, the structure of the measurement matrix is dictated by the mathematical formulation of the problem. As a result, one may reduce the high coherence issue by making a mathematical change in the formulation of the measurement matrix. We suggest a modification based on the matrix algebra that replaces the singular value decomposition[12] of the matrix B_{old}^{-1} in (13.29). Following this replacement, we arrive at the following alternative formulation for (13.29):

$$y_{\text{svd}} = \Sigma^{-1} U^T \theta_{\text{diff}} = V_{(B-1)}^T Mx + V_{(B-1)}^T \eta \rightarrow y_{\text{svd}} = A_{\text{svd}} x + n. \tag{13.36}$$

where the new measurement matrix A_{svd} is defined as $A_{\text{svd}} = V^T M$. The structure of the corresponding IEEE Bus-118 measurement matrix A_{svd} and its Grammian have been visualized in Figure 13.10(b) and (d), respectively. Although the coherence value is still unchanged, implementing the matrix decomposition step, the average pairwise correlation among the columns can be considerably reduced. Moreover, another advantage of this formulation is that the residual term η preserves its white Gaussian distribution once multiplied by a unitary matrix $V_{(B-1)}^T$, whereas in (13.29), multiplying with B_{old}^{-1}, the residual term η is changed into a colored noise which results in a roughly tougher situation for SRP solvers. On the other hand, taking a look at the structure of the corresponding Grammian of the matrix A_{svd}, the following important point is revealed: since the coherence of the measurement matrix is exactly "1," there should be at least one pair of columns which are exactly the same. The corresponding index set of such columns can be found as follows: $L_R = \{(66,67),(75,76),(85,86),(98,99),(123,124),(138,139),(141,142)\}$. One may ask, which factor makes such an issue in our measurement matrix? Take a brief look at the same-index lines in the IEEE Standard 118 Bus, we see that these lines connect the following set of buses, respectively; $N_R = \{(49,42),(49,54), (56,59),(49,66),(77,80),(89,90),(89,92)\}$. Moreover, we can observe "in lieu of a

---

[11]In the linear algebra, the Gramian matrix of a set of vectors $a_1, \ldots, a_n$ within an inner product space is defined as the Hermitian matrix of pairwise inner products. The Gramian entries are calculated by $G_{ij} = \langle a_i, a_j \rangle$. In the case of finite-dimensional real vectors in which a usual Euclidean dot product is validated, the Gramian matrix can be simply calculated from $G = A^T A$ (or $G = A^\dagger A$ in the case of complex vectors in which $\dagger$ defines the conjugate transpose), and the matrix A is formed by the vectors a_k as its columns.

[12]Suppose A is an $m \times n$ matrix in which all the components are belonging to either the set of real numbers or the set of complex numbers. The following matrix decomposition can be defined for A:

$$A = U \Sigma V^H$$

where the matrix U is an $m \times m$ unitary matrix, Σ is an $m \times n$ diagonal matrix with non-negative real numbers on the diagonal, and V is an $n \times n$ unitary matrix. Term H denotes the conjugate transpose. Such a decomposition is called a singular value decomposition of A. This matrix decomposition plays an important role in solving LS problems, especially once we face with ill-conditioned systems of equations.

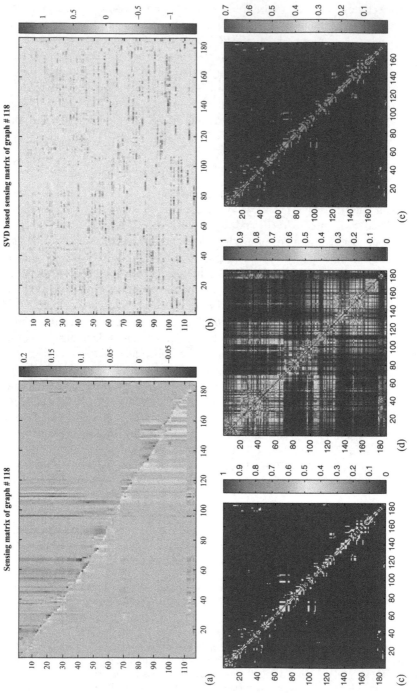

Figure 13.10 Structure of the normalized measurement matrix (a) A and (b) A_{svd} for IEEE Standard 118 Bus. Corresponding Grammian of the (c) matrix A, (d) matrix A_{svd} and (e) matrix A_{svd} (after network model modification)

unique transmission line, two distinct parallel lines are connecting each pair of these buses within the PN structure." This fact results in exactly the same pairs of columns in the incidence matrix M and as a result, produces exactly the same columns in the measurement matrix A as well [33]. To address this issue, a network model modification has been suggested in Reference 33, in which one can simply replace the corresponding positions in the nodal-admittance matrix with $b_T = b_1 + b_2$. Figure 13.10(d) shows the Grammian of the measurement matrix $A_{QR_{118bus}}$, after network model reduction procedure.

13.4.3 Addressing high coherence by support selection modification: BLOOMP algorithm

We already discussed a couple of techniques which can reduce the coherence and average correlation pattern of a measurement matrix in POI-SRP. However, a new greedy SR algorithm has been introduced in Reference 49, which can deal with highly coherent measurement matrices by modifying the support selection step of OMP; named band exclusion.

Roughly speaking, in band exclusion technique, whenever the correlation between two columns of the measurement matrix is high or they are in each other's *coherence-band*, as stated in Reference 49, one of them is not selected within the true support. BLO-based OMP (BLOOMP) can be used to deal with the high coherence issue in POI-SRP. The support selection modification step of this algorithm is as follows:

Coherence band $\alpha > 0$,
2.identify support indicator:

$$sup^l = \{argmax_j|h^l(j)|\}, j \notin B_\alpha^{(2)}(SUP^{l-1})$$

3.update the support:

$$SUP^{l+1} = LO(SUP^l \cup sup^l)$$

Local Optimization (LO):
require: *A, y,* ***Coherence band*** $\alpha > 0, SUP^0 = \cup\{sup^l\}$ *for* $i = 1:k$
repeat: for $i = 1:k$
1. $y^i = argmin_{z:supp(z)=(SUP^{i-1}\setminus\{sup^l\})\cup\{\lambda^i\}, \lambda^i \in B_\alpha(\{sup^l\})}||p - Az||_2$
2. $SUP^i = supp(y^i)$
output: SUP^k

13.4.4 Case studies and discussion

In the following set of simulations, the IEEE Standard 118-Bus system was considered as the case study. The full technical descriptions and specifications of this PN can be found in the MATPOWER toolbox. In order to implement the DC power-flow-based formulations, following a fault event the system is considered to be settled down into a quasi-stable state (also refer to References 32 and 48). Moreover,

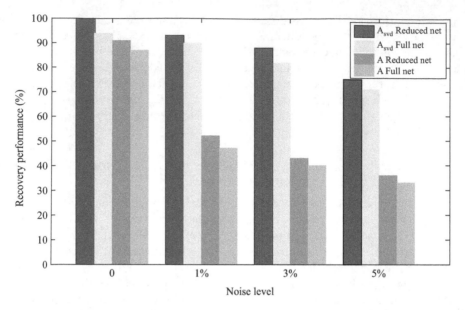

Figure 13.11   Single-line outage identification results for IEEE Standard 118-Bus system

in order to solve the corresponding power-flow equations, the MATPOWER solver assumes the PN full connectivity as a mandatory perquisite. Thus, the outages are not allowed if they result in an islanding situation for any bus within the PN.

13.4.4.1   Single-line outage

We initially investigate the line outage detection, using the aforementioned sparse POI framework, in the case of all possible single line outages. Figure 13.11 illustrates the corresponding identification results in which different noise (residual) levels, η, have been tested.[13] Figure 13.11 indicates the positive effect of the line reduction model, in which due to similar columns elimination, the OMP SRP solver illustrates a better recovery performance. Especially, in the noise-free situation, the OMP solver achieves up to 100% identification performance. On top of that, implementing the matrix decomposition step, within the POI-SRP formulation, the overall identification result is considerably improved. Moreover, within the higher noise levels the colored noise will have a destructive effect on the OMP recovery performance.

13.4.4.2   Multiple-line outages

To investigate the multiple-line outage situations, the following two different scenarios are considered: distributed and structured multiple outages. Suppose that the following three individual outages happen among the IEEE Standard 118-Bus system {transmission line index: 6, 104, and 118}. As a result, a 3-sparse

---

[13]Specifically, the standard deviation of the residual vector, η, has been set into zero (accounts for the noise-free situation) and equal to 1%, 3%, or 6%, of the average pre-event power injection.

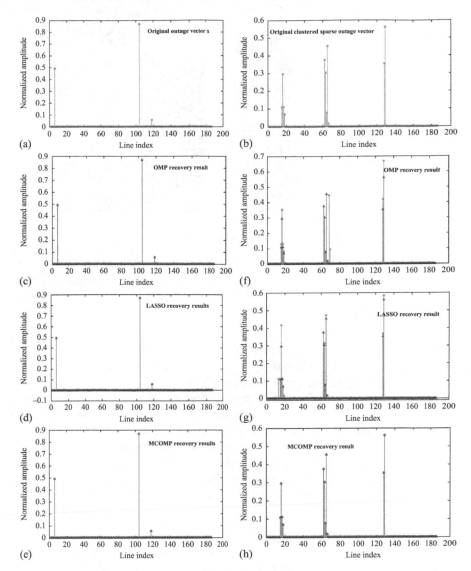

Figure 13.12   *Power-line outage identification results for 3 SRP solvers in the case of separated multiple outages ((a) and (c)–(e)). Power-line outage identification results for three SRP solvers in the case of structured multiple outages ((b) and (f)–(h))*

outage vector x_3 (Figure 13.12(a)) should be identified through the POI-SRP. Figure 13.12(c)–(e)[14] is representing the corresponding recovery results for LASSO, OMP, and MCOMP algorithms, respectively. Due to small sparsity level,

---

[14]In the rest of this section, red bars illustrate the true signal x, whereas blue bars are indicating the estimated output of any recovery algorithm.

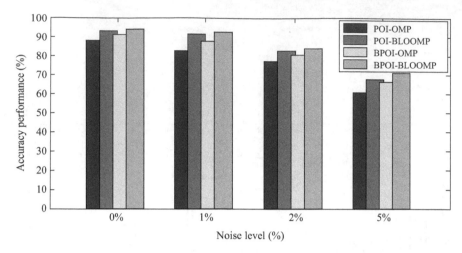

Figure 13.13   Comparison of recovery performance of OMP vs BLOOMP
method (for 118-Bus IEEE Standard system) in the case of 1,000
randomly selected 20 line outages for POI-SRP and BPOI-SRP vs
noise level

in addition to aforementioned implemented matrix modifications, all these SRP solvers are able to identify the faults locations with high accuracy.

Now consider a cascade failure situation in which following an outage in one of the feeder transmission lines of electrical buses 16, 51 and 105, the outage has been expanded into other neighbor lines as well. From sparse recover perspective, we face with a clustered sparse outage vector $x_{(11,3)}$ as illustrated in Figure 13.12(b). Figure 13.12(f)–(h) represents the recovery performance of the LASSO, OMP, and MCOMP, respectively. Although in the case of separated outages the identification performance is almost the same, exploiting the preknowledge of structural outage pattern in the sparse outage vector, MCOMP outperforms the other SRP solvers in the case of Structured-POI-SRP.

Large scale multiple line outages
In the case of single, or small-scale multiple outages, the matrix decomposition formulation in addition to network line reduction technique helps the OMP algorithm to perform adequately in solving the distributed version of POI-SRP which have been introduced in References 32 and 33. However, once the number of outages and the noise level increase, the high coherence issue is highlighted and the overall situation becomes tough. The more number of outage lines we have, the higher the probability of highly correlated columns appearance within the outage vector support. These problems can make serious combined challenges for traditional SRP solvers in the case of large-scale distributed-POI-SRP.

In order to have a comprehensive evaluation, Figure 13.13 shows the recovery performance percentage of algorithms vs noise level for 1,000 randomly chosen 10-sparse outage vectors. Figure 13.13 indicates that the POI results can be improved by applying the BPOI formulation in addition to BLOOMP technique. The standard deviation of the residual vector, η, is set to zero (account for noise-free situation) and 1%, 2%, or 5% of the average pre-event power injection. If the nonzero elements are happened to be in each other coherence-bound, we will be limited with the overall performance of the BLOOMP. To address the high coherence issue, in the case of Structure-POI-SRP, one may use an algorithm composed of BLOOMP and MCOMP that we refer to as BLOMCOMP (Algorithm 13.2); this algorithm adds the preknowledge of clustered sparsity to BLOOMP.

Figure 13.14(a) shows that when the OMP algorithm fails, in a sample case with a 20-sparse outage vector x and measurement noise set to 1%, there we have a

---

Algorithm 13.2. Band-excluded Locally Optimized MCOMP (BLOMCOMP)

---

require: matrix A, measurements p, Coherence band $\alpha > 0$, maximum cluster size m, stop
initialize: $r^0 = p, y^0 = 0, l = 0, SUP^0 = \phi$
   repeat
      1.match: $h^l = A^T r^l$
      2.identify support indicator:

$$sup^l = \left\{ argmax_j |h^l(j)| \right\}, j \notin B_\alpha^{(2)}(SUP^{l-1})$$

      3.update the support:

        $\boldsymbol{Upper\ Bound} = min((sup^l + m - 1), 1))$
        $\boldsymbol{Lower\ Bound} = max(sup^l - m + 1), K)$
        $\widehat{sup}^l = \left\{ Lower\ bound, \ldots, sup^l, \ldots, Upper\ Bound \right\}$
        $SUP^{l+1} = LO(SUP^l \cup \left\{ \widehat{sup}^l \right\})$

      4.update signal estimate:

        $y^{l+1} = argmin_{z:supp(z) \subseteq SUP^{l+1}} ||p - Az||_2$
        $r^{l+1} = p - Ay^{l+1}$
        $l = l + 1$

   Until stopping criterion met
output: $\hat{y} = y^l$
Local Optimization Procedure
require: A, y, Coherence band $\alpha > 0, SUP^0 = \cup\{\widehat{sup}^i\}$ *for* $i = 1{:}k+1$
repeat: for $i = 1{:}k+1$
   1. $y^i = argmin_{z:supp(z)=(SUP^{i-1} \setminus \{\widehat{sup}^i\}) \cup \{\lambda^i\}, \lambda^i \in B_\alpha(\{\widehat{sup}^i\})} ||p - Az||_2$
   2. $SUP^i = supp(y^i)$
output: SUP^{k+1}

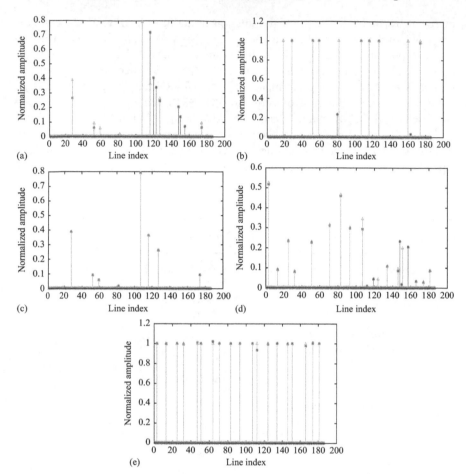

Figure 13.14   (a) and (c) OMP and BLOOMP recovery performance in POI-SRP,
for a 10-sparse outage vector (noise level 1%), respectively.
(b) OMP result for the same case in BPOI-SRP. (d) and
(e) Recovery results of BLOOMP in POI and BPOI SRPs, for a
19-sparse outage signal in 2% noise

couple of false positive identified nonzero samples. Applying the BPOI formulation, we can address the dynamic range issue of the signal, so we got a better
recovery result in Figure 13.14(b). Finally, BLOOMP has addressed almost all of
these issues, and we get a perfect reconstruction results in Figure 13.14(c); even
without using BPOI setup. In order to emphasize the performance of BPOI formulation, we have illustrated a harder case in which the sparsity level of the outage
signal x is set to 19, and the noise level is 2%. Although the BLOOMP could recover

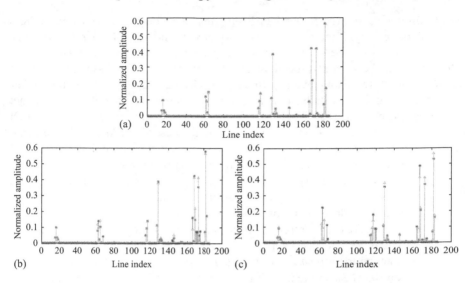

Figure 13.15   *Recovery performance of (a) BLOCOMP vs (b) MCOMP vs*
(c) OMP method for a sample (33,10)-clustered sparse outage
vector in IEEE 118-Bus

most of the nonzero elements within the true support of the outage signal x
(Figure 13.14(d)), but couple of errors are also appeared. This issue can be solved by
applying the BLOOMP in BPOI formulation (Figure 13.14(e)).

Figure 13.15 illustrates the performance of BLOCOMP (a) vs MCOMP (b) and
OMP (c), in a highly complex situation where the outage vector x to be recovered is a
(33,10)-clustered sparse signal. In structured multiple outages in addition to high
between-cluster correlation issue, MCOMP recovery performance is affected by high
within-cluster dynamic range of the nonzero elements in each particular cluster in the
signal as well. BLOCOMP can overcome this challenge utilizing the LO procedure
to peak the best center for each cluster. Although the BLOCOMP illustrates adequate
performance, we can still improve the results using BPOI formulation if needed. As
can be seen, the maximum size of the cluster should be given to the MCOMP and
BLOCOMP as a preknowledge; however, this can be easily determined from preknowledge of the local structure of the incidence matrix of the PN graph.

13.5   Sparse recovery-based smart-grid topology identification

A drawback with the aforementioned POI-SRP formulation and many other recent
POI approaches (such as those that have been introduced in References 32, 33, and
44–48) is their dependency on the nodal-admittance or grid topology matrix B as a
known factor in formulation, for example. A variety of issues may occur due to
this dependency: the first problem is related to the rank deficiency issue in the

nodal-admittance matrix B. As has been mentioned before, in order to address the rank deficiency issue, one bus is normally considered as the reference bus. However, due to cyber-attacks or other precision deficiency problems in the reference bus, the overall procedure can be affected. Another issue is related to the unreported switching events, which can change the structure of the nodal-admittance matrix B. In such a situation, if the last structural data of the PN has not been updated prior to the identification algorithm operating time instant, the POI results can be corrupted. On the other hand, to the best of our knowledge, most of the previously developed POI methods do not account for the random behavior of renewable energy sources, nonlinearity strengthened by power electronic interfaces [50–52], and the uncertainty in loads. Also, an unrealistic assumption in these formulations is the connectivity of the power grid graph, which means that power outages may not result in islanding in any part of the network.

In order to address the aforementioned imperfections, we are going to represent a totally different CS–SR-based approach for PN topology identification that can also be used for outage identification. In the new approach, the presented reformulation relies only on the measurements from system parameters and does not need any a priori information about the topology of the network as stated in References 34 and 35. Within this approach, the TI problem formulation is changed in such a way that the solution of the problem is the structure of the nodal-admittance matrix B of the PN itself (PNTI-SRP). Case studies using standard IEEE test-beds show that the proposed method represents a promising new strategy for line change, fault detection, and monitoring issues in SGs.

13.5.1 Sparse setup of topology identification problem

Given an interlocked PN of N electrical buses, let the measurements of the active power and the phase angle of node i be associated with the two following time series of M sample times:

$$p_i(t) \quad \text{for } t = 1, 2, \dots, M \tag{13.37}$$

$$\theta_i(t) \quad \text{for } t = 1, 2, \dots, M. \tag{13.38}$$

As mentioned before, in a linear DC model, the active power injected into a distinct bus i follows the superposition law (13.17); this means, for each node in the network and in each sample time t, we have

$$p_i(t) = \sum^{j} p_{ij}(t) = \sum_{j \in N_i} b_{i,j}\big(\theta_i(t) - \theta_j(t)\big). \tag{13.39}$$

As mentioned before, $b_{i,j}$ is the susceptance along the line $l_{i,j}$ under the DC model regime, and N_i is the set of neighbor buses connected directly to bus i. Since $b_{i,j} = 0$ for $j \notin N_i$, we can extend (4.3) as follows:

$$p_i(t) = \Sigma^{j} p_{ij}(t) = \Sigma_{j \in S_N^i} b_{i,j}\big(\theta_i(t) - \theta_j(t)\big) + r_i(t) + n_i(t), \tag{13.40}$$

where S_N^i is the set of all nodes in the network except node i (so the set S_N^i contains $N-1$ nodes), r_i is a possible leakage of active power in node i itself, and n_i accounts for the measurement noise. Now we ignore the time-sample notation and use the following vector notation instead:

$$p_i = A_i x_i + r_i + n_i, \tag{13.41}$$

where $p_i, r_i, n_i \in R^M$, and

$$A_i = \left[a_{1,i}^T, \ldots, a_{i-1,i}^T, a_{i+1,i}^T, \ldots, a_{K,i}^T \right] \in R^{M \times N-1} \tag{13.42}$$

$$a_{j,i}^T = \left(\theta_i(t) - \theta_j(t) \right) \in R^M \quad \text{for } t = 1, 2, \ldots, M \tag{13.43}$$

$$x_i = \left[b_{i,1}, \ldots, b_{i,i-1}, b_{i,i+1}, \ldots, b_{i,K} \right]^T \in R^{N-1}. \tag{13.44}$$

In our formulation, each column of the matrix A_i represents the difference between phase angles of node i vs each node $j \in S_N^i$ in the network for M samples of time. It has been discussed in the literature [44,53] that due to probabilistic behavior of load and renewable sources, this difference can be approximated with a Gaussian random variable [44,53]. Vector x_i is a sparse vector, with all of its components equal to zero except a small portion, which are located in those positions corresponding to each one of the neighbor nodes of node i. Replacing power vector p_i with our global notation for measurement vector y_i, and summing r_i and n_i (which we assume to be a vector of white Gaussian noise), we end up with the following equation for each bus:

$$y_i = A_i x_i + \eta_i. \tag{13.45}$$

Therefore, the PNTI problem can be viewed as the estimation of all vectors $\{x_i\}_{i=1}^N$ that best match the observed measurements $\{y_i, \theta_i\}_{i=1}^N$. To solve such a problem, one may define the following optimization problem:

$$\min_{\{\hat{x}_i\}_{i=1}^K} \sum_{i=1}^K A_i \hat{x}_i - y_{i2}^2. \tag{13.46}$$

This problem can be solved individually for each node i. If A_i is full rank, the number of measurements, M, exceeds the number of unknowns, $N-1$, and the variance of the noise vector, η_i, is small enough, this optimization problem can be solved using LS techniques. Since the value of N depends on the number of buses in the grid, a large number of measurements M will be needed to solve such a problem in the case of large PNs; this will make the calculations complex and significantly increase the time cost. In order to avoid this problem, we suggest using CS–SR techniques instead.

Due to sparsity fact in the nature of vector x_i, we can look at (13.45) or (13.46) as a SRP for an K-sparse vector $x_i \in R^{N-1}$ from a set of observations $y_i \in R^M$, where $A_i \in R^{M \times N-1}$ is the sensing matrix, $\eta_i \in R^M$ is a vector of white Gaussian noise, and K is the number of electrical buses that are directly connected to the one individual bus i for which we are solving the problem.[15] In general, we want $M \ll N$ so the problem can be solved using a reasonable set of observations in huge power grids.

After solving this problem for each sparse vector x_i corresponding to each node i, we can concatenate all of the sparse vectors together, form the weighted Laplacian matrix B, and the process is completed. Once we rewrite (13.46) in the form of P_1 (13.9) or NP_1 (13.10), this problem is termed PNTI-SRP. Next we can solve the PNTI-SRP using optimization-based sparse solvers or greedy-based algorithms. Before representing some case studies on PNTI-SRP, we are going to discuss a special feature in nodal-admittance matrix of PNs that can be used to develop a new SR formulation for PNTI [35]. This new formulation can help to increase the overall identification results using even a smaller number of measurements from system parameters.

13.5.1.1 l_1 Minimization versus reweighted l_1 minimization PNTI-SRP

As has been mentioned before, it is well discussed in literature that we can reconstruct a sparse vector with high probability from a highly overcomplete sets of linear measurements using P_1 or NP_1. Variety of solvers are available which can find the correct solutions for these two problems. However, a couple of generalized methods have been introduced so far, which are able to exploit extra information to solve the SRP more efficient and more accurate even with smaller subset of measurements; such as COMP. From Figure 13.16, it can clearly be seen that most of the nonzero elements in matrix B are concentrated closely to the main diagonal elements. This means that, in PNTI-SRP, the occurrence rate of nonzero coefficients in recovery of each sparse vector x_i (correspond to each column of nodal-admittance matrix B) is

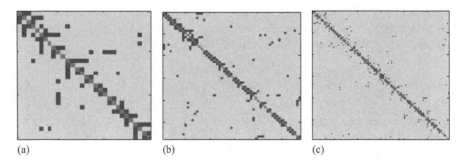

(a)   (b)   (c)

Figure 13.16   Corresponding Laplacian matrices of four of the standard IEEE test-beds, (a) IEEE Standard 30 Bus, (b) IEEE Standard 56 Bus, and (c) IEEE Standard 118 Bus (scaled between [−1, 1])

---

[15]Because $\theta_i(t) - \theta_j(t) = 0$ for $i = j$, S_N^i should include $N - 1$ nodes; in other words, we should keep the node i out of S_N^i since it produces a vector of zeros in the corresponding column of the matrix A_i. This means that we are not able to find the value of the parameter B_{ii} from the recovered vector x; however, according to the definition of the nodal-admittance matrix B, $B_{ii} = \Sigma_{j \in N_i} b_{i,j}$. Thus, after recovering the vector x_i, B_{ii} can be easily calculated.

higher among special locations. Exploiting this sparsity pattern, we can redefine P_1 as the following weighted optimization problem instead:

$$WP_1 : \hat{x} = \min_{x'} \sum_{n=1} \omega_{(n)} |x'_{(n)}| \quad \text{subject to} \quad y = Ax, \tag{13.47}$$

where $\omega_1, \omega_2, \ldots, \omega_N$ are non-negative weights. Similar to P_1, WP_1 is a convex optimization problem that can be solved as a linear program. Due to concentration of nonzero coefficients among the main diagonal elements, one may assign smaller weights near the diagonal elements to encourage those elements to be selected as the true support of the sparse signal x_i. The WP_1 can be solved within a single step using an initial preknowledge about the weights. However, in order to increase the accuracy of the final results, the problem is regularly solved through an iterative procedure called reweighted l_1 minimization (Rwl_1). The Rwl_1-based PNTI setup is detailed in Algorithm 13.3. This iterative algorithm (Rwl_1) updates the weights in each step based on the estimated sparse vector magnitudes from the previous step. In Algorithm 13.3, ε is defined as a stabilizer parameter that is used in order to obviate the effect of zero-valued components in x^l. It has been shown that in general the Rwl_1 recovery process tends to be reasonably robust to the choice of this parameter ($\varepsilon > 10^{-3}$ is suggested for practical situations). The diagonal matrix W^l is defined as follows:

$$W^l = \begin{bmatrix} \omega^l_{(1)} & \cdots & 0 \\ 0 & \ddots & 0 \\ 0 & \cdots & \omega^l_{(N)} \end{bmatrix} \tag{13.48}$$

Algorithm 13.3 Reweighted l_1-minimization based PNTI

---

require: stopping criterion, phase angle measurements, active power measurements, each of M sample times

$p_i(t), \theta_i(t) \quad \text{for } t = 1, 2, \ldots, M.$

1. form: measurement matrix $A \in R^{N \times M}$, and measurement vector y
2. Set counter l to zero and initialize $\omega^0_{(n)}$ *for $n = 1 : N$*
3. solve:

$$x^l = argmin \|W^{(l)} x\|_{l_1} \quad \text{subject to} \quad y = Ax$$

4. update weights:

$$\omega^{l+1}_{(n)} = \frac{1}{|x^l_{(n)}| + \varepsilon} \quad \text{for } n = 1 : N$$

5. Until stopping criterion met:
6. output: $\hat{x} = x^l$

13.5.1.2   Noisy reweighted l_1-minimization based PNTI

In short, to adapt the $\mathrm{R}wl_1$ algorithm in the case of noisy measurements, the following change should be made in the third step:

$$x^l = \mathrm{argmin}\|W^{(l)}x\|_{l_1} \quad \text{subject to} \quad \|y - Ax\|_{l_2} < \eta. \tag{13.49}$$

Additional discussion on WP_1 and $\mathrm{R}wl_1$, including a discussion of the recovery setup in the noisy case, is contained in Reference 54. There are a variety of optimization packages that can be used to solve P_1, NP_1, WP_1 and $\mathrm{R}wl_1$; examples include CVX in Reference 55 and NESTA in Reference 56.

13.5.2   Case studies and discussion

13.5.2.1   Setup

In this part, the identification performance of the proposed PNTI method is examined using compressive observations $(M < K)$ collected from the system parameters. An initial observation from node-wise comparison is that the identification performance is directly changed on the basis of the local pattern of sparsity. In all these simulations, we use the IEEE Standard-30 Bus, 118 Bus, and 300 Bus as case studies. These three PNs include 30 Buses, 47 power transmission lines, 118 Buses, 186 lines, and 300 Buses, 304 power transmission lines, respectively, and their detailed specifications can be found in the MATPOWER toolbox [57].

In References 44 and 53, it has been thoroughly discussed that due to the load uncertainty in the large-scale power grids both of the injected active and reactive powers can be characterized as random variables. Roughly speaking, the integrated load pattern from a large number of users, in addition to uncertainty caused by utilization of renewable resources, can be well modeled by Gaussian random variables. Moreover, following the linear relation in (13.39) the difference of the phasor angles $(\theta_i - \theta_j)$ for each sample of time can be well approximated by a Gaussian random variable. Consequently, we consider that during the observation time frame of the PNTI, the measurements of phase angle of node i for the time series of M sample times can be well estimated by a vector of Gaussian random variables, as stated in References 44 and 53. This random behavior is helpful in the recovery performance of SR techniques since it reduces the general coherence of the corresponding sensing matrix.

As a result, in order to generate the data, the system is fed by a Gaussian load, in which the MATPOWER toolbox is used for solving the power-flow equations in various demands and the resulting phase angle and active power measurements are applied as the input to the SRP. Each column of the matrix B represents one of the sparse vectors x_i which we are going to recover by solving an SRP. Figure 13.17 shows how the average coherence[16] (13.6) corresponding sensing matrix A of IEEE Standard-30 and 118 Bus systems changes while the number of measurements M increases. Following the increment in the number of measurements, this coherence

---

[16]Over 100 realizations of each PN overall 30/118 buses.

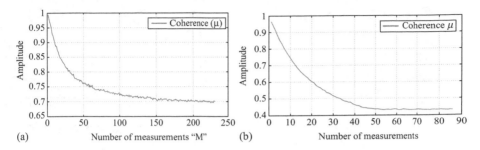

Figure 13.17 *The average value of Coherence vs the number of measurements overall 30/118 corresponding sensing matrices A of all of the nodes, where the curve is averaged over 100 realizations of each network*

measure almost approaches a lower bound, consequently. As has been mentioned before, within a smaller coherence metric, a larger value is permitted for S, which accordingly results in a broader class of signals that can be recovered. Regarding the presented sparse reformulation in Section 13.3.4.2, the structure of the sensing matrix is formed by the variation in bus phase angles. Thus, the closer the behavior of the phase angles is to being random, the stronger the SR guarantees, the greater the class of sparse vectors to be correctly estimated, and the better the overall performance of the recovery procedure.

13.5.2.2 Results

Within the network graph, optimization-based SRP solvers have different recovery performance for different nodes mainly because of the sparsity level of the corresponding signal (or in our PN, in-degree or the number of coming links to a node). Figure 13.18(a)–(e) is illustrating the recovery performance using BP (P_1), and reweighted l_1-based (WP$_1$) formulations for nodes 1, 2, 15, 3, and 130 of IEEE Standard 300-Bus, with sparsity level (in-degree), 3, 4, 6, 7, and 9, respectively. For a certain number of measurements, the corresponding success rate is computed over 100 realizations of the PN. From the sparsity level viewpoint, node 130, corresponds to the most complicated signal x_{130} to be recovered in IEEE test-bed 300 bus. As it can be seen, the successful recovery rate altered once the signal sparsity level augments. As a result, the 3-sparse signal x_1 (triple in-connection bus 1) can be recovered from a smaller set of measurements than 6-sparse signal y_{15} (6 in-connection bus 15) with a higher probability. Moreover, Figure 13.18 illustrates how the reweighted l_1-based (WP$_1$) formulations exploit the structured sparsity pattern in order to outperform the conventional BP formulation; especially in the case of nodes with higher in-degree orders. In general, the lack of measurements destructively affects the recovery performance for any SRP solver, however, the Rwl_1-based algorithm (Rwl_1), which works based on slightly different assumptions, is less affected. In the case of high in-degree order nodes such as node 3, we are limited to a certain recovery performance till the number of available measurements M reaches to

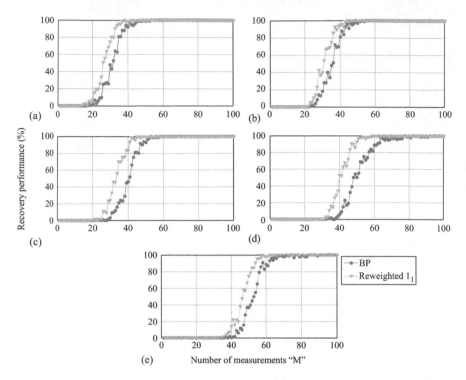

Figure 13.18 Recovery rate (%) comparison of nodes (a)–(e): 1 (in-degree 3),
2 (in-degree 4), 15 (in-degree 6), 3 (in-degree 7), 130 (in-degree 9),
respectively. Success rate is calculated over 100 realizations of the
IEEE Standard-300 Bus for a given number of measurements

an adequate value, in which the PN coherence metric arrives at a suitably small level. If this level is never to be reached within a particular PN, we would be restricted in the degree of nodes that can be recovered using this technique.

Moreover, considering Definition 13.6 for clustered sparse signals, within the corresponding nodal-admittance matrices of the standard IEEE test-beds (Figure 13.16), a subset of columns can be considered as clustered sparse signals. Along with this feature, among an interconnected system such as a PN, the parameters of different buses (specifically in neighbor buses) may share a level of solidarity. This possible correlation among the phase angles can affect the general behavior of the coherence (13.6) of the resulted sensing matrix in the PNTI-SRP and may cause a different recovery performance beyond the network nodes. In order to address the coherence issue, one may implement the BLOMCOMP algorithm (Algorithm 13.2). In order to investigate these features, the recovery performance of greedy-based algorithms: OMP, MCOMP, and BLOMCOMP have been compared in the case of IEEE Standard 118 Bus.

Figure 13.19(a) and (b) represents the recovery performance of three SRP solvers OMP, LASSO, and MCOMP for two and five sparse signals x_{22} and x_{105}

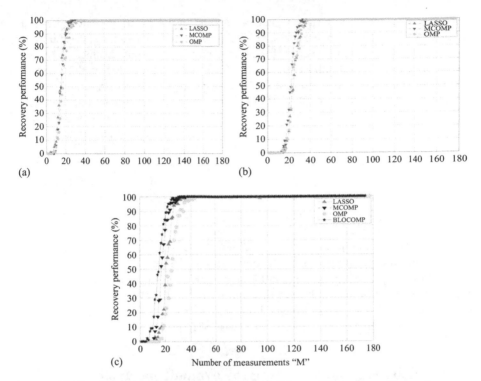

Figure 13.19 *Recovery rate (%) comparison of three SRP solvers for nodes: (a) 22: 2 in-degree and (b) 105: 5 in-degree. (c) Recovery rate comparison of four SRP solver for node 17: 6 in-degree. An initial value of m = 4 is chosen in MCOMP and BLOMCOMP. Success rate is calculated over 100 realizations of the network for a given number of measurements*

corresponding to columns 22 and 105 in the nodal-admittance matrix of IEEE Standard 118 Bus system. Comparing the corresponding result for each individual algorithm, the 2 in-connection node 22 can be recovered using a smaller number of measurements than the 5 in-connection node 105. Moreover, it can be clarified, how the presence of the clustered sparsity pattern can help the MCOMP algorithm to illustrate a better performance compare to conventional OMP and optimization-based LASSO, in the case of node 22, where the corresponding signal y_{22} is a (2,1)-clustered sparse signal, and node 105, where y_{105} is a (5,2)-clustered sparse signal. Figure 13.19 indicates how the BLOCOMP solver outperforms the three aforementioned solvers over node 17, in presence of a complicated and inter-connected local topology with a higher coherence in the corresponding sensing matrix. Finally, Figure 13.20(a) and (b) demonstrates the network-wide topology recovery error for the BP, and Rwl_1 solutions in the case of IEEE Standard-30/300 Bus, over 100 realizations of the network, respectively. For each curve, the vertical axis represents the percentage of trials in which all 30/300 columns of the corresponding nodal-admittance matrix B (and, as a result, the network-wide topology) are ideally recovered. As has been discussed, theoretically, the minimum number of measurements needed for perfect recovery M must be merely proportional to $K \log(N/K)$. In the case of the IEEE Standard-30 Bus, node 6 has the highest in-degree of 7, as a result one may expect the perfect network-wide recovery to be happened at $M = 7 \log(30/7) \approx 5$ or more. Moreover, based on our discussion in Section 13.2.2, for a wider PN with more number of buses, we may need more measurements to reach to the perfect network-wide topology identification. For instance, beyond the IEEE Standard-300 Bus, node 276 has in-degree 12, where the perfect recovery is theoretically possible when $M = 12 \log(300/12) = 16.77$. However, as it can be observed from Figure 13.20, practically, the whole network topology for these two systems can be recovered from $M \approx 20/60$ or fewer number of measurements per bus, respectively. This number is mainly affected from the coherence behavior of the sensing matrices over all of the nodes as well as the additional measurement noise. Finally, although the number of measurements needed for full recovery increases by the network size, N (number of buses), but this is not a linear relationship. For example, albeit IEEE Standard-300 Bus is 10 times bigger than IEEE Standard-30 Bus in scale, we would need almost fewer than three times more number of measurements for full recovery. This last point emphasizes on the merit of the sparse PNTI setup in the case of large scale power grids.

13.6   Modeling and analyzing the dynamic behavior of transmission lines using structured sparsity

Due to fast transitions and regarding the high mathematical modeling complexity, transmission line dynamic behavior tracking and estimation are classified as intricate monitoring challenges in SGs [58,59]. Climate situations and ambient temperature are, however, continuously affecting and changing transmission line parameter behavior. As a result, developing an accurate real-time line parameter monitoring and tracking methodology is a crucial and complicated issue in future SGs [59]. Supervisory control and data acquisition system (SCADA) is the most well-known data collection system widely used in PN monitoring, security, and so on. However, the typical sampling rate of SCADA is too slow to capture the dynamic and transient events in a PN and is mostly used in steady-state analysis. In order to analyze and track the dynamic operation of PNs, PMUs were introduced in the mid-1980s. Due to much faster sampling rates, PMU technology has found many applications in online monitoring and control analysis of dynamic events in PNs [3]. Nowadays a vast amount of work is happening in power sensor technology especially on PMU technology. In near future, this technology will change from a fully hardware structure to a software-based technology with much lower cost of units. Low-cost PMU architecture in addition to virtualizing the PMUs have been addressed in References 61–63 and the references therein.

Beyond the PMU technology recent applications, the synchronized phasor data, that is the PMU data, have been used for line parameter estimation in

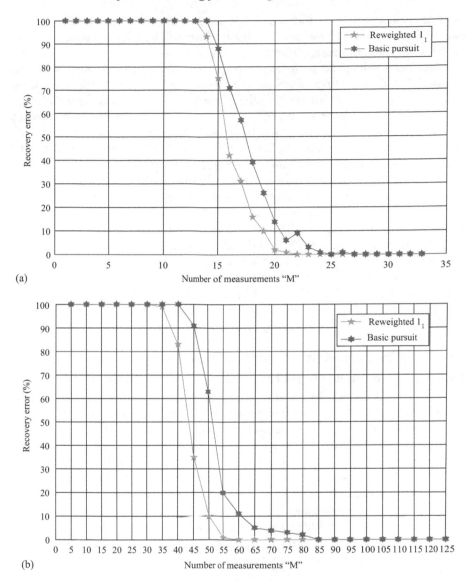

(a)

(b)

Figure 13.20   Full-network topology identification error for two SRP solvers for (a) the standard IEEE Standard-30 Bus, and (b) the standard IEEE Standard-300 Bus over 100 realizations of the network

References 60–67. More recently, a combined data approach has been developed in Reference 59 that utilized both the SCADA and PMU data in order to model the line parameter estimation problem. The major drawback with most of the aforementioned approaches is the nonlinearity and mathematical complexity of the presented formulation. Some works try to simplify the mathematical calculation; however, most of them still require nonlinear LS techniques in order to solve the parameter estimation and tracking problem. This nonlinearity will be a serious issue for online network-wide monitoring applications in the case of large-scale PNs with thousands of lines. As a result, the performance of most of these methods have been only reported for individual lines. However, the need of a general framework for modeling the overall dynamic line behaviors in the whole of the PN is still an open challenge.

In this section, in order to address the time cost and algorithmic complexity issues, we introduce a sparse-based framework for monitoring and modeling the dynamic behavior of transmission lines in SGs. First, a mathematical formulation for line parameters dynamical modeling (LPDM) is presented. Within the LPDM, a new dynamic parameter called *transmission line dynamic index coefficient* (TLDIC) is defined, based on the current wave analysis in transmission lines, where dynamical LPDM is interpreted as tracking the behavior of these DCIs in real-time. Next, using a new port-Hamilton model of the PNs (developed in Reference 68), in addition to the inherent sparse nature of PN models, we mathematically reformulate the LPDM problem as an SSRP. Finally, using an extended greedy SR method, named block orthogonal matching pursuit (BOMP) (introduced in Reference 69), the LPDM-SSRP is solved for TLDIC. The simulation results on IEEE standard testbeds indicate that the presented line dynamic behavior modeling framework can be developed quickly and accurately for modeling the overall dynamic behavior of the PN.

13.6.1   Line parameters dynamical modeling

13.6.1.1   Dynamical current wave analysis in transmission lines

We start LPDM formulation with a mathematical analysis on the electrical current waves propagation through a transmission line. At the beginning, we consider the single-phase two-wire lossless transmission line with source and load terminals as shown in Figure 13.21(a), where $Z_R(s)$ is the impedance at receiving end, Z_C is the characteristic impedance of the line and, the source bus is modeled with its Norton equivalent current source ($I_N(s)$) and Norton equivalent impedance ($Z_N(s)$). A section of length Δx of the line has been shown in Figure 13.21(b), where L represents the line inductance, C is the parallel capacitor.

It has been shown in Reference 70 that, using the Kirchhoff's current law (KCL) and Kirchhoff's voltage law (KVL) equations for the circuit in Figure 13.21(b), in addition to Laplace transform properties, for each point x along the transmission line, we have

$$\frac{d^2 I(x,s)}{dx^2} - s^2 LCI(x,s) = 0 \tag{13.50}$$

solving this second-order ODE equation, we have

$$I(x,s) = I^+(s)e^{-sx/\vartheta} + I^-(s)e^{+sx/\vartheta} \tag{13.51.1}$$

similarly, for voltage:

$$V(x,s) = V^+(s)e^{-sx/\vartheta} + V^-(s)e^{+sx/\vartheta} \tag{13.51.2}$$

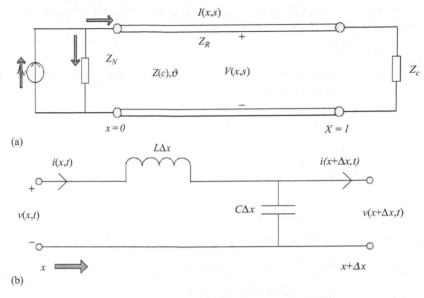

Figure 13.21   *(a) Single-phase two-wire lossless transmission line with source and load terminals. (b) A section of length Δx of the line*

where $\vartheta = (1/\sqrt{(LC)})$ m/s is about the light speed in the line and I^+, I^-, V^+ and V^-, are calculated from boundary conditions. Using (13.51) and considering (from KVL and KCL equations),

$$\frac{dV(x,s)}{dx} = -sLI(x,s) \tag{13.52}$$

we can easily find out

$$\frac{s}{\vartheta}\left[-V^+(s)e^{-\frac{sx}{\vartheta}} + V^-(s)e^{+\frac{sx}{\vartheta}}\right] = -sL\left[I^+(s)e^{-sx/\vartheta} + I^-(s)e^{+sx/\vartheta}\right] \tag{13.53}$$

Equating the coefficients $e^{+sx/\vartheta}$ and $e^{-sx/\vartheta}$ on the both side of (13.53), we can reach to the following equivalent boundary conditions for voltage in term of current:

$$\begin{cases} V^+(s) = \vartheta L I^+(s) = I^+(s)\sqrt{\dfrac{L}{C}} = I^+(s)Z_c \\ \qquad\qquad V^-(s) = -I^-(s)Z_c \end{cases} \tag{13.54}$$

where $Z_C = \sqrt{(L/C)}\,\Omega$ is the characteristic impedance of the line. Finally, we will have the following equivalent equation for voltage (13.51.2):

$$V(x,s) = Z_c\left[I^+(s)e^{-\frac{sx}{\vartheta}} - I^-(s)e^{+\frac{sx}{\vartheta}}\right]. \tag{13.55}$$

Using the general model in Figure 13.21(a) and applying boundary conditions at the receiving end, we will have

$$V(l,s) = Z_R(s)I(l,s) \tag{13.56}$$

where $Z_R(s)$ is the impedance at receiving end and l is the line length. Applying (13.51.1) and (13.55) in (13.56):

$$Z_c\left[I^+(s)e^{-(sl/\vartheta)} - I^-(s)e^{+(sl/\vartheta)}\right] = Z_R(s)\left[I^+(s)e^{-(sl/\vartheta)} + I^-(s)e^{+(sl/\vartheta)}\right] \tag{13.57}$$

Solving for $I^-(l,s)$

$$\begin{cases} I^-(l,s) = \Omega_R(s)I^+(s)e^{-2s\tau} \\ \Omega_R(s) \quad = \dfrac{((Z_c/Z_R(s)) - 1)}{((Z_c/Z_R(s)) + 1)} \quad \text{p.u.} \end{cases} \tag{13.58}$$

Using (13.58), (13.51.1), and (13.55), we have

$$\begin{cases} I(x,s) = I^+(s)\left[e^{-(sx/\vartheta)} + \Omega_R(s)e^{s((x/\vartheta)-2\tau)}\right] \\ V(x,s) = I^+(s)Z_c\left[e^{-(sx/\vartheta)} - \Omega_R(s)e^{s((x/\vartheta)-2\tau)}\right] \end{cases} \tag{13.59}$$

Applying the KCL rule in the sending end, we can reach to the following boundary condition:

$$I_N(s) - \frac{V(0,s)}{Z_N(s)} = I(0,s) \tag{13.60}$$

Using (13.59) in (13.60), we have

$$\begin{cases} I_N(s) = I^+(s)\left[1 + \dfrac{Z_c}{Z_N(s)}\right][1 - \Omega_R(s)\Omega_U(s)e^{-2s\tau}] \\ \Omega_U(s) = \dfrac{((Z_c/Z_N(s)) - 1)}{((Z_c/Z_N(s)) + 1)} \end{cases} \tag{13.61}$$

Finally, applying (13.58) and (13.61) in (13.51.1) we will have

$$I(x,s) = I_N(s)\left[\frac{Z_N(s)}{Z_N(s) + Z_c}\right]\left[\frac{e^{-(sx/\vartheta)} + \Omega_R(s)e^{s((x/\vartheta)-2\tau)}}{1 - \Omega_R(s)\Omega_U(s)e^{-2s\tau}}\right] \tag{13.62}$$

Thus, the time domain current wave equation at point x along the line is calculated from inverse Laplace transform of (13.62). Regarding the values of the set of involved impedances, this time-domain representation at the receiving bus current ($x = l$) has different forms. However, in general, we will need the Taylor expansion series to be able to factorize the denominator of (13.62), so the general time-domain representation follows the following format:

$$i(l,t) = \sum_{k=1}^{K} x_k I_g(t - \alpha_k \tau) \tag{13.63}$$

we call the sequence of coefficients, $\mathbf{x} = \{ x_k \quad for \quad k = 1, \ldots, k \}$, as TLDIC *vector*. Obviously, these index coefficients are directly dependent on the dynamic behavior of the transmission lines, and network impedances. As a result, calculating the values of TLDICs can be interpreted as an alternative approach for modeling the dynamic behavior of the transmission lines in the PN. The total number of included delayed copies, K, which appears in measurements, depends on the recording sampling rate and time constant of the line, which itself mainly depends on the line length. In order to be able to record the effect of these delayed copies in measurements, the sampling rate should be comparable to the line time constant, τ.

The common sampling rate of PMUs is around 2,880 samples/s, which means the phenomena with time constant of order around $\tau \geq (3.47/2) \times 10^{-4}$s, can be captured, without aliasing. Nowadays, the new technologies of PMU can computes electrical phasors with an accuracy better than 1 µs, reported in References 71–74. Thus, one is able to capture the dynamical behavior of transmission lines from a much shorter range of the transmission lines. In this chapter, we describe how this superposition format (13.63) helps us to present a new framework for LPDM and monitoring. A similar superposition as (13.63) can be found by replacing the lossless line model with lossy configuration. (Please also refer to Reference 70, chapter 13 for more info on dynamic transmission line modeling.)

13.6.1.2   Power network modeling under dynamical behavior

Within this chapter, we will consider the same graphical model for PNs as in Section 13.3.1. The only difference is that we call a bus a generator bus if a generator is connected to it and a load bus if it is connected to a load. There is also another category that contains interior buses, which are neither a generator nor a load bus. Following the approach, which has been used in Reference 68, we will ignore the interior buses in this study. Regarding the above description and using the port-Hamilton interconnection model of the power grid, which has been discussed in Reference 68, each bus is modeled as a super node with the configuration that has been illustrated in Figure 13.22. In other words, the input–output relation of bus i has the following format:

$$I_i(t) = \sum_{p=1}^{P} I_{i,p}(t) = \sum_{j \in \mathcal{L}_i} z_{ji}(t-1) + b_i(t) + c_{ji}(t), \tag{13.64}$$

where P is the total number of output terminals from bus i, and z_{ji} indicates the input current wave from bus j to bus i. Depending on the bus category (load-generator), b_i is the generator or load currents, which are connected to bus i; c_{ji} is the corresponding current to the capacitor of the line l_{ji} connected to bus i; and finally, $\mathcal{L}_i$, represents the set of indexes of all of the neighbor buses directly input to the bus i. We will call the cardinality of $\mathcal{L}_i$ as the in-degree of the bus i.

13.6.1.3   Mathematical formulation of LPDM

Using (13.63), the receiving current waves z_{ij}s at bus i can be modeled as a superposition of delayed copies of the sending current wave, I_{ji}, at the sending bus j

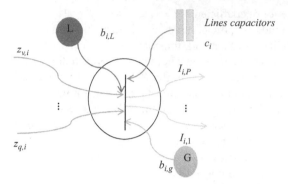

Figure 13.22   A typical bus model under PN KCL law (super-node)

(for ease of expansion we will use the discrete sample-index notation of the delay terms, so $a_k\tau = k$),

$$z_{ji}(t) = \sum_{k=1}^{K} x_{ji,k} I_{ji}(t - k + 1), \tag{13.65}$$

where K is the total number of delay terms appearing in measurements. We initially assume that all of the line lengths are beyond a specific range, which ideally makes them have the same total number of time delay terms; we term this number *line dynamic order* (LDO). In the rest of this section, we describe how we can express the TLDICs estimation, from the measurements of the current waves, in the form of an optimization problem. Replacing (13.64) in (13.65), we have

$$I_i(t) = \sum_{j \in \mathcal{L}_i} \sum_{k=1}^{K} x_{ji,k} I_{ji}(t - k) + b_i(t) + c_{ij}(t). \tag{13.66}$$

For time samples, $t = 1, \ldots, M$, setting $x_{ji} = 0$ for $j \notin \mathcal{L}_i$, assuming $I_i(t) = 0$ for $t \leq 0$ and considering no feed-through term, $z_{ji}(0) = 0$, then according to (13.66), the output of each node $I_i \in R^M$ can be written as

$$I_i = \sum_{j=1}^{N} A_{ji} x_{ji} + b_i + c_i \quad i = 1, 2, \ldots, N, \tag{13.67}$$

where N is the total number of buses in the PN, A_{ji} is an $M \times K$ Toeplitz matrix, $x_{ji} \in R^K$ and $b_i \& c_i \in R^M$. For each single node i, (13.67) can be rewritten in the following matrix-vector format:

$$I_i = A^i x^i + b_i + c_i \tag{13.68}$$

where $A^i = [A_{1i} \ldots A_{ji} \ldots A_{Ni}], x^i = [x_{1i} \ldots x_{ji} \ldots x_{Ni}]^T, I_i \in R^M, x^i \in R^{NK}$, and $A^i \in R^{M \times NK}$ (called the sensing matrix) is defined as the concatenation of

N Toeplitz matrices. As we discussed so far, the LPDM problem may be interpreted as recovering the set of TLDIC vectors $\{x^i\}_{i=1}^{N}$, for a given set of measurements, from the current wave values at each bus in the network with a proper sampling frequency. In this work, we assume the available set of measurements includes all output currents $\{I_i\}_{i=1}^{N}$ and known current terms $\{b_i\}_{i=1}^{N}$. Although, in general, the capacitor currents are much smaller than other currents and unknown, we model the c_i as a Gaussian random vector with small variance. Moreover, replacing $I_i - b_i = y_i$ and $c_i = \eta_i$ for each individual node, we arrive at $y_i = A^i x^i + \eta_i$. Therefore, the LPDM problem can be viewed as the estimation of transient sub-vectors $\hat{x}_{ji}$ that best matches the observed measurements. In order to solve such a problem, we can define the following optimization problem:

$$\min_{\{x^i\}_{i=1}^{N}} \sum_{i=1}^{N} \|A^i x^i - y_i\|_2^2. \tag{13.69}$$

This problem can be solved individually for each node i. However, while the structure of the matrix A^i is similar for all nodes, without loss of generality, we simplify the index-notation $A^i = A, x^i = x, y_i = y$ and focus on the following optimization problem:

$$\min_{\hat{x}} \|A\hat{x} - y\|_2^2. \tag{13.70}$$

Defining $L = N \times K$, we have $y \in R^M, x \in R^L$, and $A \in R^{M \times L}$. If A is full rank, the number of measurements, M, exceeds the number of unknowns, L, and the variance of the noise, η, is small enough; this optimization problem can be solved, fairly correctly, using LS techniques. Although the value of L depends on the number of PN buses and LDOs, a big number of measurements will be needed to solve such a problem in the case of large power grids; this will make the calculations to be complex and will highly increase the time cost, which is an unwanted issue, especially when we are dealing with the dynamic behavior of the network. In order to avoid this problem, we suggest using the structured SR.

13.6.2   Structured sparse LPDM

In this section, we will show how the LPDM (13.70) can be represented as a mathematical SSRP. Moreover, a brief overview on Block-sparse signal definition and BOMP greedy-based SR algorithm is presented.

Following (13.68), for each individual node, the TLDIC vector $\mathbf{x}$ can be considered as concatenation of N blocks with the following representation

$$\mathbf{x} = \left[\mathbf{x}_1^{\mathbf{T}}, \mathbf{x}_2^{\mathbf{T}}, \ldots, \mathbf{x}_N^{\mathbf{T}}\right]^{\mathbf{T}} \tag{13.71}$$

Whenever the node i is not connected to node j, the subblock $\mathbf{x}_j^{\mathbf{T}} = \mathbf{0} \in R^K$. Such a vector $\mathbf{x}$ is called S-block-sparse if $S \times K \ll L$, where $S = \|\mathcal{L}_i\|_0$, or the cardinality of the index set of neighbor nodes of node i. This signal contains S nonzero block with each of length K, so the sparsity level or number of nonzero elements of the signal is defined as $||x||_0 = S \times K = F$. Although the number of nonzero blocks in TLDIC vectors **x** is limited to the number of its connections to other nodes, which is always a small fraction of total nodes in the network, **x** can be represented as an S-block-sparse vector for all of the nodes in the PN. Obviously, whenever $M < L$ in (13.70), the null space of the sensing matrix A is nontrivial, and infinitely many candidate solutions exist; however, under certain guarantee conditions on sensing matrix A, structured SR methods will be able to efficiently find the sparse candidate solution. Here again, we can interpret the SRP as recovery of a F-sparse signal $x \in R^L$ (which means that $||x||_0 = F$) from a set of observations $\mathbf{y} = A\mathbf{x} \in R^M$ where $A \in R^{M \times L}$ is the sensing matrix with $M \ll L$. If the signal has a special structure such as block-sparse signals, the problem is called as structured SR.

13.6.2.1 Block OMP recovery algorithm for block sparse signals

Beyond the several extensities of standard greedy SR solvers, which have been adapted in order to consider the extra information in the structure of the sparse signals, the BOMP is especially introduced to exploit block-structured sparsity [69]. Due to block sparse nature of the optimization variable **x** in LPDM (13.70), we use BOMP as the recovery method in this work. The stepwise format of BOMP is represented in Algorithm 13.4 as stated in Reference 69.

Algorithm 13.4 The Block Orthogonal Matching Pursuit (BOMP)

---

require: matrix A, measurements y, stopping criterion
initialize: $r^0 = y, x^0 = 0, l = 0, SUP^0 = \phi$
repeat
1.match: $h_j^l = A_j^T r^l \quad for \ j = 1, \ldots, N$
2.identify support indicator:

$$u^l = \left\{ argmax_j ||h_j^l||_2 \right\}$$

3.update the support:

$$SUP^{l+1} = SUP^l \cup u^l$$

4.update signal estimate:

$$x^{l+1} = argmin_{z:SUP(z) \subseteq SUP^{l+1}} ||y - Az||_2$$
$$r^{l+1} = y - Ax^{l+1}$$
$$l = l + 1$$

Until stopping criterion met
Output: $\hat{x} = x^l$

13.6.2.2   BOMP guarantee condition

Definition 13.7: *The block-coherence of matrix A (with normalized columns) is defined as follows:*

$$\mu_{\text{block}}(A) = \max_{i,j \neq i} \frac{1}{K} \left\| A_i^T A_j \right\|_2 . \tag{13.72}$$

Definition 13.8: *The subblock-coherence of matrix A (with normalized columns) is defined as follows:*

$$\mu_{\text{sub-block}}(A) = \max_{n} \max_{i,j \neq i} \left| a_{n_i}^T a_{n_j} \right|, \tag{13.73}$$

where a_{n_i} and a_{n_j} are the columns of nth block of matrix A or A_n (13.68) and (13.70). Eldar et al. [69] introduced a sufficient condition to guarantee the recovery of any S-block sparse signal x from the set of measurements via BOMP method.

Theorem 13.4 [69]: *For an S-block-sparse vector x with S blocks of length K, and a vector of measurements y such that $y = Ax$, if $S \times K < \mu_T$, then the BOMP can recover the original vector x from the set of measurements y; where*

$$\mu_T = \frac{1}{2} \left(\mu_{\text{block}}^{-1} + K - (K - 1) \frac{\mu_{\text{sub-block}}}{\mu_{\text{block}}} \right). \tag{13.74}$$

Within the noisy measurements, $y = Ax + \eta$, the arguments can be extended to robust recovery in noisy settings. While $S \times K < \mu_T$ should be satisfied as guarantee condition, the maximum permitted block-sparsity level S for a vector, which can be recovered using BOMP, is limited by block-coherence of A. The smaller the block-coherence metrics, the higher the possible value of S, and the wider the sparsity level of the signal that can be recovered using BOMP.

Remark 13.4: *As it can be seen from the sparse formulation of LPDM problem (13.70) the structure of the corresponding sensing matrix A, is composed of measurements from branches flows. Since the overall behavior of loads and weather conditions in neighboring areas is similar to each other, it is highly likely to have a meaningful correlation within the current behavior in neighbor lines. Similar to PNTI-SRP, this correlation itself will result in higher coherence in corresponding sensing matrices in LPDM-SSRP. In order to deal with this correlation issue, a similar BLO-based modification (Section 13.4.3) approach can be taken into account. Since in LPDM-SSRP, we face with block-sparse signals one may develop a new BLO-based block orthogonal greedy algorithm to address this issue.*

Remark 13.5 (LDO Selection and Clustered Sparse Approach): *As has been mentioned in Section 13.6.1.2, we initially assumed that all of the lines are following the same LDO. Considering a similar LDO for all lines ($l_{(i,j)}$ for $(j \in \mathcal{L}_i)$) connected to the bus i,[17] sparse vector x is a S-block sparse signal with each of length K. However, this assumption can be ruined in realistic cases, since line lengths and physical properties can vary in a big range in PNs. This means that each nonzero block x_{ji} (of the TLDIC vector x^i) can have arbitrary number of nonzero elements.[18] From SR perspective, this will change the sparse signal x^i shape from a block to a clustered-sparse format (Figure 13.23(a) and (b), respectively). As has been mention in Section 13.4.1.3, COMP modifies the support identification step of OMP in order to address those situations where we expect a*

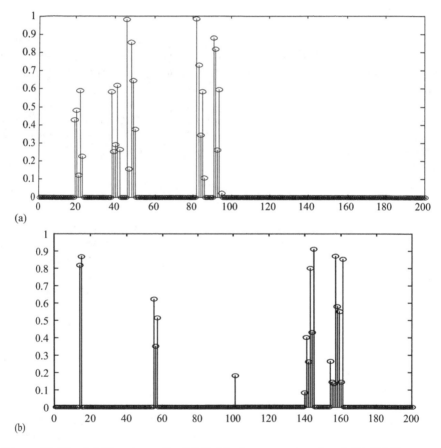

(a)

(b)

Figure 13.23   (a) Structure of a (25,5)-block-sparse signal and (b) (21,5)-clustered-sparse signal

---

[17]Or equivalently the number of affective delayed terms of $I_{ji}(t-k)$ in $I_i(t)$ to be fixed and equal to K.
[18]Or equivalently arbitrary number of affective delayed terms (K).

clustered sparsity pattern to be happened in the support of the signal. The recovery performance of COMP algorithm (Section 13.4.1.3) will be compared vs BOMP in the case of arbitrary LDOs in the LPDM problem.

13.6.3 Case studies

In this part, we examine the proposed sparse-based approach for recovering the TLDIC vectors (and as a result LPDM-SSRP) using compressive observations. As was expected, the percentage of successful recovery rate changes for different LDOs based on the pattern of sparsity. In simulations, we use the structure of the IEEE Standard 118 Bus and IEEE Standard 30 Bus, from Reference 57, as case studies to model the PN and generate the current data. Since the amplitude of the TLDICs decays fast, initially we consider a fixed LDO ($K = 2$ and 4) for all lines as a reasonable choice. The coherence measures of the IEEE Standard 30 Bus (vs number of measurements) are illustrated in Figure 13.24 for $K = 4$, where the illustrated coherence metric curves are averaged over 100 realizations of the PN. Clearly, by increment in the number of available measurements, the coherence metrics are asymptotically decreased; which means that having more measurements, we can recover the signals with higher number of non-zero blocks. However, due to interconnected nature of PN networks, this improvement will stop at some point and the coherence measures converge to a lower bound.

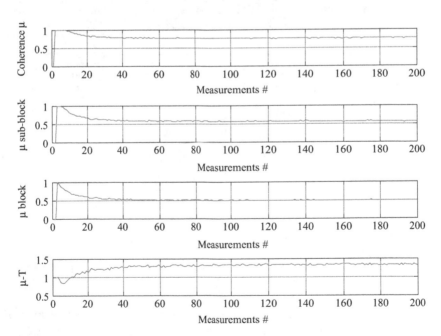

Figure 13.24   The coherence measures of IEEE Standard 30 Bus (vs the number of measurements)

In LPDM-SSRP, the sparsity level of the vector $\mathbf{x}^i$ is forced by two factors: first, the in-degree of the bus, $||\mathcal{L}_i||_0$ (number of links or lines merging into the bus i) and second, the LDO of the PN lines K. Figure 13.25(a) shows the recovery performance of the BOMP for IEEE Standard 118 Bus network, with LDO "4" for the following set of specifically assumed nodes: 74, with in-degree 2, 49, with in-degree 9, and 59, with in-degree 6. Since in this case the LDO of all lines is the same, the successful recovery

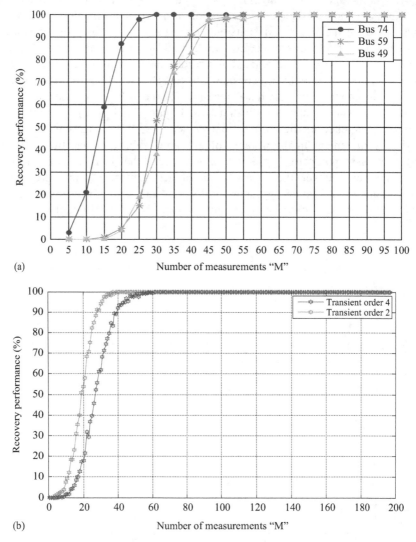

(a)

(b)

Figure 13.25   (a) TLDIC vector recovery performance of three selected buses in IEEE Standard 118 Bus with LDO "4." (b) The network-wide LPDM recovery performance of the BOMP for IEEE 30 Bus network with LDO "2" and "4"

probability alters based on the in-degree of the buses. As a result, the double in-connection node 74 is more likely to be recovered using a smaller number of measurements than the 6 in-connection node 59 as one might expect, moreover, the 6 in-degree node 59 is more likely to be recovered using a smaller number of measurements than the 9 in-degree node 49. However, the difference in recovery per-formance for these two buses is not considerable. Figure 13.25(b) shows the whole network lines recovery performance of the BOMP for IEEE-30 Bus network, with LDO "2" and "4" (over all buses), respectively. For each curve, the vertical axis represents the percentage of trials in which all TLDCI vectors corresponding to all transmission lines (and, as a result, the network-wide LPDM) are ideally recovered.

In order to address the role of the arbitrary LDOs, in next set of simulations we model the network considering the LDO of different lines to be picked randomly between 1 and 4. As a result the sparse TLDIC vector $\mathbf{x}^i$ changes its structure from a block-sparse signal into a clustered-sparse signal with maximum cluster size 4. The OMP, COMP (with cluster size 4), and BOMP (with block size 4) algorithms have been used as SRP solvers, and results have been compared in Figure 13.26.

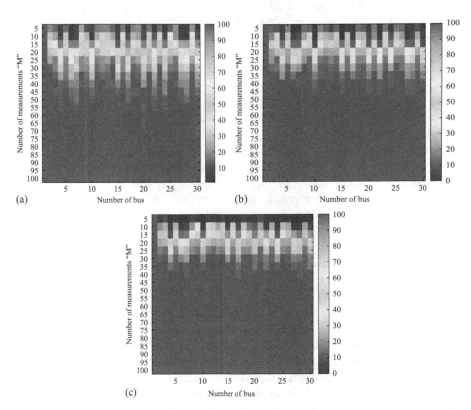

Figure 13.26   Recovery performance for (a) OMP, (b) BOMP, and (c) COMP over all of the buses of the IEEE Standard 30 Bus in the case of arbitrary LDOs

The vertical axis represents the number of measurements taken for SR of each sparse TLDIC vector **x** from the following set: {5:5:100}. The horizontal axis shows the bus number from 1 to 30. The recovery performance of each sparse signal $\mathbf{x}^i$ for $i = 1:30$, under given number of measurements have been illustrated by a color spectrum from dark blue corresponding to 0% recovery up to dark red corresponding to 100% recovery vs selected number of measurements over 100 realization of IEEE 30-Bus networks, respectively. Using the structural knowledge in sparsity pattern BOMP outperforms OMP in average over all buses with smaller number of recorded measurements. As can be seen the COMP algorithm represents better performance than BOMP and OMP when the exact LDOs cannot be estimated easily.

13.7 Conclusions

In this chapter, we aimed to familiarize the reader with possible applications of a new signal-processing theorem named CS–SR in the era of power engineering. A brief review on CS–SR mathematical concepts and formulations have been given in Section 13.2. Next in Section 13.3, the sparse structural pattern of the corresponding graphs of the PNs has been discussed and variety of the state of the art in the applications of CS–SR in SGs control and monitoring have been introduced, including system state estimation, communication protocol design in SGs, attack recognition, power-flow modeling, power signals processing, power outage identification, fault type recognition, topology identification, and transmission line modeling. Sparse-based power-line outage detection and localization, CS-based SG topology identification and structured sparse-based transmission line dynamic behavior modeling have been specifically discussed in detail with their corresponding mathematical formulations within Sections 13.4, 13.5, and 13.6, respectively. The inherent sparsity patterns in the nature of these problems have been investigated, and the performance of the corresponding sparse-based formulations has been explored through variety of IEEE standard testbeds and case studies. Moreover, in order to address a couple of practical challenges within implementation of SR-based techniques in the era of SGs, some generalized adapted SR algorithms have been detailed within each of these sparse-based frameworks.

   Due to their inherent data and structural sparse patterns, SGs and their related frontier technologies such as smart cities are a great potential era of applications for CS–SR-based frameworks. From small size renewable-based microgrids into the huge size PNs, variety of data analysis, system modeling, control, and monitoring problems can be mathematically reformulated under the domination of CS–SR theorems. Regarding their mathematical and time cost efficiency, the CS–SR-based frameworks can be considered as the next generation of frontier algorithms to address the inherent data collection, modeling, and analysis challenges within the large-scale and complex future SGs.

Acknowledgements

The authors would like to sincerely appreciate Prof. Michael Wakin (EECS Department, Colorado School of Mines) for his help and support with this work.

References

[1] Yu X, Cecati C, Dillon T, Simões MG. "The new frontier of smart grids". *IEEE Industrial Electronics Magazine*. 2011 Sep;5(3):49–63.

[2] Güngör VC, Sahin D, Kocak T, *et al.* "Smart grid technologies: Communication technologies and standards". *IEEE Transactions on Industrial Informatics*. 2011 Nov;7(4):529–39.

[3] Fang X, Misra S, Xue G, Yang D. "Smart grid—The new and improved power grid: A survey". *IEEE Communications Surveys & Tutorials*. 2012 Oct;14(4):944–80.

[4] Candès EJ, Wakin MB. "An introduction to compressive sampling". *IEEE Signal Processing Magazine*. 2008 Mar;25(2):21–30.

[5] Donoho DL. "Compressed sensing". *IEEE Transactions on Information Theory*. 2006 Apr;52(4):1289–306.

[6] Candès EJ, Romberg J, Tao T. "Robust uncertainty principles: Exact signal reconstruction from highly incomplete frequency information". *IEEE Transactions on Information Theory*. 2006 Feb;52(2):489–509.

[7] Shannon CE. "Communication in the presence of noise". *Proceedings of the IRE*. 1949 Jan;37(1):10–21.

[8] Wakin MB. "Compressive sensing fundamentals". In: Amin M, editor. *Compressive Sensing for Urban Radar*, pp. 1–47. CRC Press, Boca Raton, FL, 2014.

[9] Sanandaji BM, Vincent TL, Wakin MB. "Compressive topology identification of interconnected dynamic systems via clustered orthogonal matching pursuit". In *Decision and Control and European Control Conference (CDC-ECC), 2011 50th IEEE Conference on*, 2011 Dec 12 (pp. 174–180). Orlando, FL: IEEE.

[10] Elad M, Figueiredo MA, Ma Y. "On the role of sparse and redundant representations in image processing". *Proceedings of the IEEE*. 2010 Jun;98(6):972–82.

[11] Davenport MA, Needell D, Wakin MB. "Signal space CoSaMP for sparse recovery with redundant dictionaries". *IEEE Transactions on Information Theory*. 2013 Oct;59(10):6820–9.

[12] Tropp JA. "Greed is good: Algorithmic results for sparse approximation". *IEEE Transactions on Information Theory*. 2004 Oct;50(10):2231–42.

[13] Tibshirani R. "Regression shrinkage and selection via the lasso". *Journal of the Royal Statistical Society. Series B (Methodological)*. 1996 Jan;58(1):267–88.

[14]   Ender JH. "On compressive sensing applied to radar". *Signal Processing.* 2010 May;90(5):1402–14.

[15]   Huang H, Misra S, Tang W, Barani H, Al-Azzawi H. "Applications of compressed sensing in communications networks". arXiv preprint arXiv:1305.3002. 2013 May 14.

[16]   Cassidy B, Rae C, Solo V. "Brain activity: Connectivity, sparsity, and mutual information". *IEEE Transactions on Medical Imaging.* 2015 Apr;34 (4):846–60.

[17]   Herrmann FJ, Friedlander MP, Yilmaz Ö. "Fighting the curse of dimensionality: Compressive sensing in exploration seismology". *IEEE Signal Processing Magazine.* 2012 May;29(3):88–100.

[18]   Wai HT, Scaglione A, Leshem A. "Active sensing of social networks". arXiv:1601.05834. 2016 Jan 21.

[19]   Zhu Z, Wakin MB. "Wall clutter mitigation and target detection using Discrete Prolate Spheroidal Sequences". In *Compressed Sensing Theory and its Applications to Radar, Sonar and Remote Sensing (CoSeRa), 2015 Third International Workshop on, 2015 Jun 17* (pp. 41–45). Pisa, Italy: IEEE.

[20]   Sabbah AI, El-Mougy A, Ibnkahla M. "A survey of networking challenges and routing protocols in smart grids". *IEEE Transactions on Industrial Informatics.* 2014 Feb;10(1):210–21.

[21]   Li W, Ferdowsi M, Stevic M, Monti A, Ponci F. "Cosimulation for smart grid communications". *IEEE Transactions on Industrial Informatics.* 2014 Nov;10(4):2374–84.

[22]   Ozay M, Esnaola I, Vural FT, Kulkarni SR, Poor HV. "Sparse attack construction and state estimation in the smart grid: Centralized and distributed models". *IEEE Journal on Selected Areas in Communications.* 2013 Jul;31 (7):1306–18.

[23]   Manikandan MS, Samantaray SR, Kamwa I. "Detection and classification of power quality disturbances using sparse signal decomposition on hybrid dictionaries". *IEEE Transactions on Instrumentation and Measurement.* 2015 Jan;64(1):27–38.

[24]   Majidi M, Fadali MS, Etezadi-Amoli M, Oskuoee M. "Partial discharge pattern recognition via sparse representation and ANN". *IEEE Transactions on Dielectrics and Electrical Insulation.* 2015 Apr;22(2):1061–70.

[25]   Majidi M, Oskuoee M. "Improving pattern recognition accuracy of partial discharges by new data preprocessing methods". *Electric Power Systems Research.* 2015 Feb;119:100–10.

[26]   Arabali A, Majidi M, Fadali MS, Etezadi-Amoli M. "Line outage identification-based state estimation in a power system with multiple line outages". *Electric Power Systems Research.* 2016 Apr;133:79–86.

[27]   Majidi M, Etezadi-Amoli M, Fadali MS. "A sparse-data-driven approach for fault location in transmission networks". *IEEE Trans. Smart Grid.* 2015;99:1–9.

[28] Majidi M, Etezadi-Amoli M, Livani H, Fadali MS. "Distribution systems state estimation using sparsified voltage profile". *Electric Power Systems Research*. 2016 Jul;136:69–78.

[29] Majidi M, Etezadi-Amoli M, Sami Fadali M. "A novel method for single and simultaneous fault location in distribution networks". *IEEE Transactions on Power Systems*. 2015 Nov;30(6):3368–76.

[30] Majidi M, Arabali A, Etezadi-Amoli M. "Fault location in distribution networks by compressive sensing". *IEEE Transactions on Power Delivery*. 2015 Aug;30(4):1761–9.

[31] Zhang Z, Nguyen HD, Turitsyn K, Daniel L. "Probabilistic power flow computation via low-rank and sparse tensor recovery". arXiv preprint arXiv:1508.02489. 2015 Aug 11.

[32] Zhu H, Giannakis GB. "Sparse overcomplete representations for efficient identification of power line outages". *IEEE Transactions on Power Systems*. 2012 Nov;27(4):2215–24.

[33] Babakmehr M, Simoes MG, Al-Durra A, Harirchi F, Han Q. "Application of compressive sensing for distributed and structured power line outage detection in smart grids". In *American Control Conference (ACC), 2015*, 2015 Jul 1 (pp. 3682–3689). Chicago, IL: IEEE.

[34] Babakmehr M, Simoes MG, Wakin MB, Harirchi F. "Compressive sensing-based smart grid topology identification". *IEEE Transactions on Industrial Informatics*. 2016 Apr;12(2):532–43.

[35] Babakmehr M, Simoes MG, Wakin MB, Al Durra A, Harirchi F. "Sprase-based smart grid topology identification". *IEEE Transaction on Industry Applications*. 2016 Sep–Oct;52(5):4375–84.

[36] Babakmehr M, Simoes MG, Ammerman R. "Modeling and tracking transmission line dynamic behavior in smart grids using structured sparsity", In *54th Annual Allerton Conference on Communication, Control, and Computing*, 2016, Sep, Monticello, IL, USA.

[37] West DB. *Introduction to Graph Theory*. Upper Saddle River, NJ: Prentice Hall; 2001.

[38] Wood AJ, Wollenberg BF. *Power Generation, Operation, and Control*. New York: Wiley; 2012.

[39] Davoudi MG, Sadeh J, Kamyab E. "Time domain fault location on transmission lines using genetic algorithm". In *Environment and Electrical Engineering (EEEIC), 2012 11th International Conference on, 2012 May 18* (pp. 1087–1092). Venice, Italy: IEEE.

[40] Davoudi M, Sadeh J, Kamyab E. "Parameter-free fault location for transmission lines based on optimization". *IET Generation, Transmission & Distribution*. 2015 Aug;9(11):1061–8.

[41] Azizi S, Sanaye-Pasand M. "A straightforward method for wide-area fault location on transmission networks". *IEEE Transactions on Power Delivery*. 2015 Feb;30(1):264–72.

[42]   Chen YC, Banerjee T, Domínguez-García AD, Veeravalli VV. "Quickest line outage detection and identification". *IEEE Transactions on Power Systems*. 2016 Jan;31(1):749–58.

[43]   Wu J, Xiong J, Shi Y. "Efficient location identification of multiple line outages with limited PMUs in smart grids". *IEEE Transactions on Power Systems*. 2015 Jul;30(4):1659–68.

[44]   He M, Zhang J. "A dependency graph approach for fault detection and localization towards secure smart grid". *IEEE Transactions on Smart Grid*. 2011 Jun;2(2):342–51.

[45]   Chen JC, Li WT, Wen CK, Teng JH, Ting P. "Efficient identification method for power line outages in the smart power grid". *IEEE Transactions on Power Systems*. 2014 Jul;29(4):1788–800.

[46]   Wu J, Xiong J, Shi Y. "Efficient location identification of multiple line outages with limited PMUs in smart grids". *IEEE Transactions on Power Systems*. 2015 Jul;30(4):1659–68.

[47]   Babakmehr M, Simoes MG, Al-Durra A, Muyeen SM. "Exploiting compressive system identification for multiple line outage detection in smart grids". preprint.

[48]   Tate JE, Overbye TJ. "Line outage detection using phasor angle measurements". *IEEE Transactions on Power Systems*. 2008 Nov;23(4):1644–52.

[49]   Fannjiang A, Liao W. "Coherence pattern-guided compressive sensing with unresolved grids". *SIAM Journal on Imaging Sciences*. 2012 Feb;5(1):179–202.

[50]   Harirchi F, Simoes MG, AlDurra A, Muyeen SM. "Short transient recovery of low voltage-grid-tied DC distributed generation". In *Energy Conversion Congress and Exposition (ECCE), 2015 IEEE*, 2015 Sep 20 (pp. 1149–1155). Montreal, Canada: IEEE.

[51]   Harirchi F, Simoes MG, Babakmehr M, Al-Durra A, Muyeen SM. "Designing smart inverter with unified controller and smooth transition between grid-connected and islanding modes for microgrid application". In *Industry Applications Society Annual Meeting, 2015 IEEE*, 2015 Oct 18 (pp. 1–7). Dallas, TX: IEEE.

[52]   Simões MG, Busarello TD, Bubshait AS, Harirchi F, Pomilio JA, Blaabjerg F. "Interactive smart battery storage for a PV and wind hybrid energy management control based on conservative power theory". *International Journal of Control*. 2015;89(4):850–70.

[53]   Como G, Bernhardsson B, Rantzer A, editors. *Information and Control in Networks*. Berlin: Springer International Publishing; 2014.

[54]   Candes EJ, Wakin MB, Boyd SP. "Enhancing sparsity by reweighted l_1 minimization". *Journal of Fourier Analysis and Applications*. 2008 Dec;14(5–6):877–905.

[55]   Grant M, Boyd S. "CVX: Matlab software for disciplined convex programming", version 2.0 beta. Available at http://cvxr.com/cvx, 2013 Sep.

[56]   Becker S, Bobin J, Candès EJ. "NESTA: A fast and accurate first-order method for sparse recovery". *SIAM Journal on Imaging Sciences*. 2011 Jan;4(1):1–39.

[57] Zimmerman RD, Murillo-Sánchez CE, Thomas RJ. "MATPOWER: Steady-state operations, planning, and analysis tools for power systems research and education". *IEEE Transactions on Power Systems*. 2011 Feb;26(1):12–9.

[58] De La Ree J, Centeno V, Thorp JS, Phadke AG. "Synchronized phasor measurement applications in power systems". *IEEE Transactions on Smart Grid*. 2010 Jun;1(1):20–7.

[59] Mousavi-Seyedi SS, Aminifar F, Afsharnia S. "Parameter estimation of multiterminal transmission lines using joint PMU and SCADA data". *IEEE Transactions on Power Delivery*. 2015 Jun;30(3):1077–85.

[60] Liao Y, Kezunovic M. "Online optimal transmission line parameter estimation for relaying applications". *IEEE Transactions on Power Delivery*. 2009 Jan;24(1):96–102.

[61] Du Y, Liao Y. "On-line estimation of transmission line parameters, temperature and sag using PMU measurements". *Electric Power Systems Research*. 2012 Dec;93:39–45.

[62] Shi D, Tylavsky DJ, Logic N, Koellner KM. "Identification of short transmission-line parameters from synchrophasor measurements". In *Power Symposium, 2008. NAPS'08. 40th North American 2008*, Sep 28 (pp. 1–8). Calgary, Alberta, Canada: IEEE.

[63] Bi T, Chen J, Wu J, Yang Q. "Synchronized phasor based on-line parameter identification of overhead transmission line". In *Electric Utility Deregulation and Restructuring and Power Technologies, 2008. DRPT 2008. Third International Conference on*, 2008 Apr 6 (pp. 1657–1662). NanJing, China: IEEE.

[64] Yang J, Li W, Chen T, Xu W, Wu M. "Online estimation and application of power grid impedance matrices based on synchronised phasor measurements". *IET Generation, Transmission & Distribution*. 2010 Sep;4(9):1052.

[65] Indulkar CS, Ramalingam K. "Estimation of transmission line parameters from measurements". *International Journal of Electrical Power & Energy Systems*. 2008 Jun;30(5):337–42.

[66] Kurokawa S, Pissolato J, Tavares MC, Portela CM, Prado AJ. "A new procedure to derive transmission-line parameters: Applications and restrictions". *IEEE Transactions on Power Delivery*. 2006 Jan;21(1):492–8.

[67] Kurokawa S, Asti GA, Costa EC, Pissolato J. "Simplified procedure to estimate the resistance parameters of transmission lines". *Electrical Engineering*. 2013 Sep;95(3):221–7.

[68] Fiaz S, Zonetti D, Ortega R, Scherpen JM, van der Schaft AJ. "A port-Hamiltonian approach to power network modeling and analysis". *European Journal of Control*. 2013 Dec;19(6):477–85.

[69] Eldar YC, Kuppinger P, Bölcskei H. "Block-sparse signals: Uncertainty relations and efficient recovery". *IEEE Transactions on Signal Processing*. 2010 Jun;58(6):3042–54.

[70] Glover JD, Sarma M, Overbye T. *Power System Analysis and Design, SI Version*. Boston, MA: Cengage Learning; 2011.

[71]   Al-Hammouri AT, Nordstrom L, Chenine M, Vanfretti L, Honeth N, Leelaruji R. "Virtualization of synchronized phasor measurement units within real-time simulators for smart grid applications". In *Power and Energy Society General Meeting, 2012 IEEE*, 2012 Jul 22 (pp. 1–7). San Diego, CA: IEEE.

[72]   Laverty DM, Best RJ, Brogan P, Al Khatib I, Vanfretti L, Morrow DJ. "The OpenPMU platform for open-source phasor measurements". *IEEE Transactions on Instrumentation and Measurement*. 2013 Apr;62(4):701–9.

[73]   Chenine M, Ullberg J, Nordstrom L, Wu Y, Ericsson GN. "A framework for wide-area monitoring and control systems interoperability and cybersecurity analysis". *IEEE Transactions on Power Delivery*. 2014 Apr;29(2):633–41.

[74]   Shi D, Tylavsky DJ, Logic N. "An adaptive method for detection and correction of errors in PMU measurements". *IEEE Transactions on Smart Grid*. 2012 Dec;3(4):1575–83.

Chapter 14

Stability enhancement issues of power grid

Srinivasa Rao Kamala

This chapter gives an overview about the basic principle and operation of various shunt, series and series–shunt types of flexible alternating current transmission system (FACTS) controllers. The steady-state power transfer capability can be improved significantly in the power system with various FACTS controllers, and it is the main focus of this chapter. This can be achieved only by optimal placement of FACTS controllers in the network, otherwise they have negative impact on the system. To demonstrate the concept of power-transfer capability enhancement concept under steady-state operating condition, static VAr compensator (SVC) and thyristor-controlled series compensator (TCSC) have been considered. The simulation results were carried out for IEEE 30-bus and the New England 68-bus test systems to check the effectiveness of the algorithm.

14.1 Introduction

The loads in power system vary continuously over a day in general, and they also subjected to variations caused by weather (ambient temperature) and other unpredictable factors. Thus, the power flow in transmission line can vary even under normal, steady-state conditions. The occurrence of a contingency (due to tripping of a line, generator, load etc.) can result in a sudden increase/decrease in power flow of the transmission lines. This can result in overloading of some lines and consequently voltage collapses at the buses due to shortage of reactive power. These factors lead to the complexity of maintaining economic and secure operation of large interconnected power system network. The required safe operating margin of transmission lines can be substantially reduced through fast dynamic control over reactive and active powers, which can be done through flexible alternating current transmission system (FACTS) devices. A power electronic-based system and other static equipment that provide control of one or more AC transmission system parameters are called as FACTS controllers. The FACTS technology opens up new opportunities for controlling power flow in the transmission lines and enhances the usable capacity of present, as well as new and upgraded lines. The FACTS controllers

Research Scholar, National University of Singapore, Singapore.

Figure 14.1   SVC installation in a substation (courtesy of ABB Pte. Ltd.)

Figure 14.2   TCSC installation in a substation (courtesy of ABB Pte. Ltd.)

can enable a line to carry power closer to its thermal rating and offers continuous control of power flow or voltage changes, against daily load variations. These opportunities arise through the ability of FACTS controllers to control the interrelated parameters that govern the operation of transmission system including series impedance, shunt impedance, line current, bus voltage and load bus phase angle.

In world scenario, the FACTS controllers have been installed in various countries. To mention a few: the static var compensator (SVC) for voltage control was established in Nebraska in 1974 and a thyristor-controlled series compensator (TCSC) rating of 202 MVAr at Slatt 500 KV Substation of Bonneville Power Administration in Oregon [1]. In 1997, American Electric Power has selected its Inez substation in eastern Kentucky for the location of the world's first unified power-flow controller (UPFC) installation [2]. In 1991, a ±80-MVAr static synchronous compensator (STATCOM) extended by Kansai Electric Power Co. and Mitsubishi Motors was fixed at Inuyama Switching Station to improve the stability of a 154-kV system [3]. A ±75-MVAr STATCOM developed by Alstom Pte. Ltd., the first cascade multilevel-inverter-based STATCOM in the world, entered commercial service at National Grid Company East Claydon, England in 2001 [4]. Figures 14.1–14.4 show some of the FACTS controllers installed by various companies.

Figure 14.3   STATCOM installation in a substation (courtesy of Siemens Pte. Ltd.)

Figure 14.4   UPFC installation in a substation (courtesy of AEP Ltd.)

In academic and industrial research stand point, several papers have been published in the literature. A survey has been made on the approximate number of publications published from 1995 to 2010 by Eslami *et al.* [5], and Figure 14.5 statistically illustrates the publication distribution of FACTS devices in this period.

In this chapter, a detailed study has been given on optimal placement and optimal rating of FACTS controllers to improve the steady-state power system stability. The chapter has been organized as follows: Section 14.2 gives an introduction to power-flow control in the transmission system, transmission system with and without series and shunt compensation, the power-transfer capability enhancement of the transmission with shunt and series compensation. The detailed

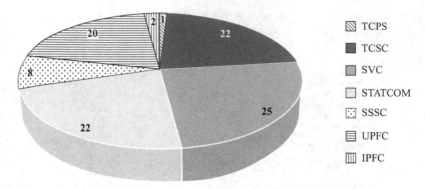

▨	TCPS
■	TCSC
▨	SVC
□	STATCOM
⋰	SSSC
▤	UPFC
▥	IPFC

Figure 14.5   Publication distribution (%) of FACTS devices from 1995 to 2011

classification of the FACTS controllers in the power system network, the power-angle characteristics of the system with FACTS controllers has been explicated in Section 14.3. The optimal placement of the FACTS controllers to improve the power-transfer capability of the system and the simulation results are given in Sections 14.4 and 14.5, respectively. Section 14.6 gives the import conclusions and follows the references in Section 14.7.

14.2 Power-flow control in transmission system

Consider a transmission system with two buses connected by a long transmission line as shown in Figure 14.6 with generator at bus-1 and load at bus-2. Assume that $\overline{V}_1$ and $\overline{V}_2$ are the voltages at the two buses, and impedance of the transmission is jX (for long transmission line $R \ll X$).

The power transferred from the generator bus to load bus is given by

$$P_{12} = \frac{V_1 V_2}{X} \sin \delta \tag{14.1}$$

In equation (14.1), the power-flow between the two buses is dependent on the magnitudes of the bus voltages (V_1 and V_2), reactance of the transmission line (X), and the phase angle difference between two buses (δ). To increase the power flow in the line connected between the buses, the following actions can be considered,

- Increase the magnitudes of $\overline{V}_1$ and $\overline{V}_2$
- Decrease the series line reactance X_L
- Control the phase-angle difference between the voltages $\overline{V}_1$ and $\overline{V}_2$

It is possible to control the mentioned parameters with various FACTS controllers, with satisfying thermal, dielectric and stability limits. The power-flow control in an AC transmission line is required to enhance power-transfer capacity and to change power flow under dynamic conditions to assure system stability and security.

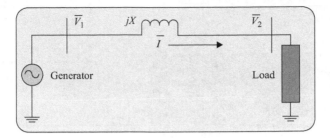

Figure 14.6 Schematic of a typical two bus system

14.2.1 Stability enhancement with shunt and series compensation

There are various applications of the FACTS controllers in the transmission system during its steady-state operating conditions. The main applications include steady-state power-transfer capability, voltage control (low and high) and power-flow control [6–9]. These applications are explained as follows:

- Power transfer is the amount of electric power that can be transferred from one area (the source area) to another (the sink area) over the interconnected transmission network in a reliable manner constrained by thermal limits, voltage magnitude limits and stability limits. FACTS devices provide better power-flow control by controlling the transmission network parameters like series impedance, shunt impedance, line current, bus voltage and load bus phase angle.
- Under steady-state conditions heavy loading and low voltage are the system limiting factors, and the corrective action is to supply reactive power so as to correct the load power factor and to compensate for the reactive power losses in transmission lines. Traditionally, mechanically switched shunt capacitors and reactors were used for voltage control. Instead, FACTS devices can be used for smooth control of reactive power thereby voltage, without time delay.

The steady-state voltage profile improvement and power-transfer capability enhancement in the system with series and shunt controllers are explained in the following sections.

14.2.2 Transmission system without compensation

The single-line diagram of a typical two-bus transmission system with one generator and load and its phasor diagram are shown in Figure 14.7.

In Figure 14.7(a), $\overline{V}_1 = V_1 \angle \delta$ is the source voltage, $\overline{V}_2 = V_2 \angle 0$ is the load voltage, $R + jX$ is the impedance of the transmission line, and $\overline{I} = I \angle - \phi$ is the current in the transmission line. In the phasor diagram (see Figure 14.7(b)), the current $\overline{I}$ can be split into two components: one in-phase $(\overline{I}_d)$ with the load voltage and other in-quadrature $(\overline{I}_q)$ to the load voltage. The reactive power drawn by

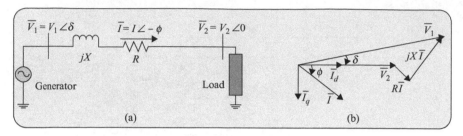

Figure 14.7   (a) Schematic of two bus system and (b) phasor diagram of system without compensation

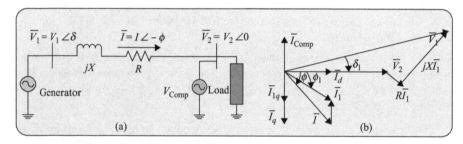

Figure 14.8   (a) Schematic of two bus system with shunt compensation and (b) phasor diagram of system with shunt compensation

the load is given as $\overline{V}_2\overline{I}_q$, and this has to be supplied by the source apart from the reactive power drop XI^2. It can be observed from the phasor diagram that as the reactive power requirement increases the current drawn by the load will increase and also the drop in the transmission line will increase. This leads to a drop in the load voltage $\overline{V}_2$ and hence a poor voltage regulation as well as poor power factor as seen from source side.

14.2.3   Transmission system with shunt compensation

Shunt compensation can be used for the system shown in Figure 14.7 to mitigate the problems due to load reactive power requirement. Figure 14.8 shows ideal shunt compensator connected at bus-2 of the transmission system and its phasor diagram. Assume that an ideal compensator exchanges only reactive power with the system.

In Figure 14.8(a), $\overline{I}_1 = I_1\angle - \phi_1$ is the current passing through the transmission line after including the compensator. The shunt compensator injects a current $I_{\text{comp}}\angle 90°$ in quadrature with the load voltage $\overline{V}_2$ into the transmission line at the load side and supplies some part of the reactive power required by the load. The amount of reactive power drop in the transmission line is reduced from I^2X to I_1^2X due to addition of shunt compensator (see Figure 14.8(b)). Similarly, real power loss in the transmission line I^2R is reduced to I_1^2R. It can be observed from the phasor diagram that, as the reactive power injected by shunt compensator increases, the current drawn from the source reduces gradually and thereby reduces the

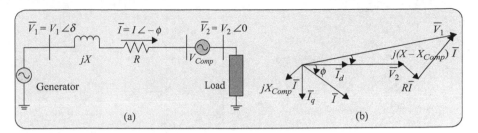

Figure 14.9   *(a) Schematic of two bus system with series compensation and*
(b) phasor diagram of system with series compensation

voltage drop in the transmission line. Thus, the injection of reactive power leads to improvement of voltage profile, power factor and reduction in transmission losses.

14.2.4 Transmission system with series compensation

Figure 14.9 shows ideal series compensator connected between bus-1 and bus-2 and its phasor diagram for the system shown in Figure 14.7. Assume that the compensator exchanges only reactive power with the transmission line.

In Figure 14.9, $V_{comp} = jX_{comp}I$ is the voltage across the ideal series compensator. Series compensator introduces a capacitive voltage drop in the transmission line, which is in quadrature with the line current $\overline{I}$ as shown in Figure 14.9(b). It can be observed from the figure that the amount of reactive power drop in the transmission line is reduced from I^2X to $I^2(X - X_{comp})$ due to the addition of series compensator. Also, the capacitive voltage increases in the line, the total voltage drop in the transmission line gradually reduces. Thus, the injection of reactive power leads to improvement of voltage profile in the system.

14.2.5 Power transfer capability enhancement with shunt compensation

To evaluate the power-transfer capability with shunt compensation, consider the system shown in Figure 14.7, with an ideal shunt compensator connected in the middle of the line as shown in Figure 14.10. Assume that the resistance of the transmission line is neglected and reactance of the line is split into two halves.

In Figure 14.10(a), the shunt compensator is represented by a sinusoidal AC voltage source $\overline{V}_m$ with an amplitude identical to that of the sending and receiving end voltages, that is $|\overline{V}_m| = |\overline{V}_1| = |\overline{V}_2| = V$. $\overline{V}_{sm}$ and $\overline{V}_{rm}$ are the voltages across the first and second half of the transmission line, and $\overline{I}_{sm}$ and $\overline{I}_{mr}$ are the line currents passing through the first and second half of the transmission line, respectively. The values of transmission line voltages and currents can be calculated as

$$\overline{V}_{sm} = j\frac{X}{2}\left(\overline{I}_{sm}\right)$$

(14.2)

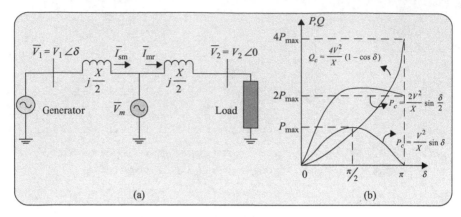

Figure 14.10   (a) Transmission system model with shunt compensation at the
midpoint and (b) the relationship between real and reactive powers
transmitted versus angle

$$\bar{I}_{sm} = \frac{4V}{X}\sin\frac{\delta}{4}\angle\frac{3\lambda}{4} \tag{14.3}$$

$$\bar{V}_{mr} = j\frac{X}{2}(\bar{I}_{mr}) \tag{14.4}$$

$$\bar{I}_{mr} = \frac{4V}{X}\sin\frac{\delta}{4}\angle\frac{\lambda}{4} \tag{14.5}$$

For the assumed loss less system, the real power is the same at each terminal of the
line. The real power transmitted P from source to the load without compensation,
the real and reactive powers injected by the shunt compensator are given in
equations (14.6)–(14.8).

$$P = \text{Real}(VI^*) = \frac{V^2}{X}\sin\delta \tag{14.6}$$

$$P_C = 2\frac{V^2}{X}\sin\frac{\delta}{2} \tag{14.7}$$

$$Q_C = 4\frac{V^2}{X}\left(1-\cos\frac{\delta}{4}\right) \tag{14.8}$$

where P_C and Q_C are the real power transferred from source and reactive power
injected by the shunt compensator, respectively. It can be observed from the
equations that the magnitude of real power transferred from source to the load is
doubled. The power-angle characteristics are shown in Figure 14.10(b), in which
P_{max} is the maximum value of the real power transmitted. It can be observed from
power-angle characteristics that the midpoint shunt compensation can significantly
increase the transmittable power.

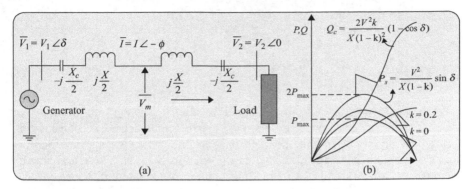

Figure 14.11 *(a) Transmission system model with series compensation along the line and (b) the relationship between real and reactive powers transmitted versus angle*

14.2.6 Power transfer capability enhancement with series compensation

To measure the power-transfer capability with series compensation, consider the system shown in Figure 14.7 but with series capacitance connected in series with the transmission line as shown in Figure 14.11.

In Figure 14.11(a), $\overline{V}_m$ is the midpoint voltage and $\overline{I} = I\angle - \phi$ is the current passing through the transmission line. X is the total reactance of the line and resistance of line is neglected. The basic idea behind series compensation is to decrease the overall effective series impedance of transmission line and this series compensator is represented by two ideal capacitive reactances of value $X_C/2$ each, which is placed in series with the transmission line. The effective transmission line reactance X_{eff} after including the series compensator is given in following equations.

$$X_{\mathrm{eff}} = (1 - k)X \tag{14.9}$$

where k is the degree of series compensation, that is,

$$k = \frac{X_C}{X} \tag{14.10}$$

The series compensator introduces a voltage V_C in the transmission line which is out of phase with the line voltage, there by reduces the total reactive voltage drop which leads to improvement of power transfer from source to load. The real power P and reactive power Q transmitted from source to the load without series compensator can be derived as equations (14.11) and (14.12). The total reactive power injected by the series compensator is given by Q_C and given in (14.13).

$$P = \frac{V^2}{X}\sin\delta \tag{14.11}$$

$$Q = \frac{V^2}{X}(1 - \cos \delta) \tag{14.12}$$

$$Q_C = \frac{2V^2}{X}\frac{k}{(1 - k)^2}(1 - \cos \delta) \tag{14.13}$$

It can be easily observed from the above equations that, the magnitude of real power transferred from source to the load depends on the degree of compensation. Figure 14.11(b) represents the power versus angle characteristics of the transmission system with compensation that is the relation between active and reactive powers supplied from source to load with different amounts of degree of compensation k. In Figure 14.11(b), $k = 0$ curves correspond to the system without compensation. The $k = 0.2$ and $k = 0.4$ curves represent the system with series compensation with degree of compensation 0.2 and 0.4. It can be noticed from power-angle characteristics that series compensation can significantly increase the transmittable power at the expense of a rapidly increasing reactive power demand in the transmission line.

14.3   Classification of FACTS controllers

The classification of FACTS devices has been done on the basis of mechanical and power electronic switching. The mechanical switch-based devices have been developed based on fixed or mechanical switches with passive elements like R, L and C together with transformers.

The power electronics switch-based devices contain power electronic valves or converters which are used to switch the elements in smaller steps or with switching patterns within a cycle. The power electronics-based controllers are further classified as shunt, series, combined shunt–series and combined series–series controllers.

- Shunt controllers
  - Shunt controllers are connected in shunt with the transmission line. They can be of variable impedance, variable source converter type or a combination of both. SVC and STATCOM are two commonly used shunt controllers
- Series controllers
  - Series controllers are connected in series with the transmission line. The series controller could be variable impedance, such as capacitor, reactor or a power electronic-based variable source of main frequency. Thyristor switched series compensator (TSSC), TCSC and static synchronous series compensator (SSSC) are the commonly used series controllers
- Shunt–series controllers
  - Combined shunt–series controllers could be a combination of separate shunt and series controllers, which are controlled in a coordinated manner. The shunt and series parts of the combined shunt and series controllers

inject current in shunt and voltage in series, respectively. UPFC is the best example for this category

- Series–series controllers
    - Combined series–series controllers could be a combination of separate series controllers, which are controlled in a coordinated manner. These controllers have the capability to directly transfer real power between the transmission lines through the common DC-link, along with independently controllable reactive series compensation of each individual line. The interline power flow controller (IPFC) is the important type of combined series–series controller

In this section, principle and operational characteristics (the transmitted power versus angle characteristics) of various FACTS controllers are discussed and the extensive elaborations on FACTS controllers can be found given in References 6–9.

14.3.1 Shunt controllers

Shunt controller is used to enhance the electrical characteristics of the transmission line to increase the steady-state transmittable power and to control the voltage profile along the line. The shunt controller injects or absorbs reactive power into or from the bus when the current injected by the controller is in quadrature with the bus voltage. In the following sections, the principle and operating characteristics of SVC and STATCOM are discussed.

14.3.1.1 Static VAr compensator

A shunt-connected SVC, whose output is adjusted to exchange capacitive or inductive current so as to maintain or control-specific parameters of the electrical power system (typically bus voltage). Fixed capacitor–thyristor-controlled reactor (FC–TCR) and thyristor-switched capacitor–thyristor-controlled reactor (TSC–TCR) are the two main types of configurations to represent SVC. The performance of SVC is determined by its constituents: TCR and TSC.

14.3.1.2 Thyristor-controlled reactor

TCR is a shunt-connected device, whose effective reactance is varied in a continuous manner by partial-conduction control of the thyristor value. A single-phase TCR and its operational waveforms are shown in Figure 14.12. TCR consists of a fixed reactor of inductance L, and a bidirectional thyristor valve.

In Figure 14.12(a), $v(t)$ and $i(t)$ are the voltage applied and current passing through the TCR. The thyristors T_1 and T_2 form a bidirectional thyristor valve. The valve will automatically come to blocking mode immediately after the AC current $i(t)$ crosses zero, unless the gate signal is applied. The current in the reactor can be controlled from maximum to zero by the method of firing delay angle (α) control. The applied voltage $v(t)$ and the reactor current $i_L(t)$, at zero delay angle (switch fully closed) and at an arbitrary a delay angle, are shown in Figure 14.12(b). When $\alpha = 0$, the valve closes at the crest of the applied voltage and the resulting current in the reactor. When the gating of the valve is delayed by an angle α with respect to

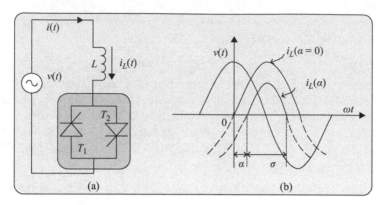

Figure 14.12   (a) Schematic of thyristor-controlled reactor and (b) operating waveforms

the crest of the voltage $v(t)$, the current $i_L(t)$ in the reactor can be expressed as follows:

$$v(t) = V_m \cos(\omega t) \tag{14.14}$$

$$i_L(t) = \frac{1}{L} \int_\alpha^{\omega t} v(t) dt = \frac{V_m}{\omega L} (\sin \omega t - \sin \alpha) \tag{14.15}$$

The delay angle (α) defines the conduction angle (σ), and it is evident from the equation (14.15) that the magnitude of the current $i_L(t)$ in the TCR can be varied from maximum value to zero continuously by controlling the α. In a practical valve contains many thyristors (10–20) are connected in series to meet the required blocking voltage levels at a given power rating.

14.3.1.3   Thyristor-switched capacitor

A shunt-connected, TSC whose effective reactance is varied in a stepwise manner by full- or zero-conduction operation of the thyristor value. A single-phase TSC and its operating waveforms are shown in Figure 14.13. It consists of a capacitor, a bidirectional thyristor valve and a relatively small surge current limiting reactor.

In Figure 14.13(a), $v(t)$ and $i(t)$ are the voltage applied and current passing through the TSC, and $v_L(t)$ is the voltage across the current limiting reactor. Under steady-state conditions, with a sinusoidal applied AC voltage source, $v(t) = V \sin(\omega t)$ across TSC, the current in the branch can be derived as

$$i(\omega t) = V \frac{X_C^2}{X_C^2 - X_L^2} \omega C \cdot \cos(\omega t) \tag{14.16}$$

The TSC branch can be switched off at any current zero, at this time the capacitor voltage will reach its peak value $v_C(t) = V(X_C^2)/(X_C^2 - X_L^2)$ as shown in Figure 14.13(b). The disconnected capacitor stays charged to this voltage and, consequently, the voltage across the non-conducting thyristor valve varies between zero

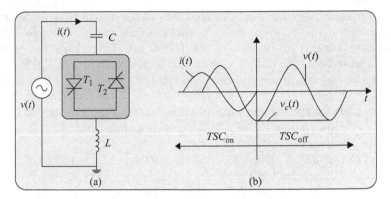

Figure 14.13 (a) *Schematic diagram of thyristor-switched capacitor and* (b) *operating waveforms*

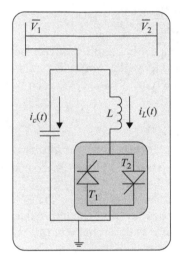

Figure 14.14 *Schematic diagram of FC–TCR type SVC*

and the peak-to-peak value of the applied AC voltage. If the voltage across the disconnected capacitor remained unchanged, the TSC bank could be switched in again, without any transient, at the appropriate peak of the applied AC voltage. But, normally the voltage across the capacitor does not remain constant during the time when the thyristor is switched off, but it is discharged after disconnection. Thus, the reconnection of the capacitor may have to be executed at some residual capacitor voltage between zero and $VX_C^2/(X_C^2 - X_L^2)$. To approximate continuous current variations, several TSC branches in parallel will be used.

14.3.1.4 FC–TCR type static var compensator
The schematic of FC–TCR type SVC connected in system at the bus is shown in Figure 14.14. The SVC consists of a FC and TCR connected in parallel, and it is

connected in shunt with the bus. SVC can be operated in inductive mode or capacitive mode by the firing angle delay control as explained in Section 14.3.1.2. At the maximum capacitive var output, the TCR is off. To decrease the capacitive output, the current in the reactor is increased by decreasing delay angle. At zero var output, the capacitive and inductive currents become equal and thus capacitive and reactive vars cancel out. With further decrease of delay angle, the inductive current becomes larger than the capacitive current, resulting in a net inductive var output. At a zero-delay angle, the TCR conducts current over full 180° interval, resulting in maximum inductive var output.

In Figure 14.14(a), $\overline{V}_1$ and $\overline{V}_2$ are the bus voltages and $\overline{i}$, $\overline{i}_C$ and $\overline{i}_L(\alpha)$ represent the currents passing through the circuit, fixed capacitor and TCR, respectively. The equivalent admittance B_{SVC} is the parallel combination of capacitive admittance and the TCR admittance. The value of the admittance of SVC at fundamental frequency is given below

$$B_{SVC} = \frac{X_C(2\sigma + \sin 2\alpha) - \pi X_L}{\pi X_C X_L} \tag{14.17}$$

where X_C and X_L are the capacitive and inductive reactances, respectively and σ, α are the conduction and firing angles of the thyristors. The reactive power, Q_{SVC}, injected by the SVC into the transmission system can be given as equation (14.18).

$$Q_{SVC} = B_{SVC} V^2 \tag{14.18}$$

14.3.1.5   TSC–TCR type static var compensator

The TSC–TCR type SVC is shown in Figure 14.15, and it consists of n number of TSC branches and one TCR branch.

In Figure 14.15, $v(t)$ is the voltage across the SVC, and the total capacitive output range of TSC branches are divided into n intervals. In the first interval, the output of the var generator is controllable in the zero to Q_{Cmax}/n range, where Q_{Cmax} is the total rating provided by all TSC branches. In this interval, one

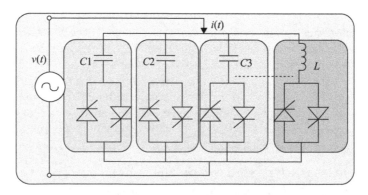

Figure 14.15   Basic TSC–TCR type static var compensator

capacitor bank is switched so that, the current in the TCR is set by the appropriate firing delay angle so that the sum of the var output of the TSC (negative) and that of the TCR (positive) equals the capacitive output required. In the second, third, ..., and nth intervals, the output is controllable in the Q_{Cmax}/n to $2Q_{Cmax}/n$, $2Q_{Cmax}/n$ to $3Q_{Cmax}/n$, ... and $(n-1)Q_{Cmax}/n$ to Q_{Cmax} range by switching in the second, third, etc. and nth capacitor bank and using the TCR to absorb the surplus capacitive VArs. By switching the capacitor banks on and off within one cycle of the applied AC voltage, the maximum surplus capacitive VAr in the total output range can be restricted to that produced by one capacitor bank, and thus the TCR should have the same VAr rating as the TSC. The reduction in the reactor size, increased flexibility of control and better performance under system fault conditions are the main advantages of TSC–TCR type SVC over FC–TCR type SVC.

To explain the transmitted power versus angle characteristics of system with SVC, assume the SVC connected in the middle of the transmission line shown in Figure 14.16(a), and its phasor diagram is shown in Figure 14.16(b). Assume that the end line voltages, and the midline voltage all have the same magnitude V. The phasor angle of $\overline{V}_2$ is set to zero, and it can be used as reference value for the other phasors.

$$\overline{V}_2 = Ve^{j0}, \quad \overline{V}_1 = Ve^{j\delta}, \quad \overline{V}_S = Ve^{j\delta/4} \tag{14.19}$$

By applying Kirchoff's laws in the circuit shown in Figure 14.16(a), $\overline{I}_2$ can be derived as

$$\overline{I}_2 = \frac{4V}{X} \sin(\delta/4)\, e^{j\delta/4} \tag{14.20}$$

and the resultant transmitted power is

$$P = \text{Real}(\overline{V}_1\, \overline{I}_1^*) = \text{Real}(\overline{V}_2\, \overline{I}_2^*) \tag{14.21}$$

$$P = \text{Real}\left(\frac{4V^2}{X} \sin(\delta/4)(\cos(\delta/4) - j\sin(\delta/4))\right) = \frac{2V^2}{X} \sin(\delta/2) \tag{14.22}$$

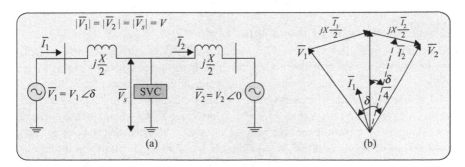

Figure 14.16   *(a) Two machine system with SVC in the middle and (b) phasor diagram of the system*

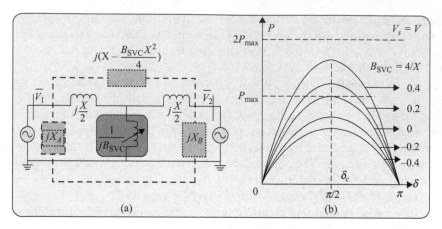

Figure 14.17   (a) Equivalent network of the two machine system and
(b) transmitted power versus angle characteristics

As the transmitted power without the SVC is $(V^2/X)\sin(\delta)$, and the maximum transmittable power is doubled from V^2/X to $2V^2/X$.

In the above discussions, it was assumed that the SVC can provide the voltage $|\overline{V}_S| = V$ at any transmission angle. If we look at the SVC as adjustable susceptance B_{SVC}, then there will be upper and lower limits to the value of SVC susceptance. The larger the transmission angle then larger the necessary susceptance, because at the same voltage a higher current has to be provided.

In Figure 14.17(a), the impedance of the system is shown by dotted lines, and this can be transformed by Y–D transformation as shown. After simplification, we get the transmitted power in terms of susceptance as given in (14.23).

$$P = \frac{V_1 V_2}{X - (X^2 B_{SVC}/4)} \tag{14.23}$$

The transmitted power versus transmission angle characteristic is shown in Figure 14.17(b) with various susceptance values.

14.3.1.6   Static synchronous compensator

STATCOM is operated as a shunt-connected SVC whose output current (capacitive or inductive) can be controlled independently. It provides voltage support by generating or absorbing reactive power at the point of common coupling.

The basic voltage–source converter (VSC) employed STATCOM is shown in Figure 14.18(a). The charged capacitor C_{DC} provides a DC voltage to the converter. By varying the amplitude of the output voltage V, the reactive power exchange between the converter and the AC system can be controlled. If the amplitude of the output voltage $\overline{V}$ is increased above that of the AC system $\overline{V}_t$, a leading current is produced by STATCOM, that is, the STATCOM is seen as a capacitor by the AC system, and reactive power is generated. Decreasing the amplitude of the output

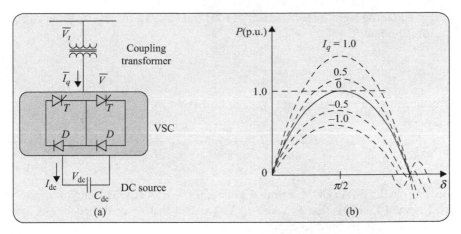

Figure 14.18   (a) Schematic of static synchronous compensator (STATCOM) and (b) transmitted power versus transmission angle characteristics

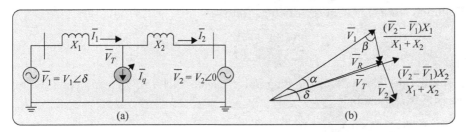

Figure 14.19   (a) Two machine system with STATCOM and (b) phasor diagram of the system

voltage below that of the AC system, a lagging current results and the STATCOM is seen as an inductor. In this case, reactive power is absorbed. If the amplitudes are equal no power exchange takes place. Instead of a capacitor a battery can also be used as DC energy. In this case, the converter can control both reactive and active power exchange with the AC system.

To explain the transmitted power versus angle characteristics of system with STATCOM, let us assume the STATCOM connected to the transmission line shown in Figure 14.19(a), and its phasor diagram is shown in Figure 14.19(b), where $\overline{V}_1$ and $\overline{V}_2$ are the voltages across the source and the load, X_1 and X_2 are the line reluctances, $\overline{V}_T$ and $\overline{I}_q$ are the voltage across and current passing through the STATCOM.

Alloying the Kirchoff's laws for the above figure, we get

$$\overline{I}_2 = \frac{\overline{V}_T - \overline{V}_2}{jX_2} = \frac{(\overline{V}_1 - j\overline{I}_1 X_1) - \overline{V}_2}{jX_2} \tag{14.24}$$

$$\overline{I}_2 = \overline{I}_1 - \overline{I}_q \tag{14.25}$$

By equalling right-hand terms of (14.25) and (14.24), the value of the current $\overline{I}_1$ is obtained as

$$\overline{I}_1 = \frac{\overline{V}_T - \overline{V}_2}{j(X_1 + X_2)} + I_q \frac{X_2}{(X_1 + X_2)} \tag{14.26}$$

from the value of $\overline{I}_1$, the voltage $\overline{V}_T$ can be derived as

$$\overline{V}_T = \overline{V}_1 - j\overline{I}_1 X_1 = \overline{V}_1 - \frac{(\overline{V}_1 - \overline{V}_2)X_1}{(X_1 + X_2)} - jI_q \frac{X_1 X_2}{X_1 + X_2} = \overline{V}_R - j\overline{I}_q \frac{X_1 X_2}{X_1 + X_2} \tag{14.27}$$

where $\overline{V}_R$ is the STATCOM terminal voltage if the STATCOM is out of operation (see Figure 14.19(b)), that is when $\overline{I}_{q=0}$. The fact that $\overline{I}_q$ is shifted by 90° with regard to $\overline{V}_R$ can be used to express $\overline{I}_q$ as

$$\overline{I}_q = jI_q \frac{\overline{V}_R}{V_R} \tag{14.28}$$

Equation (14.27) is then rewritten as follows:

$$\overline{V}_T = \overline{V}_R + I_q \frac{\overline{V}_R}{V_R} \cdot \frac{X_1 X_2}{X_1 + X_2} = \overline{V}_R \left(1 + \frac{I_q}{V_R} \cdot \frac{X_1 X_2}{X_1 + X_2}\right) \tag{14.29}$$

Applying the sine law to the diagram in Figure 14.19(b), we can obtain the following two equations:

$$\frac{\sin \beta}{V_2} = \frac{\sin \delta}{|\overline{V}_1 - \overline{V}_2|} \tag{14.30}$$

and

$$\frac{\sin \alpha}{|\overline{V}_1 - \overline{V}_2|(X_1/(X_1 + X_2))} = \frac{\sin \beta}{V_R} \tag{14.31}$$

from which value of $\sin \alpha$ is derived

$$\sin \alpha = \frac{V_2 \sin \delta X_1}{V_R(X_1 + X_2)} \tag{14.32}$$

The transmitted active power can be given as

$$P = \frac{V_T V_1}{X_1} \sin \alpha = \frac{V_1 V_2 \sin \delta}{X_1 + X_2} \frac{V_T}{V_R} \tag{14.33}$$

To eliminate the term V_R the cosine law is applied to Figure 14.19(b) and we can obtain

$$V_R = |\overline{V}_R| = \left|\frac{\overline{V}_1 X_2 + \overline{V}_2 X_1}{X_1 + X_2}\right| = \frac{\sqrt{V_1^2 X_2^2 + V_2^2 X_1^2 + 2V_1 V_2 X_1 X_2 \cos \delta}}{X_1 + X_2} \tag{14.34}$$

Substituting the value of V_R and (14.29) into (14.33), we can get the value of the transmitted active power as

$$P = \frac{V_1 V_2 \sin \delta}{X_1 + X_2} \left(1 + \frac{I_q}{V_R} \cdot \frac{X_1 X_2}{X_1 + X_2}\right) \tag{14.35}$$

The resultant transmitted power versus angle characteristics are given in Figure 14.18(b).

14.3.1.7 Comparisons of shunt compensators

- SVC and STATCOM are very similar in their functional compensation capability, but the basic operating principles are fundamentally different. A STATCOM functions as a shunt-connected synchronous voltage source, whereas a SVC operates as a shunt-connected, controlled reactive admittance. This difference accounts for the STATCOM's superior functional characteristics, better performance and greater application flexibility than those attainable with a SVC.

- In the linear operating range, the *V–I* characteristics (see Figure 14.20(b)) and functional compensation capability of the STATCOM and the SVC are similar. Concerning the non-linear operating range, the STATCOM is able to control its output current over the rated maximum capacitive or inductive range independently of AC system voltage, whereas the maximum attainable compensating current of the SVC decreases linearly with AC voltage. Thus, the STATCOM is more effective than the SVC in providing voltage support under large system disturbances during which the voltage excursions would be well outside of the linear operating range of the compensator. The ability of the STATCOM to maintain full capacitive output current at low system voltage also makes it more effective than the SVC in improving the transient stability.

- The attainable response time and the bandwidth of the closed voltage regulation loop of the STATCOM are also significantly better than those of the SVC.

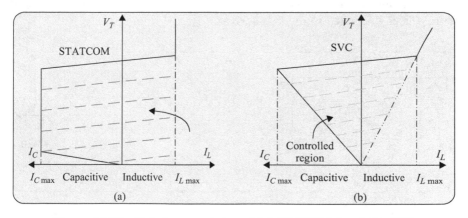

Figure 14.20 V–I characteristics of (a) STATCOM and (b) SVC

In situations in which it is necessary to provide active power compensation, the STATCOM is able to interface a suitable energy storage (large capacitor, battery etc.) from which it can draw active power at its DC terminal and deliver it as AC power to the system, but the SVC does not have this capability.

14.3.2   Series controllers

The variable series compensation is highly effective in both controlling power flow in the line and in improving stability. The series controller could be variable impedance, such as capacitor, reactor or a power electronic-based variable source of main frequency. With series compensation, the overall effective series transmission impedance from the sending end to the receiving end can be arbitrarily decreased thereby influencing the power flow ($P = (V^2/X) \sin \delta$). This capability to control power flow can effectively be used to increase the transient stability limit and to provide power oscillation damping.

Thyristor-switched series reactor, TCSC and SSSC are the main series controllers. The operational principle and characteristics of these controllers are explained in the following section.

14.3.2.1   Thyristor-switched series capacitor

A TSSC consists of a capacitor shunted by a thyristor valve to provide a smoothly variable series capacitive reactance.

The schematic of a TSSC is shown in Figure 1.21(a). The capacitor is inserted into the line if the corresponding thyristor valve is turned off; otherwise, it is bypassed. A thyristor valve is turned off in an instance when the current crosses zero. Thus, the capacitor can be inserted into the line by the thyristor valve only at the zero crossings of the line current. On the other hand, the thyristor valve should be turned on for bypass only when the capacitor voltage is zero in order to minimize the initial surge current in the valve and the corresponding circuit transient.

Therefore, if the capacitor is once inserted into the line, it will be charged by the line current from zero to maximum during the first half-cycle and discharged from maximum to zero during the successive half-cycle until it can be bypassed again. This is illustrated in Figure 14.21(b).

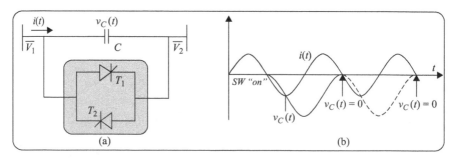

Figure 14.21   (a) The schematic of a TSSC and (b) associated voltage and current wave forms

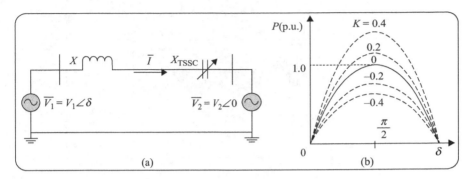

Figure 14.22   (a) Two systems with TSSC and (b) the transmitted power versus angle characteristics

The TSSC can be considered as a controllable reactance in series with the line reactance as shown in Figure 14.22(a). The ratio of the inserted TSSC reactance to the line reactance is a measure for the compensation degree of the line given by

$$K = \frac{X_{\text{TSSC}}}{X} \tag{14.36}$$

The transmitted active power is calculated from the general formula for transmitted active power on a line and is given as

$$P = \frac{V_1 V_2}{X + X_{\text{TSSC}}} \sin \delta \tag{14.37}$$

Thus, the transmitted active power versus angle characteristic for a TSSC is shown in Figure 14.22(b). It can be seen that the value of K determines the maximal transmittable power.

14.3.2.2   Thyristor-controlled series capacitor

A TCSC consists of a series capacitor bank shunted by a thyristor-controlled reactor in order to provide a smoothly variable series capacitive reactance.

Figure 14.23 represents the basic TCSC connected in series with a transmission line connected between the two buses 1 and 2. R and X are the resistance and reactance of the transmission line. TCSC consists of the series compensating capacitor C shunted by a TCR. The currents passing through the capacitor and TCR circuits in terms of delay angle α are represented by $i_C(\alpha)$ and $i_{\text{TCR}}(\alpha)$, respectively, and i is the current in the transmission line. The TCR acts as a continuously vari- able reactive impedance at fundamental system frequency, and controlled by delay angle α as explained in Section 14.3.2.1.

Basically, TCSC is operated in three different modes: bypassed mode, thyristor valve blocked mode and vernier control mode and they are explained as following:

Bypassed mode: Figure 14.24 represents the bypassed mode of TCSC. In this mode of operation, the thyristor valves T1 and T2 as shown in Figure 14.24 are gated for 108° conduction (in each direction), and the current flow in the reactor is

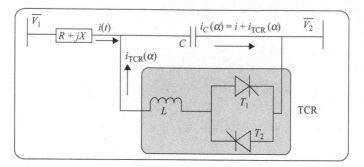

Figure 14.23   Schematic diagram of TCSC

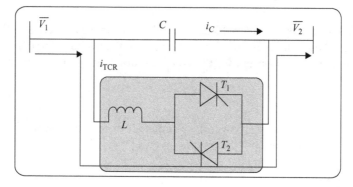

Figure 14.24   Bypassed mode of TCSC

continuous and sinusoidal. The net reactance of the module is slightly inductive as the susceptance of the reactor is larger than that of the capacitor. During this mode, most of the line current is flowing through the reactor and thyristor valves with some current flowing through the capacitor. This mode is used mainly for protecting the capacitor against over voltages.

Thyristor valve blocked mode: Figure 14.25 represents the thyristor valve blocked mode of TCSC. In this operating mode, no current flows through the valves with the blocking of gate pulses. Here, the TCSC reactance is the same as that of the fixed capacitor, and there is no difference in the performance of TCSC in this mode with that of a fixed capacitor.

Vernier control mode: Figure 14.26 represents the vernier control mode of TCSC. In this operating mode, the thyristor valves are gated in the region of $(\alpha_{min} < \alpha < 90°)$ such that they conduct for the part of a cycle. The effective value of TCSC reactance (in the capacitive region) increases as the conduction angle increases from zero. α_{min} is the value of α corresponding to the parallel resonance of TCR and the capacitor (at fundamental frequency). In the inductive vernier mode, the TCSC (inductive) reactance increases as the conduction angle (α) reduced from 108°. The direction of currents through the capacitor i_C and TCSC,

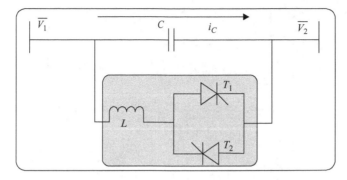

Figure 14.25 Thyristor valve blocked mode of TCSC

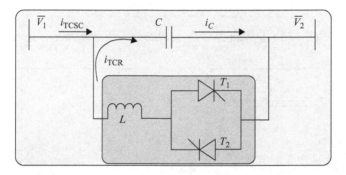

Figure 14.26 Vernier control mode of TCSC

i_{TCSC} are as shown in Figure 14.26. Generally, vernier control is used only in the capacitive region and not in the inductive region.

The steady-state impedance of the TCSC circuit (see Figure 14.23) is a parallel LC circuit, consisting of a fixed capacitive impedance X_C and a thyristor-controlled reactor impedance $X_{\mathrm{TCR}}(\alpha)$, the effective value of reactance $X_{\mathrm{TCSC}}(\alpha)$ is given below:

$$X_{\mathrm{TCSC}}(\alpha) = \frac{X_C\, X_{\mathrm{TCR}}(\alpha)}{X_{\mathrm{TCR}}(\alpha) - X_C} \tag{14.38}$$

where $X_{\mathrm{TCR}}(\alpha)$ is the reactance of the TCR at the fundamental frequency and is given as following:

$$X_{\mathrm{TCR}}(\alpha) = X_L \frac{\pi}{(\pi - 2\alpha - \sin \alpha)} \tag{14.39}$$

where, $X_L = \omega\, L$, and α is the delay angle.

The impedance versus delay angle characteristics of TCSC are shown in Figure 14.27. In Figure 14.27, $\alpha_{L\mathrm{lim}}$ and $\alpha_{C\mathrm{lim}}$ represents the limits of the delay angle corresponding to inductive and capacitive regions of TCSC. TCSC has two

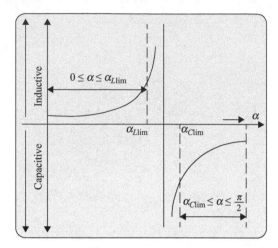

Figure 14.27   The impedance versus delay angle characteristics of TCSC

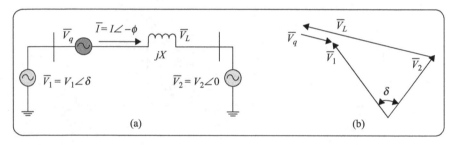

Figure 14.28   (a) Synchronous voltage source for compensation and (b) phasor diagram of the system with SSSC

operating ranges inductive and capacitive depending on the value of α varied from 0 to 90°.

As a TCSC is based on the same idea as the TSSC, namely to introduce additional reactances, the characteristics of the transmitted power versus transmission angle looks alike, the one of the TSSC in Figure 14.22.

14.3.2.3   Static synchronous series compensator

A SSSC is static synchronous generator operated without an external electric energy source as a series compensator whose output voltage is in quadrature with, and controllable independently of, the line current for the purpose of increasing/decreasing the overall reactive voltage drop across the line and thereby controlling the transmitted power.

A SSSC is a VSC-based series compensator. The principle operation of SSSC can be explained with Figure 14.28(a); it shows a two-bus system with SSSC connected in the line. The phasor diagram (see Figure 14.28(b)) shows that the voltage source increases the magnitude of the voltage across the inductance, that is the line, and therefore also increases the magnitude of the current $\overline{I}$ resulting in an increase in the power flow. This corresponds to the effect of a capacitor placed in series. By making the output voltage of the synchronous voltage source $\overline{V}_q$ a function of the current $\overline{I}$, the same compensation as provided by the series capacitor is accomplished.

However, with a voltage source, it is possible to maintain a constant compensating voltage in the presence of variable line current because the voltage can be controlled independently of the current, that is the voltage source can also decrease the voltage across the line inductance having the same effect as if the reactive line impedance was increased. Thus, the SSSC can decrease as well as increase the power flow to the same degree, simply by reversing the polarity of the injected AC voltage. The series reactive compensation scheme, using a switching power converter as a synchronous voltage, is termed SSSC. The relation between transmitted power versus angle with SSSC in the system is as follows.

The SSSC injects the compensating voltage in series with the line irrespective of the line current, and the value of the current $\overline{I}$ is given by

$$\overline{I} = \frac{\overline{V}_1 - \overline{V}_q - \overline{V}_2}{jX} = \frac{1}{jX}\left((\overline{V}_1 - \overline{V}_2) - V_q \cdot \frac{\overline{V}_1 - \overline{V}_2}{|\overline{V}_1 - \overline{V}_2|} \right) \tag{14.40}$$

by simplifying the above equation we get

$$\overline{I} = \frac{j(\overline{V}_2 - \overline{V}_1)}{X}\left(1 - \frac{V_q}{|\overline{V}_1 - \overline{V}_2|} \right) \tag{14.41}$$

The term $(\overline{V}_1 - \overline{V}_2)$ represents the phasor difference between $\overline{V}_1$ and $\overline{V}_2$. Without source this would be the voltage drop on reactance X. The injected voltage $\overline{V}_q$ phasor has the same direction as it is a reactive voltage source. This direction is determined by the term $(\overline{V}_1 - \overline{V}_2)/|\overline{V}_1 - \overline{V}_2|$. Choosing $\overline{V}_2$ as reference phasor, that is $\overline{V}_2 = V_2$, $\overline{V}_1 = \overline{V}_1(\cos\delta + j\sin\delta)$, the transmission characteristic can be obtained from the following equation:

$$P_1 = P_2 = P = \mathrm{Real}(\overline{V}_1 I^*) = \overline{V}_2\,\mathrm{Real}(\overline{I}^*) \tag{14.42}$$

and

$$|\overline{V}_1 - \overline{V}_2| = \sqrt{V_1^2 + V_2^2 - 2V_1 V_2 \cos\delta} \tag{14.43}$$

The following formula results for the transmitted active power,

$$P = \frac{\overline{V}_1 \overline{V}_2 \sin\delta}{X}\left(1 - \frac{V_q}{\sqrt{V_1^2 + V_2^2 - 2V_1 V_2 \cos\delta}} \right) \tag{14.44}$$

Therefore, the transmitted power P is a function of the injected voltage V_q. The transmitted power versus transmission angle characteristics are given in Figure 14.29.

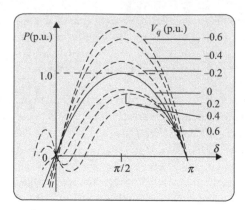

Figure 14.29　Transmitted power versus transmission angle provided by the SSSC

14.3.2.4　Comparison of series compensators

The SSSC is a voltage source type and the TSSC and TCSC are variable impedance type series compensators. Resulting from the different structures, there are essential differences in characteristics and features of these devices:

- The SSSC is capable of internally generating a controllable compensating voltage over an identical capacitive and inductive range independently of the magnitude of the line current. The compensating voltage of the TSSC over a given control range is proportional to the line current. The TCSC can maintain maximum compensating voltage with decreasing line current over a control range determined by the current boosting capability of the thyristor-controlled reactor.

- The SSSC has the ability to interface with an external DC power supply to provide compensation for the line resistance by the injection of active power as well as for the line reactance by the injection of reactive power. The variable impedance type series compensators cannot exchange active power with the transmission line and can only provide reactive compensation.

- The SSSC with an energy storage increases the effectiveness of power oscillation damping by modulating the series reactive compensation to increase and decrease the transmitted power and by concurrently injecting an alternating virtual positive and negative real impedance to absorb and supply active power from the line in sympathy with the prevalent machine swings. The variable impedance type compensators can damp power oscillation only by modulated reactive compensation affecting the transmitted power.

Series reactive compensation can be highly effective in controlling power flow in the line and in improving the dynamic behaviour of the power system. But certain problems related to the transmission angle cannot be handled by series compensation. For example, the prevailing transmission angle may not be compatible with the transmission requirements of a given line or it may vary with daily or seasonal system loads over too large a range to maintain acceptable power flow in some lines.

14.3.3 Combined FACTS controllers

In the preceding sections, shunt controllers and series controllers have been considered. They both have different influences on the line. In this section, two of these controllers, namely UPFC and the IPFC are discussed.

14.3.3.1 Unified power flow controller

A UPFC is a combination of STATCOM and an SSSC, which are coupled via a common DC link, to allow bidirectional flow of active power between the series output terminals of the SSSC and the shunt output terminals of the STATCOM, and are controlled to provide concurrent active and reactive series line compensation without an external electric energy source.

The UPFC is conceptually a synchronous voltage source with controllable magnitude V_{pq} and angle placed in series with the line (see Figure 14.30(a)). The voltage source exchanges both active and reactive power with the transmission system. But the voltage source can only produce reactive power, the active power has to be supplied to it by a power supply or a sink. A UPFC consists of two VSCs which are placed back-to-back and operated from a common DC link. This implementation is shown in Figure 14.31.

The active power can freely flow in either direction between the AC terminals of the converters and each converter can generate or absorb reactive energy independently. Converter 2 injects the voltage V_{pq}, which is controllable in magnitude and phase (ρ), in series with the line and therefore acts as the voltage source. The reactive power exchanged at the AC terminal is generated by the converter internally. Opposed to this, the active power is converted into DC power which appears at the DC link as a positive or negative active power demand. This DC power demand is converted back to AC power by converter-1 and coupled to the transmission line bus via the supply transformer. In addition, converter-1 can also exchange reactive power with the line, if necessary and provide independent shunt reactive compensation for the line.

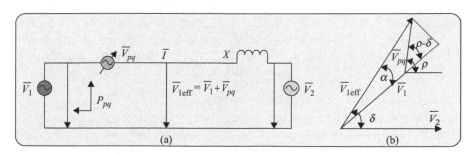

Figure 14.30 (a) UPFC in two machine system and (b) phasor diagram of the system with UPFC

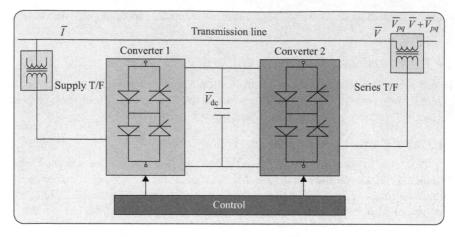

Figure 14.31   Implementation of a UPFC in the system

For the system given in Figure 14.30(a), the transmitted active power can be calculated as

$$P_1 = P_2 = P = \text{Real}\left(\overline{V}_2 \cdot \overline{I}^*\right) \tag{14.45}$$

with,

$$\overline{I} = \frac{\overline{V}_{1\text{eff}} - \overline{V}_2}{jX} = \frac{\overline{V}_1 \angle \delta + \overline{V}_{pq} \angle \rho}{jX} \tag{14.46}$$

results in

$$P = \text{Real}\left(V_2 \frac{V_1(\cos \delta - j\sin \delta) + V_{pq}(\cos \rho - j\sin \rho) - V_2}{-jX}\right) \frac{V_1 V_2}{X} \sin \delta \tag{14.47}$$

$$P = \frac{V_1 V_2}{X} \sin \delta + \frac{V_2 V_{pq}}{X} \sin \rho \tag{14.48}$$

For a maximal influence of V_{pq} on the transmitted power, the angle ρ is equal to 90°. In Figure 14.32, the corresponding transmitted power versus transmission angle characteristic is shown. Thus, the transmission characteristic is shifted up and down depending on the magnitude of the voltage of the UPFC.

14.3.3.2   Interline power flow controller

The IPFC is the combination of two or more SSSCs that are coupled via a common DC link to facilitate a bi-directional flow of active power between the AC terminals of the SSSCs and are controlled to provide independent reactive compensation for the adjustment of active power flow in each line and maintain the desired distribution of reactive power flow among the lines.

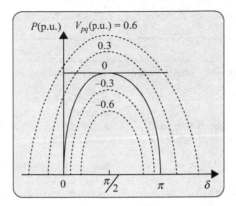

Figure 14.32 Transmitted power versus transmission angle for UPFC ($\rho = 90°$)

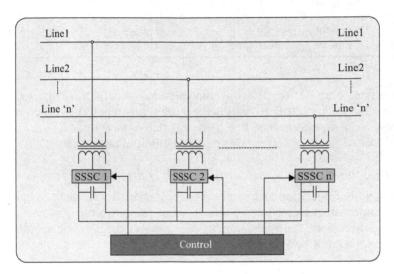

Figure 14.33 Schematic of IPFC in the system

The IPFC addresses the problem of compensating a number of transmission lines at a given substation. It is possible to transfer the active power between different transmission lines with IPFC and the following objectives can be met with IPFC,

- equalize both active and reactive power flow between the lines,
- reduce the burden of overloaded lines by active power transfer,
- compensate against resistive line voltage drops and the corresponding reactive power demand and increase the effectiveness of the overall compensating system for dynamic disturbances.

The general form of a IPFC is shown in Figure 14.33. It employs a number of DC-to-AC converters, namely SSSC, each providing series compensation for a

Table 14.1   Applications of various FACTS controllers to power system

Objective	Reason	Corrective action	FACTS required
Voltage limits	Low voltage at heavy load	Supply reactive power	SVC, STATCOM
		Reduction of line reactance	TCSC
	High voltage at low load	Absorbs reactive power	SVC, STATCOM
	High voltage following an outage	Absorbs reactive power	SVC, STATCOM
	Low voltage following an outage	Supply reactive power	SVC, STATCOM
Thermal limits	Transmission circuit overload	Increases line capacity	TCSC, SSSC, UPFC
Load flow	Power distribution on parallel lines	Adjust line reactances	TCSC, SSSC, UPFC
		Adjust phase angle	UPFC, SSSC
	Load flow reversal	Adjust phase angle	UPFC, SSSC
Stability	Limited transmission power	Decrease line reactance	TCSC, SSSC

different line. With this scheme, the converters do not only provide series reactive compensation but can also be controlled to supply active power to the common DC link from its own transmission line. Like this, active power can be provided from the overloaded lines for active power compensation in other lines.

14.3.4   Summary of application of FACTS controllers

The principle of operation and power angle characteristics of shunt compensators, series compensators as well as combinations of these two types of compensators have been discussed. The application of these devices depends on the problem that has to be solved has been summarized in Table 14.1.

14.4   Optimal placement of FACTS controllers

In order to get the maximum benefits from the FACTS controllers, it is very important to optimally place the devices in the power system with appropriate rating; otherwise, it will have negative impacts on the transmission system. There are several methods suggested in the literature for finding the optimal locations of various FACTS controllers, such as TCSC, thyristor-controlled phase-angle regulator (TCPAR), SVC and UPFC to minimize the power loss and to maximize the power transfer capability.

Singh *et al.* [10] suggested an approach based on the sensitivity analysis on FATCS device control parameters with respect to reduction in real power-flow performance index to enhance the security of the power system. Gerbex *et al.* [11] advised genetic algorithm to seek the optimal location, controller type and device rating of multi-type FACTS controllers: SVC, TCSC, TCVR and TCPST in the power system. Preedavichit and Srivastava [12] used loss sensitivity index to determine the optimal placement of TCSC, TCPAR and SVC to minimize the total system real power loss. Verma *et al.* [13] advised the real power flow performance index used to determine the suitable locations of TCSC and TCPAR for total transfer capability (TTC) to enhance the TTC. Leonidaki *et al.* [14] used repeated power-flow method combined with the sensitivity index of the loading margin to transmission line impedances to determine the location of TCSC for maximizing TTC. Wu *et al.* [15] proposed that a group search optimizer with multiple producer method is used to optimally place TCSC, SVC and thyristor-controlled phase shifting transformer to minimize the real power loss. Idris *et al.* [16] suggested bees algorithm to determine the optimal allocation of SVC and TCSC devices for maximizing the available transfer capability. Mori and Goto [17] explain a Tabu Search algorithm-based technique to optimally locate the multi-type FACTS controllers to minimize the power system loss and to improve the loadability in a power system network. Rajaraman *et al.* [18] have used continuation power flow for obtaining the size and locations of series compensators to increase the loadability limit of the system. Kazemi *et al.* [19] proposed eigen-vector analysis for optimizing location, size and control modes of SVC and TCSC in order to achieve the maximum loadability.

The methods cited above has their own advantages as well as disadvantages. A simple and reliable method for optimal location of FACTS considering the sensitivity of the system with respect to the control parameter of FACTS controller is given in Reference 12, and it has been taken as reference in this chapter to explain the concept of power transfer capability enhancement. In Reference 12, sensitivity index-based methods calculated the sensitivity of total transmission loss (P_{Loss}) with respect to the control parameters of SVC and TCSC for optimal placement. The control parameter is the line net series reactance (X_{Line}) in a line for TCSC and reactive power injection (Q) at a bus for SVC. Sensitivity index-based methods are best suitable for optimally placing SVC and TCSC when the system is under steady-state condition. After fixing the location, particle swarm optimization (PSO) algorithm is used to find the optimal ratings of SVC and TCSC to minimize the loss and maximize the power transfer capability.

14.4.1   Optimal placement of static var compensator

The SVC and its steady-state model are shown in Figure 14.34(a) and (b), respectively. SVC is a shunt-compensating device that should be placed at a particular bus in the system. A sensitivity index-based method [12] was adopted for optimal placement of SVC to minimize the power loss. The loss sensitivity index is the sensitivity of the system power loss with respect to the reactive power injected by SVC at a bus.

In a power system with n number of buses, the total power loss of the system in terms of the injected real and reactive powers at each bus can be expressed as following [20]:

$$P_{\text{Loss}} = \sum_{i=1}^{n}\sum_{j=1}^{n}\left[\alpha_{ij}\left(P_iP_j + Q_iQ_j\right) + \beta_{ij}\left(Q_iP_j - P_iQ_j\right)\right] \tag{14.49}$$

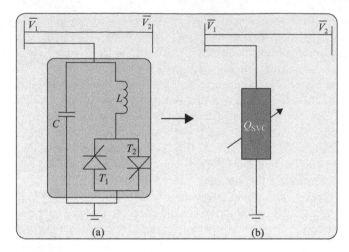

Figure 14.34   (a) Schematic of SVC in the system and (b) steady-state model of SVC

where,

$$a_{ij} = \frac{r_{ij}}{V_i V_j} \cos(\delta_i - \delta_j) \tag{14.50}$$

$$\beta_{ij} = \frac{r_{ij}}{V_i V_j} \sin(\delta_i - \delta_j) \tag{14.51}$$

where α, β represent the loss coefficients. P_i, P_j and Q_i, Q_j represent the real power and reactive power injections at buses i and j, respectively. V_i, V_j and δ_i, δ_j represent the voltages and phase angles at the buses i and j, respectively. r_{ij} is the resistance of the transmission line that is connected between the buses i and j. If it is assumed that SVC is placed at all the buses injecting certain amount of reactive power then the sensitivity of the power system loss with respect to reactive power Q_i, injected at the ith bus due to SVC is given as

$$\frac{\partial P_{\text{Loss}}}{\partial Q_i} = 2 \sum_{j=1}^{n} \left(a_{ij} Q_j + \beta_{ij} P_j \right) \tag{14.52}$$

At ith bus, if the sensitivity is negative, it means that with increase in the injected reactive power at ith bus the system loss will decrease and if sensitivity is positive then losses will increase. Hence, the SVC has to be placed at a bus which has most negative sensitivity index.

The optimal placement of SVC for minimizing the power loss is the best location for maximizing the TTC as well, because the best location of SVC for minimizing the loss improves the voltage profiles in the total system which leads to the enhancement of TTC.

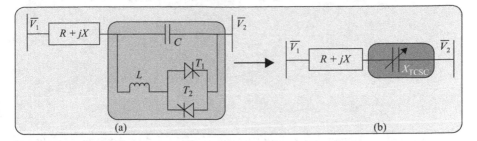

Figure 14.35   (a) Schematic of TCSC in the system and (b) equivalent steady-state model of TCSC

14.4.2   Optimal placement of thyristor-controlled series capacitor

The TCSC and its steady-state model are shown in Figure 14.35(a) and (b), respectively. TCSC is a series compensating device that should be placed in a particular transmission line in the system.

A sensitivity index-based method [12] has been taken for optimal placement of TCSC to minimize the power loss. Loss sensitivity index is the sensitivity of the system loss with respect to the reactance of the transmission line at which TCSC is placed. It can be computed from the power loss equation [20] of the system which is having *l* number of lines is given by

$$P_{\text{Loss}} = \sum_{\text{Line}=1}^{l} P_{\text{Line}} \tag{14.53}$$

where

$$P_{\text{Line}} = \left(\frac{V_i^2 + V_j^2 - 2V_iV_j \cos(\delta_i - \delta_j)}{r_n^2 + x_n^2} \right) * r_n \tag{14.54}$$

where P_{Loss} is the total power loss, and P_{Line} is the loss in a particular transmission line connected between the buses *i* and *j*. V_i and V_j represent the magnitudes of the voltages at the buses *i* and *j*, respectively. δ_i and δ_j are phase angles at the buses *i* and *j*. r_n is the resistance of the line that is connected between the buses *i* and *j*. x_n is the total reactance of the transmission line in which TCSC is placed. If it is assumed that TCSC is placed in all the lines, and it is injecting reactive power into the lines, then the sensitivity of the power system loss with respect to controlled parameter, that is reactance of the transmission line in which TCSC is placed is given as

$$\frac{\partial P_{\text{Loss}}}{\partial x_n} = 2G_nB_n\left(V_i^2 + V_j^2 - 2V_iV_j \cos(\delta_i - \delta_j)\right) \tag{14.55}$$

where,

$$x_n = X_{\text{TCSC}} + X_{\text{Line}} \tag{14.56}$$

$$G_n = \frac{r_n}{r_n^2 + x_n^2} \tag{14.57}$$

$$B_n = \frac{-x_n}{r_n^2 + x_n^2} \tag{14.58}$$

where X_{TCSC} is the reactance offered by the TCSC. x_{ij} and x_n are the reactance of the transmission line that was connected between the buses i and j without and with TCSC. r_n, G_n and B_n are the resistance, conductance and susceptance of the transmission line. At ith line, if the sensitivity is positive, it means that with increase in the injected reactive power in ith line the system loss will decrease, and if sensitivity is negative, then losses will increase. Hence, TCSC must be placed in a line which is having the most positive sensitive index. TCSC should not be placed between two generator buses, even though the line sensitivity of the line is high.

The optimal placement of TCSC for minimizing the power loss is the best location to maximize the TTC as well, because the best location of TCSC for minimizing the loss improves the line power flows in the total system which will lead to the increment of TTC.

14.4.3   Total transfer capability

Transfer capability of an interconnected electric power system is defined as the maximum amount of power that can be transferred reliably from generating stations to the load centres without violating the system operating conditions. According to the report approved by North American Electric Reliability Corporation [21], TTC can be defined as the quantity of electric power that can be transferred over the interconnected transmission path reliably without violating the predefined set of conditions of the system. The main reasons to calculate the TTC are

- estimation of TTC can be used as a rough indicator of system stability
- to improve reliability and economic efficiency of the power markets
- for providing a quantitative basis for assessing transmission reservations to facilitate energy markets.

The problem formulation to calculate the TTC is explained as following. To calculate the value of TTC, the injected real powers P_{Gi} at source area and the load demands P_{Di} and Q_{Di} at sink area are increased, simultaneously to maximize the value of loading factor λ in (14.59) subjected to equality and inequality constraints [22] given in (14.60)–(14.66). Constraints are handled using the penalty function method given in Reference 12, and the optimal value of λ is calculated using PSO algorithm.

$$P_{Gi}(\lambda) = P_{Gi0}(1 + \lambda)$$
$$P_{Di}(\lambda) = P_{Di0}(1 + \lambda) \tag{14.59}$$
$$Q_{Di}(\lambda) = Q_{Di0}(1 + \lambda)$$

where $P_{Gi}(\lambda)$ is the real power generation at the ith bus in terms of loading factor. $P_{Di}(\lambda)$ and $Q_{Di}(\lambda)$ are the real and reactive power demands at the ith bus in terms of loading factor λ. P_{Gi0}, P_{Di0} and Q_{Di0} are the real power generation, real and reactive power loads at the bus i, corresponding to the base case ($\lambda = 0$).

Equality constraints: These constraints are real and reactive power balance equations.

$$P_{Gi} - P_{Di} - \sum_{j=1}^{N} V_i V_j \left[G_{ij} \cos(\delta_i - \delta_j) + B_{ij} \sin(\delta_i - \delta_j) \right] = 0 \quad i \in N_0$$

(14.60)

$$Q_{Gi} - Q_{Di} - \sum_{j=1}^{N} V_i V_j \left[G_{ij} \sin(\delta_i - \delta_j) - B_{ij} \cos(\delta_i - \delta_j) \right] = 0 \quad i \in N_{PQ}$$

(14.61)

where V_i, V_j and δ_i, δ_j are voltage magnitudes and angles at buses i and j. P_{Gi}, P_{Di}, Q_{Gi} and Q_{Di} are the real power generation, the real power demand, the reactive power generation and the reactive power demand at ith bus, respectively. B_{ij} is the susceptance of the line connected between ith and jth bus. N, N_0, N_{PQ} and N_G are the total number of buses, total number of buses excluding slack bus, number of load buses and number of generator buses in the system.

Inequality constraints: These constraints represent the system operating condition limits.

- Generation constraints: Generator voltages, real power outputs and reactive power outputs are restricted by their lower and upper bounds as given below

$$V_{Gi}^{\min} \leq V_{Gi} \leq V_{Gi}^{\max}, \quad i = 1, \ldots, N_G$$

(14.62)

$$P_{Gi}^{\min} \leq P_{Gi} \leq P_{Gi}^{\max}, \quad i = 1, \ldots, N_G$$

(14.63)

$$Q_{Gi}^{\min} \leq Q_{Gi} \leq Q_{Gi}^{\max}, \quad i = 1, \ldots, N_G$$

(14.64)

where $V_{Gi}^{\min}$, $V_{Gi}^{\max}$ are the minimum and maximum limits of voltages at the generator buses. $P_{Gi}^{\min}$, $P_{Gi}^{\max}$, $Q_{Gi}^{\min}$ and $Q_{Gi}^{\max}$ are the real and reactive power generation limits at the generator buses.

- Security constraints: These include the constraints of voltage magnitudes at load buses and transmission line loadings as given below

$$V_{Li}^{\min} \leq V_{Li} \leq V_{Li}^{\max}, \quad i = 1, \ldots, N_{PQ}$$

(14.65)

$$S_{li} \leq S_{li}^{\max}, \quad i = 1, \ldots, nl$$

(14.66)

where N_{PQ} and nl are the total number of load buses and total number of transmission lines, respectively. $V_{Li}^{\min}$, $V_{Li}^{\max}$ are the minimum and maximum limits of the voltages magnitudes at load buses. $S_{li}^{\max}$ is the maximum complex power that can be transmitted in the transmission line without affecting the system security.

The TTC value is equal to the difference between the sum of all real power loads at sink area corresponding to $\lambda_{\min}$ and $\lambda = 0$ as given in (14.67):

$$\text{TTC} = \sum_{i=1}^{N_{PQ}} P_{Di}(\lambda_{\max}) - \sum_{i=1}^{N_{PQ}} P_{Di}(\lambda = 0) \tag{14.67}$$

14.4.4   Optimal rating of static var compensator

To calculate optimal rating of the SVC to minimize system loss and to maximize the TTC, optimization function consisting of real power loss and TTC are taken as the objectives as given in (14.68). As we are minimizing the total objective function, we need to take the negative sign for TTC. Real and reactive power generation limits at voltages buses, bus voltage magnitude limits, line thermal limits and transformer tap setting limits given in (14.62)–(14.66) are taken as the operational constraints. The objective function given in (14.68) is a function of the reactive power injected by SVC, that is Q_{SVC} placed at the optimal location selected through the sensitivity index given in (14.52). Hence, Q_{SVC} is the control variable for optimization, and its limits are given in (14.69). PSO technique is used to solve the objective function given in (14.68) to find the optimal rating of the SVC.

$$F = \min(w1 * P_{\text{Loss}} + w2 * (-\text{TTC})) \tag{14.68}$$

$$Q_{\text{SVC}} \le Q_{\max} \tag{14.69}$$

where,

$$P_{\text{Loss}} = \sum_{i=1}^{n} \sum_{j=1}^{n} \left[\alpha_{ij} \left(P_i P_j + Q_i Q_j \right) + \beta_{ij} \left(Q_i P_j - P_i Q_j \right) \right] \tag{14.70}$$

$$\text{TTC} = \sum_{i=1}^{N_{PQ}} P_{Di}(\lambda_{\max}) - \sum_{i=1}^{N_{PQ}} P_{Di}(\lambda = 0) \tag{14.71}$$

where $w1$ and $w2$ are the weighting factors to set relative importance to the objectives real power loss and TTC, respectively. The values of $w1$ and $w2$ are given in such a way that the total value $w1 + w2$ should be equal to 1. In the present work, equal importance is given to each individual objective function, that is minimization of loss and maximization of TTC and are taken as $w1 = w2 = 0.5$.

14.4.5   Optimal rating of thyristor-controlled series compensator

The optimal rating of the TCSC to minimize system loss and to maximize the TTC, an objective function consisting of loss minimization and TTC maximization is formulated as given in (14.72). Real and reactive power generation limits at voltages buses, bus voltage magnitude limits, line thermal limits and transformer tap setting limits given in (14.62)–(14.66) are taken as the operational constraints. The objective function given in (14.72) is optimized with respect to the reactance of

TCSC placed optimally through loss index given in (14.55). The limits on the control variable X_{TCSC} (pu) are given in (14.73). PSO technique is used to solve the objective function given in equation (14.72) to find the rating of the TCSC.

$$F = \min(w1 * P_{Loss} + w2 * (-TTC)) \tag{14.72}$$

$$-0.75X_{Line} \leq X_{TCSC} \leq -0.25X_{Line} \tag{14.73}$$

where,

$$P_{Loss} = \sum_{i=1}^{n} \sum_{j=1}^{n} \left[\alpha_{ij}(P_iP_j + Q_iQ_j) + \beta_{ij}(Q_iP_j - P_iQ_j) \right] \tag{14.74}$$

$$TTC = \sum_{i=1}^{N_{PQ}} P_{Di}(\lambda_{max}) - \sum_{i=1}^{N_{PQ}} P_{Di}(\lambda = 0) \tag{14.75}$$

where X_{TCSC} and X_{Line} are the reactances offered by the TCSC and transmission line, respectively. $w1$ and $w2$ are the weighting factors corresponding to the objective functions loss minimization and TTC maximization and taken as $w1 = w2 = 0.5$.

14.4.6 Particle swarm optimization

PSO, a part of swarm intelligence family, is known to effectively solve the large-scale nonlinear optimization problems. PSO was introduced by Kennedy and Eberhart in 1995 [23]. PSO is a population-based optimization technique that is originally inspired by the sociological behaviour associated with bird flocking and fish schooling.

The PSO algorithm consists of a population of particles called "swarm" in which all the particles move in the search space to get the solution. Each particle represents a potential solution to the optimization problem in search space. The velocity and position of ith particle in kth iteration are represented as

$$V_i^k = v_{i1}^k \tag{14.76}$$

$$X_i^k = x_{i1}^k \tag{14.77}$$

then in the next iteration, the velocity and positions are evaluated using

$$V_i^{k+1} = X_i^k + c_1 * \text{rand}() * (pbest - X_i^k) + c_2 * \text{rand}() * (gbest - X_i^k) \tag{14.78}$$

$$X_i^{k+1} = X_i^k + V_i^{k+1} \tag{14.79}$$

where,

$$pbest^{k+1} = \begin{cases} pbest & \text{iff} \quad f(X_{k+1}) \geq f(pbest^k) \\ X^{k+1} & \text{iff} \quad f(X_{k+1}) < f(pbest^k) \end{cases} \tag{14.80}$$

$$gbest^{k+1} = \begin{cases} pbest & \text{iff} \quad f(pbest^k) \leq f(gbest^k) \\ gbest & \text{iff} \quad f(pbest^k) > f(gbest) \end{cases} \tag{14.81}$$

where *pbest* and *gbest* are the particle best position and global best position of swarm. A limit to the velocity, that is $[-V_{max}, V_{max}]$, is given to decrease the possibility of the particle out of the search space is given. The particle having the position limits $[-X_{max}, X_{max}]$ can be related with velocity limits as $V_{max} = kX_{max}$, where $0 < k \leq 1$. The velocity equation (14.78) contains three terms which are getting updated after each iteration. The first term represents the information of about its current velocity, that is change in position in the search space dimensions. The second term represents its own cognition, that is its own experience. The third term represents the social interaction with other particles. Thus, there is an adaptation process going while updating the velocity equation after each iteration instead of replacing old particles to new particles. The three terms in (14.78) can be explained with the help of Figure 14.36 [24]. In Figure 14.36, the thick line represents the direction in which particles move finally, the dashed line represents the direction of the particle towards its local best position. Along with the local best, the global best position also shifts with iterations, but not represented in figure. As the iteration process progresses, the fitness value approaches near to global minimum.

To increase the rate of convergence of the above-mentioned PSO algorithm, an inertia weight w has been introduced in the velocity update equation. The inertia weight is a scaling factor that combines with memory representation (previous velocity) of the velocity equation. After the inertia weight is added, the equation is modified to as following:

$$V_i^{k+1} = wV_i^k + c_1 * \text{rand}() * \left(pbest_i - X_i^k\right) + c_2 * \text{rand}() * \left(gbest_i - X_i^k\right) \quad (14.82)$$

where w is the inertia weight. The inertia weight specifies how much of the previous velocity is to be retained for the next iteration. In general, a linearly decreasing inertia weight is used. In the algorithm, the inertia weight is set to

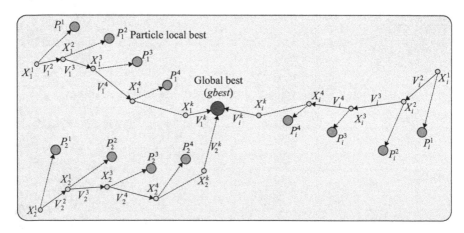

Figure 14.36   Illustration of particle movements in PSO

decrease from 0.95 to 0. As a result of this setting, the particles can explore a large area at the start of the simulation when the inertia weight is large and to refine the search later by using a decreasing inertia weight. When the inertia weight is large, the oscillations are found. Thus, the particle oscillations around the *gbest* are damped by using small inertia weight, thus assisting the particles of the swarm to converge to the global minimum. The changes introduced by the velocity update equation represents acceleration; thus, the constants c_1 and c_2 are called as acceleration coefficients. The acceleration constants c_1 and c_2 can be understood as a balance between exploration (searching for good solution) and exploitation (using someone else's success), respectively [25]. This can be also understood as the balance between individuality and sociality. Ideally, individuals prefer being individualistic yet they want to know what others have achieved.

Compared with traditional optimization algorithms, PSO does not need the information of the derivative of functions to be optimized. Unlike classical optimization techniques like gradient search methods, the PSO-based optimization does not get stuck up at a local minimum but will converge to a global minimum most of the cases, if at all it exists. In addition to that, the advantage of PSO is that it is so simple in terms of mathematical expression and understanding. PSO is very fast in converging to a solution because of its mathematical simplicity. Due to the above-mentioned advantages, PSO has been used to the present optimization problem.

14.5   Simulation results

The methodology of optimal placement of SVC and TCSC to minimize losses and maximize TTC has been studied for two systems: IEEE 30-bus system [26] and the New-England 16-machine, 68-bus system [27] as two test case scenarios. The simulations for the test systems have been carried out in two stages. The first stage is the optimal placement of SVC and TCSC using sensitivity-based methods [12]. The second stage is finding the optimal rating of SVC and TCSC to minimize the loss and maximizing the TTC using PSO. The simulation results are taken from References 28 and 29.

14.5.1   *Optimal placement and parameter settings of SVC and TCSC in IEEE 30-bus system*

The IEEE 30-bus test system [26] represents a portion of the American Electric Power System (in the Midwestern US) as of December, 1961. The data were provided by Iraj Dabbagchi of AEP and formatted by Rich Christie at the University of Washington in August 1993. The IEEE 30-bus test system has 6 generator buses and 24 load buses. It has 41 transmission lines and 4 tap changing transformers. Bus-1 is taken as the slack bus. The six generators are placed at buses 1, 2, 5, 8, 11 and 13, respectively. Four transformers with off-nominal tap ratio are placed in the lines between buses 6–9, 6–10, 4–12 and 27–28.

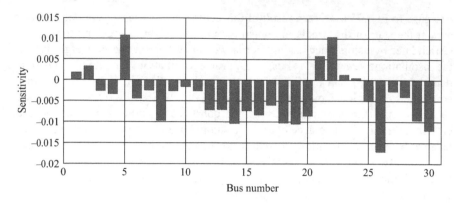

Figure 14.37   Power-loss sensitivity index with respect to SVC

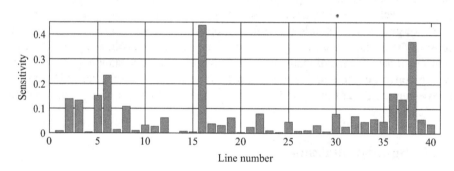

Figure 14.38   Power-loss sensitivity indices with respect to TCSC

14.5.1.1   Optimal placement of static var compensator

The sensitivity-based method [12], as explained in Section 14.4.1, is considered for optimal placement of SVC to minimize the transmission system power loss.

Figure 14.37 shows the bus sensitivity index versus bus number plot for the SVC in IEEE 30-bus. It can be observed from Figure 14.37 that 26th bus is having the most negative sensitivity index. So 26th bus is the best locations to place SVC in the system.

14.5.1.2   Optimal placement of thyristor-controlled series capacitor

Sensitivity-based method [12], as explained in Section 14.4.2, is considered for optimal placement of TCSC to minimize the transmission system power loss.

Figure 14.38 shows the power loss sensitivity index of the system with respect to line number plot for the SVC in IEEE 30-bus. It can be found from Figure 14.38 that 16th line connected between the buses 12 and 13 is the most positive sensitive line, and it is the best location to place the TCSC to minimize the power loss in the system.

Table 14.2 PSO parameters

Parameter	Value
No. of particles	100
Maximum inertia weight	0.9
Minimum inertia weight	0.4
$c1$ and $c2$	1.49
No. of iterations	50

14.5.1.3 Total transfer capability

The objective function, to find TTC, is given in (14.83) optimized with loading factor (λ) as the control variable. The equality and inequality constraints given in (14.60)–(14.66) are considered.

$$\text{TTC} = \sum_{i=1}^{N_{\text{PQ}}} P_{Di}(\lambda_{\max}) - \sum_{i=1}^{N_{\text{PQ}}} P_{Di}(\lambda = 0) \tag{14.83}$$

PSO has been applied to the minimization problem, and Table 14.2 gives the PSO parameters used in minimization problem with the objective function given in (14.83). The IEEE 30-bus system can deliver 229.20 MW of real power (with a loading factor λ equal to 1.1785) from generating stations to all load centres without SVC and TCSC in the system and without violating the system constraints.

14.5.1.4 Optimal rating of static var compensator

The objective function consisting of power loss minimization and TTC maximization given in (14.84), as explained in Section 14.4.4, is considered for finding the optimal rating of the SVC. The equality and inequality constraints are given in (14.60)–(14.66).

$$F = \min(w1 * P_{\text{Loss}} + w2 * (-\text{TTC})) \tag{14.84}$$

where P_{Loss} is the total power loss in the system, and $w1$ and $w2$ are the weighting factors to give relative importance to the individual objectives, and in the present case equal importance is given to both objectives, that is minimizing losses and maximizing TTC. Hence, $w1$ and $w2$ are considered as 0.5. It can be observed that TTC is taken with a negative sign in (14.84) as the total objective function is minimization, and TTC has to be maximized. PSO has been applied to the constrained minimization problem.

Table 14.3 gives the simulation results for the system. It can be observed from Table 14.3 that the real power loss is reduced from 2.45 to 2.40 MW, that is loss reduced by 50 kW. TTC is increased from 229.20 to 243.29 MW, that is increased by 14.39 MW approximately by placing 3.457 MVAr rated SVC at 26th bus for IEEE 30-bus system.

Table 14.3   Power loss and TTC of system with and without SVC

Parameter	Without SVC	With SVC3.457 (MVAr)
P_{Loss} (MW)	2.45	2.40
TTC (MW)	229.20	243.29

Table 14.4   Power loss and TTC of system with and without TCSC

Parameter	Without TCSC	With TCSC 0.0146 pu
P_{Loss} (MW)	2.45	2.36
TTC (MW)	229.20	249.56

14.5.1.5   Optimal rating of thyristor-controlled series capacitor

The objective function consisting of power-loss minimization and TTC maximization given in (14.85), as explained in Section 14.4.5, is considered for optimal rating of the TCSC. The equality and inequality constraints are given in (14.60)–(14.66).

$$F = \min\left(w1 * \sum_{Lines=1}^{l} P_{Loss} + w2 * (-TTC) \right) \tag{14.85}$$

where $w1$ and $w2$ are the weighting factors to give relative importance to the individual objectives, and the values of $w1$ and $w2$ are taken as 0.5. Table 14.4 gives the simulation results for IEEE 30-bus test system. It can be observed from Table 14.4 that the real power loss is reduced from 2.45 to 2.36 MW that is reduced by 90 kW. TTC is increased from 229.20 to 249.56 MW that is, increased by 20.36 MW approximately by placing 0.0146 pu. of TCSC in 16th line connected between buses 12 and 13 in IEEE 30-bus system.

14.5.2   Optimal placement and parameter settings of SVC and TCSC in the New England 16-machine, 68-bus system

The 68-bus system [27] is a reduced order equivalent of the inter-connected New England test system (NETS) and New York power system (NYPS), with five geographical regions out of which NETS and NYPS are represented by a group of generators, whereas the power imports from each of the three other neighbouring areas are approximated by equivalent generator models. The New England 16-machine, 68-bus test system, has 68 buses of which 16 are generator buses, and 52 are load buses. It has 86 transmission lines and 16 off-nominal tap ratio transformers.

14.5.2.1   Optimal placement of static var compensator

Optimal placement of SVC in test system can be calculated as explained in Section 14.5.1.1.

Figure 14.39 depicts the bus sensitivity index versus bus number plot for the SVC in the New England 16-machine, 68-bus test system. It can be found from the

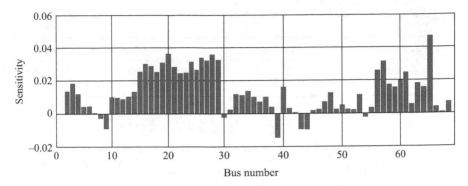

Figure 14.39   Power-loss sensitivity index with respect to SVC

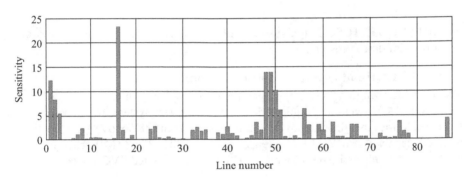

Figure 14.40   Power-loss sensitivity indices with respect to TCSC

figure that 39th bus is having the most negative sensitivity index. So 39th bus is the best location to place SVC in the system.

14.5.2.2   Optimal placement of thyristor-controlled series capacitor

Optimal placement of TCSC in test system can be calculated as explained in Section 14.5.1.2.

Figure 14.40 shows the power-loss sensitivity index of the system with respect to line number plot for the SVC in the New England 16-machine, 68-bus test system. It can be found from Figure 14.40 that 16th line connected between the buses 8 and 9 is the most positive sensitive line, and it is the best location to place the TCSC to minimize the power loss in the system.

14.5.2.3   Total-transfer capability

The objective function, to find TTC, is given in (14.83) optimized with loading factor (λ) as the control variable. The equality and inequality constraints given in (14.60)–(14.66) are considered.

The New England 16-machine 68-bus can deliver 26,592 MW of real power (with a loading factor λ equal to 1.1785) from generating stations to all load centres

Table 14.5   Power loss and TTC of system with and without SVC

Parameter	Without SVC	With SVC 362.32 (MVAr)
P_{Loss} (MW)	288.56	284.41
TTC (MW)	26,582	27,687

Table 14.6   Power loss and TTC of system with and without TCSC

Parameter	Without TCSC	With TCSC 0.0239 pu
P_{Loss} (MW)	288.56	273.14
TTC (MW)	26,582	27,711

without SVC and TCSC in the system and without violating the system constraints from the simulation results.

14.5.2.4   Optimal rating of static var compensator
The optimal rating of SVC can be calculated as explained in Section 14.5.1.4.

Table 14.5 gives the simulation results for the test system. It can be observed from Table 14.5 that the real power loss is reduced from 288.56 to 284.41 MW that is, loss reduced by 4.15 MW. TTC increased from 26,582 to 27,687 MW that is increased by 1,100 MW approximately by placing 362.32 MVAr rated SVC at bus 39.

14.5.2.5   Optimal rating of thyristor-controlled series capacitor
The optimal rating of TCSC can be calculated as explained in Section 14.5.1.5.

Table 14.6 gives the simulation results for the test system. It can be observed from Table 14.6 that the real power loss is reduced from 288.56 to 273.14 MW, that is loss reduced by 15.42 MW. TTC is increased from 26,582 to 27,711 MW, that is increased by 1,129 MW approximately by placing 0.023955 pu of TCSC in 16th line connected between the buses 8 and 9 in the New England 16-machine, 68-bus test system.

14.6   Conclusions

This chapter covered the effect of series and shunt compensations in the transmission system. The various types of FACTS controllers and the system transmitted power versus angle characteristics are explained in detail. SVC and TCSC have been considered to improve the TTC and to minimize the power loss of the system under steady-state condition. An objective function has been formulated to find the optimal rating of the controllers to maximize the TTC and to minimize of real power losses. The problem has been solved in two stages: Stage-1: optimal placement of the controllers based on sensitivity index method, and Stage-2: PSO algorithm has been applied to solve objective function to find the optimal ratings.

The SVC of 3.457 MVAr capacity has been placed optimally at 26th bus, in IEEE 30-bus system loss has been reduced from 2.45 to 2.40 kW and TTC improves from 229.20 to 243.39 MW and by placing SVC of capacity 362.32 MVAr at 39th bus, in the New England 16-machine, 68-bus system loss has reduced from 288.56 to 284.41 kW and TTC improves from 26,582 to 27,687 MW.

The TCSC of 0.0146 pu capacity in 16th line, in IEEE 30-bus system and 0.0239 pu in 16th line, in the New England 16-machine 68-bus system, the loss has reduced from 2.45 to 2.36 kW and TTC improves from 229.20 to 229.20 MW in IEEE 30-bus system and loss has reduced from 288.56 to 273.14 kW and TTC improves from 26,582 to 27,711 MW in the New England 16-machine, 68-bus system respectively, and this shows the effectiveness of the algorithm.

References

[1] S. Mukhopadhyay, A.K. Tripathy, V.K. Prasher and K.K. Arya. "Application of FACTS in Indian power system", *IEEE Transmission and Distribution Conference and Exhibition: Asia Pacific. PES*, vol. 1, 2002.

[2] C. Schauder, M. Gernhardt, E. Stacey, *et al.* "Operation of ±100 MVAR STATCOM", *IEEE Trans. PWRD*, pp. 1805–1811, 1997.

[3] D. Povh, "FACTS controller in deregulated systems", *Power Systems Symposium*, Rio de Janeiro, Brazil, May 1998.

[4] B. Fardanesh, A. Edris, B. Shperling, *et al.* "NYPA convertible static compensator validation of controls and steady state characteristics", *CIGRE*, France, August 2002.

[5] M. Eslami, S. Hussain, M. Azah, K. Mohammad, "A survey on flexible AC transmission systems (FACTS)", *Przeglad Elektrotechniczny*, 2012.

[6] N.G. Hingorani, L. Gyugyi, *Understanding FACTS: Concepts and Technology of Flexible AC Transmission Systems*, IEEE Press, New-York, 2000.

[7] K. Padiyar, *FACTS Controllers in Power Transmission and Distribution*, Delhi, India: New Age International, 2007.

[8] R.M. Mathur, R.K. Varma, *Thyristor-Based Facts Controllers for Electrical Transmission Systems*, IEEE Press, Piscataway, NJ, 2002.

[9] Y.H. Song, *Flexible AC Transmission Systems (FACTS)*, The Institution of Electrical Engineers, London, 1999.

[10] S.N. Singh, "Location of FACTS devices for enhancing power systems security", *The 2001 Large Engineering Systems Conference on Electrical Power Engineering (LESCOPE)*, Halifax, Nova Scotia, Canada, pp. 162–166, 2001.

[11] S. Gerbex, R. Cherkaoui, A. Germond, "Optimal location of multitype FACTS devices in a power system by means of genetic algorithms", *IEEE Transaction Power System*, vol. 16, no. 3, pp. 537–544, 2001.

[12] P. Preedavichit, S. Srivastava, "Optimal reactive power dispatch considering facts devices", *Electric Power Systems Research*, vol. 46, no. 3, pp. 251–257, 1998.

[13] K. Verma, S. Singh, H. Gupta, "Facts devices location for enhancement of total transfer capability", *IEEE Power Engineering Society Winter Meeting*, vol. 2., pp. 522–527, 2001.

[14] E. Leonidaki, N. Hatziargyriou, G. Manos, B. Papadias, "A systematic approach for effective location of series compensation to increase available transfer capability", *IEEE Porto Power Tech Proceedings*, vol. 2, pp. 6–11, 2001.

[15] Q. Wu, Z. Lu, M. Li, T. Ji, "Optimal placement of facts devices by a group search optimizer with multiple producer", *IEEE Congress on Evolutionary Computation*, pp. 1033–1039, 2008.

[16] R. Idris, A. Khairuddin, M. Mustafa, "Optimal allocation of facts devices in deregulated electricity market using bees algorithm", *WSEAS Transactions on Power Systems*, vol. 5, no. 2, pp. 108–119, 2010.

[17] H. Mori, Y. Goto, "A parallel tabu search based method for determining optimal allocation of facts in power systems", *International Conference on Power System Technology, PowerCon*, vol. 2, pp. 1077–1082, 2000.

[18] R. Rajaraman, F. Alvarado, A. Maniaci, R. Camfield, S. Jalali, "Determination of location and amount of series compensation to increase power transfer capability", *IEEE Transactions on Power Systems*, vol. 13, no. 2, pp. 294–299, 1998.

[19] A. Kazemi, B. Badrzadeh, "Modeling and simulation of SVC and TCSC to study their limits on maximum loadability point", *International Journal of Electrical Power & Energy Systems*, vol. 26, no. 5, pp. 381–388, 2004.

[20] J. Dhillon, D. Kothari, *Power System Optimization*, PHI, New Delhi, 2004.

[21] T. Force, "Available transfer capability definitions and determination", *North American Electric Reliability Council*, Princeton, New Jersey, 1996.

[22] C. Taylor, *Power System Voltage Stability*, McGraw-Hill Companies, New York, 1994.

[23] J. Kennedy, R. Eberhart, "Particle swarm optimization", *IEEE International Conference on Neural Networks*, vol. 4, pp. 1942–1948, 1995.

[24] J.-B. Park, Ki-Song Lee, Joong-Rin Shin, Kwang Y. Lee, "A particle swarm optimization for economic dispatch with nonsmooth cost functions", *IEEE Transactions on Power Systems*, vol. 20, pp. 34–42, 2005.

[25] K.E. Parsopoulos, M.N. Vrahatis, "Particle swarm optimization method for constrained optimization problems", *Intelligent Technologies Theory and Application: New Trends in Intelligent Technologies*, vol. 76, pp. 214–220, 2002.

[26] IEEE 30-bus system data,1993 [online], available at https://goo.gl/RDRu4T.

[27] B. Pal, B. Chaudhuri, *Robust Control in Power Systems*, Springer, Berlin, 2005.

[28] K.S. Rao, B. Kalyan Kumar, "Placement of SVC for minimizing losses and maximizing total transfer capability using particle swarm optimization", *IET Conference on Renewable Power Generation (RPG 2011)*, 2011.

[29] K. S. Rao, B. Kalyan Kumar, M.K. Mishra, "Optimal placement and parameter settings of TCSC to minimize the loss and maximize the total transfer capability using particle swarm optimization", *World Congress on Engineering and Technology*, Shanghai, China, 2011.

Chapter 15

Security analysis of smart grid

Charalambos Konstantinou[1] and Michail Maniatakos[2]

Legacy power grids were typically designed with reliability as the main goal. With the transition to the smart grid, security concerns have begun to arise; due to the computational and communication capabilities of the integrated elements, smart grid technologies are becoming vulnerable to cyber-attacks. This chapter aims to enumerate existing threat vectors in the various layers of smart-grid architecture and provide insights of how security techniques should be implemented in order to ensure smart grid resiliency.

15.1 Introduction

The electric power grid is increasingly integrating advanced digital technologies into its existing infrastructure. These information and communication technologies (ICT) operate as the control layer on top of the physical energy grid aiming to manage and automate stable operation of power system processes. As a result, the grid is evolving towards a smart grid in order to ensure more reliable and efficient delivery of electricity. Although the ICT integration is of paramount importance for the smart-grid transformation, it comes at the cost of exposing the entire power system network to new security challenges. Specifically, since the ICT-based systems expand the threat landscape, the smart grid is effectively becoming more prone to cyber intrusions and attacks.

The attack incident against the Ukrainian power system in late 2015 shows the grid vulnerability and how real the threat is. Specifically, on December 23, 2015, media reported a cyber-attack on the Ukrainian grid that left thousands of people in the Ivano-Frankivsk region without electricity [1]. The attackers compromised control systems and infected software with malicious code. As a result, they were able to trip breakers, cause a power outage, and prevent the utility from detecting the attack. This real-world example demonstrates the potential implications and damage that a sophisticated cyber-attack can cause on the critical infrastructure.

[1]Electrical and Computer Engineering, New York University, Tandon School of Engineering
[2]Electrical and Computer Engineering, New York University Abu Dhabi, Abu Dhabi, UAE

The main smart-grid components (generation, transmission, distribution, and consumption) are equipped with cyber systems, as seen in Figure 15.1, including communication networks, control automation systems, and intelligent electronic devices (IEDs). Bulk generation and distributed energy resources (DER) are equipped with programmable logic controllers (PLCs) and other distributed control systems (DCS) enabling automation and adjustment of the power generation level. At the transmission level, substations, power lines, and towers support supervisory control and data acquisition (SCADA) and energy management systems (EMS), wide area monitoring systems with phasor management units (PMUs), dynamic line rating sensors, and others. Distribution systems deal with the increasing occurrence of decentralized generation and focus on the control, monitoring, and automation of power distribution through the utilization of remote terminal units (RTUs). The introduction of smart meters at the consumption level has driven advanced metering infrastructure (AMI) to be an imperative component of smart grid by providing a two-way communication scheme between utilities and customers. Modernizing the grid through the integration of ICT technologies at every level of the system increases both the complexity of the grid and the exposure to potential attacks.

The threat landscape of smart grid is in a constant state of evolution due the inadequate level of security measures. In order to assess the security of physical processes, it is necessary to identify the vulnerability sources. Embedded elements, such as PLCs, RTUs, data concentrators, and others, are hardware designs that control a cyber physical process. Attacks on the hardware level of these systems aim to gain access to critical information, cause denial of important services, and generally lead to various kinds of security failures. ICT and control systems employ a variety of software platforms. Vulnerabilities in these interfaces may range from simple software errors to poor management of authentication credentials. Every smart-grid implementation contains a proliferation of hardware equipment and software, interacting on a network communication layer. Regardless of network topology, vulnerabilities within the network can be introduced from the different composing elements (e.g., insecure communication protocols). Furthermore, security issues may emerge on the operational and process layer of smart grid. For instance, attackers could inject falsified data to operational critical routines in order to hamper the state of the system.

Since smart-grid security threats are constantly evolving, proper defenses require advanced cyber security mechanisms. Achieving a secure smart grid is a difficult and complex task. It is essential to understand potential vulnerabilities and weaknesses as well as identify the challenges arising from the incorporation of ICT systems in every level of smart grid. Mitigation and defense mechanisms can be further designed to detect and ultimately prevent malicious attempts to attack the smart grid. Also, initiatives, guidelines, and reference models are valuable, and it is highly recommended that these standards are studied in detail. The security of the grid can be further improved through initiatives that raise awareness and provide training among grid operators, utilities, services providers, manufacturers, and end consumers.

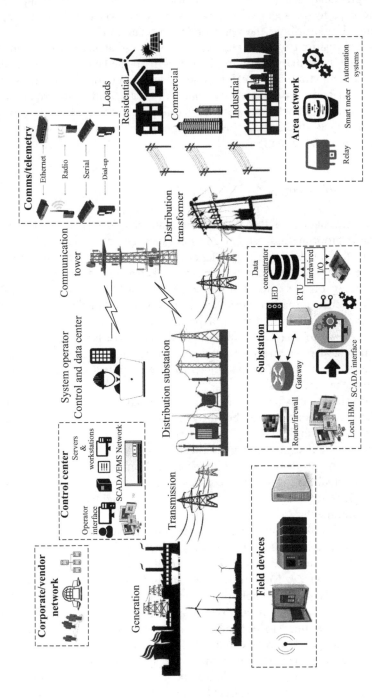

Figure 15.1 The smart-grid structure equipped with cyber systems (dashed line boxes)

This chapter analyzes the challenges that smart grids and their supporting infrastructure pose to the security of power system components. To that end, the analysis begins by describing the nature of threats due to the various motives of adversaries. The main part of the chapter deals with the attack strategies and the protection mechanisms that can be implemented in all the layers of smart grid. In order to give a complete picture, efforts and guidelines aiming to mitigate vulnerabilities are described. The chapter also includes a case study with experimental results demonstrating attack and defense techniques. Finally, the future of the smart-grid security is discussed by examining the trends in cyber security considerations and industry directions.

15.2 Nature of threats

Understanding the nature of threat actors in a smart grid environment helps to determine the appropriate assessment and mitigation strategies that will take into consideration the sensitive grid dependencies and connectivity. Threats are typically defined as the "possible actions that can be taken against a system" [2]. These actions, depending on the impact they may cause, can create significant safety issues to people, energy market participants, operations and maintenance routines, damage equipment, and even trigger a power outage. The merge of ICT with industrial control systems (ICS) expands the threat landscape and increases the exposure to potential risks and vulnerabilities. Therefore, as the number of threat agents increases, the nature of threat actors must be examined in detail in order to determine their characteristics.

Smart-grid threats can be classified into two broad categories: inadvertent or deliberate. Inadvertent threats may be further categorized on the basis of the origin of failure such as natural disasters, equipment, and safety failures. Due to the diversity and complexity of smart grid, the intent of deliberate malicious threats again smart-grid components is also important for determining proper risk assessment techniques. Deliberate threats may be further divided into active or passive threats. Figure 15.2 summarizes the existing threat types and their subcategories in smart-grid environments.

Threats resulting from inadvertent actions could be attributed to failures, natural disasters, and carelessness in establishing security policies. Failures can result from equipment breakdowns, poorly designed protocols, and noncompliance with safety codes. For instance, Modbus, a common SCADA protocol, was originally designed for use only within simple process control networks to enable low-speed serial communications between clients and servers. Hence, threats could lead to vulnerability exploitations since Modbus cannot address security concerns [3]. Natural threats such as severe weather conditions, earthquakes, hurricanes, and other phenomena (e.g., vegetation, animals) are also an ongoing threat to grid infrastructure. Smart-grid planning should encounter the behavior of the system subjected to such natural hazardous events. Events related with careless actions can result in incorrect operation and control of the grid. Unintentional errors from

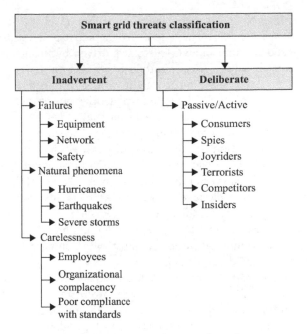

Figure 15.2   Security threats classification

inadequately trained employees can allow unauthorized personnel to acquire authentication credentials. In addition, unfamiliarity and noncompliance with security procedures, standards, and policies can cause accidental risks to the smart-grid implementations. Finally, individuals and companies often become complacent when it comes to security. Managers can be satisfied with mediocre security performance and do not work to improve the environment by raising security awareness and eliminating the potential threats.

In addition to inadvertent threats, deliberate security threats can cause physical damage to the power system operation with huge financial, social, legal, and political impact. Deliberate threats can be classified as passive and active. Passive threats arise from attempts that aim to learn or make use of information that can affect the performance of the system. Although such threats can cause harm to the smart grid, their target is not to alter the operational state of the system but to gain insight to its inner workings and operational details (e.g., insecure communication links may allow adversaries to eavesdrop the traffic and deduce practical/useful information between communication parties). On the other hand, active threats could affect the operation or resources of the power system. The exploitation of active threats has the potential to impact more victims and have more far-ranging effects than any other type of threat. The growing number of network inter-connections in smart-grid implementations results in the creation of new active threats and greatly increases the number of entry points for attacks. The outcome of orchestrating the exploitation of these exposed threats results in damaging grid components causing operation disruptions and performance degradation.

Deliberate threats vectors may arise from actions taken by security hunters who view the security mechanisms deployed in smart-grid infrastructure as a puzzle to be cracked for amusement. Customers driven by financial motives pose also a type of threat in smart grid. Fraudulent activities in AMI systems include hacking into smart meters targeting to falsify billing information and eventually manipulating energy usage data. The race to modernize power grid systems and, at the same time, keeps the cost low increased industrial espionage. Spies, malicious insiders, and even disgruntled employees operating on behalf of utility competitors can gather critical information of power system services. In addition, independent groups and nation-wide terrorists often view grid infrastructure as an attractive target, which if attacked, can result in extensive financial and political implications besides the direct physical damage. The scale of incidents resulting from both passive and active threat actors indicates that adversaries target valuable information on their way to attack a critical system. In 2013, almost one-third of utilities reported losing valuable internal information with 46% citing cyber criminals and competitors as the probable culprit [4].

In order to implement defense strategies against malicious adversaries, it is important to examine all smart-grid elements from an attacker's perspective. Adversaries may follow different attack methodologies depending on their motives, capabilities, and resources. The awareness regarding vulnerabilities exploitation approaches will lead to the development of detection and protection schemes able to capture the various threats across all smart-grid layers.

15.3   Attacking the smart grid

Identifying the potential threats and vulnerabilities at all levels, ranging from operation management procedures to hardware specific aspects, is the first step towards protecting the smart grid. The various paths that an attacker could utilize in order to exploit grid vulnerabilities and circumvent security features should be taken into consideration when configuring networks and devices.

For adversaries to successfully achieve their malicious objectives, the attack target must be identified first, as presented in Figure 15.3. Possible targets and attack vectors include service provider systems and their customer information interfaces, wide area measurement routines in EMS, smart meters in AMI, and others. The dependencies between the smart-grid interconnections often allow attack exploits to propagate within the system and reach additional targets. For instance, a SCADA server may be connected to an industrial Ethernet *switch*[1] which is further connected with *routers* that are accessible from the utility's corporate network. In the scenario that the SCADA server allows write access to the system settings through the slave controllers, a breached device could modify the information sent from other units to the server.

---

[1]A glossary of terms used in this chapter (presented in *italic* throughout the text) appears in Appendix A.

Figure 15.3 Steps of attack methodologies

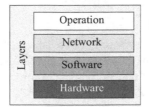

Figure 15.4 Layered smart-grid architecture

The identification and enumeration of possible attack targets follows the planning of the attack methodology. This methodology aims to enumerate the steps needed to exploit vulnerabilities specific to a device or related to the network design. For example, software vulnerabilities such as non-inclusion of device memory checks can provide an adversary with the ability to disrupt the operation of grid units like PLCs and PMUs. Attackers capable of violating these boundaries can alter the way the software program operates or trigger execution of malicious code. In addition, widely adopted serial or internet protocol (IP)-based protocols have vulnerabilities that may result in *denial-of-services (DoS)* attacks [3].

The inclusion of legacy systems into grid technologies provides opportunities to an attacker to seamlessly exploit existing vulnerabilities. To achieve this, vulnerability databases such as the National Vulnerability Database, the Open Source Vulnerability Database, and others can be utilized. For industrial systems in particular, the ICS Cyber Emergency Response Team (ICS-CERT) provides alert notifications related to critical infrastructure threats. The security warnings provide information about threats, activity, exploits, and the potential impact.

A vulnerability is exploited by attackers to compromise the target system. Nowadays, most of the adversaries employ readily available intrusion tools and exploit scripts. For example, operation Night Dragon in 2010 utilized common hacking tools aiming to extract financial information and specific project details about oil, gas, and energy companies [5]. Depending upon the vulnerabilities of the system, the methodology could follow different approaches in order to remain undetected and effective as long as possible. A cyber security firm discovered that it takes, on average, 205 days before target companies identify security breaches [6]. Due to the complexity and interconnections within smart-grid environments, many attack methodologies use a multilayer threat model, that is the attack strategy could target one or multiple layers of smart grid presented in Figure 15.4. Such attacks leverage vulnerabilities in different layers in order to affect as many infrastructure domains as possible (markets, consumers, distribution, transmission, generation, etc.), increase the severity of the damage and the speed of contagion on the infected equipment. For instance, databases can be connected to computers or other databases with web-enabled applications located at the business network. An attacker able to gain access to the database on the business network could thereafter exploit the communication channel between the two networks and thus bypass the security mechanisms protecting the control systems environment.

15.3.1   Operation layer

The delivery of electricity to end users in a reliable and secure manner relies on the interaction and interoperability across all the layers that form the smart grid. Power systems operational behavior determines the balance between supply and demand and relies on the dynamic process characteristics of the designed smart-grid model. Attacks on the operation layer of the grid aim to hamper the efficiency of the controlled process and therefore degrade the performance of operational routines. Such operation-centric attacks may inject spurious data through specially crafted commands and modify run-time process variables or control logic in order to disturb the operation state and ultimately cause denial of critical services. Besides crashing or halting an industrial process, attack strategies in the operation layer of smart grid aim to conceal their digital footprints and action. Consequently, it is often infeasible for the system operator to determine whether or not the variations in the system process are nominal effects of an expected operation or an attack indicator.

A prominent example of an operation-aware attack is Stuxnet. The malware is believed to be a state-developed cyber weapon, and it is considered the most sophisticated malware targeting industrial operations [7]. Stuxnet, presented in Figure 15.5, could spread stealthily between Windows computers running a Siemens specific programming software for PLCs (Step7). It has been estimated that the malware infected around 100k systems worldwide. Capable of reconfiguring the operation of PLCs, Stuxnet infiltrated a uranium-enriching plant and accessed the control mechanisms of the centrifuges spin speed. The manipulation of the spin speed of centrifuges allowed the attackers to cease the operation of nuclear plants several times. Since the malware has been identified in mid-June 2010, ICS-targeting malware like Flame, Duqu and Gauss have appeared, sharing many code similarities to Stuxnet [8]. Operation-centric cyber-attacks have been also uncovered before the appearance of Stuxnet. For instance, in 2001, hackers installed a *rootkit*, a special category of malicious computer software, to the network of California Independent System Operator [9]. The rootkit caused rolling blackouts throughout the state, affecting over 400k utility customers.

In the context of smart grid, *false data injection (FDI)* attacks can manipulate the system state estimation and have a significant impact on the power distribution. An adversary able to obtain access to SCADA could alter the measurement data send from the RTUs to the master station. Bad Data Detection (BDD) system as

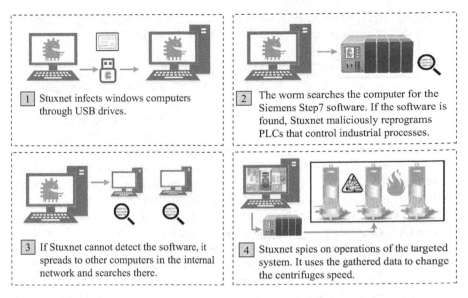

1. Stuxnet infects windows computers through USB drives.

2. The worm searches the computer for the Siemens Step7 software. If the software is found, Stuxnet maliciously reprograms PLCs that control industrial processes.

3. If Stuxnet cannot detect the software, it spreads to other computers in the internal network and searches there.

4. Stuxnet spies on operations of the targeted system. It uses the gathered data to change the centrifuges speed.

Figure 15.5   Attack strategy of Stuxnet

part of the state estimation module identifies and eliminates data attributed to topology errors and measurement abnormalities. The goal of an adversary is to corrupt the measurement data from a subset of RTUs D_S. As a result, the measured data $z \in \mathbb{R}_{m \times 1}$ will become $z_a = z + a$, where $a \in A$ is the attack vector and A is the set of feasible attack vectors:

$$A \triangleq \{a \in \mathbb{R}_m : a_i = 0, \forall i \notin D_S\} \tag{15.1}$$

where $H \in \mathbb{R}_{m \times n}$ is the nonlinear function vector determined according to the physical structure of the power system, and it provides the relationship between state variables and measured values. The malicious vector a is called the FDI attack that can pass the BDD scheme if and only if a can be expressed as a linear combination of the columns of H, that is $a = Hc$. It has been shown that data can be falsified in order to introduce errors in certain state variables without being detected by the BDD system [10]. This class of attacks could compromise signals in the electricity market or even mask the outage of lines. Also, the impact from FDI attacks could be the same as removing the attacked RTUs from the network [11].

As in every layer of smart-grid architecture, an authorized employee or in general a legitimate user can access privileged system's resources to perform malicious actions. Particularly in the operation layer, insider's knowledge of operational intrusion detection and protection mechanisms allows an attack methodology to effortlessly bypass protection settings to achieve its malicious intents. The Maroochy attack in 2000, although not targeting directly smart-grid equipment, demonstrated the impact of a malicious insider attack on ICS. Specifically, a disgruntled employee used inside knowledge to attack a sewage treatment plant and pump 800k l of sewage in a river in Queensland, Australia [12].

15.3.2 Network layer

The communication and interaction of smart-grid systems is achieved via the network layer. Due to the various entry points in the network layer, there are a large number of ICS-related vulnerabilities connected with the operation of smart grid at this level. Such entry points include control networks and protocols that link the SCADA systems to lower level control equipment. Firewalls and modems can also be listed as entry point candidates. Firewalls aim to protect certain network levels by applying filtering policies on the monitored communication packets. Modems enable devices to communicate by converting data in order to be transmitted over a modulated carrier wave. Similarly, communications systems and routers transfer messages between two networks. In addition, entry points include industrial network protocols that link field sensors and other devices to control units (fieldbus protocols). By eavesdropping on the exchanged protocol data, adversaries can manipulate the integrity of communication packets. Remote access points are also ideal candidates for attack entry points to the system since they can remotely query network data and configure smart-grid technologies. The poor configuration and the lack of patch management procedures remains, however, the main security risk to emerging threats [13]. Finally, even though most of the smart-grid systems are not directly connected to untrusted networks, portable media can be transferred inside the trusted perimeter by personnel. As a result, malware can infect system components and propagate to the critical field equipment (e.g., Stuxnet).

Implementation and design flaws in industrial communication protocols are frequently the source of smart grid related vulnerabilities. In 2013, 22% of ICS/SCADA vulnerabilities appeared in communication protocols [14]. The most widely used protocols in the electric sector are International Electrotechnical Commission (IEC) 60870-5, Distributed Network Protocol (DNP3), Modbus, FOUNDATION Fieldbus, and Inter-Control Center Communications Protocol [15]. Moreover, smart grid equipment vendors often include support of proprietary communication protocols in their products. In both cases, a number of these protocols offer security through obscurity[2] measures, whereas others have known vulnerabilities or they are insecure by design. Modbus and DNP3, two widely used protocols in SCADA systems, have several identified vulnerabilities [16]. Figure 15.6 illustrates the percentage of vulnerabilities identified in industrial communication protocols in 2013 [14]. The absence of integrity, confidentiality, antireplay and authentication check features in these protocols is the main source of their inherit vulnerabilities. The lack of these mechanisms allows a variety of network attacks such as *man-in-the-middle (MitM)* attacks, *replay attacks*, denial of control and monitor services, identity *spoofing* attacks, and others.

---

[2]Security through obscurity is the use of secrecy in the design or implementation of particular protocol modules (e.g., how the authentication password gets encrypted) to provide security.

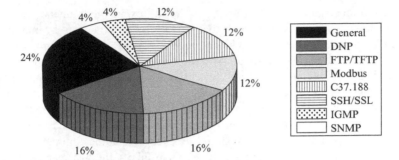

Figure 15.6   ICS/SCADA protocol vulnerabilities (2013)

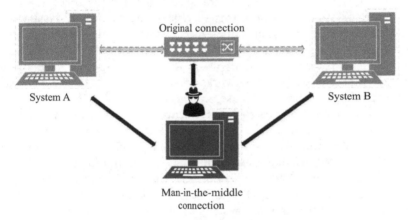

Figure 15.7   An illustration of the man-in-the-middle attack

In MitM attacks, an adversary intercepts the communication between two systems, A and B, as shown in Figure 15.7. The attacker "splits" the original communication link into two connections, one between system A (e.g., SCADA server) and the attacker, and the other between the attacker and system B (e.g., slave RTU). Although the two systems believe that they directly communicate with each other, the adversary acts a proxy who eavesdrops on the connection. The execution of a MitM attack by the unauthorized party requires to modify the exchanged data or inject new traffic into the communication channel. In the scenario that the communication protocol transfers encrypted data or supports authentication mechanisms, MitM attacks are still feasible by eavesdropping for key exchanges. The MitM party may establish two distinct key exchanges with the two systems, masquerading as A to B, and vice versa, allowing the MitM to decrypt and then re-encrypt the messages passed between A and B. However, industrial protocol connections often do not support authentication and encryption mechanisms; therefore, MitM attacks could allow access to entire smart-grid networks with minimal effort.

Since industrial control traffic is often exchanged unencrypted, the communication packets via a valid data transmission link can be maliciously repeated or delayed. Replay attacks intercept the data and retransmit a desired process command into a protocol stream received by an ICS component. For instance, if an attacker captures the network traffic between the commands from master SCADA stations requesting tripping of relay controllers, then these packets can be replayed to perform the same trip task. In cases which the transmitted traffic is in plain text, custom packets can be crafted to perform other tasks ultimately altering the behavior of the entire system. If the receiver in the communication path is a PMU, setpoint registers can be overwritten to falsify the phasor measurements transferred to the transmission system operation center.

Identity spoofing attacks allow attackers to impersonate authorized users. Consequently, spoofed messages are transmitted through the smart-grid network and appear to be originated from a trusted system. For example, if the attacker can manipulate a network address or routing mechanisms, identify spoofing attacks can further trigger MitM and replay attacks to the grid network [17].

Firewalls and *virtual private networks (VPNs)* are essential for the network perimeter. If these devices and networks are poorly configured, the risk of cyber intrusions could significantly increase. Improper firewall setup can allow attackers to inject a large number of packets into the network that may cause congestion and limit the network's availability. VPNs create encrypted connections to ensure secure and confidential data transmission between a client device and a server device. These VPN links, however, secure only the connection tunnel and not the client or the server unit. VPN connection hijacking can be avoided by integrating VPN communications into suitable firewalls. In addition, when designing a network architecture for a smart grid system, special attention should be paid towards separating the industrial from the corporate network. In cases in which the networks must be connected, only minimal connections should be allowed and the connection must be through a firewall and a *demilitarized zone*, that is "a network area (a subnetwork) that sits between an internal network and an external network" [18].

Although power utilities are taking steps to establish better protection against attacks on the smart-grid infrastructure, many of the products currently being deployed in power systems have not been designed with security in mind. Commercial-off-the-shelf designs typically use common technologies that have both known and unknown security vulnerabilities. The extensive use of such devices led to the porting of serial protocols over transmission control protocol/internet protocol (TCP/IP) for interoperability reasons. However, this tactic has expanded the attack surface by including *TCP/IP attacks*. Several industrial processes are connected directly to the Internet via TCP/IP as evident by the Shodan search engine [19]. In early 2016, ICS Radar, built on top of the Shodan engine, registers more than 14k Internet-facing systems that use DNP3 and Modbus protocol over TCP/IP. This facilitates reconnaissance and ICS targeting. For example, a study utilized Shodan to first identify Internet-connected PLCs and then acquire their code via specially crafted network requests [20].

15.3.3 Software layer

The variety of applications and platforms within the software layer increase the smart grid's attack surface. Terminal systems and human–machine interface (HMI) suites interacting with the industrial processes may include sections of code that contain errors or implement poor *access control* mechanisms. For example, *memory corruption* errors between 1988 and 2012 demonstrate the prevalent impact of such vulnerabilities: about 35% of "critical" common vulnerabilities and exposures (CVEs) were memory corruptions [21]. In addition, software interfaces and terminals are frequently based on commodity operating systems (OSs). These vulnerabilities could be exploited by adversaries to gain administrative access to key system components. Attackers can then modify the legitimate *control flow* and alter the behavior of software platforms responsible for controlling the system.

The highest percentage of vulnerabilities identified in industrial control products include improper input validation by ICS software, poor management of credentials, and authentication-related vulnerabilities [22]. The analysis of ICS-related software flaws from 2010 until 2013 shows that improper input validation, at 61% of reported vulnerabilities, is by far the most commonly identified vulnerability [23]. Improper input validations arise in scenarios in which the data or control flow of a software platform are affected by the incorrect or nonexistent validations of input packets. Hence, an attacker able to craft the software input in a form that is not properly handled by the designed platform can cause modifications to the software control flow or even achieve arbitrary code execution.

Besides attacks applicable to the smart-grid software itself, additional attacks are feasible, particularly in cases which web browsers are used on the same system as part of multipurpose workstations or as part of the HMI platform. For instance, *cross-site scripting (XSS)* attacks could enable malicious adversaries to inject client-side script into web pages viewed by other users. Thus, attackers could disclose host files, steal session cookies, install further malware, and alter the browser exchanged data. The prevalence of XSS attacks is highlighted from an analysis report that indicates that from 1988 to 2012, 13% of all CVE security advisories were related with XSS code injection vulnerabilities [21].

Sophisticated malware often target both firmware and software. Firmware includes instructions and data that reside between the hardware and software of devices. It is programmed into the read-only memory of a system. The functionality of firmware ranges from booting the hardware and providing run-time services to loading an OS. In order to meet the real-time constraints related to the behavior of smart grid operations, firmware-driven systems typically employ real-time OSs. Similarly to software vulnerabilities, firmware flaws pose a severe threat to the security of smart grid. For instance, malicious firmware images can be distributed from a central system in AMI to smart meters [24]. Firmware vulnerabilities in wireless access points and recloser controllers could even cause serious erosion of the power system's stability margin [25]. In addition to the research studies, real-world examples clearly indicate that power systems are vulnerable to firmware-related threats. In late 2015, attackers were able to cause a blackout in the

Ivano-Frankivsk region of Ukraine by overwriting the legitimate firmware on critical devices at the substations, leaving them unresponsive to any remote commands from operators [26]. Specifically, the adversaries wrote malicious firmware to replace the legitimate firmware on serial-to-Ethernet converters at more than a dozen substations, rendering the converters thereafter inoperable and unrecoverable, unable to receive commands. A case study presenting the impact of modifying the firmware of such critical controllers is examined in Section 15.5.

Data acquisition in smart grid begins at the RTU or PLC level. The meter readings and device status reports from such equipment is transferred to the SCADA master as required by the network model. These special purpose field devices may include vulnerabilities in the software layer allowing attackers to exploit them either remotely or locally. As a consequence, malicious activity could compromise the device integrity and availability. PLCs deployed in many complex control processes run ladder-logic[3] program code containing numerous intentional or unintentional errors. Furthermore, IEDs as "smart" sensors/actuators are often identified with vulnerabilities contributing to MitM and DoS attacks [28].

In the past, several real-world examples have shown that the grid as an ICS is exposed to various software threats that can lead to severe consequences. In March 2008, the Hatch nuclear power generation plant in Georgia, USA, was forced into an emergency shutdown for 48 h [29]. The shutdown is attributed to a software update on a single computer, charge of the chemical process monitoring and diagnosis data of one of the plant's primary control systems. After applying the update, the computer rebooted and the synchronization software routine reset the data on the control network. Safety control systems misinterpreted the reset action causing a sudden drop in the reactor's water reservoirs and initiated an automatic shutdown. The 48-h disruption of the nuclear plant operation is a striking indication of the impact of malicious software patches and their resultant financial costs. Considering power transmission and distribution, in 2009 cyber spies accomplished to penetrate part of the US electricity grid, installing malicious software capable of disrupting power supplies [30]. According to US public administration officials, the malicious software had the ability to interfere with critical control systems and leave thousands of people without electricity at any given point during its existence.

Regardless of the attacker's incentive, the exploitation of the underlying vulnerabilities at the software layer is simplified and accelerated by leveraging attack frameworks. Malicious adversaries typically use common tools and tactics in order to attack computer programs and identify the vulnerabilities of grid components, such as Metasploit [31] and Nessus [32]. In addition, attackers seek opportunities in underground markets for *zero-day* exploit sales [33]. The "purchase" of zero-day exploits anonymously minimizes the effort invested by adversaries. Since these

---

[3]Ladder logic is widely used to program PLCs, and it represents a software program by a graphical diagram based on the circuit diagrams of relay logic hardware. PLCs can be also programmed with both graphical and textual programming languages. IEC 61131 standard for instance, defines two graphical (ladder diagram, function block diagram) and two textual (structured text, instruction list) PLC programming language formats [27].

exploits are unknown to software and security developers, they are expected to remain unpatched for extended periods of time, increasing the attack surface of the smart-grid software layer.

15.3.4 Hardware layer

All the electronic equipment and physical elements such as microcontrollers, data storage disks, and microprocessors constitute the hardware layer of a computer system. Attack vectors in this layer rely on routes or methods used to get into hardware and physically corrupt or manipulate a combination of hardware objects. This requires detailed knowledge of the specifics of the system and its architecture. In the cyber security context of smart grid, the role of hardware layer is essential, as all the other layers rely and build upon it to perform all operations. Hardware-based components in smart-grid include embedded systems such as PLCs, RTUs, IEDs, SCADA servers, workstations, relays, and communication routers. These hardware modules are susceptible to both invasive and noninvasive hardware attacks. For example, hardware backdoors can be exploited by adversaries to enable remote control of the target device. The activation of hardware attacks can rely on a specific timing or functional condition which when activated will degrade the performance of the system or even disable the circuit logic [34].

Due to supply-chain globalization, security in the hardware development cycle is a major concern. Attackers can inject malicious hardware logic at any stage of the supply chain including the fabrication phase. Attack vectors that rely on hardware *trojans* can introduce potential risks in the implementation of vital security functions resulting in hardware reliability problems [35]. Besides vulnerabilities arising from supply chain, smart-grid systems are prone to attacks targeting specific hardware modules initially designed to accelerate certain procedures such as testing, verification, memory expansion, and others. For instance, unauthorized users can use the *Joint Test Action Group* interface, an industry standard employed for testing integrated circuits, for reverse engineering and intellectual property [36]. Peripherals also introduce hardware vulnerabilities. Malicious universal serial bus (USB) drives, for example, can redirect communications by changing network settings or destroy the circuit board [37]. Memory units, expansion cards, and communication ports pose as well a security threat [38]. Real incidents reveal that hardware backdoors exist in critical equipment. Specifically, a backdoor discovered in widely used routers allows information to be transmitted to unauthorized users [39].

Fault injection attacks deliberately inject hardware faults during the normal operation of a device in order to manipulate the computation results. One method of injecting faults into a hardware unit is clock glitching, that is the attacker gradually decreases the clock period via glitch injection in order to make the hardware circuit fail. A number of existing false injection attacks are destructive to the circuit and require physical proximity to the hardware device. However, they are effective in disrupting the integrity of smart-grid devices and leaking confidential information [40]. Depending on the fault injection method, faults can be transient or permanent. Also the accuracy of false injection attacks varies. Low accuracy fault injection attacks can be realized using electromagnetic (EM) pulses, by falsifying the environment temperature of the device, manipulating the system power supply and clock, and others. On the other hand, high accuracy fault injection attacks require costly and more sophisticated tools (e.g., focused ion beams) which are able to achieve a granularity of single bit-flips.

Information can be extracted from the hardware of smart-grid devices through side channels [41]. Side-channel attacks (SCAs) are known to be quite effective on hardware, and they are related to adversary actions that take advantage of the physical module information leakages. The leaked information (mostly unintentionally) from a particular module to its environment can be derived from data such as power consumption, EM radiation, timing information, acoustic vibrations, and others. In general, there are various information sources leaking information from hardware modules that can consequently be exploited by malicious attackers. These characteristics are directly visible from measurement traces and could simply compromise the confidentiality of the data computation. SCAs based on reverse engineering techniques are easy to implement and ultimately entail powerful attacks, mainly against cryptographic primitives. Most of the times, it is assumed that cryptographic implementations are ideal "black-boxes" and thus their internals cannot be observed or interfered by malicious activities. It is unrealistic, however, to consider that these blocks provide perfectly secure solutions. Cryptographic primitives rely on both software and hardware and hence this interaction can instigated and monitored by adversaries. For instance, a malicious user could monitor the power traces of cryptographic computations to gather information regarding the algorithm rounds and thereafter use this valuable information to deduce the number of internal secret keys (Figure 15.8).

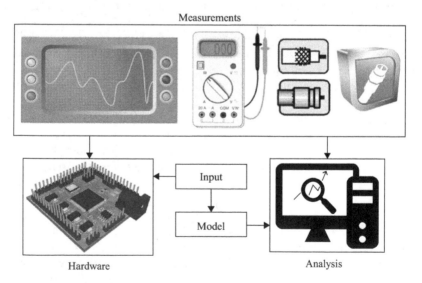

Figure 15.8   Side-channel attack methodology: it compares observations of the side-channel measurements (e.g., timing information, power dissipation, etc.) with model estimations of the leakage

15.4 Smart-grid security enhancement

The outcomes of cyber-attacks against the smart grid range from electricity theft to widespread blackouts. Towards protecting the grid against such threats, security standards, guidelines, and regulatory documents play a key role in smart-grid cyber security. These publications provide guidance in understanding smart-grid inter-operability, interconnections, architectural designs and layers in order to achieve seamless, secure, and reliable operation of the electric power system.

In general, availability, integrity, and confidentiality are the core information objectives that provide assurance of security requirements for smart grids. The comprehensive security requirements of a smart grid, however, are different from traditional information technology (IT) security due to the constraints imposed by the scale, complexity, and distributed nature of the grid. Therefore, when designing protection solutions against smart-grid threats, first it is essential to be aware of the fundamental differences between the security requirements in those two domains. The following list provides a basic understanding as to why smart-grid security is different than any other traditional security:

- Lifetime cycle: the average life of the equipment and systems deployed within the grid exceeds 10 years. In comparison with traditional IT systems, the development of secure power systems is inherently difficult. Many grid technologies, due to their long life cycle and improper patch management operational procedures, are running systems with published vulnerabilities.
- Performance and availability: availability in smart-grid systems is critical since power outages are not acceptable. Furthermore, grid interactions and processes are continuous in nature making smart-grid applications time-critical. For example, cryptography as a security control mechanism used in smart grid may cause unacceptable communication overhead and computational cost. Also, patching often requires downtime which for power systems may result in both operational and economic consequences.
- Proprietary control and communication protocols: in smart grid systems, there is a high amount of proprietary information due to the unique standards and protocols. Moreover, many designs use specialized software and hardware. These proprietary or customized system implementations and interfaces are often designed to support-specific processes and may not have built-in any security capabilities. In addition, proprietary systems are not exposed to the scrutiny of public auditing.
- Risk management goals: human safety is foremost, followed by protection of process. Other major risks include regulatory compliance, loss of equipment or production, environmental impacts, and others.
- Component location: power systems consist of geographically disperse and isolated resources, for which gaining physical access may require extensive effort. For the same reason, many of these components may be exposed to physical attacks.

- Security mechanisms: the protection of central servers and field components is significant. In addition, "fail-closed" security mechanisms are restricted to avoid prevention of accessing a system in critical situations.
- Legacy systems: every domain of electric grid includes implementations with legacy communication, control, and monitoring systems. These legacy technologies were not necessarily developed with resilience against cyber-attacks as a primal concern.

15.4.1   Policies and standards

Federal, state, and local agencies and organizations involved in the electric industry should cooperate for the creation of smart grid security plan templates. The templates can be adapted to each particular situation, accelerating the response of both personnel and infrastructure components in attack incidents. The plan guidelines can include details regarding crisis management, operational, cyber and physical security recommendations, technical information, roles and responsibilities, and others. In addition, smart-grid stakeholders should establish channels for information sharing. Such initiatives can assist both the government and the private sector to develop situational awareness in support of smart-grid security. For example, the S.754 "Cybersecurity Information Sharing Act of 2015" aims to enhance information sharing between government and private entities [42]. Furthermore, vulnerability databases can help in effectively dealing with cyber-attacks. Regulators, customers, and especially utilities can have a reference for publicly disclosed vulnerabilities and the required actions for mitigating the threat. Training, education, and awareness programs and events are also necessary. Such programs must link availability and safety with good cyber security practices. Every person involved in any smart grid related activity should be trained for security awareness as a way to mitigate the threats arising from the increased use of ICT.

The different needs of smart-grid technologies in terms of security have been recognized by governments, agencies, and institutes. Several published security guidelines provide guidance and recommendations for a resilient smart grid. Table 15.1 presents some of the most frequently used security roadmaps. Several other reports exist in literature which provide the appropriate measurements to be taken in order to meet industrial systems security challenges [43–46].

Due to the large complexity of the smart grid, different models are used for different studies. Each model represents a particular view of the system aiming to solve a specific issue. For example, power systems commonly adopt a graph-theoretic model in which each link (e.g., distribution line) is the connection between two nodes (e.g., smart meters). The data and measurements of the system are then used in an estimation model (e.g., power-flow model) to solve the formulated problem. The smart grid also endorses the use of common information models (CIMs) as a key to achieve interoperability objectives, that is to supply a common format for information exchange to support power system management. For example, CIMs such as the IEC-61970 and IEC-61968 focus on transmission and distribution systems, respectively, defining the required structures, information,

Table 15.1   Mission-critical systems security: roadmaps and security guidelines

Policies/documents	Issues addressed
ENISA: Smart Grid Security, Recommendations for Europe and Member States [47]	Makes ten recommendations to the public and private sector regarding the implementation of smart grids
NIST SP 800-82: Guide to Industrial Control Systems (ICS) Security [48]	Identifies cyber security concerns within ICS systems, including SCADA and DCS systems
ISA99 (ANSI/ISA-62443): Industrial Automation and Control Systems Security [49]	Includes standards, recommended practices, technical reports, and related information for implementing manufacturing and control systems securely
NIST-IR 7628 Rev. 1: Guidelines for Smart Grid Cybersecurity [50]	Provides a comprehensive overview for cyber security strategy practices, taking into account privacy and vulnerability classes
NERC/DOE: High-Impact, Low-Frequency Event Risk to the North American Bulk Power System [51]	Determines the appropriate balance of prevention, resilience, and restoration in the American power system
NERC: Critical Infrastructure Protection (CIP) [52]	Requires the identification and protection of all cyber assets supporting the power grid
MIT: The Future of the Electric Grid [43]	Provide a detailed portrait of the US electric grid including the challenges and opportunities it is likely to face over the next years
GAO-11-117: Electric Grid Modernization [44]	Identifies challenges for securing smart grid systems
EOP: A Policy Framework for the 21st Century Grid: Enabling Our Secure Energy Future [45]	Outlines fundamental concerns related to the security of electric grid
DHS (ICS-CERT): Cross-Sector Roadmap for Cybersecurity of Control Systems [46]	Contains a set of priorities that address specific control systems needs over the next decade

and relationships that facilitate grid support and harmonization development [53]. In order to address current and future security threats, the electric-power industry should also foster and leverage research on smart-grid information modeling as a way to forge ahead with grid modernization.

15.4.2   Multilayer protection for smart grid

Since threats are constantly evolving, protection demands advanced cyber security mechanisms. The development of a secure smart grid should adopt fundamental security techniques for defending against cyber-attack methodologies at all layers of the smart grid. In the *operation layer*, integration of energy and information technologies are necessary to ensure resiliency and permit two-way flow for communication and control. The first step to achieve these goals is to perform vulnerability assessments in order to identify security weaknesses and potential risks with the industrial operation. Due to the real-world consequences of smart-grid operations, vulnerability assessment must be performed regularly to ensure that the grid elements, composing the grid infrastructure and interface with the system perimeter, are secure. In addition, the assessment should take into consideration the sensitive smart-grid dependencies and connectivity and also account for all possible operating conditions of each smart-grid component. The steps of vulnerability assessment methodologies include document analysis, mission and asset prioritization, vulnerability extrapolation, design of an assessment environment, testing and impact assessment, vulnerability remediation, validation testing, and continuous monitoring [54].

Operation-centric access points in the smart grid should be hardened to limit access only to authorized processes and personnel (role-based access control). One of the basic concepts in system hardening is applying the principle of "least privilege": only points necessary for reliable operation should be enabled. Unnecessary applications should be removed, and access should be limited to the minimal level. Furthermore, practices such as separation of services are recommended to prevent cross-contamination when a smart-grid system becomes compromised.

Similarly, in the *network layer*, the identity between communication parties should be verified with proper authentication procedures. It is also important to secure the communication link between grid elements (e.g., smart meters and gateways) without compromising the communication performance. In the past, smart-grid systems used to be isolated or "air-gapped", that is having a physical air gap with no common system crossing that gap. The Stuxnet case demonstrated that an ultimate air gap does not work in real environments, thus VPNs, point-to-point firewall rules, and others must be applied for proper network segmentation. In addition, industrial protocols are required to protect communication networks in smart grids. For instance, IEC-61850 communication standard defines data formats and interoperability technologies for layered substation communication architectures [55]. Due to the concerns related with the security of the protocol, IEC-62351 is a support standard defining how to secure IEC-61850 communications [56].

Network Intrusion Detection System (IDS) technologies should be included in smart-grid network designs to augment host or network-based defenses and monitor system activities for malicious attempts. IDS modules installed distributively along the network hierarchy can profile network communications in order to flag any suspicious traffic and identify deviations from correct behaviors [57]. The role of an IDS is typically passive, that is it gathers network information with the purpose only to identify and alert anomalous/suspicious activity. Thus, IDS technologies have limitations in protecting network systems. Intrusion prevention systems (IPS) on the other hand are considered active security solutions since they can protect the system from outside and inside network attacks. IPS designs have all the capabilities of an IDS and can also use attack signatures and deep packet inspection to prevent specific types of attacks. Figure 15.9 illustrates the utilization of IPS technology for networks in smart grid environment to keep intruders off the network devices.

The main causes of incidents in industrial networks are attributed to errors and malware attacks in the *software layer* [58]. The negative effects of software-related

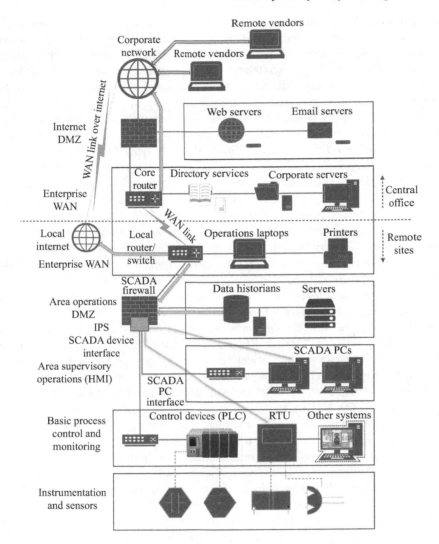

Figure 15.9   The use of intrusion prevention system (IPS) technology in different levels of smart-grid networks

problems could have several implications in the control operation of smart-grid systems. Therefore, implementing software security controls such as code integrity checking and antivirus software can minimize or even eliminate the exposure and propagation of malicious software activities. Smart-grid software-based systems must be also updated and patched frequently to avoid running code that is known to have software vulnerabilities. Furthermore, embedded systems in smart grid typically run software supplied merely by the manufacturer. Thus, manufacturers should incorporate software validation procedures in their products. For example, validation keys can be located in a secure memory storage of the device to protect users from installing rogue copies of the software. In addition, application *white-listing* solutions can prevent unauthorized software program execution without impacting the performance of the system. On one hand, in contrast with antivirus and other software packages, whitelisting does not require to be updated regularly. On the other hand, every time software applications are patched their signature changes; it is therefore required that administrators update the whitelist database.

Providing security countermeasures to firmware malicious actions should be one of the initial steps for protecting not only embedded systems, but any kind of computing system within the smart grid. If the underlying firmware is not trusted than any other mechanisms implemented at the OS or application level cannot be trusted. Considering the limited resources of an embedded system (e.g., computation resources, memory usage, communication bandwidth, etc.) as well as the requirement to defend legacy systems within the grid, malicious firmware detection techniques must remain low-cost and overcome the challenges and constraints imposed by the security model. For example, low-level hardware events such as *Hardware Performance Counters (HPCs)* can be utilized to monitor and model the behavior of firmware images [59].

At the *hardware layer* and therefore at the device level, hardware-assisted functionality can make systems more resilient and able to reduce recovery time in case of a malicious attack. Hardware support of cryptographic capabilities can empower devices like RTUs and IEDs. Due to the limited storage space within the hardware of field devices, lightweight authentication protocols can assist in securing the communication amongst various components at different points of the smart grid. A Trusted Platform Module (TPM) can also be used to provide hardware-based trust for the smart grid equipment [60]. A TPM is a cryptographic chip designed to provide greater levels of security (than was previously possible) within the hardware layer of a device. Currently, TPM technology is part of most the computer designs and is manufactured by nearly all chip producers, including Intel, Atmel, STMicroelectronics, Broadcom, Toshiba, and others. The main functions of a TPM are (a) to verify the integrity of the software running on the platform, that is it provides to the challenger that the attester executes legitimate software packages with the assistance of a remote trusted third party (remote attestation), (b) data sealing, that is encrypted data are sealed to a specific TPM platform and a particular system configuration, and (c) data binding, that is user data are encrypted using a capable of migrating secret key.

Besides TPM authentication of general purpose computing platforms, security extensions to already deployed processors can enable protection of both peripherals and memory. For instance, TrustZone technology, a security extension in ARM processors [61], employs two virtual processors (called normal and secure) backed by hardware-based access control to provide isolation and enforce tighter *digital rights management*. In addition, hardware-enforced virtualization can reduce hardware dependencies, separate memory access, isolate execution environments, reduce the costs of maintaining hardware computing platforms, and most importantly, limit the damage from a successful attack within the virtual machine. Virtualization can also assist in restoring the system to a previous snapshot.

The ability to return back to a previous point in time is useful during forensic investigations after a successful breach into the system.

15.4.3 Security assessment environment

It is important to understand that realistic impact assessment of cyber-attacks on electricity service requires appropriate representation of the studied system. The impact analysis should identify dependencies between various complex systems (both cyber and physical) showcasing the characteristics of interdependent networks. Hence, for proper vulnerability evaluation assessment, testbeds provide an ideal development environment. A testbed typically includes interdependent software, physical/hardware equipment, and network components for studying, understanding, and improving smart grid security without causing real-world failures. A large number of testbeds exist for a variety of purposes and industries. In 2014, over 35 smart-grid applicable testbeds had been developed in the United States [62]. In order to expand the vulnerability impact analysis to nation-scale, the hardware testbed needs to be combined with an adequate simulation model. For that purpose, hardware-in-the-loop (HIL) testbeds are attractive as assessment environments [63]. HIL methodologies connect hardware equipment to a host computer that runs the simulation model.

A HIL testbed must model the operation and evaluate the security of electronic equipment employed in smart grid domains. The configuration setup outline of a sample testbed is presented in Figure 15.10. It includes a data acquisition that processes the information regarding the testbed network structure (sectionalized areas through relay controllers). The acquired data are used in a power-flow simulation environment to calculate the optimal electricity flow through the network. The testbed also contains a real-time automation controller (RTAC) that collects data from smart meters and relay controllers. The information collected via the RTAC is transmitted to a long-term evolution (LTE) router that shares it with a

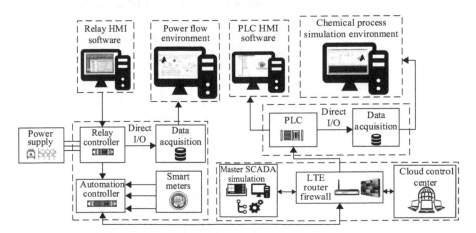

Figure 15.10 Diagram of a sample smart-grid testbed

4G/3G/2G connection to the SCADA control center. This sample testbed can address vulnerabilities for each smart-grid layer and across layers. Besides the security objectives, it can also be used to ensure the interoperability of components and capture the information flow throughout the layers of the grid architecture. Moreover, since the structure of the environment can be reconfigurable, the testbed offers flexibility to examine several scenarios and emulate different conditions of the real smart grid environment.

15.5 Case study

In this section, a case study is presented focusing on both attack and defense mechanisms applicable to the firmware layer of smart-grid systems. Particularly, the study illustrates the capabilities of cyber-attacks in the smart-grid context and examines the security and reliability of the network. Preparing for the future smart grids, studies are essential to evaluate security requirements and practices in the power industry. In addition, by mapping attacks onto proposed protection models, the following study can demonstrate the effectiveness of security countermeasures.

The challenge: The digitization of smart-grid technologies leads to the formation of power systems substantially dependent on embedded devices. Firmware in embedded systems, such as microprocessor-based relays, controls the hardware of the device. Therefore, firmware attacks can bypass access control and security modules. An attacker capable of maliciously modifying the firmware can introduce backdoors, control the operation of a device, and have unrestricted access to the system components. In the smart-grid scenario, one of the attack challenges is to alter the relay firmware in order to open and close circuit breakers at undesirable time, inducing catastrophic damage to machines or even leading to cascading systems failure [25]. Given the critical role of firmware in these embedded systems, implementation of effective security controls against malicious actions is essential [59]. The design of such countermeasures needs to take into consideration the resources constrains in embedded devices (e.g., computation resources, power consumption, memory usage, etc.).

The impact: Firmware reverse engineering can reveal information of the system features and unlock hidden functionalities. In this study, the firmware analysis findings of a commercial relay controller are applied to corrupt the circuit breaker (CB) status signal in a developed testbed. Two firmware attack vectors are developed based on the aurora-type vulnerability and the relay inability to sense a fault and initiate a trip to the breaker.

Every relay has a deliberate operational delay to avoid any protection activity during transients. These delays leave an open window of opportunity for defective operation [64]. The out-of-sync closing of breakers via the connected relays results in the aurora vulnerability by changing the operating frequency of the generator and causing frequency difference between the machine and the grid. To meet the requirement of repeatedly sending trip and reclose commands to the generator relay, the first firmware modification disables the communication port of the relay controller so that there is no transmission of digital data to the master SCADA system (DoS attack). While the relay is offline, the relay reboot address is injected into specific firmware locations in order to cause the relay to restart resulting in an aurora-type event.

Protective relays are designed to handle power network faults (e.g., short-circuits). This involves detecting the presence of faults, isolating them by tripping the breaker connected to the relay, and reclosing circuits automatically. Failure to sense and clear the fault may start a chain reaction to the system. The second attack vector modifies the relay protection profiles specifying the operation of the relay control. This is accomplished by altering the calibration control mechanisms encompassed in the firmware initialization process. In order to keep the modifications minimal, changes are only performed for the overcurrent protection parameters. For example, even if the phase and ground minimum trip currents set in the relay software are 400 A and 280 A, respectively, the calibration registers are modified to be always programmed to the relay maximum trip settings that is 3,200 A for phase and 1,600 A for ground minimum trip currents.

The proposed attack scenarios are applied on a HIL testbed. The breaker control signal is transmitted to the simulation environment modeling the Institute of Electrical and Electronics Engineers (IEEE) 14 benchmark. The contingency ranking of the system specifies that the most critical generators are G5 and G3 [65]. The developed bus system, however, is able to handle both $N - 1$ and $N - 2$ contingencies, that is the generator rotor angles are transiently stable after causing G5 and G3 breakers to trip.

In the first scenario, the modification of aurora-type event is simulated by intentionally opening the CBs at $t = 1$ s and reclose/trip every 15 cycles (0.25 s). When the breaker opens and closes once (scenario 1a), the out-of-phase generators are imposed to torque pulsation to remain in synchronism with the grid. When the attack is repeated two times, there is a voltage collapse due to the limited power transfer capability (scenario 1b). The graphs for this case are shown in Figure 15.11. In the fault-clearing failure scenario, the firmware modifications are simulated by applying a short-circuit three phase fault to the system. When the fault current flows above the preset overcurrent value, the corresponding relay instead of initiating the status signal to open the corresponding breakers, the breakers remain close and the system fails to clear the fault. Figure 15.12 shows the inability of the breaker to clear faults, which leads to voltage instability responsible for network collapse.

The solution: Since a firmware program is composed of a sequence of various types of instructions and data, the program behavior can be uniquely characterized by the total occurrences of hardware events during its execution as well as the relationship between the counts of different monitored events. The low-level hardware events can be efficiently measured using HPCs. Although HPCs are typically used for performance tuning, HPCs can be utilized for security purposes with no extra hardware cost [59]. HPCs events include retired instructions, branches, returns, and others. When firmware is maliciously modified, the counts of the executed HPCs events will be different from those during legitimate firmware runs.

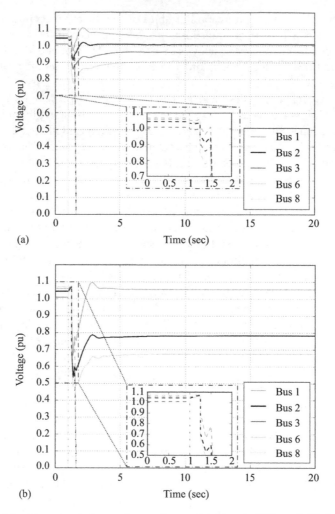

*Figure 15.11   Generators bus voltage due to (a) N − 1 (G5) and (b) N − 2
(G5, G3) generator aurora-type contingencies [scenario 1a
(solid line), scenario 1b (dotted line)]*

The high-level structure of the HPC-based security module is shown in
Figure 15.13. A legacy *bootloader* is extended to include three components: (a) an
insertion module that places checkpoints to the monitored firmware, (b) an HPC
handler that drives the HPCs, and (c) a database that stores valid HPC-based signatures. All these components are stored in write-protected non-volatile memory.
This prevents attacks from compromising the security module while still allowing
authorized updates. Once the execution reaches a checkpoint, the control flow is
intercepted and redirected to the core module. The core then communicates with the
HPC handler and the HPC-based signature database. Specifically, the event counts

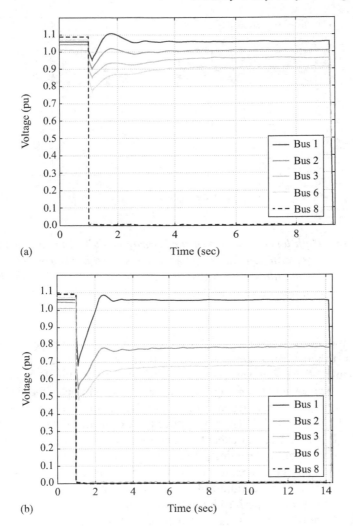

*Figure 15.12   Generators bus voltage due to (a) N − 1 (G5) and (b) N − 2
(G5, G3) generator fault-clearing failure contingencies*

for the previous check window are read and compared with the corresponding signatures in the database. Then the HPCs are reset for the next check window and the
execution of the monitored firmware continues. The HPCs keep counting the
occurrences of the hardware events until the next checkpoint is reached.

The benefits: the effectiveness of the designed security mechanism is tested
with two real-world firmwares of embedded systems on two different platforms.[4]

---

[4]Samsung Exynos Arndale (ARM Cortex-A15, 6 HPCs, 70 events) and Freescale MPC8308RDB
(PowerPC e300c3, 4 HPCs, 40 events).

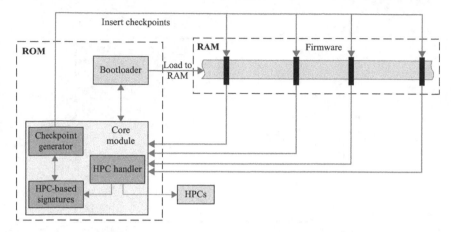

Figure 15.13   High-level structure of the security module. The core consists of
three components: an insertion module that inserts checkpoints to
the monitored firmware, an HPC handler that drives the HPCs,
and a database that stores valid HPC-based signatures

The first embedded system is an ARM-based wireless access point and access gateway. In this device, a DoS attack is performed which targets the task scheduling module of the firmware. When a task with a specific ID running it will occupy the processor without being switched out, thus will impact the availability of other tasks. Owing to their repeatability, four events are selected to be monitor: INSTRUCTION executed, BRANCH instruction executed, LOAD instruction executed, and STORE instruction executed. There are seven valid paths in the original scheduling subroutine. At runtime, the HPC-based signature is compared with a subset of the valid signatures to check if a match occurs. The results are shown in Table 15.2(a) for three randomly selected check windows. The minimum deviation for a malicious path to be detected is 8.7% in check window 3 (among the events whose occurrences have changed). In this case, the monitored LOAD event defines the minimum noise detection threshold N. For example, a detection threshold of 5% is adequate to identify the malicious modifications in every chosen check window.

The second embedded system in this study is a commercial relay controller. The PowerPC-based controller firmware is modified to implement a MitM attack that sniffs Ethernet packets. The attack targets the Ethernet packet receiving subroutine in order to capture the packets of data flowing across the network. The modification intercepts the control flow and copies the received packets to a specific memory location. As a result, an attacker can retrieve the critical information in the received Ethernet packets. Due to the hardware events occurrences, the selected events are similar to the ARM-based platform. The HPC-based signature of the malicious path exhibits large deviations when compared with the signatures of the five valid paths in the Ethernet packet receiving task. The generated

Table 15.2(a) *Detection capability using HPCs. The numbers are the event count deviations (%) of the malicious path from all the valid paths in monitored subroutines for (a) the wireless access gateway and (b) the recloser controller. For each path, the bold number indicates the largest deviation among all events. The tested path (malicious) is not matched to any valid path indicating a successful detection*

Path	Hardware event			
	Instruction	**Branch**	**Load**	**Store**
Check window 1				
1	27.3	33.3	37.5	**50.0**
2	77.1	71.4	**81.5**	50.0
3	73.3	71.4	**76.2**	60.0
4	51.5	**60.0**	58.3	0.0
5	65.9	50.0	**75.0**	60.0
6	69.8	71.4	**76.2**	33.3
7	62.8	**71.4**	66.7	33.3
Check window 2				
1	77.8	33.3	**150.0**	0.0
2	44.8	60.0	16.7	**66.7**
Check window 3				
1	33.8	**175.0**	31.6	40.0
2	**16.5**	15.8	8.7	12.0

Table 15.2(b)

Path	Hardware event			
	Instruction	**Branch**	**Load**	**Store**
Check window 1				
1	65.0	**266.7**	78.6	250.0
2	10.0	10.0	0.0	**55.6**
3	41.4	**83.3**	16.7	33.3
4	6.5	22.2	4.2	**47.4**
5	5.7	10.0	7.4	**33.3**
Check window 2				
1	**95.8**	76.8	62.1	30.0
2	19.5	51.1	**70.6**	30.0
3	**65.0**	46.7	37.5	44.0
Check window 3				
1	30.3	12.0	16.7	**47.4**

signature is compared with the valid ones included in the checking window. The results are presented in Table 15.2(b) for three randomly chosen check windows. The smallest deviation in this scenario is 4.2% due to the LOAD event counts of check window 1. Selecting the appropriate threshold N, for instance 4%, the

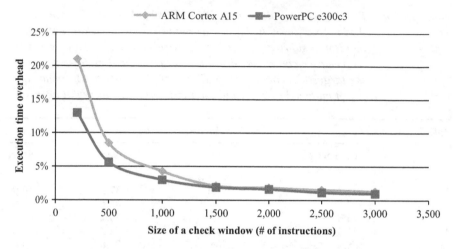

Figure 15.14   The execution time overhead with different sizes of check windows in terms of number of total instructions

implemented security mechanism differentiates between valid and malicious paths in order to detect the packet sniffing.

Figure 15.14 shows the execution time overhead on the monitored firmwares when different check window sizes are applied. For instance, a check window size of 500 instructions leads to an average execution time overhead of 8.48% on the ARM Cortex-A15 platform and 5.62% on the PowerPC e300c3 platform. For the test cases, the performance overhead for the scenario that includes all the sub-routine paths is 14.2% and 7.3% for the ARM and PowerPC case, respectively. The storage overhead of the security module is mainly for storing the components instructions (less than 10 kB) and the known valid HPC-based signatures. For example, an HPC vector in a check window that counts five hardware events requires 10 B storage for the signature of a valid path. If ten check windows are applied and there are ten valid paths in each window, then the required storage is 1 KB. In the scenario in which the firmware image size is 1 MB then the storage overhead (10 KB + 1 KB) is translated to approximately 1% of the firmware code size.

15.6   Future cyber security considerations

Successfully implementation of secure smart-grid systems requires careful attention to the future cyber security landscape. Smart-grid technologies will need to take into consideration trends in malware activity, cyber espionage, protocol vulnerabilities, network dependencies, and others. In addition, the multitude of new research studies and innovative technological developments can enhance observability in system operation and control.

One step towards improving cyber security is to create quantitative metrics for attack incidents. The metrics can be used to assess the dependency among information and communication objects, help to evaluate the impact of incidents, and prioritize the adoption of appropriate security countermeasures. Future cyber security considerations of the smart grid should also consider the introduction of security breach notification frameworks. Early notification warnings about disclosed vulnerabilities along with improved information sharing mechanisms can assist utilities to coordinate and implement strong security controls to mitigate the identified risks.

Smart-grid investments must allow the integration of new products and services that can provide comprehensive and coherent protection against cyber threats. In addition, "smart" equipment can enable demand participation and therefore improve both observability and controllability of the grid networks. Moreover, the optimal placement and sizing of DER have the potential to strengthen energy security and provide greater stability and efficiency to the smart grid. Transmission and distribution automation deployments are expected to receive more attention in the future. Such systems can assist in maintaining a reliable and secure electricity infrastructure by optimizing the utilization of assets and efficiently leverage smart grid instruments in the field. In addition, big data analytics will play a vital role in future implementations. The aggregation and analysis of smart-grid operational information (e.g., logs/databases, alarms/events, etc.) can assist in identifying threats and breaches as well as predict and prevent equipment failure.

Due to the rapid smart-grid modernization, the nature of human interaction with electric systems is expected to result in dramatic changes (as demonstrated in the past with cell phone communications and Internet). This social transformation will define new levels of security and privacy for data. Hence, to successfully maintain the security of both customers and the system, current and future smart-grid technologies need to be supervised and controlled by appropriate policies and regulations that will address the upcoming cultural change. In addition, it is necessary as we advance toward a smart grid to adopt a cross-disciplinary approach that instills greater coordination and interaction among government, electric utilities, and customers.

15.7 Conclusions

Over the last decade, the grid infrastructure has faced a transition from legacy systems to new technologies. The increased dependency of these components on cyber resources has an immediate impact to the exposure of the grid to potential vulnerabilities. During this, transition multilayer security approaches can help all involved stakeholders to defend against malicious actions and assess the security designs deployed in the smart grid. The attack methodologies and countermeasures discussed in this chapter aim to reflect the needs for current and future frameworks towards reducing the risks to an acceptable secure level and providing fine-grained security solutions.

A.   Appendix—Glossary

- *Access control mechanism*: It is a set of controls enforcing security policies that restrict access to particular resources. A security policy is a statement of what is, and is not, allowed.
- *Bootloader*: It is responsible for locating and loading the OS or firmware. If the device is a non-OS based system, then the firmware execution tasks are run in an infinite loop.
- *Control flow*: It is the order in which instructions, functions, or statements of a program are executed or evaluated.
- *Cross-site scripting (XSS)*: Attack is a type of attack based on web-application vulnerabilities. It involves client-side code injection in which malicious scripts are injected into web pages executed on the user's web browser (client-side).
- *Demilitarized zone (DMZ)*: It is a *network area (a subnetwork) that sits between an internal network and an external network* [18]. DMZs allow connections from the internal and the external network to the DMZ network. However, connections from the DMZ are only permitted to the external network.
- *Denial-of-service (DoS)*: Attack is a type against the availability of a service, machine, or network, that is DoS prevent legitimate users from accessing information or services.
- *Digital rights management (DRM)*: Technologies impose access control mechanisms and restrictions that control the usage of proprietary hardware and copyrighted works.
- *Gateway*: It is a hardware-based device that serves as a network point allowing entrance to another communication network.
- *Hardware performance counters (HPCs)*: These are special-purpose registers built into the performance monitoring unit of a modern microprocessor. The number of available HPCs as well as the number of hardware events vary from one processor model to another.
- *Joint Test Action Group (JTAG)*: Interface is an IEEE 1149.1 industry standard used for testing ICs using boundary scanning [66]. JTAG is widely used to communicate with microcontrollers and perform operations such as single-step execution and breakpoint insertion.
- *Memory corruption*: Errors that occur from violations of memory contents due to noninclusion of memory boundaries checks (e.g., buffer overflow).
- *Rootkit*: It is a special category of malicious computer software that can enable privilege access to a computing system or areas of its software, while, at the same, time hides its existence (or the existence of other software modules or both) in user-level objects.
- *Router*: It is a "traffic directing" hardware device that determines the next point within a network to which a data packet should be forwarded until it reaches its destination.

- *Switch*: It is a device that utilizes packet switching to process the incoming data from multiple ports and forwards that data to the intended destination (bridge between network points).
- *TCP/IP attacks*: These refer to any type of attack that exploits vulnerabilities in the transmission control protocol/internet protocol (TCP/IP) protocol suite.
- *Trojan*: In the context of software, a trojan is a malicious program that hides within other seemingly harmless programs that misrepresent themselves to appear interesting or useful in order to persuade and eventually trick a victim user to install it. Similarly, hardware trojans can be described as malicious and deliberately stealthy modifications to the circuitry of an integrated circuit.
- *Virtual private network (VPN)*: It is a technology that expands a private network over a less secure public network. It creates virtual network connections using virtual tunneling protocols, dedicated links, or traffic encryption.
- *Zero-day vulnerability*: It is a vulnerability that is unknown to the vendor. Such undisclosed security hole is known as a "zero-day" because once the vulnerability becomes known, the developer has zero days to patch and thus fix the flaw.

References

[1] D. Trivellato, D. Murphy. *Lights Out! Who's Next? How to Anticipate the Next "cyber-blackout"*, (Security Matters, 2016). Available at http://www.secmatters.com/sites/www.secmatters.com/files/documents/whitepaper_ukraine_EU.pdf.

[2] American National Standard (ANSI). *ANSI/ISA99.00.012007 Security for Industrial Automation and Control Systems, Part 1: Terminology, Concepts, and Models*. International Society of Automation (ISA), (2007).

[3] Y. Mo, T. H.-J. Kim, K. Brancik, *et al*. "Cyber–physical security of a smart grid infrastructure", *Proceedings of the IEEE*, **100**(1), pp. 195–209, (2012).

[4] PricewaterhouseCoopers (PWC). *Power & Utilities: Key Findings from the Global State of Information Security Survey*, (2013).

[5] McAfee. *Global Energy Cyberattacks: Night Dragon*, (2011).

[6] FireEye. *M-Trends 2015: A View from the Front Lines*, (2015).

[7] N. Falliere, L. O. Murchu, E. Chien. "W32. stuxnet dossier", *White Paper, Symantec Corp.*, (Cupertino, CA: 2011). Available from https://www.symantec.com/content/en/us/enterprise/media/security_response/whitepapers/w32_stuxnet_dossier.pdf.

[8] B. Bencsáth, G. Pék, L. Buttyán, M. Felegyhazi. "The cousins of stuxnet: Duqu, flame, and gauss", *Future Internet*, **4**(4), pp. 971–1003, (2012).

[9] SANS Institute. *Can Hackers Turn Your Lights Off? The Vulnerability of the US Power Grid to Electronic Attack*, (2001).

[10] Y. Liu, P. Ning, M. Reiter. "False data injection attacks against state estimation in electric power grids", *ACM Transactions on Information and System Security*, **14**(1), pp. 1–33, (2011).

[11] O. Kosut, L. Jia, R. Thomas, L. Tong. "Limiting false data attacks on power system state estimation", *Information Sciences and Systems (CISS), 2010 44th Annual Conference on*, pp. 1–6, (2010).

[12] J. Slay, M. Miller. *Lessons Learned from the Maroochy Water Breach*, (Springer, Berlin, 2008).

[13] C. Nan, I. Eusgeld, W. Kröger. "Hidden vulnerabilities due to inter-dependencies between two systems", *Critical Information Infrastructures Security*, pp. 252–263, (Springer, Berlin, 2013).

[14] A. Sarwate. *Vulnerability Analysis of 2013 SCADA Issues*, (Redwood City, CA: Qualys Inc., 2007).

[15] Sandia. *Control System Devices: Architecture and Supply Channels Overview*, (Albuquerque, New Mexico and Livermore, CA: Sandia National Laboratories 2010). Available from http://energy.sandia.gov/wp-content/gallery/uploads/JCSW_Report_Final.pdf.

[16] S. East, J. Butts, M. Papa, S. Shenoi. "A taxonomy of attacks on the dnp3 protocol", *Critical Infrastructure Protection III*, volume 311, pp. 67–81, (Springer, Berlin, 2009).

[17] U. Premaratne, J. Samarabandu, T. Sidhu, R. Beresh, J.-C. Tan. "An intrusion detection system for iec61850 automated substations", *IEEE Transactions on Power Delivery*, 25(4), pp. 2376–2383, (2010).

[18] Industrial Control Systems Cyber Emergency Response Team, U.S. Department of Homeland Security (ICS-CERT). *Control System Security DMZ*, (2016). Available from https://ics-cert.us-cert.gov/Control_System_Security_DMZ-Definition.html.

[19] SHODAN *Search Engine for Internet-Connected Devices*, (2016). Available from https://www.shodan.io/.

[20] P. M. Williams. "Distinguishing internet-facing ICS devices using PLC programming information", Technical Report, DTIC Document, (2014).

[21] Y. Younan. "25 years of vulnerabilities: 1988–2012", Technical Report, Sourcefire Vulnerability Research, (2013).

[22] Department of Homeland Security, Control Systems Security Program, National Cyber Security Division. *Common Cybersecurity Vulnerabilities in Industrial Control Systems*, (2011). Available from https://ics-cert.us-cert.gov/sites/default/files/recommended_practices/DHS_Common_Cybersecurity_Vulnerabilities_ICS_2010.pdf.

[23] Industrial Control Systems Cyber Emergency Response Team, U.S. Department of Homeland Security (ICS-CERT). *Year in Review*, (2014). Available at https://ics-cert.us-cert.gov/sites/default/files/Annual_Reports/Year_in_Review_FY2014_Final.pdf.

[24] CRitical Infrastructure Security AnaLysIS. *Crisalis Project EU, Deliverable D2.2 Final Requirement Definition*, (2013).

[25] C. Konstantinou, M. Maniatakos. "Impact of firmware modification attacks on power systems field devices", *Sixth IEEE International Conference on Smart Grid Communications (SmartGridComm)*, pp. 283–288, (Miami, FL: IEEE, 2015).

[26] K. Zetter. *Inside the Cunning, Unprecedented Hack of Ukraine's Power Grid*, (San Francisco, CA: Wired, 2016). Available from https://www.wired.com/2016/03/inside-cunning-unprecedented-hack-ukraines-power-grid/.

[27] K. H. John, M. Tiegelkamp. *IEC 61131-3: Programming Industrial Automation Systems Concepts and Programming Languages, Requirements for Programming Systems, Decision-Making Aids*, 2nd edition, (Springer Publishing Company, Incorporated, Berlin, 2010).

[28] J. Weiss. *Protecting Industrial Control Systems from Electronic Threats*, (New York, NY: Momentum Press, 2010).

[29] F. Skopik, P. Smith. *Smart Grid Security: Innovative Solutions for a Modernized Grid*, (Waltham, MA: Elsevier Science, 2015).

[30] The Wall Street Journal. *Electricity Grid in U.S. Penetrated By Spies*, (2009).

[31] D. Maynor. *Metasploit Toolkit for Penetration Testing, Exploit Development, and Vulnerability Research*, (Waltham, MA: Elsevier, 2011).

[32] N. I. Daud, A. Bakar, K. Azmi, M. Hasan, M. Shafeq. "A case study on web application vulnerability scanning tools", *Science and Information Conference (SAI), 2014*, pp. 595–600, (London, UK: IEEE, 2014).

[33] A. Greenberg. *New Dark-Web Market Is Selling Zero-Day Exploits to Hackers*, (San Francisco, CA: Wired, 2015). Available from https://www.wired.com/2015/04/therealdeal-zero-day-exploits/.

[34] R. Karri, J. Rajendran, K. Rosenfeld, M. Tehranipoor. "Trustworthy hardware: Identifying and classifying hardware trojans", *Computer*, **43**(10), pp. 39–46, (2010).

[35] N. G. Tsoutsos, C. Konstantinou, M. Maniatakos. "Advanced techniques for designing stealthy hardware trojans", *Proceedings of the 51st Design Automation Conference*, pp. 1–4, (San Francisco, CA, 2014).

[36] M. Breeuwsma. "Forensic imaging of embedded systems using JTAG (boundary-scan)", *Digital Investigation*, 3(1), pp. 32–42, (2006).

[37] D. Schneider. *USB Flash Drives Are More Dangerous Than You Think*, (New York, NY: IEEE Spectrum, 2014). Available from http://spectrum. ieee.org/tech-talk/computing/embedded-systems/usb-flash-drives-are-more-dangerous-than-you-think

[38] S. Skorobogatov. "Flash memory 'bumping' attacks", *Proceedings of Cryptographic Hardware and Embedded Systems*, pp. 158–172, (Springer, Berlin, 2010).

[39] J. Duffy. *ISP Routers Have Backdoors That Expose User Data*, (Framingham, MA: Network World, 2010).

[40] H. Bar-El, H. Choukri, D. Naccache, M. Tunstall, C. Whelan. "The sorcerer's apprentice guide to fault attacks", *Proceedings of The IEEE*, **94**(2), pp. 370–382, (2006).

[41] D. Agrawal, B. Archambeault, J. R. Rao, P. Rohatgi. "The EM sidechannel (s)", *Cryptographic Hardware and Embedded Systems–CHES 2002*, pp. 29–45, (Springer, Berlin, 2003).

[42]   CISA S.754 – The Cybersecurity Information Sharing Act (CISA S. 754 [114th Congress]), United States federal law. Available from https://www. congress.gov/bill/114th-congress/senate-bill/754.

[43]   Massachusetts Institute of Technology (MIT). *The Future of the Electric Grid*, (2011).

[44]   United States Government Accountability Office. *GAO-11-117, Electric Grid Modernization*, (2011).

[45]   Executive Office of the President of the U.S. *A Policy Framework for the 21st Century Grid: Enabling Our Secure Energy Future*, (2011).

[46]   ICS-CERT. *Cross-Sector Roadmap for Cybersecurity of Control Systems*, (2011).

[47]   European Network and Information Security Agency (ENISA). *Smart Grid Security – Recommendations for Europe and Member States*, (2012).

[48]   National Institute of Standards and Technology (NIST). *Nist Special Publication 800-82, Guide to Industrial Control Systems (ICS) Security*, (2011).

[49]   International Society of Automation (ISA). *ISA99 Security Guidelines and User Resources for Industrial Automation and Control Systems*, (2015).

[50]   National Institute of Standards and Technology (NIST). *Introduction to NISTIR 7628 Guidelines for Smart Grid Cyber Security*, (2010).

[51]   North American Electric Reliability Corporation (NERC). *High-Impact, Low-Frequency Event Risk to the North American Bulk Power System*, (2010).

[52]   North American Electric Reliability Corporation (NERC). *Critical Infrastructure Protection (CIP) standards*, (2007).

[53]   J. Hughes. "Harmonization of IEC 61970, 61968, and 61850 models", *Electric Power Research Initiative (EPRI) Rep.*, (2006).

[54]   S. McLaughlin, C. Konstantinou, X. Wang, *et al.* "The cybersecurity landscape in industrial control systems", *Proceedings of The IEEE*, **104**(5), pp. 1039–1057, (May 2016).

[55]   D. Baigent, M. Adamiak, R. Mackiewicz, G. M. G. M. SISCO. "IEC 61850 communication networks and systems in substations: an overview for users", *SISCO Systems*, (2004).

[56]   S. Fries, H. J. Hof, M. Seewald. "Enhancing IEC 62351 to improve security for energy automation in smart grid environments", *Internet and Web Applications and Services (ICIW), 2010 Fifth International Conference on*, pp. 135–142, (Barcelona, Spain: IEEE, 2010).

[57]   Y. Zhang, L. Wang, W. Sun, R. C. G. II, M. Alam. "Distributed intrusion detection system in a multi-layer network architecture of smart grids", *IEEE Transactions on Smart Grid*, **2**(4), pp. 796–808, (2011).

[58]   Kaspersky. Lab. *Cybethreats to ICS Systems: You Don't Have to be a Target to Become a Victim*, (Woburn, MA: 2014). Available from http://media. kaspersky.com/en/business-security/critical-infrastructure-protection/Cyber_ A4_Leaflet_eng_web.pdf

[59]   X. Wang, C. Konstantinou, M. Maniatakos, R. Karri. "Confirm: detecting firmware modifications in embedded systems using hardware performance counters", *Proceedings of the 34th IEEE/ACM International Conference on Computer-Aided Design*, pp. 544–551, (Austin, TX: 2015).

[60]   J. D. Osborn, D. C. Challener. "Trusted platform module evolution", *Johns Hopkins APL Technical Digest*, **32**(2), p. 536, (2013).

[61]   T. Alves, D. Felton. "Trustzone: integrated hardware and software security", *ARM White Paper*, **3**(4), pp. 18–24, (2004).

[62]   NIST. *Measurement Challenges and Opportunities for Developing Smart Grid Testbeds*, (2014).

[63]   A. Keliris, C. Konstantinou, N. G. Tsoutsos, R. Baiad, M. Maniatakos. "Enabling multi-layer cyber-security assessment of industrial control systems through hardware-in-the-loop testbeds", *2016 21st Asia and South Pacific Design Automation Conference (ASP-DAC)*, pp. 511–518, (Macau: IEEE, 2016).

[64]   M. Zeller. "Myth or reality does the aurora vulnerability pose a risk to my generator?", *Protective Relay Engineers, 2011 64th Annual Conference for*, pp. 130–136, (College Station, TX: IEEE, 2011).

[65]   A. Srivastava, T. Morris, T. Ernster, C. Vellaithurai, S. Pan, U. Adhikari. "Modeling cyber-physical vulnerability of the smart grid with incomplete information", *IEEE Transactions on Smart Grid*, **4**(1), pp. 235–244, (2013).

[66]   "IEEE standard test access port and boundary scan architecture", *IEEE Std 1149.1-2001*, pp. 1–212, (IEEE, 2001).

Chapter 16

Smart grid security policies and regulations

Robert Czechowski

Smart grid is a concept and also a way to mitigate the deficiencies in infrastructure and to counteract the effects of the growing demand for electricity. One of the ways to increase management efficiency in power network is to use the latest communication solutions, in which digital communication technologies and information and communication technologies are applied. These solutions provide reduced energy consumption, daily load leveling, and reduce losses. Also by automatically balancing energy, they allow for improving the efficiency of its transmission. Such solutions can directly manifest in increase in the efficiency of the entire power system.

16.1 Introduction

Providing energy with the use of smart power grids is associated with the control information flow, allowing both to continuously monitor demand and to control automation power systems by acting on the individual components of the system. This will allow flexible demand shaping and the adjustment of supply to the daily demand. In conjunction with the increasing use of energy-saving devices and optimized processes, this will lead to an increase in the objectives of the climate-energy package the European Union [1]. The introduction of Smart Power Grid—SPG—as a modern standard of network management will bring changes to the existing patterns of energy consumption and increase awareness among its customers about the possibilities of its more efficient use. These projections both pertain to individual (consumers and households) and collective (public institutions) consumers. Despite many concerns with the modernization of the network, better and more focused management of the network through the use of information and communication protection measures and security of database systems will increase the safety and efficiency, which directly can be reflected by lower costs and ultimately benefit the final consumer. An important advantage of the smart grid is the ability to adapt it to the existing energy system to improve management efficiency through integration of distributed generation, installation of renewable

Wroclaw University of Science and Technology, Department of Electrical Power Engineering, Wybrzeze Wyspianskiego 27, 50-370 Wroclaw, Poland

energy sources, energy storage systems, and as a result increase the effectiveness and efficiency of the entire power system.

16.1.1   The development of information and communication technology in electrical power systems

The development of telecommunication networks cooperating with virtually all industries has been observed in recent decades. It is increasingly used in the comprehensive management of transmission and distribution of electricity systems. This development aims at increasing the integration of the network with a power system, in which the network becomes an integral part of the object, playing a role of supervision of the technological process SCADA (supervisory control and data acquisition), PLC transmission (Power Line Communication), or encryption and transfer of control commands using open-communication standards, such as PRIME (Prime Alliance). The most important function in the system will be performed by the transmission and distribution networks. In particular, the latter ones due to their specificity in the first place will be subject to digital modernization in order to create an intelligent backbone network (smart grid). Equally important function will be played by intelligent solutions, mainly those from the group of smart metering. They play an increasingly important role in ensuring the safety, reliability, and operation of power system and management of intelligent devices included in the "last mile," so-called low-voltage network [2]. Modernization of distribution networks and replacement of traditional electricity meters by smart meters, which is the technical aspect to be faced by a modern network, is not everything. The key role that cannot be ignored with this type of investment is also to ensure the security of electronic distribution networks, which will require knowledge of many issues that are not known by experts in the field of electrical power, for example, specialists in the field of protections. With the implementation of automatic measuring equipment, the traditional network by its structure will resemble modern telecommunication ICT networks (Information and Communication Technologies). Implementation of intelligent networks will require cooperation of not only electricians who will take care of installation works, but completely new professionals of widely understood computer science, ranging from network administrators, professionals in the field of IT security, database and data warehouse administrators to analysts of processes management, and business layers. Another task that the representatives of these professional groups must face is the ability to design such systems in the future, practically from scratch based on the experience gained during the upgrade of earlier models of power systems, which are not the most recent. Infrastructure built according to the new concept of smart grid will bring distribution network operators not only the measured or statistical data, which the supplier will be able to use to improve the quality of services or increase profits, but new challenges to the security. Unfortunately, it will be reflected in the need to search for new specialists and conduct specialized training in this field. The integration of power and information technology systems in few years will introduce the need to modernize curricula at the faculties of electrical universities, including the issue of the appropriate textbooks that would take the subject into account. There will also be a change in the look at the threats that are faced by any network on such a large scale and approach to security policy, which will have to be verified in terms of new project assumptions and potential lurking dangers [3].

16.1.2　Participants, roles, and functions of SPG

In the modernized model of the smart grid, energy consumer no longer play only a passive role of electricity consumers, but it consciously and actively manages it and its consumption. Allowing customers to control energy consumption will reduce the maximum peak electricity consumption during the hours of the highest demand. Energy management by consumers, and in the process of transformation of the network—the prosumer, will involve a conscious saving through efficient use of the devices with the highest demand for power (e.g., heating) in off-peak or evening hours. The target shape of the consumer becomes a prosumer, that is, a producer of energy in microscale, which will allow for the simultaneous production of energy for own needs and injecting the produced surplus to the system operator. Investment in knowledge about smart grid, making consumer aware and the development of intelligent technology and related solutions will translate into better use of energy and related cost savings (and thus lower energy bills). The opportunity to earn on the sale of energy produced—in the case of presumption from the so-called distributed generation—will become widespread and natural. Even if, consumers would not want to transmit excess power generated by them to the operators network, then such a solution will be reflected by reduction in the energy consumed from it, resulting in a decrease of demand in the greatest peak. The development of SPGs requires a conscious and active customer and consumer of energy, contributing also to the development of the information and low-emission society, and in the perspective of local government units—the development of sustainable energy community (smart communities) [4]. However, the presence of ICT solutions in intelligent networks introduces high risk of the emergence of a new type of threat—cyber-attacks. In order to prevent them, there must be a development of standards (also security policy) and uniform solutions that need to comprehensively cover the security of the exchange of information, ensure the privacy of users and customers of the whole power system. Developed standards and uniform solution must include the security of the power system, exchange of information, and ensure the privacy of users. Smart grid will be required to develop and specify many standardized solutions. The highest priority has the introduction of areas, such as advanced measuring infrastructure, cooperation between the network and the customer, supervision over transmission and distribution networks, automation of networks and the integration of distributed energy sources, energy storage systems, data exchange, transfer and record, and their safety. Interactions between smart grid entities via secure communication media and the flow of electricity. Description of connections: secure connections, the flow of electricity [5].

All of the above components of the system can be subject to analysis as a result of which we obtain a reference model comprising the specified logical interfaces,

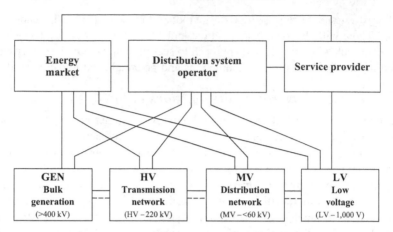

*Figure 16.1   Interactions between smart grid entities via secure communication
media and the flow of electricity. Description of connections: solid
line—secure connections, dashed line—the flow of electricity*

connecting the various actors with the specification of the location of these components in the network. Simplified components of the logical architecture for
intelligent power grid and its basic components are shown in Figure 16.1. The
target of the logical reference model of intelligent network is to show the pattern of
interaction between all its components without specifying and defining specific
media of data transmission and communication interfaces. This model shows the
short-term relationships that can naturally change, which is very likely that this will
lead to the evolution of the concept of intelligent networks. The presented system is
a big generalization and does not discuss all subcomponents, connections, and
relationships between them. However, the model can serve as a tool to identify and
organize the priorities of basic components and relationships, to inform on the
required level of security and the necessity to ensure the safety of individual
actors [6]. Intelligent electricity networks will require specification and development of many standardized solutions and administrative proceedings. The highest
priority is to introduce them to the areas such as

- advanced measurement infrastructure,
- cooperation between network and customer,
- supervision of transmission and distribution networks,
- automation of network and integration of distributed energy sources,
- energy storage systems,
- systems of automatic restoration after a crash,
- digital safety of media transmission,
- automatic reading system,
- illegal consumption of electricity,
- intelligent household installations,
- dynamic energy balancing,

- intelligent prepayment systems, and
- safety of transfer, recording, data collection, and processing.

In the near future, many companies will plan the modernization of the network, or prepare to implement smart grid and a promising, more efficient network of smart metering. Differentiation between hardware and system expectations is due to the various types of national standards, legislation, applicable solutions, and substantial financial resources, which are always a heavy burden even for a large operator. The integration of different devices and telecommunication solutions can result in an additional and completely unnecessary cost of implementations. To facilitate and accelerate the creation of SPGs, it is recommended to implement international standards in the first stage, and then enforce security policy. Normalization will further facilitate the tender procedures of hardware and software, reducing the cost of purchase. Uniform and popular solutions prevent premature aging of the system, facilitate the update, ensure scalability of deployments, and ensure future seamless integration of new systems with the systems already functioning. With this solution in each of the individual segments of the system, it will be possible to perform modernization, eventual repairs facilities, and communication between applications regardless of location and provider. The presence of ICT solutions for intelligent network introduces the danger of the emergence of cyberattacks. This will happen through the implementation of a slightly modified solutions used for years in the telecommunications and information technology. Their weaknesses had been discovered and published long ago. Introduced standards of regulation and security policy must cover all participants and functions of the power system to ensure the safety of the entire system of information privacy [5].

16.1.3   Structure, models, and dynamics of SPG

Smart grid is a modern power grid that uses digital technology to manage and monitor the transport of electricity from all generation sources through transmission and distribution networks to customers, while providing access to advanced system features such as the ability to observe the current state of network parameters. The problem of protecting enterprise network must be considered at all levels of its organizational structure and its technological components. Safety policy must be clear and transparent so that everyone in company can be familiar with it and apply it. Complicated procedures and rules often give opposite result to the expected. Therefore, it is worth remembering that by implementing security policy, one should skillfully estimate the required high safety level of difficulty of its future use. The basic step in organizing the system of protection is to recognize the protection area. First of all, one needs to determine which of the components comprise a network, what its diagram looks like, which are network access points, and who ultimately has to use them. In addition, it is necessary to assess the value and critical information, to assess what information resources are so important to protect them, and where they are located. Staff training in information security is a priority, though it is often neglected. Continuous cooperation is necessary between personnel in charge of information security and smart devices with all organizational

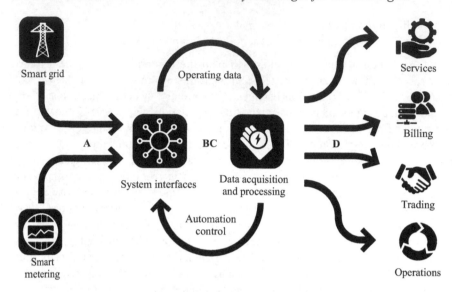

Figure 16.2　*Diagram of the basic functionalities in the smart grid and smart metering. Legend (sections): (a) infrastructure of Intelligent Power Grid zone (SPG)—communication devices smart grid and smart metering, (b) zone of technical management of GSM and broadband BPL, (c) WAN Zone—Internet connections, and (d) ICT infrastructure of electricity supplier*

units of the company. The model of SPG monitoring and surveillance (Figure 16.2) is not fundamentally different from other distributed systems supervising industrial processes. A simplified model can be divided into three distinct zones (with the point of view of their function in the system). The first zone (Figure 16.2, smart grid and smart metering) or infrastructure to which the operator can connect to tele-technical devices of the low or medium voltage, by connecting via the most frequently used private network (Access Point Name) of the provider of energy, previously purchased from suppliers of packet data (called GSM—Global System for Mobile Communications). Subsequently, in the absence of a remote connection, access to the communication interfaces of these devices shall be achieved by the use of local connectors, that is, Ethernet, serial interface, opto-connectors, interfaces, universal serial bus, etc. Such a solution should take place only in case of failure that prevents remote configuration and diagnostics [7].

The second zone (Figure 16.2, the interfaces of the system and the acquisition and processing of data) is the access to network usually of unknown topology and configuration. This network is typically a leased network or the Internet (which can be seen more and more frequently). Designers often use a public network for assembling virtual connections bridges between their own and thus trusted tele-communications equipment. On the one hand, this solution has the advantage that we do not need to invest substantial financial resources to build own network (which is associated with the difficulties of execution and lead time), but also a high level of security. Access to such networks often also occurs with the use of authorized access to the transmission medium based on authentication using a certificate or a certificate with login and password. The advantage of this solution is the ability to access the network and electricity from virtually anywhere on the supervisory network, using a VPN tunnel (VPN—Virtual Private Network). That method of remote access is widely used and provides a high level of security of transmitted information. Another zone (Figure 16.2, the management system operator's services, settlement, trading, and operations) describes the local access directly from center authorized by the system operator, which, from the point of view of system administrators, is the most safe and recommended variant of SPG management [8].

Management and monitoring of large power systems is not an easy task, so administrators often divide them into individual and independent parts in order to avoid confusion due to information overload. Keeping up with the basic safety requirements of digital communication, that is, confidentiality, completeness, timeliness, and effectiveness of the communication process, is its key requirement. In some cases, lack of or interruptions in communication can cause disturbance of the stability of the entire system. The claim that the efficiency of information transfer depends only on the distance between the sender and the recipient is not right. Information passed from node to node can still get stuck somewhere along the way, or in the node, which cannot cope with its fast processing. This problem is particularly noticeable in low-voltage networks, due to the significant density of network interfaces (electricity meters). Because of that, the final meters (terminals) and intermediate counters (switches), which are agents in the transfer of information, can be of strategic importance and their impact on the quality and level of efficiency of data transmission should be determined. The rate of transmission of information in communications networks using the existing electrical wiring (called PLC) depends on several parameters such as the type of network and communication technologies, the source of information, intermediation of physical environment, network motifs, external fault, and distortion tolerance or partial destruction of the network. The information obtained on the shaped network topology can be represented as a directed graph (Figure 16.3). The key parameter is the level of attendance and the stability of the connection (Figure 16.3, lane 1) and the existence of the largest coherent component for the surrounding neighborhood (Figure 16.3(a)). These parameters are crucial for the functioning of complex power grids. It is worth recalling that information on spontaneous creation of the flow path (Figure 16.3, lanes 1, 2, and 3), between the energy meters, concentrators, and logging units can provide the system operator with previously unknown information about the topology and the existence of external interference, allowing in the future more appropriate choice of transmission technology and selecting equipment with better communication parameters. Due to their function of sensor network, smart metering devices generate large amounts of data, so with their mismanagement this amount of data may adversely affect network performance and security of the entire system [9].

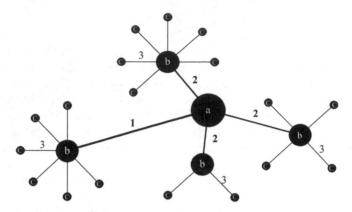

Figure 16.3   *Diagram of the hierarchy of smart metering devices on the network. Legend: (a) concentrator, (b) smart meter work as switch, and (c) smart meter work as terminal*

With the implementation of smart grid in the next stage, there will be need to upgrade the measurement infrastructure and the construction of new transmission capacity supported by ICT. Security is both the problem of technical protection measures and management skills, resources, and information. High level of security is the result of good organization and proper policy introduced at all organizational levels. Modern power systems are facing major challenges, such as aging infrastructure, increase in energy demand, and the integration of a growing number of renewable energy sources, which in the light of security is a very big challenge. The technologies used must be compatible with each other. Smart grid not only offers opportunities to achieve the above objectives, but also the possibility of a flexible development in the future with the use of newer technology [5].

16.2   Smart grids—security management

The concept of security in the context of electrical power systems should be understood as a state of certainty, lack of threat to the stability of system operation, as well as a sense of security and the existence of protection against dangers. One can therefore distinguish two aspects of security: internal and external. Internal security means stability and harmony of an entity (be it a setup, an object, or a system), while external security means lack of threat from other entities and forces of nature. The term *security* corresponds to the Latin *securitas* and is synonymous with safety, implying lack of physical threat. In the broadest sense, it can be said to include the satisfaction of needs such as existence, survival, totality, identity, independence, peace, and certainty of development. The concept of "security" is closely and inextricably linked with another notion inseparable from it, namely "threat." The implementation of smart grid solutions in the area of distribution networks is a substantially more difficult task than in the case of transmission networks. Distribution networks are equipped with a lot more elements responsible for the control and management, and they are therefore subject to higher demands, including those for information technology and telecommunication solutions. In implementing the smart grid strategy for distribution networks, the solutions available must be selected so as to maximize the benefits, while simultaneously remaining cost-effective. The statement according to which the system implementation and responsible management will bring about concrete advantages for both the power-system operator and the customers becomes unquestionable [5]. In accordance with the current provisions and legal regulations, an energy company is defined as an enterprise engaged in the distribution of electricity, responsible for network traffic within the distribution system, current and long-term security of this system, operation, maintenance and repairs of the distribution network, and its necessary expansion, including connections to other power systems [10]. Security management comprises a set of general actions and procedures to be taken in the event of an emergency or a potentially dangerous situation. Most commonly, security models assume the following form: model of threats—security policy—protection mechanisms. Security policy is understood as relating to the documents which contain an explicit and concise definition of who, how, and under what conditions should steps be taken aimed at prevention, leveling threats, or minimizing their effects and using the protection mechanisms available. Security policy may also include certain statements and assumptions as to which users (and processes) of the system may be authorized to access different information and system data.

Documentation also plays a role in specifying the system security requirements as well as the assessment of whether these requirements have been met. By the same token, the system specification is also verified in terms of overall functionality of the security mechanisms. Security policy may form part of the system specification and, similarly to it, its main function is to establish the rules of communication. A security policy comes down to a concise development of protection procedures and documents, where the company's entire executive staff agree on the objectives and protection techniques [11].

16.2.1   Security of power system operator

From the standpoint of the Transmission System Operator and Distribution System Operator, security issues regarding the network management center are not substantially different from the rules and safety procedures that apply to other equally large companies operating in other market sectors. In this context, power systems are governed by the same requirements as in processing of any data. On the one hand, these are generally accepted and currently applicable requirements for security-related applications, but on the other hand the requirements imposed by dedicated measuring and control devices differ from those of other companies. In spite of the fact that, in terms of organization, management of the company's main office (management center) is very much similar to the management model of other companies or corporations implementing ICT solutions, the management of power networks is fundamentally different.

Laws and regulations in individual countries include more or less precise provisions regarding the accessibility, confidentiality, and integrity of data processed within the Internet network. The most numerous are the provisions that relate to confidentiality, and they are associated with a variety of sanctions if not adhered to. Therefore, data security—in the broad sense of Internet technologies—is most often limited to confidentiality, that is, to protection against unauthorized access. Protection must be granted to the content itself (i.e., information and data) and optionally other assets of the computer system that allow or restrict access to contents of shared, or otherwise developed, IT resources. For this reason, as far as practice is concerned, data security also covers, under the current legal regulations, all procedures and measures used to achieve the predetermined goal [12].

Managing a smart grid must be conducted in its entirety by IT specialists, network, and system administrators employed by the operator. Outside-resource-using is not a good option, especially if it refers to the provision of ICT services. This would involve entrusting the provision of services to a third party and would thus make it necessary to let external entities in on crucial security aspects, the act which is in conflict with the very idea of security policy as a whole. This type of solution could bring the opposite effect to the one intended. These issues can be solved by investing in one's own personnel and by establishing a special advanced metering infrastructure (AMI)-dedicated team consisting of electricians and IT specialists, possibly well-trained in both fields [13].

Despite being an updated version of existing energy networks, smart grids will be subjected to the same elementary requirements that are expected from computer networks. In order to ensure a basic level of security, the following conditions must be satisfied:

- confidentiality—ensuring that information is accessible only to authorized persons,
- integrity—ensuring the accuracy and completeness of information and processing methods, and
- accessibility—ensuring that authorized persons have access to information and related assets when needed [13].

In the case of breach or noncompliance with the above crucial standards of security management for the AMI systems, the following rules should be applied:

- any change to the system configuration requires verification for compliance with security policy;
- failure to meet the standards of system security should trigger physical disconnection of the system from the network; and
- decisions to connect or disconnect the system should be made by authorized persons only.

The overall security strategy for smart grids involves both common requirements as well as requirements specific to a particular part of the infrastructure (Figure 16.4). The main task of the cyber-security strategy is to make employees aware of the existence of potential threats and prevent them from occurring.

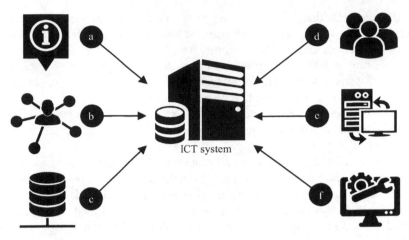

*Figure 16.4 The basic elements of a company's computer system include IOS—
information organization system; (a) set of applicable organizational
solutions, applicable instructions, and formulas; (b) relations
between the sets; (c) information resources; (d) system users;
(e) description of information system and its information resources;
and (f) set of technical tools used in the processes of downloading
(e.g., AMI smart metering systems), sending, processing, and storing
of information [14]*

However, in the event of a cyber-attack on the power system, a strategy for
response, attack mitigation, and system recovery should also be developed.
Implementation of the cyber-security strategy requires the determination and
application in smart grids a risk assessment process in terms of cyber-security. The
risk here involves the possible occurrence of a hazardous incident or event, as well
as the consequences it entails. Organizational risk may relate to several types of
threats (e.g., investment, budgetary, program management, legal liability, security,
inventory, and information systems). The process of smart grid risk assessment is
based on the existing methods of risk assessment developed by the private and
public sector, and it involves identification of consequences, vulnerability to
attacks and threats, and purpose of risk assessment in relation to the smart grid.
Since smart grids include systems from IT, telecommunications, and power systems, the risk assessment process should concern all three of these sectors and their
interaction with the smart grid and smart metering. Generally speaking, the priority
objectives of the IT system security are confidentiality, integrity, and accessibility.
In industrial control systems—which include power-supply systems—the priorities
for security are accessibility, integrity, and confidentiality, in that order.

16.2.2 Security of distribution systems

Due to their topology and distributed infrastructure, information networks utilized
in the smart grid are of an open structure. This is because, apart from the controlled

network segments, that is, systems of electric power transmission and electricity distribution, they also integrate distributed energy sources interacting with the network as well as electricity consumers and prosumers. Devoid of permanent protection, this architecture further increases vulnerability to various threats. It is estimated that the vast majority of attacks on ICT systems, curiously, comes from within the network. Thus, such half-open system requires specific protection against both external and internal threats. Smart grids are therefore subject to specific demands in order to provide a specific productivity and performance, plus a high level of security [14].

Migration of the current power network model to the smart grid model is associated with the increasingly direct involvement of IT and telecommunication solutions. These sectors have the existing cyber-security standards for dealing with vulnerabilities in systems, as well as software solutions aimed at detecting the potential and already-known system vulnerabilities. The same gaps should be assessed in the context of the smart grid infrastructure. Apart from that, smart grids will come with additional weaknesses resulting from their complexity, a large number of users, and the importance of time for their operational requirements. The traditional understanding of cyber-security assumes it is a form of protection necessary to ensure confidentiality, integrity, and accessibility of an electronic information communication system. The definition of cyber-security needs to be more extensive when discussing smart grids. Cyber-security of smart distribution networks includes both technologies and processes of the energy and cyber systems in respect to IT-systems operation and management, as well as electric power transmission. These technologies, along with the processes related to them, provide appropriate protection to the confidentiality, integrity, and accessibility of the smart grid cybernetic infrastructure. For distribution systems, for instance, these are communication systems, overload security, short-circuits detectors, programable devices, actuators (operating elements), controllers, and power supply.

An electric power system may belong to the most critical systems of strategic importance to the functioning of the entire country. Incapacity or destruction of such systems would have a debilitating impact on national security or economic and social well-being of a particular society and its neighbors, both in physical and cyberspace. Protection of key infrastructures includes

- physical security, accounting for all foreseeable threats with regard to human error, system protection against physical damage, and physical interference, for example, in connection systems and natural disasters,
- cybersecurity, accounting for security policy which, in addition to the organizational concept of security oversight, includes the relevant regulations, research, training, etc. [15].

Currently, the operation of the electric power network and efficient control of its operation depends on many other systems, ICT networks, software, and other communication technologies. The ability to anticipate threats and abnormal behavior of the system is also significant. A very good practice when developing a security policy is to put oneself in the attacker's shoes and bridge the gaps

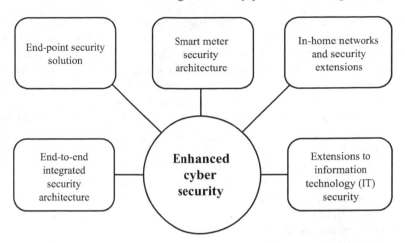

Figure 16.5   Distribution networks security

identified at the design stage. This will not only help avoid the most common errors before the implementation of the solution, but will save unnecessary expenditure in the future [13].

Power failures and violations within the security of distribution systems are becoming increasingly common, and they can result in a substantial financial loss, reputation loss, high repair costs, or even the company going bankrupt. Smart grids are increasingly gaining strategic significance in terms of energy security. Basic components of a security model are common to all segments of the smart grid (see Figure 16.5).

In order to ensure safety of the process of distribution networks management, one should be guided by the principle of allocating the minimum and necessary permissions to employees with regard to the application supervising the distribution process. Access to active devices of the distribution network should be granted based on the hierarchy of powers of persons managing the entire power system. Access to the resources of the power system should be limited only to those persons whose access is necessary for the system's proper operation. It is also important to determine the following:

- level of acceptable risk,
- access-control mechanism and tracking system events,
- identification and authorization mechanism, and
- method of registering and logging changes to the system (configuration and data modification changes) [4].

Smart power distribution systems implement two types of systems within their structure—previously described ICT systems and operational technology (OT) systems. The former are mainly responsible for data processing and transmission of digital information, whereas the latter involve industrial control through devices and systems, rarely using information technology. Thanks to the immense development and rapid integration with power systems, ICT systems have become of great significance in recent years. Initially, they were merely a project and their potential role in the energy sector was only hypothesized. Over time, however, not only has their existence, but also the functioning of these systems, grown to determine the stability and security of the entire power system. Therefore, their security is a priority, but its lack might challenge the security of a system, which is critical and important for the proper functioning of the entire country. When designing the systems, distribution system administrators should, to the greatest extent possible, seek to isolate the information infrastructure (ICT) from the control system (OT). This solution is related to the protection of the system and information networks, but also of visualization systems. In addition, it involves SCADA. Quite frequently—for various reasons—such complete separation is not possible, while the provision of high-level security is not supported by the increasingly popular IP technology implemented within these systems. Upon the loss of autonomy of the control systems provided by physical separation from the external network, these networks become increasingly exposed to cyber-attacks. Power-systems operators should rely on the same security measures for SCADA systems as they do in case of any other ICT infrastructure they use.

The most effective way that could ensure the security of power systems is a combination of selected techniques, that is, intruder detection, network technologies enabling encrypted communication, principles and rules for filtering traffic based on internal firewalls, monitoring system logs and the network, use of updated software, and compliance with the security policy. Ensuring the protection of critical systems depends not only on the existing solutions and codes of conduct, but also on the ability to anticipate potential threats and swift leveling of their effects, which will in turn make them unusable [16].

16.2.3   Security of smart metering networks

In its basic functionality, the AMI provides metering of all endpoints and intermediate points using electricity meters and concentrators, as well as automation of communication between them. Any violations and manipulations generally have very little impact on the operation of the power system as a whole in the presence of such functionality. Nevertheless, it would be highly problematic to not only manipulate, but also lower the meter reading. Equally problematic is the possibility of leaving many customers without electricity through massive meter outage (switching the meters transmitter) [4]. In addition to the most basic functions of outage and real-time reading, the AMI has many other functions, such as reception control when changing time zones or displaying prices. The remotely controlled automation systems can also enable or disable specific receivers through integration, for example, with a home area network. Manipulating these functions on a large scale can cause or simulate overload of the power system, or make life difficult for a particular consumer, for example, by exposing them to costs they would normally not be subject to without the interference of third parties [3]. The security of SMNs through the use of the same equipment in quantities of hundreds of thousands may also come with the risk of learning by unauthorized persons the techniques of their protection and thus finding out how to access a significant part of the metering system. Security gaps in this network segment may lead to massive and successful attacks being conducted. It is worth noting that communication between the meters and the data concentrators should be, at the very least, encrypted.

Implementation of smart metering will contribute to increased attacks on the network due to the emergence of a new attack-prone object of a very specific and thus far unknown architecture that will pose a challenge, especially to specialists in the field of computer networks and server services. An especially dangerous target of attack is ICT systems that contain relevant statistical data or personal information in one place, that is, an attack targeting a computer database in the network operator center. If some network security measures get broken during this time, especially in devices responsible for the communication between the meters and the concentrators, it will become impossible to replace them, and sharing an effective method of breaking the algorithms will not only undermine the entire system, but will also entail additional capital expenditures—this is because there is no easy and cheap way to replace the software of these devices in terms of increasing security when authorizing access to data and device control [17]. The only alternative to providing constant high-level security of these devices is to update the firmware and use authentication and encryption on the basis of the device's unique ID, serial number, password, or hash, known only to the operator. Using the meter's ID, the network operator can generate a unique code (a key designed to communicate with one unit only), allowing its authorization. Another factor contributing to security involves nonstandard communication devices which are not available on the market to an average telecommunications enthusiast.

One of the main elements of the smart grid operation is smart metering—a system for measurement, collection, and analysis of energy consumption, comprising electricity meters, concentrators, communication media, and software. This system is based on metrology (collection and processing of data), telecommunications and computer networks (data transfer), as well as information technology and hardware IT (processing, storage, and presentation of data).

The main goals of introducing smart metering are

- designing a demand management system,
- rationalizing electricity consumption,
- developing a competitive electricity market through introduction of billings based on actual consumption profile,
- making it easier to change suppliers,
- providing information about the current energy consumption to enable energy savings and improve the efficiency of its consumption,
- reducing the impact of power generation on the environment,
- making it obligatory to use electronic meters that enable the transmission of price signals to energy consumers, and
- introducing national standards for technical characteristics, installation, and reading of electricity meters [18].

Current power systems are facing a very difficult challenge of protecting their own infrastructure and ensuring the continuity of their services. This is especially true for low-voltage networks on which the consumers directly depend on. Oftentimes, there are cases where customers (consumers) themselves are a source of potential problems, and sometimes even threats. A major issue for the power system operator is financial losses resulting from electricity being stolen. This practice of electricity theft is due to the very accessibility of the electric line and measurement devices. In order to prevent this sort of behavior, or at least significantly reduce its impact, the operators employ increasingly sophisticated techniques to protect their infrastructure. As shown by experiences of the power system operators, scammers too, resort to increasingly sophisticated methods, and often even to reverse engineering—unfortunately for the purpose of illegal power consumption or providing illegal services involving modifications within meter parameters. The currently installed electricity meters are typically placed in communal lockers located near the building entrance. This solution is, on the one hand, very convenient for energy suppliers and building administrators, but it tends to pose a certain threat coming from consumers residing there. Despite the meters being physically protected by their enclosure in a steel locker, access to them is not much of an inconvenience, especially for people equipped with a basic tool. For this reason, electricity meters should come with additional mechanisms that would protect them from having their cover, main terminal cover, or optical connector removed [19]. According to energy suppliers, the eclectic creativity of the so-called customers knows no bounds. Speaking from many years of experience, they ensure that there are nearly 300 different techniques that make it possible to steal electricity. Meter readers and inspectors visiting customers are alert to these types of ideas. There are also cases in which customers—as a result of an attempt to interfere—destroy the meter so that it can no longer function properly and instead displays overestimated numbers. Illegal electricity consumption is usually (although not always) suggested by scratches, damaged plastic welds on the meter housing, or broken seals. These problems relate to old installations using analog energy meters. Issues connected with electricity theft can be presented in the following categories: (1) direct hooking from line (most common), (2) bypassing the energy meter—short-circuiting or manipulation of the terminals and subsequent interference with the physical mechanisms protecting the meter, (3) bypassing the energy meter, and (4) interfering with memory and chip of the meter's motherboard. There is also another form of theft that involves collusion with the employee supervising the process of reading and billing. One of the least detectable electricity theft techniques with the use of digital technology is interference with the digital meter's software. This method comes down to logical swapping of variables to which we can gain insight through numerical codes, known as Object Identification System. These variables are responsible for storing the numbers based on which the energy supplier bills the consumer. Due to the extremely complex and time-consuming process of breaking a number of security features and the highly specialized knowledge it requires, this method of fraud is extremely rare.

16.2.4 Consumer security

Unsecured smart grids implemented today may lead to a disaster in the future. Someone who is able to bidirectionally transmit data in metering and billing systems may, to some extent, control prepaid meters and their outage mechanisms (by first breaking the keylock and encryption). Such person can also change the tariff assigned to the meter and make other changes burdensome to recipients, some incurring additional costs.

The commercially available electricity meters allow to identify individual domestic appliances with high-power demand. It is also possible to detect the histogram of current electricity consumption in appliances such as ovens, electric kettles, water heaters, stoves, or irons (Figure 16.6). The above information is not particularly troublesome for the customers and raises no major concerns, provided that the operator uses this information only for statistical purposes or in order to offer the customer a better tariff plan. Concerns arise in case of its fraudulent use, for example, to rob someone's apartment when they are not at home. The second important aspect that prompts people to be suspicious toward smart metering is the possible profiling of customers based on their lifestyle and habits. From a formal point of view, personal data are data that uniquely identify a particular person, whereas smart metering makes this uniqueness debatable. The ambiguity of the status of these data is due to the identification of the place of residence only—used by an unspecified number of consumers, rather than a specific user. In some cases, it is possible to extract customer behavior patterns and assign them to specific groups of customers based on their preferences or habits.

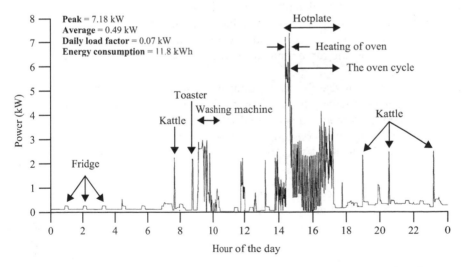

Figure 16.6   An example of household power consumption with the identification of the appliances used [20]

Intelligent home gateways used by some customers (designed to integrate the electric power system with automated household devices) may also pose a serious threat to the smooth operation of the user's devices and the functional stability of a large part of the power system [21]. Not only can intelligent home systems establish communication with the electricity meter to read it from afar, they can also be used to control home automation and household appliances. The most common solutions for remote control are switching on heating and air-conditioning, control of window blinds, switching on the washing machine, or other appliances controlled by programable logic controllers.

In the case of an AMI-targeted attack, it is emphasized that some solutions do not have the possibility of direct load control, and the attacker may only send to the customers the wrong price for electricity (or set a more expensive time zone), and solely on the basis of such information some devices may be switched off. After such load reduction and system stabilization, the attacker may send new information regarding the price (or applicable time zone), which will prompt smart meters to mass-launch multiple receivers for many customers. It is currently difficult to accurately predict the effects of a sharp increase in peak load on a large part of the power grid. Distribution companies and regulatory agencies are not interested in AMI implementations as long as the infrastructure does not support operation of the power system, and there is no economic justification for making such investment [22].

ICT systems dedicated to monitoring solutions, using the processing of personal data, should be designed with default assumptions and solutions ensuring privacy protection of their future users [23]. Smart grids and smart metering, which uniquely identify individual devices and their application, can reveal customer profiles and create new threats to their privacy, such as

- identity theft and disclosure of personal behavior patterns,
- collecting and grouping customers based on behavior patterns,
- possible disclosure of controlled devices in one's home or apartment,
- real-time consumption monitoring—risk of disclosing that a recipient is not at home, and
- manipulation of energy prices transmitted to the meter; for example, transmitting a significantly lowered energy price during peak hours and displaying it to many customers may potentially trigger an important behavioral change in terms of electricity consumption, where a significant increase in this consumption by many defrauded customers can be dangerous to the network in general.

Some consumers are concerned with the lack of control over collecting, processing, accessing, and using sensitive personal data. The problem is, of course, a little broader than that and refers to unauthorized collection, use, and disclosure of information obtained by inference from the metadata. Therefore, a comprehensive security strategy, covering the transmission of information and the security of personal and telemetric data, needs to be developed.

16.3   Sources and classification of threats

Smart grids coordinate the needs and capabilities of energy producers, transmission and distribution network operators, users, and other entities on the electricity market in order to ensure security, stability and efficiency of a system, decrease costs, and minimize the negative impact these systems have on the environment. Security mechanisms and utilization of latest digital technologies intended for digital security assurance can be verified by persons other than those formally related to the operator.

It should be, therefore, assumed that SPGs will sooner or later be under attack, which is only a matter of time, so the possibility to ensure security should be looked into as soon as the design phase: for management software, inventory of computer equipment, company data, staff (including the list of ICT/AMI specialists), documentation of metering equipment, that is for example, access to enterprise resource planning and critical data of the company: data about contractors, commercial information, data creating a risk of loss of positive image, ways of unauthorized access, the so-called information security policy [24]. What also bears significance is protection of the most important documents and papers related to logical and physical project of a data communications network, rules of data storage and processing, addressing network interfaces, ways to access a mobile operator's network or the system's security policy itself.

Ensuring proper functionality of these networks, their security, and protection from attacks of cybercriminals for years will become a serious issue, which is why introduction of real-time smart management should be considered. Its main features include the following:

- possibility of energy production by consumers, its transfer to the operator's network and sale,
- two-way energy and information flow,
- decentralized energy production by means of micro-generation,
- communication between all market entities,
- increased efficiency of energy transmission and distribution by means of network monitoring,
- monitoring of peak-hour power,
- monitoring of transmission networks' physical parameters (including environmental parameters: ambient temperature, wind strength, degree of sunlight),
- current metering and balancing of energy,
- reduction of energy demand by means of its conscious consumption,
- data registration and visualization of technological processes,
- gathering data on energy consumption by specific users and its prioritization,
- sending control signals to devices and their remote configuration, and
- efficient system management via all the system and network components, sending control signals to devices and their remote configuration.

Increased scope of automation and communication of smart grids will surely bring numerous benefits, but also certain risks—due to availability of ICT technology in a new, hitherto unknown (for such solutions) industry branch. There will definitely be individuals willing to test their skills, which will result in increased susceptibility of these networks to attacks [25]. Utilized solutions have to ensure sufficient security to ensure that even if an attack against one of the network's components succeeds, breaking through subsequent security measures does not entail a snowball (cascading) system collapse and cause decreased trust for further devices or services. When designing a secure power grid, one must assume that such a grid will sooner or later be attacked by a cybercriminal who is familiar with commonly utilized ICT system security measures and has sufficient practical skills and knowledge allowing for bypassing them, as well as properly authorizing his or her access to the network in order to get familiar with it, modify, destroy, and steal data [2].

Power systems in the coming years will be faced with a significant challenge in terms of security due to their increasing popularity. Moreover, their dissemination will be supported by various informational campaigns conducted by energy distributors, as well as campaigns spreading awareness about possibilities of more effective energy consumption and management of one's own devices.

16.3.1 Sources of threats

The growth of automation—especially with IT devices included—inevitably increases susceptibility to actions that might disrupt its operation. It is both possible to store important information—especially of administrative nature—related to both energy trading or company development strategy and plans. A significantly more serious threat is an attack entailing blocked functionality of a specific network fragment or service (e.g., access to a database server). What can be especially dangerous is blocked operation of a network transmitting real-time information related to security measures and system control. However, one of the greatest failures in terms of captured data is theft and subsequent sale of an entire database. Although there are mechanisms preventing data being downloaded in bulk by the system operator, there is no perfect, unbreakable security, and the effectiveness of an attack depends only on the skills and often underestimated imagination of the attacker. Network intrusions can also be performed by authorized users from outside the system, which is quite a serious problem due to the increased difficulty of detecting them. The most common threats to IT systems include [3]:

- blocked access to a service,
- intrusion into an IT system's infrastructure,
- data loss,
- data theft,
- disclosure of confidential data,
- falsifying information,
- software-code theft,
- hardware theft, and
- damage to computer systems.

Sharing an ICT power network to meet other needs (not necessarily related to network control and monitoring) is a potential source of danger. It is necessary to separate information transmitted for the needs of power supply from exterior traffic in a network. This solution is most often implemented by means of logical segmentation mechanisms, that is, private virtual networks or distinct addressing of an internet protocol. Moreover, administration-office traffic should also be separated from traffic related to remote supervision over power facilities. The most commonly encountered problems in terms of improper design and management of network architecture are:

- lack of proper security architecture,
- errors in information security management,
- software errors,
- human errors and intentional actions,
- implementation errors,
- closed and poorly tested software,
- errors of system design and security management model,
- usage of obsolete or poorly tested technologies,
- not accounting for information security issues,
- insufficient security monitoring,
- no cyclically performed security audits, and
- lack of sufficient technical knowledge (including training of employees).

No clear separation of these networks can, for example, cause a potential intrusion into a power-plants control system or a distribution system, via access through an administrative network, or cause blocked operation and deletion of data from a SCADA system. Naturally, the scenarios presented above are rather pessimistic and operators are often perfectly aware of threats that can result from ignoring these simple security rules. The most commonly encountered threats are based on:

- vulnerabilities of operating systems which are potential targets of hackers' attacks,
- opening potentially dangerous e-mail attachments from strangers or untrusted senders,
- lack of installed antivirus programs on customer computers or lack of regular virus definition database updates,
- lack of defined entry rules and configured firewalls on a networks' contact edges,
- allowing third parties to access some of the system's functionalities,
- excessive granting of access rights and lack of awareness about threats in people responsible for security issues of a network or system, and
- IT chaos resulting from lack or incompleteness of documentation.

As experience teaches, not even very well-secure systems (like banks whose specific nature makes them the most secure) are unbreachable. Using any security measures is definitely better than not using them at all, because even if they do not completely prevent intrusions, they make it significantly harder for unauthorized people with average knowledge and skills to gain access to a smart grid. It is worth noting that even poor security measures significantly reduce the possibility of a successful attack by people who should not have any access at all. It is much more difficult to mount a successful defense against people with a lot of experience who have performed such attacks on networks with similar structure and principle of operation. In case of an attack by a "highly trained specialist," successful defense depends on a number of mechanisms with varied principles of operation, which will buy enough time for intrusion-prevention systems, intrusion-detection systems to effectively work and for the administrator to react.

16.3.2   Motives behind attacks

It is worth noting that not every attack is driven by intention to destroy, steal data, or derive financial benefits from breaching a security system. Some hackers are motivated by their desire to learn the construction and functionality of a power system, which gives them great satisfaction with their skills and knowledge. Hackers who have successfully broken through security also often leave information (usually in the form of a file) in the system about the used vulnerability in hopes of the administrator taking note and implement preventative measures. From time to time, they check if the information was noticed and the vulnerability fixed. It should not be assumed that such a thing will happen every time. There is no shortage of people driven by a desire to destroy, steal, and blackmail. To illustrate the effects of attacks, we can recall the spectacular attack against Sony Network [26] in 2011 when 77 million user accounts were stolen. The attack lasted for 2 days and caused downtime of the service for 23 days. The hackers also acquired data of 12 thousand credit cards. It is estimated that the attack might have cost the company as much as 171 million dollars. The presented example helps realize just how catastrophic neglecting security measures can be. It is especially important in critical systems, as an attack will result in not only service downtime, but also another tragic outcome. To sum up, attacks against smart grids are caused by:

- desire to lower the costs of electricity,
- excessive interest in novel technologies,
- self-realization and search for new challenges,
- putting computer security to a test,
- desire to acquire power over a network,
- mischief, curiosity, revenge, and vindictiveness,
- trivial mischief or curiosity,
- expected financial benefits,
- sabotage by disgruntled or underappreciated employees,
- intellectual challenge,
- desire to match famous hackers,
- attack against the widely understood system, and
- intentional terrorism.

Intrusion into SPGs can also be caused by a desire of an individual to prove themselves before other hackers or win a bet. Performing a successful attack can also often be a test of one's experience and a condition of admittance to an elite hacker group [2]. A whole different kind of threat (equally dangerous) is the previously mentioned attack by people (employees) from inside the network, the so-called insiders. Such attacks, while performed by people without as extensive knowledge and skills as hacker specialized in breaching security, come to fruition due to the attacker's priceless knowledge of the system's architecture and being undertaken from inside the network. There are many motives for attacks, but the most commonly encountered and dangerous group of potential attackers with sufficient knowledge and motivation consists of disgruntled and underappreciated (also fired) administrative level employees, especially IT technicians and security engineers.

16.3.3 Scale and scope of action

Unlike typical mechanical sabotage attacks, an attack against an electronic grid of energy distribution can be performed with little resources, in a coordinated and extremely precise way. Moreover, it can be initiated via a public network from distant places and undertaken as a coordinated action from many locations at once. Such an attack can target multiple places simultaneously, which can contribute to faster discovery of the entire defense system's weak points. In order to maintain a high level of security, it is necessary to utilize predefined procedures and security policy. They become more and more like a traditional corporate network, which means that similar security means can be used, including intrusion detection, access control, and event monitoring systems. Concentrators connected to ethernet switches or routers fitted with packet data transmission interface, using the very common TCP/IP protocol, are at the highest risk of attacks by use of packet data transmission [2].

That is why one should consider unintended events and focus on all possible threats which most commonly include the following:

- man-made deliberate threat—incidents that are either enabled or deliberately caused by human beings with malicious intent, for example, disgruntled employees, hackers, nation-states, organized crime, terrorists, and industrial spies,
- man-made unintentional threat—incidents that are enabled or caused by human beings without malicious intent, for example, careless users and operators/administrators that bypass the security controls, and
- natural threat—non-manmade incidents caused by biological, geological, seismic, hydrologic, or meteorological conditions or processes in the natural environment, for example, earthquakes, floods, fires, and hurricanes.

Attacks against a power grid and system can assume various forms, take different amounts of time to undertake and prepare, originate from various places or utilize different resources. Transformation of a current network structure into a smart grid necessitates introduction of several new security solutions borrowed from already utilized solutions. Typical problems of today's computing include hacking, data theft, and even cyberterrorism, which will sooner or later also affect power grids. Most of attack forms and kinds can be presented as follows,

- by attack location:
  - attack against an AMI device (mainly energy meters),
  - attack against a data transmission medium, intermediate devices (active and passive), and
  - attack against an operator's data center (extortion and passwords and access to services by means of various techniques even bordering on social engineering, attack against access control servers, databases, warehouses, and entitlements).

- by the target and scale of a potential attack:
  - attack against a single client [27], and
  - attack against the entire system's operation or its significant portion [22].

Implementation of smart grids via installation of remote reading meters, active network elements, construction of new IT systems including data on energy usage gives power engineers many brand new security-related problems. A complex system of multi-layered security requires an overall concept of information security assurance in every area of its functionality. Security in smart grids can be divided as follows:

- by continuity and security of services:
  - ensuring continuity of electricity supply as guaranteed by the contract binding the supplier and the client (also concerns cases of two-way energy flow-smart grids with a prosumer),
  - ensuring confidentiality of information about clients and security of statistical data they generate, like "amount of usage," peak-hours of energy demand or complete lack of thereof, and
  - security related to management of energy distribution and protection of telemetry and personal data in data centers,

- by security class:
  - protection against unauthorized access to data transmission digital media and physical security of the devices themselves in intermediate stations,
  - security of telemetric terminal devices against unauthorized access, disruption of transmission or total blockade of their operation, and
  - protection against unauthorized access to data acquisition systems' databases,

- by policy:
  - data access policy—user authorization, permission management,
  - management security policy—rules and principles of conducting investment processes, and
  - system security policy—reaction to incidents, management of confidential information like passwords, cryptographic keys.

Introduction of smart software (a very important part of distribution networks) will contribute to escalation of attacks against the network due to there being a new object prone to attack, one with a very specific, previously unknown architecture. These systems will pose a challenge, especially for specialists in the areas of computer networks and server services. Especially, dangerous attack targets include ICT systems containing essential statistical data or personal data in one place, meaning an attack against a computer database in a network operator's center. If certain security measures in a low-voltage networks are also breached, especially related to devices responsible for communication and access to concentrators, there will not be any possibility to replace them, as there is not effective methods allowing for relatively cheap and quick replacement of all these devices' software. Moreover, understanding and dissemination of an effective method to breach security algorithms will not only undermine the entire security system, but also entail further investments [17]. Such actions can involve uploading malicious software. It necessitates proper encryption mechanisms and advanced methods of these devices' authorization. Sadly, these aspects are often neglected by novice installers and system administrators, which puts the system at risk of serious consequences already at the implementation phase. A significant portion of secured systems is related to monitoring environments in which one should draw attention to possible attacks against means of communication and devices monitoring the state of distribution network operation. Many of such monitoring systems are at significant risk of physical damage, as well as attempts at acquiring unauthorized access to other system components. In both metering and monitoring systems, we must provide undeniable evidence in order to prove undertaken tampering, which can be a difficult task in case of devices belonging to the OT group. The opponent could gain advantage by not only falsifying information (e.g., by repeating old control communications or changing settings), but also suggesting that it was someone else that did it [28]. The above problem is but an introduction to a broader issue of cyberterrorism and attempts at disrupting the system's stability. These problems are increasingly often becoming a serious issue for companies dealing with distribution and trade of electric energy.

16.3.4 Cyberterrorism

In general, the cybersecurity of the power system covers all issues related to protection of digital information, signals control that technological processes and ICT security of devices and media, responsible for the correct operation of power distribution systems and the management of the entire power system. To counter potential threats consists in preventing cyber incidents, preparing for them, protection and response to them, and to minimize their effects. Until recently, energy industry has focused exclusively on increasing the reliability of power systems. But in recent years, one can clearly see a trend to ever more daring reaching for ICT solutions to improve the efficiency of the network and to increase the safety of its operation. This can be seen even by analyzing the detailed terms and conditions of public contracts and attached to them lists of technical requirements to be fulfilled by the equipment or services concerned. The continuing integration of power systems and data communication systems results in very rapidly increasing manageability and efficiency of the systems themselves, however, it also increases the vulnerability of these systems and networks against cyber-attacks, the effects of which are difficult to imagine. Reports of government institutions and companies involved in the provision of ICT-security solutions, are becoming increasingly frequent. This means that every year there is a growing interest of hackers in various industries, which is not encouraging to those responsible for the cybernetic protection of these systems. In particular, the failure of the power system may cause a cascade of failures in other parts of the system or network. It can result not only in the exclusion of a portion of the system, but also a consequence for electricity consumers, such as financial losses due to downtime, destruction of products, or stopping technological process. Unforeseen interruptions in production, paralysis of the financial, data or transport system, and a number of other basic services that require continuous power supply can entail not only a multimillion dollar losses, or even endanger life or health of people who indirectly depend on the uninterrupted supply of electricity. One should be aware that a cyber-attack can be easily carried out against any party responsible for generation, transmission, and distribution of energy. Currently, there is no need to conduct an open and costly armed conflict to paralyze the opponents energy infrastructure and introduce information confusion, which makes it more difficult to restore the system to full efficiency. Much more effective and cheaper solution is a thoughtful and precise attack which will result in the destabilization of the system.

In order to deeper present the magnitude of cyberterrorism and the extent to which the effectiveness of information transfer and its safety are important for the proper functioning of power systems, some spectacular cases of attacks in recent years are shown. One of the more interesting system failures is a failure, which occurred August 14, 2003 in one in the most industrialized countries in the world. Initially, a small negligence of one of the system operators and the subsequent unfortunate combination of events led to the biggest blackout in the US history. The failure led to the exclusion of 265 power plants (531 power units) in the United States and Canada [29]. An interesting fact is that the development of the accident went on mainly due to problems with the transfer of the right information in the right time and place.

Moreover, IT infrastructure breakdowns were not the result of terrorist or hacker activities, but of random events (errors), the absence of key alarms and weak project design. The next two attacks, which particularly were engraved in such a short history of cyber-attacks and cybersecurity, and resulted in lot of material losses, were the events in Iran and Germany. The first of them was an attack on Iran's nuclear program in 2009 (scientific center in Natanaz) [30]. In order to carry out this attack, Stuxnet (Malware—malicious software) [31] was used, which caused great damage to the nuclear program, taking over the control of operation of centrifuges used to enrich uranium leading to their overload and ultimately forcing them to work at full capacity, hence their rapid destruction occurred. From the moment of infection in Natanaz proliferated failures occurred as well as work interruptions and production of enriched uranium has decreased. However, Iranian experts were unable to discover the causes of these problems and massively bought the new centrifuges abroad.

It was only in June 2010, when the virus has been detected by the experts from the Belarusian company of computer security who described its destructive properties as a weapon that destructs industrial installations. According to the most widespread opinion, the attack was jointly prepared by the secret services of Israel and the United States. It should be added that this resort was completely separated from the external network and infecting the system was due to connection of the previously left data carrier at the resort. Procedures and awareness of danger failed. It was created probably in 2004–07, first detected until June 2010 and by September 29, 2010 its presence was detected at over 100 thousand computers. This software works in cyberspace today. Similar in operation to Stuxnet: Duqu [32], Gauss [33] (both discovered in September 2011), Mahdi (February 2012) [34], and Flame (May 2012) [35], capture data from the keyboard, periodically perform screenshots, record video and audio with microphones and cameras installed on infected computers, intercept cookies and passwords from Internet browsers, instant messaging, e-mail, they steal configuration files and access data to banks, social networks, and send the stolen information in the background outside.

Currently, cyber-attacks are widespread, well-organized, and most importantly— highly effective. All the previously mentioned software has self-destructive modules, so that it blurs the traces of its activity. Another attack—very known for its destructive effect—although lacking detail, which caused destruction was an attack in August 2015 against a German steelworks. The German Federal Office for Information Security (Ger. BSI—Bundesamt für Sicherheit in der Informationstechnik), made public a report [36] containing information that unidentified hackers caused the failure of the furnace in one of the German steel mills. The place and time of the attack is kept secret, but the BSI report mentions that because of the attack "it was not possible to extinguish the furnace in a controlled manner," which resulted in "serious damage to the system." However, we should be aware that the majority of attacks on critical infrastructure are kept secret. Interestingly, the know-how of the attacker was very evident, not only in terms of knowledge of the techniques of IT security, but also on knowledge of the applied control of industrial and manufacturing processes [37]. This case proves that more and more attackers are high-class specialists with not only extensive knowledge in the field of cybersecurity but also about the technological process of attacked object. Another fairly well-known operation was a campaign "Night Dragon" [38], which started in the beginning of 2009. Experts of one of the most popular security companies in the network—MacAfee, have identified it as a well-organized, directed at specific targets, using advanced software and methods of social engineering. The aim of the operation was to steal sensitive data such as operational details, balances of finance, and the results of research, from the largest energy and petrochemical companies. According to experts, the methods of attacks and the tools used clearly indicate Chinese hackers. This information is not a surprise, given the commitment of China in cyber-espionage and the growing interest in the energy sector, particularly oil and gas imports from that country [39]. The last of the known attacks was conducted on December 23, 2015, after which hundreds of thousands of residents of Ivanofrankovsk circuit in Ukraine had been left without power supply for few hours. Some sources say that there were over one million customers without electricity. According to experts, the cause of power failure was a "hacker attack" on power network operators Prykarpattyaoblenergo and Kyivoblenergo, which resulted in the exclusion of 30 transformer stations [40]. The analysis of the company ESET shows that other energy companies in Ukraine were attacked at that time. The above-described cyber-attack is attributed to the Moscow group of hackers called Sandworm [41].

Naturally, the examples described above are only those selected and those the existence of which was published. However, they prove that there are different methods that lead to failure in the power sector. Types of attacks evolve over time, attackers could cripple power grids, using the so-called effect of cascading collapse of the system [42]. It consists in the fact that instead of concentrating on the most important and best defended elements of the energy infrastructure, the attack would be aimed at substations, which are usually poorly protected. Their exclusion by cyber-attacks forces the load transfer to other stations, which leads to overloading and, consequently, to deny access to electricity for significant amount of people. It is extremely difficult to prevent such attacks due to lack of time and appropriate measures to ensure the defense of all the elements of the energy infrastructure.

Current statistics [43] represent a remarkable rapidly increasing number of attacks and incidents. As opposed to 2010, which saw 50 cyber-attacks, their number already increased to 138 in 2015. The average time to detect the attack was 170 days, time to remove the effects—45 days and the cost of removing an average of 1.6 million USD. The Trend Micro report [44] presents an example of prices of various services offered on the Russian underground forums of economic underground [44]:

- 50 USD—source code Trojan horse backdoor Trojan [45],
- 30–70 USD—a one-day DoS attacks,
- 35/40 USD—ZeuS virus [46] installation on the host (of buyer/seller), and
- 30–80 USD—the basic version of the software to hide infected files or malicious malware from security scanners.

However, one should be aware that the majority of attacks on critical infrastructure is kept secret. Hence, the interest of hackers in electricity networks, where damage cannot be concealed [40]. According to Microsoft Corporation, the average time to detect a sophisticated cyber-attack in 2014 was 200 days [47]. This may mean that unnoticed hackers were able to penetrate the telecommunication infrastructure of any power system, gather necessary information to carry out cyber-attack, and already the subject attacked and infected to this day does not know about it. In recent years, we have had unprecedented forms of military action: cyber-attack on the Libyan networks (Arab Spring 2011) [48], the attacks on Google (2012) [49], an attack on the International Atomic Energy Agency (2013) [50], the installation of malware on the computers of energy companies in the United States and Europe: Dragonfly [51], BlackEnergy [52], and Sandworm [53].

Current solutions to ensure an acceptable level of safety are only tools, the use of which does not present any security guarantees. Moreover, they do not assure that the system will be safe in the coming years. Each security system is only as strong as its weakest link. This requires a comprehensive understanding of potentially dangerous threats and weaknesses of the system. In order to protect effectively, security systems need to engage all employees and participants in the system in all its segments, and therefore system must consider also factors of social engineering, which most often are the weakest element of the security system. In addition, we observe a constant evolution of threats, which are extremely difficult to predict. Due to the fact that the attack scenario is being prepared under the specific purpose, there is a lack of comprehensive protection techniques that prevent the implementation of appropriate preventive measures. Although there are many proven technologies and security methods, we can very quickly decide to use the solution before verification of safety requirements and transgressive analysis of potential hazards specific to the system. The first step in planning the implementation of security in electricity systems should be consulting a planned project with specialists in cybersecurity, reconciliation of solution, and then its implementation [54].

16.3.5 Blackouts and backbone network failures

Long-term and large-area system failure is the blackest possible scenarios of power system operation. The failures that additionally deprive consumers of electricity at the same time include the collapse of the system (Blackout), and are particularly significant and painful consequences. This is particularly the case when the source of initially small failure involves more network segments, so-called cascading system failure [55]. The various phases of large-scale failure of the system are shown in Figure 16.7. In summary, large-area power failure is a long-term loss of or voltage deviation in distribution power system. The failure is defined as a loss of voltage in the electricity network over a large area and at a specific time. Any large-scale failure has different reasons. Often an attack is the cause of a blackout but not always. Network failures may also arise as a result of a sequence of several fortuitous events, network failure, power off, extreme weather conditions, but also may owe its genesis to human error. Other adverse factors causing system failures

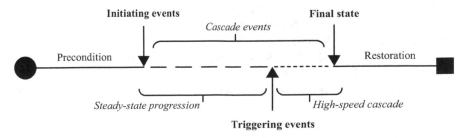

Figure 16.7 Large-scale blackout phases

networks are weather conditions, especially during the summer when the critical value of basic technical parameters of the system are exceeded.

Increased demand for electricity is related to higher energy consumption of electrical appliances (cooling in particular), than in other seasons. It is estimated that with the massive launch of air-conditioning and refrigeration equipment, these devices increase the maximum peak of electricity consumption of electricity up to 2% [23]. In addition, high temperatures in summer and extended overhang of power lines effectively limit the maximum power transmission. Whatever the cause of failure, it is subject to classification. Its size and time of occurrence is decisive in its classification and decide whether its results allow classifying it as a blackout. To call the failure of the system a blackout, thus to gain the interest of safety experts and to be subjected to a thorough analysis, the failure must meet all the following conditions:

• Interruption in the power supply cannot be planned by the distributor or manufacturer of energy,
• failure must reach at least 1,000 customers and take at least 1 h, and
• duration of the outage must be at least 1 million h per number of customers.

In other words, the failure of the power system can be called a blackout, if 1,000 recipients are without power for a minimum of 1,000 h. But if the failure touches fewer than 1,000 customers, such an event, irrespective of the duration, cannot be called a blackout. This failure also is not classified to the level of a blackout when it touches a million customers, if the duration of the outage is less than 1 h. The event will not be taken into account (regardless of the number of recipients). For example, 10,000 customers affected by the 100 h or 100,000 within 10 h will be classified as a blackout failure. By far the majority of blackouts, the largest previous failures were caused by nature or human error, rather than by terrorists. The biggest blackouts in 2015 were:

• Pakistan, January 26, 2015—80% of the population in Pakistan have been deprived of power (about 140 million people) because of unspecified technical error that occurred in a power plant in Sindh [56],
• Kuwait, February 11, 2015—a technical problem in one of the five power plants in Kuwait caused a lack of power in many buildings, street lighting, and the airport. As informed by local authorities, power was restored in a few hours [57],
• The Netherlands, March 27, 2015—a technical problem in one of the major energy networks in North Holland caused lack of power for 1 million customers for at least 1 h [58],
• Turkey, March 31, 2015—due to technical faults, more than 90% of energy consumers in Turkey (70 million) were without power. Regions unaffected by the failure were Hakkari and Vin provinces, as electricity was routed from Iran [59],
• USA, Washington, Spokane County, November 17, 2015—as a result of a powerful hurricane that destroyed power lines, 161 thousand customers were without electricity [60], and

- Crimea, November 21, 2015—military action involving the blowing up of transformers caused a power outage, leaving 1.2 million customers without electricity [61].

Cybersecurity of power systems covers issues related to the automation and communication that affect the management tools functioning of these systems. It also includes the tasks of prevention, mitigation, and facing other anti-cyber events. Protection of power systems is largely based on devices and protection that may significantly improve the reliability of the power system in the event of a large failure. Cascading failures are mostly related to the problems of ensuring the proper transmission of digital information in the right place and in the right time. The most commonly reported large-scale failures were caused by unintentional actions, errors, lack of early alert, and poor exchange of information [62].

16.4 Regulations and future of smart power grids

Supplying electrical power using intelligent network increasingly involves the flow of digital information, both allowing continuous monitoring of demand and control of the network, which in turn is reflected directly to its greater efficiency. This will allow flexible shaping demand and supply to daily demand. In conjunction with the increased use of energy-efficient and intelligent solutions and processes, it will increase the energy efficiency of the entire system. To some extent, this solution will also reduce the risks associated with the occurrence of accidents or intentional attacks. In order to improve energy efficiency and the development of competitive markets for fuels and energy, they include in particular:

- the use of Demand Side Management, stimulated by daily diversification in electricity prices due to the introduction of intraday market and the transmission of price signals to customers using smart meters and
- the abolition of barriers to switching suppliers by introducing standards for technical features, installation, and reading of smart electricity meters.

An important advantage of smart grid power is the ability to adapt it to the existing system in order to integrate distributed generation, the possibility of connecting renewable sources of energy, introduction energy storage system (including mobile systems), and improving energy efficiency. The priority objective of the development and modernization of the SPG should be the objectives of climate-energy package of the EU [63]. The introduction of SPG on a larger scale will bring changes in the existing patterns of energy consumption, both in terms of individual, as well as collective entities. Despite many concerns related the modernization of the network and the fear of recipients about the confidentiality of data, better and more focused management of the network will increase its security, which directly translates into lower costs and improving the quality of services provided not only for the final consumer of energy [64].

The new concept and methodology of designing safe power systems is a complex problem whose foundations go back to the current standards and regulations of the European Parliament and the internal state regulations which are often more stringent. Safety requirements of power systems, transmission of digital information, and the privacy of recipients often are regulated independently by the individual operators of the system, however, they bear the burden of compulsory general guidelines imposed by the superior authorities. Very often the design specification and concept architecture for intelligent power grids focuses on detailed guidance on the design of new systems and processes occurring in their network management. The operators of transmission and distribution systems are aware of the role of systems and information technology and business practices, which have their impact on the smooth functioning of the power system. This is described for example by Directive 2005/89/EC of the European Parliament and of the Council of January 18, 2006 on measures to safeguard security of electricity supply and infrastructure investment [65]. Subsequent directive contains information about the need for a state to take appropriate steps aimed at creating (. . .) intelligent network (. . .) in order to guarantee the safe operation of the electricity system as it is accommodated to the further development of electricity production from renewable energy sources, including interconnection between Member States and between Member States and other countries. These regulations are described in the Directive 2009/28/EC of the European Parliament and of the Council of April 23, 2009 [66]. Data protection should be built into smart metering systems from the outset as part of the compliance of life cycle and include easy-to-use technologies, strengthening the protection of privacy. Recommended is also an approach to privacy protection through its inclusion in the specifications of design of physical infrastructure, technology and business practices, and it is suggested that it is mandatory, including the obligation to minimize and processed data and their removal (Privacy by Design) [67].

The conclusions of the European Council of February 4, 2011 determine the directions of legal action in the field of standardization: (1) the adoption of technical standards for measuring equipment, designs legal solutions in the fourth quarter of 2011. (2) The adoption of technical standards for smart grids and smart meters, legal solutions draft by the end of 2012. (3) The adoption of technical standards for charging systems for electric vehicles, legal solutions drafts till mid-2012. Another recommendation of the European Commission—2012/148/EU of March 9, 2012 on preparations for the broadcasting of smart metering systems suggests [67] that the protection of measurement data should be ensured already in the design phase and means implementation—including the latest technical developments—both in determining the resources needed to process the data, and at the moment of the process, appropriate measures and technical and organization procedures so that data processing complies with the requirements of Directive 95/46/EC and guarantee protection of the rights of the data subject [68]. The primary objectives of European energy policy to 2020: (1) 20% reduction in greenhouse gas emissions compared to 1990 levels, (2) 20% reduction in energy consumption, (3) 20% share of energy from renewable sources in EU energy consumption by 2020. These objectives will be the objectives of the EU—not necessarily all Member States.

Recently, there are very important structural changes in energy companies. The transformation of the electricity grid is reflected also on the tele-information solutions, because the solutions should be, and successively are centralized and standardized. This applies to management, controlling, accounting, and reporting systems. IT solutions providers reported in 2015 increased (by 70%) sales of products and services dedicated to power generation (including SPG) compared to the previous year—according to the analysis published in this year's edition of the TOP200 Computerworld report [69].

16.5 Conclusions

The development of digital technology and tele-communication systems observed in recent years meant that companies are increasingly willing to reach and integrate the technology in their own power infrastructures. This is closely related to the dynamic growth of the network services and decreasing costs of implementing the latest technologies. According to the report Internet of Everything Cisco [70], the number of devices connected to the Internet exceeded the number of people inhabiting the Earth in 2009. It is estimated that in 2020, this number will reach astronomical 50 billion, which is an example that shows how popular and widely used digital technologies are, and how they become an indispensable part of our lives.

Undoubtedly, the benefits of effective digital communication and technologies use in power systems are very promising and exceed capital expenditures made. However, their implementation will result in increased vulnerability to cyber-attacks and system failures having their base in IT solutions. The use of these solutions and the inclusion of modern methods of cybersecurity in the design and operation of ICT systems and automation equipment will limit the effects of deliberate attacks and accidental failures of power system objects, which will also help to increase the safety of its operation. The first and also the most important element to ensure the protection of cyberspace of a company is to develop and practice the adopted security policy. Through closely described procedures and continuous training of employees, many problems can be avoided, which are often a source of serious consequences of failure (including large-scale failures). The security policy described in the chapter is closely connected with the observance of the principles of information exchange, as well as the use of technical measures to avoid the weaknesses of the system, vulnerable to failures and attacks, at the stage of management. The security policy should include: advanced intelligent data collection and analysis, intelligent monitoring, increased access control, serious consideration of effective user training, education of management staff, and modernization of the IT architecture. However, special role in the shaping of a European energy policy is played by Books and Directives published by the European Commission. They contain detailed rules for the energy strategy and descriptions of energy security topics of the European Union [71].

Acknowledgments

This chapter was realized within NCBR project: ERA-NET, No 1/SMARTGRIDS/ 2014, acronym SALVAGE: *Cyber-Physical Security for the Low-Voltage Grids*. Author also would like to thank you Mr. Professor Eugeniusz Rosolowski[1] for first review, corrections, and received good advice during writing the chapter.

References

[1] Ministry of the Environment. Pack challenge – Poland implements the climate and energy package – Summary, Original title: *Ministerstwo Środowiska, Pakiet wyzwań – Polska wdraża pakiet klimatyczno-energetyczny – podsumowanie*, Online source: https://archiwum.mos.gov.pl/artykul/archiwum/ 7_aktualnosci/10373_pakiet_wyzwan_polska_wdraza_pakiet_klimatyczno_ energetyczny_podsumowanie.html, (access date: 12.02.2016), Warszawa, 2013.

[2] Flick T., Morehouse J., *Securing the Smart Grid. Next Generation Power Grid Security*, Burlington: Elsevier Inc., 2011.

[3] Wilczyński A., Tymorek A., *The role and characteristics of information systems in power energy systems*, Energy Market, Original title: *Rola i cechy systemów informacyjnych w elektroenergetyce, Rynek Energii*, 2 (87), s. 154–158, 2010.

[4] Billewicz K., *Smart Metering. Intelligent measuring system*, The Institute of Electrical Power Engineering Wroclaw University of Technology, Original title: *Inteligentny system pomiarowy*, Instytut Energoelektryki Politechnika Wrocławska, Wydawnictwo Naukowe PWN, 2012.

[5] Zaleski P., Hajdrowski K., *Roadmap Smart Grid technology by 2050*, ENEA Operator Sp. z o.o., *Article was based on studies, Original title: Mapa dro- gowa technologii Smart Grid do roku 2050*, ENEA Operator Sp. z o.o., Artykuł powsta, na podstawie opracowania: Technology Roadmap, Smart Grids, Global Status and Vision to 2050; International Energy Agency, (last update: February 3 2010), 2011.

[6] SGiP, *Introduction to NISTIR 7628 Guidelines for Smart Grid Cyber Security*, The Smart Grid Interoperability Panel Cyber Security Working Group, 2010.

[7] Bush S.F., *Smart Grid, Communication-Enabled Intelligence for the Electric Power Grid, Complexity Teory*, Chichester: IEEE Press, John Wiley & Sons, Ltd., 2014.

[8] Gellings C.W., *The Smart Grid: Enabling Energy Efficincy and Damand Response*, Energy Management Today, Lilburn: The Fairmont Press, Inc., 2015.

[9] Carcelle X., *Power Line Communications in Practice, Effects of Interference on the Electrical Network*, London: Artech House, 2006.

[10] The Law of 4 March 2005, *Amending the Act – Energy Law and the Environmental Protection Law*, Laws No. 62, item. 552. Original title: Ustawa z dnia 4 marca 2005 r. o zmianie ustawy – Prawo energetyczne oraz ustawy – Prawo ochrony środowiska, Dz. U. Nr 62, poz. 552, 2005.

[11] Anderson R.J., *Security Engineering: A Guide to Building Dependable Distibuted Systems, Security Policy Model*, New York, NY: John Wiley & Sons, Inc., 2005.

[12] Anderson R.J., *Security Engineering: A Guide to Building Dependable Distibuted Systems, Data Security*, New York, NY: John Wiley & Sons, Inc., 2005.

[13] Zurakowski Z., *Protection of critical infrastructures in the Lower Silesia*, Department of Scientific Papers ETI Technical University of Gdansk, No 6, Original title: *Zabezpieczenia infrastruktur krytycznych Dolnego Śląska*, Zeszyty Naukowe Wydziału ETI Politechniki Gdańskiej, nr 6, 2008.

[14] Lis R., Wilczyński A., Data protection in intelligent power systems on example ZigBee, Energy Market, No 6, Original title: Ochrona danych w inteligentnych systemach elektroenergetycznych na przykładzie ZigBee, *Rynek Energii*, 1 (98), s. 84–88, 2012.

[15] Żurakowski Z., *Safety and Security Issues in Electric Power Industry*, 19th International Conference Safecomp 2000, Rotterdam, The Netherlands, 2000.

[16] Lydziński D., *ICT and OT security in the energy sector*, Smart Grids Poland, No 1/2016 (15), Original title: *Bezpieczeństwo ICT i OT w sektorze energetycznym*, Smart Grids Polska, Nr 1/2016 (15), ISSN 2084-6959, 2016.

[17] Kearney A.T., GmbH, *Technology Report, Home Network Infrastructure (ISD) within the Smart Grid*, Original title: Raport Technologiczny, Infrastruktura Sieci Domowej (ISD) w ramach Inteligentnych Sieci/HAN within Smart Grids, 2012.

[18] The Ministry of Economy, Energy Department, *Prospects for Development of Smart Metering System in Poland*, Original title: Ministerstwo Gospodarki Departament Energetyki, Perspektywy rozwoju systemu inteligentnego opomiarowania w Polsce, 2009.

[19] Czechowski R., Kosek A.M., The Most Frequent Energy Theft Techniques and Hazards in Present Power Energy Consumption. *Cyber Security in Smart Metering Low-Voltage Network*, CPS Week Conference, Viena, 2016.

[20] Wood G., Newborough M., Dynamic energy-consumption indicators for domestic appliances: environment, behaviour and design, *Energy and Buildings* 35, s. 821–841, 2003.

[21] Sorebo G.N., Echols M.C., *Smart Grid Security: An End-to-End View of Security in the New Electrical Grid, Defence-in-Depth and Other Security Solutions*, Boca Raton: CRC Press, 2012.

[22] Parks R.C., *Advanced Metering Infrastructure – Security Considerations*, Sandia Report, Sandia National Laboratories, 2007.

[23]   Billewicz K., Smart Grids. *The intelligent Energy Power Networks*, Original title: Inteligentne sieci elektroenergetyczne, www.printshop.org.pl, ISBN: 83-935366-5-8, 2015.

[24]   Billewicz K., *The Issues of Information Security in Smart Grids*, Electrical Power Institute, Wroclaw University of Technology, *Problematyka bezpieczeństwa informatycznego w inteligentnych sieciach*, Instytut energoelektryki, Politechnika Wrocławska, 2012.

[25]   Ball P., *Critical mass. Original title: Masa krytyczna*, Wydawnictwo Insignis, Kraków, 2007.

[26]   DAGMA Sp. z o.o., IT Security Office solutions of ESET distributor in Poland. *How hackers stormed Steam, Xbox Live and PlayStation Network – the top 5 attacks*, Original title: *Biuro bezpieczeństwa IT Dystrybutor rozwiazań ESET w Polsce, Jak hakerzy szturmowali Steam, Xbox Live i PlayStation Network – 5 najwiekszych ataków*, 2015.

[27]   EPiC – *Electronic Privacy information Center, Concerning Privacy and Smart Grid Technology, The Smart Grid and Privacy*, Online access: epic. org/privacy/smartgrid/smartgrid.html, 2014.

[28]   Anderson R.J., *Security Engineering: A Guide to Building Dependable Distibuted Systems, Monitoring Systems*, New York, NY: John Wiley & Sons, Inc., 2005.

[29]   Ozaist G., *Egyptian darkness in the USA*, Polish Energy, No 5/2012, Original title: *Egipskie ciemności w USA*, Polska Energia 5/2012, 2012.

[30]   Hanford S., Stuxnet: *Exploiting Trust Relationships and Expected Behavior, Cisco Blog Security*, Online source: http://blogs.cisco.com/security/ stuxnet_exploiting_trust_relationships_and_expected_behavior,   (access date: 19.03.2016), 2010.

[31]   Piaseczny J., *Insidious Stuxnet*, Weekly overview, Online source: *Podstepny Stuxnet*, Tygodnik przeglad, Online source: http://www.tygodnikprzeglad.pl/ podstepny-stuxnet (access date: 19.03.2016), 2011.

[32]   Grazka K., *Duqu – similar to Stuxnet virus attack Europe*, Original title: Duqu – podobny do Stuxnet wirus atakuje Europe, Symantec, Online source: http://www.benchmark.pl/aktualnosci/Duqu_-_podobny_do_Stuxnet_wirus_ atakuje_Europe-37250.html, (access date: 19.03.2016), 2011.

[33]   Kaspersky Lab, Kaspersky Lab detects "Gauss" – new complex cyber-threat, which target is online monitoring banking accounts in the Middle East, Original title: *Kaspersky Lab wykrywa "Gaussa" – nowe złożone cyber-zagrożenie, którego celem jest monitorowanie kont bankowości online na Bliskim Wschodzie*, Online source: http://www.kaspersky.pl/about.html? s=news&newsid=1773, (access date: 19.03.2016), 2012.

[34]   Zetter K., Wired.com Security, Mahdi the Messiah *Found Infecting Systems in Iran, Israel*, Online source: http://www.wired.com/2012/07/mahdi/, (access date: 19.03.2016), 2012.

[35]   Gostev A., *Secure List The Flame: Questions and Answers*, Online source: https://securelist.com/blog/incidents/34344/the-flame-questions-and-answers-51/, (access date: 19.03.2016), 2012.

[36] Wired.com Security, *The IT Security Situation in Germany in 2014, Original title: Die Lage der IT-Sicherheit in Deutschland 2014*, Online source: http://www.wired.com/wp-content/uploads/2015/01/Lagebericht2014.pdf, (access date: 19.03.2016), 2015.

[37] Zetter K., Wired.com Security, *A Cyberattack has Caused Confirmed Physical Damage for the Second Time Ever*, Online source: http://www.wired.com/2015/01/german-steel-mill-hack-destruction/, (access date: 19.03.2016), 2015.

[38] McAfee Foundstone Professional Services and McAfee Labs, *Global Energy Cyberattacks: "Night Dragon"*, By McAfee Foundstone Professional Services and McAfee Labs, Online source: http://www.mcafee.com/us/resources/white-papers/wp-global-energy-cyberattacks-night-dragon.pdf, (access date: 19.03.2016), 2011.

[39] Kozłowski A., *Cyber security of energy infrastructure*, Portal Geopolityka.org, Original title: *Cyberbezpieczeństwo infrastruktury energetycznej*, Portal Geopolityka.org, Online source: http://www.geopolityka.org/analizy/andrzej-kozlowski-cyberbezpieczenstwo-infrastruktury-energetycznej, (access date: 20.03.2016), 2014.

[40] Wasowski M., *Is it possible to cause a blackout using one e-mail?*, Original title: *Czy można wywołać blackout za pomocą jednego e-maila?*, Smart Grids Polska, 1/2016(15), ISSN 2084-6959, 2016.

[41] Auchard E., Finkle J., Reuters Technology, *Ukraine utility cyber attack wider than reported: experts*, Online source: http://www.reuters.com/article/us-ukraine-crisis-malware-idUSKBN0UI23S20160104, (access date: 19.03.2016), 2016.

[42] Trzaska Z., *Security and Vulnerability of Power Systems under Terrorist Threat*, Original title: Bezpieczeństwo systemu elektroenergetycznego i jego podatność na atak terrorystyczny, Online source: http://elektroenergetyka.pl/upload/file/2006/1/elektroenergetyka_nr_06_01_1.pdf, wwww.e-energetyka.pl, p. 1–7, (access date: 20.03.2016), 2006.

[43] Hawlett Packard Enterprise, *Annual Study Reveals Average Cost of Cyber Crime Per Organization Escalates to 15 Million USD*, Online source: http://www8.hp.com/us/en/hpe-news/press-release.html?id=2097584#.Vv6IGaSLT1s, (access date: 25.03.2016), 2015.

[44] Goncharov M., Raport Trend Micro, *A TrendLabs reaearch Papert, Russian Underground 2.0*, Online source: https://www.trendmicro.com/cloud-content/us/pdfs/security-intelligence/wp-russian-underground-2.0.pdf (access date: 19.03.2016), 2015.

[45] Goncharov M., Raport Trend Micro, *Cybercriminal Underground Economy Series, Russian Underground Revisited*, Online source: http://www.trendmicro.com/cloud-content/us/pdfs/security-intelligence/white-papers/wp-russian-underground-revisited.pdf, (access date: 19.03.2016), 2014.

[46] Symantec.com, *Zeus Trojan (Zbot) – Virus Definition*, Online source: https://www.symantec.com/security_response/writeup.jsp?docid=2010-011016-3514-99, (access date: 19.03.2016), 2010.

[47]   Microsoft Corporation, *Microsoft Advanced Threat Analytics, Changing nature of cyber-security attacks*, Online source: http://download.microsoft. com/download/C/F/6/CF62335F-C46B-4D84-B0C9-363A89B0C5E6/Micro- soft_advanced_threat_analytics_datasheet.pdf, (access date: 19.03.2016), 2015.

[48]   Schmitt E., Shanker T., The New York Times, *U.S. Debated Cyberwarfare in Attack Plan on Libya*, Online source: http://www.nytimes.com/2011/10/18/ world/africa/cyber-warfare-against-libya-was-debated-by-us.html, (access date: 20.03.2016), 2011.

[49]   Majdan K., Security in the network, the largest attack on Google. China fear the sentiments of Tian'anmen, Original title: *Bezpieczeństwo w sieci, Naj- wiekszy atak na Google. Chiny boja sie sentymentów z Tian'anmen*, Online source: http://wyborcza.biz/biznes/1,147883,16089789, Najwiekszy_atak_na_ Google_Chiny_boja_sie_sentymentow.html?disableRedirects=true, (access date: 20.03.2016), 2014.

[50]   Stevenson A., UK technology news, reviews and analysis, V3.co.uk, *Inter- national Atomic Energy Agency Hit by Unknown Malware Attack*, Online source: http://www.v3.co.uk/v3-uk/news/2306372/international-atomic- energy-agency-hit-by-unknown-malware-attack, (access date: 20.03.2016), 2013.

[51]   Symantec.com, *Security Response – Dragonfly: Western Energy Companies Under Sabotage Threat*, Online source: http://www.symantec.com/connect/ blogs/dragonfly-western-energy-companies-under-sabotage-threat, (access date: 20.03.2016), 2014.

[52]   Kovacs E., *Security Week, Internet and enterprise security news, insights & analysis, BlackEnergy Malware Used in Ukraine Power Grid Attacks*, Online source: http://www.securityweek.com/blackenergy-group-uses- destructive-plugin-ukraine-attacks, (access date: 20.03.2016), 2016.

[53]   Tara E., InfoSecurity Magazine, *Sandworm Team Could Be Behind Ukraine Power Grid Attack*, Online source: http://www.infosecurity-magazine.com/ news/sandworm-team-ukraine-power-grid/, (access date: 20.03.2016), 2016.

[54]   An Exploration of Utility Cyber Security Essentials, *Part One of a Two-Part Series on Smart Grid Security*. NES Networked Energy Services, 2015.

[55]   Lu W., Bésanger Y., Zamaï E., Radu D., Blackouts: Description, Analysis and Classification, Proceedings of the 6th *WSEAS International Conference on Power Systems*, Lisbon, Portugal, 2006.

[56]   Sky News, *Militant Attack Plunges Pakistan Into Darkness*, Online source: http://news.sky.com/story/1414477/militant-attack-plunges-pakistan-into- darkness, (access date: 22.03.2016), 2015.

[57]   Mail Online Wires, *Most of Kuwait hit by Power Blackout*, Online source: http://www.dailymail.co.uk/wires/afp/article-2949576/Most-Kuwait-hit-power- blackout.html, (access date: 22.03.2016), 2015.

[58]   Escritt T., Reuters, *UPDATE 5-Power Returns to Amsterdam after Outage Hits a Million Homes*, Online source: http://uk.reuters.com/article/dutch- power-outages-idUKL6N0WT1DI20150327, (access date: 22.03.2016), 2015.

[59] RT Question More, *Turkey Struck by Biggest Power Cut in 15 years, Investigation Underway*, Online source: https://www.rt.com/news/245529-massive-power-outage-turkey/, (access date: 22.03.2016), 2015.

[60] FOX News Weather Center, *Tens of Thousands Shivering without Power in Washington City*, Online source: http://www.foxnews.com/weather/2015/11/24/tens-thousands-shivering-without-power-in-washington-city/, (access date: 22.03.2016), 2015.

[61] RT Question more, State of emergency, *Blackout in Russias Crimea after Transmission Towers in Ukraine Blown Up*, Online source: https://www.rt.com/news/323012-crimea-blackout-lines-blown-up/, (access date: 22.03.2016), 2015.

[62] National Institute of Standards and Technology, *U.S. Department of Commerce, Second Draft NISTIR 7628 Smart Grid Cyber Security Strategy and Requirements, Chapter 3* – High Level Security Reqirements, 2010.

[63] EUROPEAN COMMISSION, Brussels, *Climate Action, 2020 Climate & Energy Package, European/Commission Climate Action/EU Action/Strategies/ 2020*, Online source: http://ec.europa.eu/clima/policies/strategies/2020/index_en.htm, (access date: 22.03.2016), last update 2016.

[64] Smart Grid ABC, Training e-learning: Intelligent Energy at Home and in the Municipality. The educational campaign to promote Smart Grids in Poland, Original title: *Abc Inteligentnych Sieci, Szkolenie e-learningowe: Inteligentna Energia w Domu i w Gminie*. Kampania edukacyjna na rzecz promowania Inteligentnych Sieci Energetycznych w Polsce, Online source: www.ise.ews21.pl, (access date: 22.03.2016), 2014.

[65] *Energy Regulatory Office, Law/Law Community/Directives*, Online source: http://www.ure.gov.pl/pl/prawo/prawo-wspolnotowe/dyrektywy/4352,Dz-U-UE-L-063322.html, (last upadate 2011, access date: 22.03.2016), 2011.

[66] Energy Regulatory Office, Dz.U. UE L 09.140.16, Directive of the European Parliament and Council Directive 2009/28/EC of 23 April 2009. On the promotion of the use of energy from renewable sources and amending and subsequently repealing Directives 2001/77/EC and 2003/30/EC, Original title: *Urzad Regulacji energetyki, Dz.U. UE L 09.140.16, Dyrektywa Parlamentu Europejskiego i Rady 2009/28/WE z dnia 23 kwietnia 2009 r. w sprawie promowania stosowania energii ze źródeł odnawialnych zmieniajaca i w nastepstwie uchylajaca dyrektywy 2001/77/WE oraz 2003/30/WE*, Online source: http://www.ure.gov.pl/pl/prawo/prawo-wspolnotowe/dyrektywy/4925,DzU-UE-L-0914016.html, (last update 2009, access date: 22.03.2016), 2009.

[67] European Commission, Brussels, European Commission/Energy/Topics/ Markets and consumers/Smart grids and meters/*Smart Grids Task Force*, Online source: https://ec.europa.eu/energy/en/topics/markets-and-consumers/smart-grids-and-meters/smart-grids-task-force, (access date: 22.03.2016), 2016.

[68] EUR-Lex Access to European Union Law, *2012/148/EU: Commission Recommendation of 9 March 2012 on preparations for the roll-out of smart*

metering systems, Online source: http://eur-lex.europa.eu/legal-content/EN/ ALL/?uri=CELEX:32012H0148, (access date: 22.03.2016), 2012.

[69]   Computer Word Top 200, *The Development of the Energy Sector from the Perspective of IT Providers*, Original title: Rozwój sektora energetycznego z perspektywy dostawców IT, Online source: http://www.computerworld.pl/ news/397567/Rozwoj.sektora.energetycznego.z.perspektywy.dostawcow.IT. html, 2014.

[70]   Bradley J., Reberger C., Dixit A., Gupta V., Cisco Systems, Inc., *The Public-Sector Impact of the Internet of Everything, Cisco Internet of Everything – Full Report*, Online source: http://share.cisco.com/internet-of-things.html, (access date: 24.03.2016), 2013.

[71]   RSA, The Security Division of EMC, *Getting Ahead of Advanced Threats. Achieving Intelligence-Driven Information Security*, Online source: https:// www.emc.com/collateral/industry-overview/ciso-rpt-2.pdf, (access date: 24.03.2016), 2012.

Index

Printed in the USA
CPSIA information can be obtained
at www.ICGtesting.com
JSHW011507221024
72173JS00005B/1231